NATO ASI Series

Advanced Science Institutes Series

A series presenting the results of activities sponsored by the NATO Science Committee, which aims at the dissemination of advanced scientific and technological knowledge, with a view to strengthening links between scientific communities.

The Series is published by an international board of publishers in conjunction with the NATO Scientific Affairs Division.

A Life Sciences	Plenum Publishing Corporation
B Physics	London and New York
C Mathematical and Physical Sciences	Kluwer Academic Publishers
D Behavioural and Social Sciences	Dordrecht, Boston and London
E Applied Sciences	
F Computer and Systems Sciences	Springer-Verlag
G Ecological Sciences	Berlin Heidelberg New York Barcelona
H Cell Biology	Budapest Hong Kong London Milan
I Global Environmental Change	Paris Santa Clara Singapore Tokyo

Partnership Sub-Series

1. Disarmament Technologies	Kluwer Academic Publishers
2. Environment	Springer-Verlag /
	Kluwer Academic Publishers
3. High Technology	Kluwer Academic Publishers
4. Science and Technology Policy	Kluwer Academic Publishers
5. Computer Networking	Kluwer Academic Publishers

The Partnership Sub-Series incorporates activities undertaken in collaboration with NATO's Cooperation Partners, the countries of the CIS and Central and Eastern Europe, in Priority Areas of concern to those countries.

NATO-PCO Database

The electronic index to the NATO ASI Series provides full bibliographical references (with keywords and/or abstracts) to about 50 000 contributions from international scientists published in all sections of the NATO ASI Series. Access to the NATO-PCO Database is possible via the CD-ROM "NATO Science & Technology Disk" with user-friendly retrieval software in English, French and German (© WTV GmbH and DATAWARE Technologies Inc. 1992).

The CD-ROM can be ordered through any member of the Board of Publishers or through NATO-PCO, B-3090 Overijse, Belgium.

Springer
Berlin
Heidelberg
New York
Barcelona
Hong Kong
London
Milan
Paris
Singapore
Tokyo

Computational Models
of Speech Pattern Processing

Edited by

Keith Ponting

Speech Research Unit
DERA Malvern
St. Andrew's Road
Great Malvern, Worcs. WR14 4DT
United Kingdom

Published in cooperation with NATO Scientific Affairs Division

Proceedings of the NATO Advanced Study Institute
on Computational Models of Speech Pattern Processing,
held in St. Helier, Jersey, U.K., July 7-18, 1997

Library of Congress Cataloging-in-Publication Data applied for

```
Computational models of speech pattern processing / edited by K.M.
  Ponting.
      p.   cm. -- (NATO ASI series. Series F, Computer and systems
  sciences ; no. 169)
    "Proceedings of the NATO Advanced Study Institute on Computational
  Models of Speech Pattern Processing, held in St. Helier, Jersey,
  U.K., July 7-18, 1997."--T.p. verso.
    Includes bibliographical references and index.
```
    ISBN-13: 978-3-642-64250-0   e-ISBN-13: 978-3-642-60087-6
    DOI: 10.1007/978-3-642-60087-6
```
    1. Speech processing systems--Mathematical models. 2. Automatic
  speech recognition.    I. Ponting, K. M.   II. NATO Advanced Study
  Institute on Computational Models of Speecch PAttern Processing :
  1997 : Saint Helier, Channel Islands)   III. Series.
  TK7882.S65C62   1999
  006.4'54--DC21                                          98-55123
                                                          CIP
```

ACM Computing Classification (1998): I.2.7, I.5, I.7.2, H.5, G.3

ISBN-13: 978-3-642-64250-0 Springer-Verlag Berlin Heidelberg New York

© Springer-Verlag Berlin Heidelberg 1999
Softcover reprint of the hardcover 1st edition 1999

Typesetting: Camera-ready by editor
Printed on acid-free paper
SPIN: 10648737   45/3142 – 5 4 3 2 1 0

To Rosamund
for her love, support
and encouragement

Foreword

Keith M. Ponting

Speech Research Unit, DERA Malvern
St. Andrew's Road, Great Malvern, Worcs. WR14 3PS, UK
email: ponting@signal.dera.gov.uk

"Computational Models of Speech Pattern Processing" was the title of a NATO Advanced Study Institute held in St. Helier, Jersey in July 1997. The purpose of the meeting was to discuss and disseminate the latest techniques in the field of speech science and technology with a view to working towards a unifying theory of *speech pattern processing*. With that in mind, a broad range of techniques and applications were included in the programme.

This volume contains tutorial papers from the invited lecturers and a selection of the contributed papers presented at the meeting. The first paper, by Roger Moore, sets the scene. Subsequent papers have been divided into two groups, broadly corresponding to those approaching the theme from the acoustic angle and those working from the linguistic end, although the boundary is by no means clear cut.

The acoustic perspective begins with a review by Pols of work on human perception and a description by Young of the current state of the art in very large vocabulary automatic speech recognition. Extensions to those techniques include the modelling and use of longer span correlations described by Ostendorf, the use of connectionist and hybrid models reviewed by Haton and Deng's work on auditory representations. Papers by Ponting and Furui cover robust approaches and speaker characteristics. Finally, segmental, supra-segmental and articulatory modelling are described by Ostendorf, Noeth and Deng.

The linguistic perspective begins with a review by Connine of work in psycholinguistics. This is followed by papers from Ney and deMori covering language modelling and adaptation, from Robert Moore describing the use of natural language knowledge sources in automatic speech recognition and from Gorin describing an automated operator service trained on and using natural spoken dialogue. Multilingual systems and related issues are discussed by Noeth and Chollet. The section concludes with papers by Neel and Waibel covering a range of systems, both speech-driven and multi-modal, in which speech technology is used.

The editor would like to thank the organising committee (J. P. Haton and R. de Mori) for their assistance in the planning and running of the meeting; the various authors for their efforts in producing a wide-ranging view of the state-of-the-art in this book; and all the attendees for their contribution to a stimulating meeting.

Among many other sessions there was a lively final discussion. The remainder of this introduction summarises the more important points made, with thanks to Roger Moore, who chaired the session, and to all those who contributed and whose points are reproduced unattributed here.

The discussion ranged from whether research should aim for insight or performance through a number of dangers and pitfalls for the researcher and topics "hot" at the time of the meeting to possible directions for future work.

1. Insight vs. performance

Research in any field may aim either to increase insight into the behaviour of systems, or to improve some measure of performance, or, ideally, both.

In the field of automatic speech recognition (ASR) there has been significant emphasis on performance as measured by error rates, typified by the (D)ARPA competitive evaluations of large vocabulary speaker independent continuous ASR systems. Impressive as some of the results have been, the current state-of-the-art in ASR is at least one order of magnitude worse than the performance which humans achieve.

Hidden Markov model (HMM) systems are based on crude piecewise constant approximations to a continuous dynamical system and a frame-based acoustic analysis. The resulting systems have millions of parameters, in conflict with the principle that scientific models should be parsimonious. It should be possible to do better – there is evidence that the language competency humans are born with is encoded in around 5,000 bytes worth of information on the human genome.

Systems which introduce more structure into the models, such as those described in Deng's and Holmes' papers in this volume, those based on asynchronous HMMs [3], or using recognition based on synthesis [1] have an attractiveness which is yet to be proved in terms of raw "performance".

There has also been significant progress in modelling speech production which has yet to be integrated into ASR, partly because of the difficulties in incorporating the ideas into statistical models and partly because the ASR community is strongly driven by the need to perform well in frequent benchmark tests.

Over the years considerable effort has been put into formant-based speech analysis, but formants are an internal property of vocal tract resonances rather than a direct model of the system. They are also difficult to extract and even more context dependent than cepstral representations. They are not even a necessary representation, as intelligibility in the speech signal remains even if all amplitude information is discarded and only phase is retained.

The "best" systems also use a crude statistical approach for their language models (LMs), with N-gram models trained on vast amounts of data outperforming more structured model-based systems. The relative success of N-grams shows how much there is still to learn about linguistics. It may be that N-grams and structured models actually model different aspects of language, so that a composite approach could be used, with structured components predicting what might occur in a whole utterance, while N-gram models cover the local dependencies.

One challenge for research is how to extract and use more information from these huge data sets. Another is how to build and use speaker dependent LMs, an approach encouraged by evidence for gender differences in linguistic performance and by techniques for identification of authors from their texts.

Human language is enormously complex, subtle and hard to pin down. It may well be that, in order to be able to make use of more structured models, it will prove necessary to incorporate full *understanding* of natural language, making use of all the varied knowledge sources which humans unconsciously bring to bear.

A complete solution of this kind would also be a solution to many other tasks in Artificial Intelligence (AI), so that the problem could be said to be "AI complete".

This tradeoff of data against structure reflects a general pattern recognition issue. Nearest neighbour systems are structure free, yet provide optimum performance given infinite data. However in order to achieve portable, flexible systems, which are adaptable across different domains, it is necessary to have structure which can generalise and to be able to train and and adapt such systems without manual intervention.

The primary reason for the apparent triumph of the simple systems is the power of the associated training algorithms which allow large amounts of training data to be used. For acoustic modelling, maximum likelihood linear regression (MLLR) provides a mechanism for adaptation on limited data, but such a mechanism is not yet available for language models.

2. Dangers

There are a number of dangers inherent in the drive to produce working systems.

Labelling algorithms with the names of the problems they are designed to overcome can lead to over-general conclusions. One example is "Vocal Tract Length Normalisation" algorithms which are simply methods for spectral warping. When several such attempts to model some phenomemon fail, it is all too easy to conclude "it is not worth trying to model that" rather than "my models didn't work".

For the acoustic modelling community, mixture distributions provide a real danger, as they mop up variability of many different kinds in a way which does not really help understanding of what is happening. This can take surprising forms – a number of labs have reported mixture components helping to overcome annotation errors!

A final area is the popular "fudge" factor. This is used to "balance" acoustic and language model probabilities in order to optimise overall recognition performance, but there is no theory or mathematical solution to justify or determine its value. Several areas worth investigating here were suggested:

- the speech and language model distributions have very different amounts of noise – maybe the weights could be derived from the variances of these distributions;
- experimentally the optimum values for these factors are lower for small vocabulary speaker dependent than for an equivalent large vocabulary system;
- likewise lower values are optimum for faster speech or for smaller numbers of features, reinforcing the suggestion that these factors are related to correlations introduced by the modelling paradigm.

3. Hot topics

As in all areas of research there are hot topics of the moment. One covered in Cerisara's contribution (but not covered in the main lectures due to the long lead time in planning the event) was that of asynchronous hidden Markov models [3]. Related to decomposition [4] these models allow, for example, for events in different parts of the spectrum to overlap in time. There are interesting parallels to "non-linear phonology" which has been used for speech synthesis [2].

A related area is that of multi-level hidden state models, for which the expectation-maximisation (EM) training algorithm becomes significantly more complicated. Such models require even more careful setting up in order that the hill-climbing algorithm may arrive at a (local) optimum with useful structure. Success in stochastic context free grammars, for example, has come from using annotated training data or a sufficiently good seed.

Robustness is a well-established goal in the acoustic modelling community, but less so in the language modelling community, where there is a historical tradition of exploring "competence" (ideal/perfect systems) first and "performance" (constrained by memory, time etc.) later. Even in the ASR community, robustness tends to be incorporated into the established models rather than, for example, incorporating prior knowledge of possible or likely events in a Bayesian fashion.

Adaptation has already been mentioned in this discussion, and as systems come closer to exploitation its importance becomes more obvious and more all-pervasive.

Maximum entropy modelling has produced only indifferent results. This may be due to applying it to something which is already fairly well handled rather than to something which cannot be handled by other mechanisms.

The use of entropy measures provides some interesting comparisons. Entropy measures the amount of uncertainty. To transmit a message using a vocabulary of 20,000 words and no language model at two words per second would require 29 bits per second. Current modems can transmit three orders of magnitude more information in the 3–4 kHz bandwidth required for a typical telephone system. Therefore there is either a lot of additional information or large amounts of redundancy or both in the speech signal.

With the use of a typical trigram language model the perplexity for a 20,000 word task reduces to around 200 and the corresponding bit rate to 15 bits per second. For a connected digit task, the bit rate is around 8 bits per second, suggesting that current acoustic models which perform pretty well on connected digit tasks are able to extract around 8 bits per second from the acoustic channel, but systems need to be able to obtain around 6 extra bits from somewhere.

4. Towards the future

Looking towards the future, where should research effort be concentrated? The basic HMM algorithms have been around for many years although processing power is now vastly greater, much larger databases are available and there have been some improvements to the search mechanisms. Most progress has come from refinements in statistical models based on small improvements in structure.

4.1 Integrating knowledge sources

For automatic speech recognition there are many sources of knowledge used at different levels. Major challenges are in:

- interfacing and integrating these sources;
- balancing the fidelity of the representation with ease of modelling.

The acoustic–phonetic level is common across most applications, and therefore relatively well understood. Even here there is much to be done, as people do better than machines at recognising out-of-context acoustic fragments. Likewise humans "tune in" to a new speaker very much more rapidly than even the best adaptive systems.

Phonological rules exist and are exploited to a limited extent via context dependent models at the phonetic level. Wider than triphone contexts may well be necessary to make proper use of such information.

The linguistic level to date has almost always operated in "micro-worlds", with great difficulty in jumping from one to another.

4.2 A unified theory?

Discussion of the possibility of a unified theory of speech and speech pattern processing identified, among other desireable properties, that it should provide at least the following:

- coverage of ASR, speech synthesis, talker identification, speech coding, dialect, language variation;
- a system performance model;
- a probabilistic, trainable set of parameters
- mechanisms for adaptation;

Bayes rule is one possible candidate. It is widely used in speech recognition and is very useful for developing the technology. However its use assumes known distributions and parameters, which are not available. Significant work is required to be satisfy the criteria and make the use of Bayes' rule truly Bayesian. Therefore, though useful as a way of modelling, Bayes rule as currently used is not a mathematical truth and does not provide the basis for a sound theory.

Another way of approaching this is to work towards better models with more structure and fewer free parameters, which can therefore be trained on less data. In current systems most of the variation is treated as random, meaning that huge amounts of data are required to estimate parameters and the resulting systems can be described as "statistics masking ignorance".

The system performance model is an important avenue. If it were possible to make and verify predictions based on some model, then this would make the approach to speech technology more scientific, so that every experiment would contribute something, rather than research being dominated by engineering. Being able to ask the question "Will this system recognise that utterance?" would lead to a theory of linguistic communication and methods for creating interpretable utterances.

If this could be achieved, there would also be some prospect of improving the current 1 in 10 "hit rate" for good ideas.

Because speech and language are so important to us, there is a danger of trying to cover so much ground that speech is turned into an effectively insoluble "AI complete" problem. However there is much scope for bringing together what we already know. It is the editor's hope that this volume will in some small way encourage progress towards the goal of better "computational models of speech pattern processing".

References

[1] R. Brierton and N. Sedgwick. Talker enrolment for speech recognition by synthesis. In *Proc. Eurospeech*, volume 3, pages 1619–1622, Berlin, 1993.

[2] J. K. Local. Phonological structure, parametric phonetic interpretation and natural-sounding synthesis. In Keller, editor, *Fundamentals of Speech Synthesis and Speech recognition*, pages 253–269. Wiley, Chichester, 1994.

[3] M. J. Tomlinson, M. J. Russell, R. K. Moore, A. P. Buckland, and M. A. Fawley. Modelling asynchrony in speech using elementary single-signal decomposition. In *Proc. IEEE ICASSP*, volume II, pages 1247–1250, 1997.

[4] A. P. Varga and R. K. Moore. Hidden Markov model decomposition of speech and noise. In *Proc. IEEE ICASSP*, pages 845–848, Albuquerque, 1990. IEEE.

Table of Contents

History Integration into Semantic Classification

Multilingual Speech Recognition

Toward ALISP: A proposal for Automatic Language Independent Speech Processing.

Multimodal Interfaces for Multimedia Information Agents

List of Contributors

A. *Batliner*
Lehrstuhl für Mustererkennung
(Informatik 5)
Universität Erlangen–Nürnberg
Martensstr. 3
D-91058 Erlangen
Germany

F. *Bimbot*
CNRS URA-820, ENST
Signal Department
46 rue Barrault
F-75634 Paris Cedex 13
France

M. J. *Caraty*
LIP6 - Université Pierre et
Marie Curie - CNRS
4, place Jussieu
F-75252 Paris Cedex 5
France

C. Cerisara
CRIN-CNRS & INRIA Lorraine
BP 239
F-54506 Vandœuvre-lès-Nancy
France

J. *Černocký*
FEI VUT Brno
Institute of Radioelectronics
Purkynova 118
CZ-61200 Brno
Czech Republic

M. Cettolo
IRST - Istituto per la Ricerca
Scientifica e Tecnologica
I-38050 Povo Trento
Italy

G. Chollet
CNRS URA-820, ENST
Signal Department
46 rue Barrault
F-75634 Paris Cedex 13
France

C. M. Connine
Psychology Department
SUNY Binghampton
PO Box 6000
Binghamton, NY 13902-6000
USA

A. *Constantinescu*
CNRS URA-820, ENST
Signal Department
46 rue Barrault
F-75634 Paris Cedex 13
France

A. *Corazza*
IRST - Istituto per la Ricerca
Scientifica e Tecnologica
I-38050 Povo Trento
Italy

T. Deelman
Psychology Department
SUNY Binghampton
PO Box 6000
Binghamton, NY 13902-6000
USA

S. Deligne
IBM
T.J. Watson Research Center
Yorktown Heights
USA

R. DeMori
University of Avignon
BP 1228
F-84911 Avignon Cedex 9
France

Li Deng
Department of Electrical and
    Computer Engineering
University of Waterloo
Waterloo
Ontario
Canada N2L 3G1

L. Docío-Fernández
E.T.S.I. de Telecomunicación
Dpto. Tecnologías de
    las Comunicaciones
Campus Universitario de Vigo
E-36200 Vigo (Pontevedra)
Spain

M. El-Bèze
LIA, Université d'Avignon
CERI - BP1228
F-84911 Avignon Cedex 9
France

M. Federico
IRST-Istituto per la Ricerca
    Scientifica e Tecnologica
I-38100 Trento
Italy

L. Fissore
CSELT - Centro Studi E
    Laboratori Telecomunicazioni
Via G. Reiss Romoli 274
I-10148 Torino
Italy

S. Furui
Tokyo Institute of Technology
Department of Computer Science
2-12-1, Ookayama, Meguro-ku
Tokyo, 152 Japan

C. García-Mateo
E.T.S.I. de Telecomunicación
Dpto. Tecnologías de
    las Comunicaciones
Campus Universitario de Vigo
E-36200 Vigo (Pontevedra)
Spain

A. L. Gorin
AT&T Laboratories – Research
180 Park Avenue
Florham Park
New Jersey 07932-0971
USA

S. Harbeck
Lehrstuhl für Mustererkennung
    (Informatik 5)
Universität Erlangen–Nürnberg
Martensstr. 3
D-91058 Erlangen
Germany

J. P. Haton
LORIA/Université Henri Poincaré
Nancy 1
BP 239
F-54506 Vandœuvre-lès-Nancy
France

M. C. Haton
LORIA/Université Henri Poincaré
Nancy 1
BP 239
F-54506 Vandœuvre-lès-Nancy
France

W. J. Holmes
Speech Research Unit
DERA Malvern
St. Andrew's Road
Great Malvern
Worcs. WR14 3PS
UK

A. Kannan
Nuance Communications
1380 Willow Road
Menlo Park
CA 94025
USA

A. Kießling
Speech Processing
Ericsson Eurolab Deutschland GmbH
R&D Radiocommunication
Nordostpark 12
D-90411 Nürnberg
Germany

R. Kompe
Sony International (Europe) GmbH
European Research and Development
Stuttgart (ERDS)
Advanced Developments
Stuttgarter Str. 106
D-70736 Fellbach
Germany

F. Lefèvre
LIP6 - Université Pierre et
        Marie Curie - CNRS
4, place Jussieu
F-75252 Paris Cedex 5
France

R. S. McGowan
Sensimetrics Corporation
48 Grove Street
Somerville
MA 02144
USA

W. M. Minker
LIMSI-CNRS
B.P. 133
F-91403 Orsay Cedex
France

C. Montacié
LIP6 - Université Pierre et
        Marie Curie - CNRS
4, place Jussieu
F-75252 Paris Cedex 5
France

R. C. Moore
Microsoft Research
One Microsoft Way
Redmond
WA 98052
USA

R. K. Moore
Speech Research Unit
DERA Malvern
St. Andrew's Road
Great Malvern
Worcs. WR14 3PS
UK

F. D. Néel
LIMSI-CNRS
B.P. 133
F-91403 Orsay Cedex
France

H. Ney
Lehrstuhl für Informatik VI
RWTH Aachen
 – University of Technology
D-52056 Aachen
Germany

H. Niemann
Lehrstuhl für Mustererkennung
    (Informatik 5)
Universität Erlangen–Nürnberg
Martensstr. 3
D-91058 Erlangen
Germany

E. Nöth
Lehrstuhl für Mustererkennung
    (Informatik 5)
Universität Erlangen–Nürnberg
Martensstr. 3
D-91058 Erlangen
Germany

M. Ostendorf
Electrical and Computer
    Engineering Department
Boston University
8 St. Mary's St.
Boston
MA 02215
USA

A. M. Peinado
Dpto. de Electrónica y
    Tecnología de Computadores
Universidad de Granada,
E-18071 Granada
Spain

L. C. W. Pols
Institute of Phonetic Sciences / IFOTT,
University of Amsterdam
Herengracht 338
NL-1016 CG Amsterdam
The Netherlands

K. M. Ponting
Speech Research Unit
DERA Malvern
St. Andrew's Road
Great Malvern
Worcs. WR14 3PS
UK

G. Riccardi
AT&T Laboratories – Research
180 Park Avenue
Florham Park
New Jersey 07932-0971
USA

L. Rodríguez Liñares
Dpto. de Tecnoloxías
    das Comunicacións
Universidade de Vigo
E-36200 Vigo (Pontevedra)
Spain

O. Ronen
Orith Ronen
SRI International
333 Ravenswood Avenue
Menlo Park
CA 94025
USA

A. J. Rubio
Dpto. de Electrónica y
    Tecnología de Computadores
Universidad de Granada,
E-18071 Granada
Spain

J. C. Segura
Dpto. de Electrónica y
    Tecnología de Computadores
Universidad de Granada,
E-18071 Granada
Spain

T. Spriet
LIA, Université d'Avignon
CERI - BP1228
F-84911 Avignon cedex 9
France

B. Suhm
Interactive Systems Laboratories
Carnegie Mellon University
5000 Forbes Avenue
Pittsburgh
PA 15217
USA

Á. de la Torre
Dpto. de Electrónica y
    Tecnología de Computadores
Universidad de Granada,
E-18071 Granada
Spain

C. Vair
CSELT - Centro Studi E
    Laboratori Telecomunicazioni
Via G. Reiss Romoli 274
I-10148 Torino
Italy

M. T. Vo
Interactive Systems Laboratories
Carnegie Mellon University
5000 Forbes Avenue
Pittsburgh
PA 15217
USA

A. Waibel
Interactive Systems Laboratories
Carnegie Mellon University
5000 Forbes Avenue
Pittsburgh
PA 15217
USA

J. H. Wright
AT&T Laboratories – Research
180 Park Avenue
Florham Park
New Jersey 07932-0971
USA

J. Yang
Interactive Systems Laboratories
Carnegie Mellon University
5000 Forbes Avenue
Pittsburgh
PA 15217
USA

S. J. Young
Engineering Dept.
Cambridge University
Trumpington Street
Cambridge CB2 1PZ
UK

Speech Pattern Processing

Roger K. Moore

DERA Speech Research Unit, St. Andrews Rd, Malvern, Worcs, WR14 3PS, UK

Summary. The study of 'speech' is a fragmented multi-disciplinary area of science which sits somewhere between acoustics, linguistics, engineering and psychology. The one unifying force, which links all of the practitioners in the field, is the study and exploitation of speech *patterning*. The ability to process speech patterns is thus central to the capabilities of all forms of speech communication, whether performed by a human or by a machine. To create a truly unified 'theory of speech pattern processing' it is necessary to focus on the positive contributions that can be made by both speech science and speech technology. Great strides have already been made using a paradigm based on stochastic modelling, and the prospects for further significant developments are good - as long as communications between *all* sectors of the R&D community are suitably encouraged.

Key words: Speech perception and production, speech technology, pattern processing

1. The State-of-the-Art in Speech

The ability to process speech - both on input and output - has by now reached a very high degree of sophistication. Performance on the speech input side is near perfect sentence recognition from a very wide range of talkers at signal-to-noise ratios as low as −3dB. On the other hand, performance on the speech output side is the generation of natural and intelligible speech with clear talker individuality at speaking rates of up to 550 words per minute. Vocabulary sizes range between 10,000 and 150,000 words and, by the year 2000, a very large number of systems will be deployed – probably around 6,406,000,000.

The success of these systems is due, at least in part, to the use of neural networks as the fundamental computational paradigm implemented within a highly optimised bio-mechanical and electro-chemical system architecture. Their main advantage, however, may derive from the fact that it is around 1,000,000 years since the first such systems appeared. It is not clear, therefore, whether the impressive capabilities of any given system are due to it having been exposed to just 2 or 1,000,002 years of speech!

Such high-specification speech processing systems are, of course, not artificial but *human*. Over a considerable period of time the human brain has evolved a highly sophisticated audio-visual signalling system which serves both the communicative and social needs between individuals in a community. It is only within the last 100 years that the sounds of speech have become a serious subject for scientific investigation, and it is only within the last 40 years that serious attempts have been made to mimic these capabilities using artificial means.

2. Speech Patterning

As a scientific field, speech sits at the cross-roads between acoustics, linguistics, engineering and psychology. However, speech is not a scientific discipline in its own right, rather it is a fragmented multi-disciplinary area in which many different fields intersect. Nevertheless, one unifying concept which links all these different interests is the study and exploitation of the fundamental <u>patterning</u> in speech:

> *"Speech production and perception are essentially processes which relate a series of acoustic and visual events to corresponding cognitive activity. Speech mediates the expression and communication of ideas, concepts and information between different physical entities through a regularity of behaviour which is shared (and hence understood) by the participants. It is the regularity of behaviour - the patterning - which is the central object of study in <u>all</u> areas of speech research."* - Roger Moore [1]

The study of speech patterning is thus relevant to four main areas of speech communication: spoken interaction between human beings, speech input into machines (referred to as 'automatic speech recognition'), speech output from machines (referred to 'speech synthesis') and combined interactive speech input/output (referred to as 'dialogue systems') - see Fig. 1.

Fig. 1. Speech communication

Of these four areas in speech communications, the three topics in 'speech technology' have a clear requirement for a scientific understanding of the patterning in speech; the ability to recognise or generate speech automatically is critically dependent on formal methods for processing speech patterns. Similarly, interactive dialogue systems not only require an understanding of the patterns in speech, but also depend on the invocation of a productive relationship between speech and other communication modes in the generic human-machine interface (see Fig. 2).

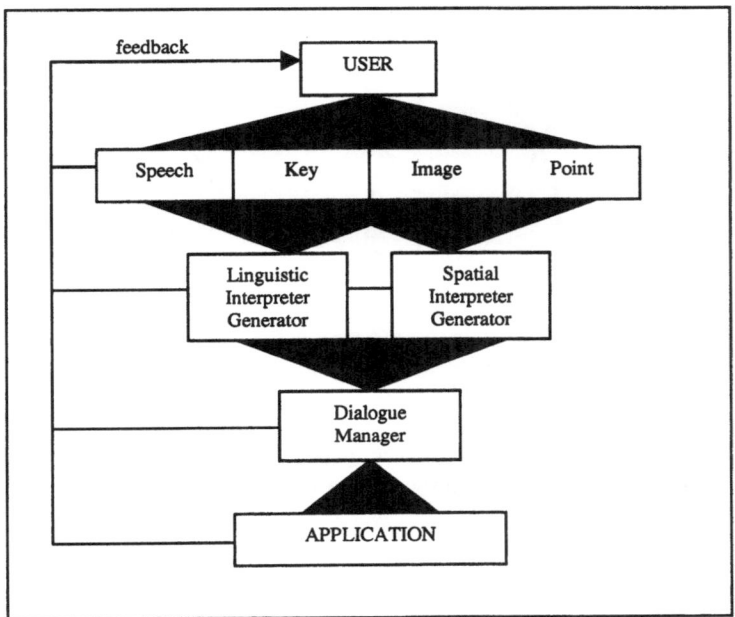

Fig. 2. Speech in the generic human-computer interface

3. Speech Pattern Processing

The ability to process speech patterns is central to the capabilities of all forms of speech communication, albeit by a human or by a machine:

> *"...it is highly likely that, in order to achieve high accuracy many talker, large vocabulary automatic speech recognition in a harsh environment, and high quality, high intelligibility, variable talker speech synthesis, it is necessary to establish a central theory of speech pattern processing"*
> *"Such a theory should be mathematically rigorous, computationally tractable and should make effective use of available information about the structure and use of human speech and language"* - Roger Moore & John Bridle [2]

As a field of study, 'speech pattern processing' sits between linguistics, psychology, engineering and artificial intelligence (see Fig. 3).

Of course Fig. 3 suggests that there exists a coherent scientific discipline which is based upon some agreed underlying principles. In reality a unified view of speech (and speech pattern processing) has yet to emerge. Thirty years ago, the reason for

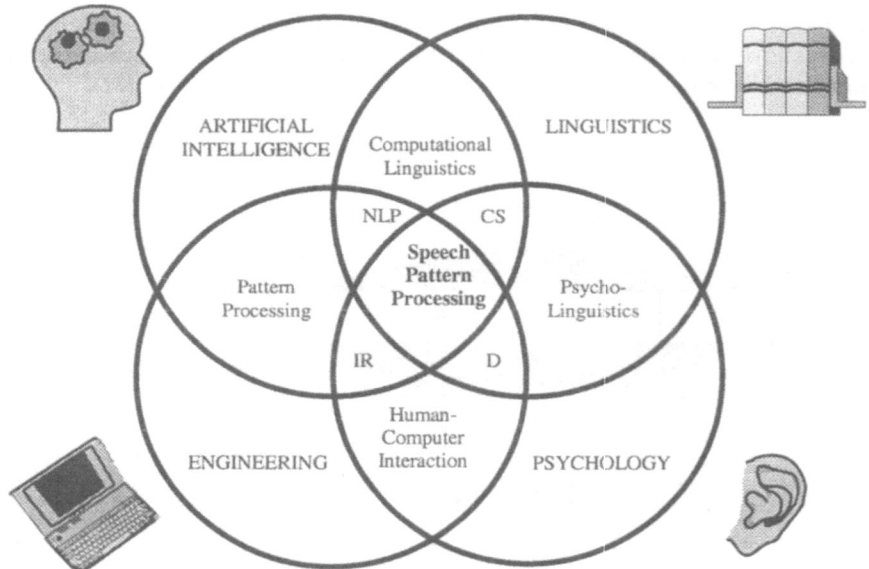

Fig. 3. The position of 'speech pattern processing' (SPP) as a scientific discipline. NLP denotes 'natural language processing', CS denotes 'cognitive science', IR denotes 'information retrieval' and D denotes 'dialogue'.

this derived from the inherently fantastical nature of talking to machines and from the consequent ad-hoc engineering approaches that emerged:

> *"We all believe that a science of speech is possible, despite the scarcity in the field of people who behave like scientists and of results that look like science."* - John Pierce [3]

As a counter to these simple engineering fixes, twenty years ago the field was dominated by supposedly more well-founded linguistic theories using knowledge-based techniques:

> *"...significant advances have been made in the speech understanding field. Yet if one spends $3M for five years and the best system turns out to be a one-man year PhD thesis, not all is well."* - Dennis Klatt [4]

Such was the (lack of) success of the linguistically motivated approach that ten years ago such expertise was even considered to be positively damaging:

> *"Every time we fire a phonetician/linguist the performance of our system goes up!"* - Fred Jelinek [5]

The reality, therefore, is that there is no 'unified theory' of speech - or of speech pattern processing. Rather, there is a large and disparate set of partial theories serving the special interests of the different contributory disciplines identified in Fig. 3.

4. Whither a Unified Theory?

It is interesting to speculate what a unified theory of speech pattern processing might look like, but first it is necessary to consider more deeply why one doesn't exist already. One reason is the contrast between the of phonetics and linguistics (so called 'speech science') on the one hand, and speech technology on the other hand. Speech science favours descriptive but under-specified models, explicit levels of representation and layered processing. However, speech technology favours functional but over-simplified models, implicit representations and integrated processing. The consequence of these opposing views (and needs) is that neither community respects the other's expertise, and communication barriers are drawn up which mitigate against ˙ effective debate.

Any attempt to bridge this gap must focus on the positive contributions that each group can make. For example, the computational disciplines are rich in *algorithms* (mathematically-based processes for manipulating data in a non-heuristic manner), the descriptive disciplines are rich in *knowledge* (in terms of information about the behaviour of the human process) and the experimental disciplines are rich in *methodology* (which is essential for progress in an empirical science). What is needed is for each to accommodate the expertise of the other in a productive and coherent manner.

4.1 Towards a Theory

As a minimum, any theory of speech pattern processing should encompass:

- *information* about the regularities of speech (that is, constraints in the form of descriptive knowledge about speech patterns and linguistic structures in addition to corpora of recorded speech material and linguistic data),
- *representation* of the information (that is, the encoding of the constraints), and
- *computation* using algorithms for constraint satisfaction.

The aim should be to move automatically between one level of representation and another. However, low level representations tend to be continuous, numeric and physical, whereas high level representations tend to be discrete, symbolic and abstract. This apparent contradiction can be overcome by using a 'stochastic model-based' approach in which the discrete structure and form of a model is used to represent high-level symbolic information and the stochastic properties are used to represent the low-level continuous information. In this regard, 'hidden Markov models' (HMMs) have been a particularly successful stochastic modelling approach to automatic speech recognition [6], and have also been usefully applied to speech synthesis [7].

4.2 Practical Issues

In order to construct a theory based on a speech pattern modelling paradigm, it is necessary to consider in detail issues which relate to that information which is given, known, or assumed *a-priori*, as opposed to that which is measured, learnt or estimated *a-postiori*. The former, being based on 'knowledge', is likely to be comprehensive but sparse and perhaps inaccurate. The latter, being based on 'data', is likely to be dense and accurate but perhaps over-specific. These differing characteristics must be accommodated appropriately.

It is also necessary to resolve representational issues relating to alternative signal transformations and to different modelling formalisms. Suitable representations - the 'units' – are highly likely to reflect phonetic and linguistic priors; explicit units arise from patterning in the data, and implicit units arise from patterning in the models. However, it is also important to consider the implications of different representations in the provision of an *interface* between the characteristics of the patterning in the signal and the assumptions inherent to a particular modelling paradigm. For example, the key to the success of the cepstral representation in automatic speech recognition is the fact that it orthogonalises the data – which fits perfectly with a conventional hidden Markov model in which the statistical output distribution covariance matrices are assumed to be diagonal.

By far the most important concern in any theory of speech pattern processing is that the process of 'recognition' should be expressible in computational terms. This means that it is necessary to be able to define (mathematically) the concept of 'goodness of fit' between a model and *any* incoming data. And this needs to be supplemented with efficient 'search strategies' in order to find the best fit, and 'parameter estimation' techniques to calibrate the models.

Finally, models must be able to be 'validated' against previously unseen data in order to be able to assess their performance (their ability to *generalise*) independently of the data with which they have been calibrated.

5. What We Know

Considerable progress has been made, particularly in the field of automatic speech recognition, in recent years, and much of this can be attributed to the application of 'Bayes Theorem' from which derives the notion of hidden Markov models [8]. Powerful algorithms exist for the HMM formalism in the form of 'Viterbi decoding' for search (i.e. recognition) [9] and the 'forward-backward' algorithm for parameter estimation (i.e model training) [10]. This means that the process of recognition can be viewed as *the most probable interpretation of some input data, given a set of (probabilistic) models which might have generated that data.*

The consequence of these theoretically-motivated developments is that there is now a much greater understanding of some key elements of speech pattern processing. For example, some of the most important concepts are as follows:

- Recognition (and synthesis) can be viewed as an integrated search through all constraints.
- An integrated architecture can be decomposed into an equivalent layered architecture if a lattice-style interface is invoked between each layer.
- Ambiguity can be resolved sequentially and continuously.

- Modelling 'ignorance' is as important as modelling knowledge [11].
- Generalisation is achieved by approximation, interpolation and extrapolation.
- Statistical modelling provides a convenient (though not unique) framework for generalising, i.e. modelling uncertainty.
- Every piece of information about speech is a constraint.
- Measured data is more valuable than hypothesised data.
- The patterning in speech facilitates parameter sharing.
- Parameter sharing gives rise to the emergence of processing units.
- Adaptation, normalisation and recognition are essentially the same process.

6. Some Things We Don't Know

Despite the greater understanding that has been achieved since the introduction of speech pattern modelling into speech technology systems, levels of performance - though impressive - are still at least an order of magnitude worse than that exhibited by a human being [12].

In practice most of the current difficulties in speech technology systems are caused by: the richness of the user environment, ergonomic problems, the gap between a sub-language and full language, the individuality of talkers (in themselves and in reaction to the environment), not treating speech as a communicative signal (speaking clearer can make things worse!), and a reliance on full enrolment.

This view was reflected in a recent email survey of the speech community [13] which contained questions relating to the communicative nature of speech, suitable mechanisms for rapid-adaptation and long-term learning, and how to accommodate disfluency and individuality.

7. The Way Forward

The challenges for spoken language systems cover a wide range of topics [14]. However, real (non-incremental) progress will come from further development of a unified multi-disciplinary approach to speech pattern modelling. First, the mathematics of HMMs has already been extended to accommodate simultaneous asynchronous events [15]. Hence by removing the 'single synchronous signal assumption', it is now possible to model speech in the presence of other structured signals [16]. These ideas can now be applied to composite models for speech itself, and already some relevant work is under way [17][18]. Second, the mathematics of HMMs is being extended to encompass dynamic segmental behaviour, i.e. breaking down the 'independence assumption' [19]. This means that it is highly likely that future modelling paradigms will be less concerned with the highly variable surface structure of speech, and be more reflective of putative underlying generative processes [20]. Third, the mathematics of HMMs should be extended to accommodate doubly stochastic processes, i.e. to remove the 'stationarity assumption'. This will facilitate the modelling of key conditional dependencies such as individual talker characteristics [21].

All these advances should facilitate a unification of HMMs with non-linear phonology, an integration of speech recognition and speech synthesis and a coming together of speech science and technology [22]. The key is to develop a multidisciplinary environment in which suitable alternative theories of speech pattern processing can mature to serve the interests of the whole speech R&D community.

Conclusion

'Speech pattern processing' is nothing less than the beginnings of a *Grand Unified Theory of Speech* in which an explanation and reproduction of human levels of performance would be the minimum target. At present, there appear to be no intellectual limits to progress towards these goals since current methods and models are quite crude, and some clear forward directions can be identified. The real challenge is to achieve the necessary degree of communication between all of the different scientific disciplines involved in spoken language R&D.

References

[1]   Moore, R. K.: Whither a theory of speech pattern processing?, *Proceedings. European Conference on Speech Communication and Technology*, 43-47, (1993).
[2]   Moore, R. K. and Bridle, J.: Speech research at RSRE, *Proceedings UK Institute of Acoustics Conference on Speech and Hearing*, (1986).
[3]   Pierce, J. R.: Whither speech recognition? *Journal of the Acoustical Society of America*, 46, 1049-1051, (1969).
[4]   Klatt, D.: Review of the ARPA speech understanding project, *Journal of the Acoustical Society of America*, 62, 1345-1366, (1977).
[5]   Jelinek, F.: public comment, *IEEE ASSPS Workshop on Frontiers of Speech Recognition*, (1985).
[6]   Moore, R. K.: Recognition – the stochastic modelling approach, *Speech Processing*, McGraw-Hill, 223-255, (1992).
[7]   Donovan, R. E. and Woodland, P.C.: Automatic speech synthesis parameter estimation using HMMs, *Proceedings IEEE International Conference on Acoustics, Speech and Signal Processing*, 640-643, (1995).
[8]   Jelinek, F., Mercer, R. L. and Bahl, L. R.: Continuous speech recognition: statistical methods, *Handbook of Statistics*, 2, 549-573, (1982).
[9]   Bellman, R. E.: *Dynamic Programming*, Princeton University Press, NJ, (1957).
[10] Baum, L. E.: An inequality and associated maximisation technique in statistical estimation for probabilistic functions of a Markov process, *Inequalities*, 3, 1-8, (1972).
[11] Makhoul, J. and Schwartz, R.: Ignorance based modelling, *Invariance and Variability in Speech Processing*, Erlbaum, (1984).
[12] Lippmann, R. P.: Speech recognition by machines and humans, *Speech Communication*, 22, 1-15, (1997).
[13] Moore, R. K.: Twenty things we still don't know about speech, *Proceedings CRIM/FORWISS Workshop on Progress and Prospects of Speech Research and Technology*, (1994).

[14] Cole R., Hirschman L., Atlas L., Beckman M., Biermann A., Bush M., Clements M., Cohen J., Garcia O., Hanson B., Hermansky H., Levinson S., McKeown K., Morgan N., Novick D., Ostendorf M., Oviatt S., Price P., Silverman H., Spitz J., Waibel A., Weinstein C., Zahorian S. and Zue V.: The challenge of spoken language systems: research directions for the nineties, *IEEE Transactions on Speech and Audio Processing*, 3, 1-21, (1995).

[15] Moore, R. K.: Signal decomposition using hidden Markov models, *RSRE Memorandum* No.3931, (1986).

[16] Varga, A. P. and Moore, R. K.: Hidden Markov model decomposition of speech and noise, *Proceedings IEEE International Conference on Acoustics, Speech and Signal Processing*, 640-643, 845-848, (1990).

[17] Bourlard, H. and Dupont, S.: A new ASR approach based on independent processing and recombination of partial frequency bands, *Proceedings International Conference on Spoken Language Processing*, 426-429, (1996).

[18] Tomlinson, M. J., Russell, M. J., Moore, R. K., Buckland, A. P. and Fawley, M. A.: Modelling asynchrony in speech using elementary single-signal decomposition, *Proceedings IEEE International Conference on Acoustics, Speech and Signal Processing*, 1247-1250, (1997).

[19] Ostendorf, M., Digilakis, V. and Kimball, O. A.: From HMMs to segment models: A unified view of stochastic modelling for speech recognition, *IEEE Transactions. on Speech and Audio Processing*, 4, 360-378, (1996).

[20] Russell, M. J.: A segmental HMM for speech pattern modelling, *Proceedings IEEE International Conference on Acoustics, Speech and Signal Processing*, 640-643, (1993).

[21] Russell, M. J.: Progress towards speech models that model speech, *Proceedings IEEE Workshop on Automatic Speech Recognition and Understanding*, 115-123, (1997).

[22] Moore, R. K.: Critique: The potential role of speech production models in automatic speech recognition, *Journal of the Acoustical Society of America*, 99, 1710-1713, (1996).

Psycho-acoustics and Speech Perception

Louis C.W. Pols

Institute of Phonetic Sciences / IFOTT, University of Amsterdam
Herengracht 338, 1016 CG Amsterdam, The Netherlands
(e-mail pols@fon.hum.uva.nl)

Summary. Computational models of speech pattern processing might be able to benefit a lot from sound and speech perception by humans. Psycho-acoustics has given us insight into the limits and the capabilities of peripheral hearing for, mainly, simple stationary sounds. Threshold phenomena and temporal and spectral resolution for such stimuli are a first indication of how the front end of a recognizer should be modeled, and what level of precision is required in rule synthesis. Much less is known about the ear's sensitivity to dynamic events with complex signals, such as formant-like transitions. Once the signal becomes a syllable or a meaningful word or sentence, our ear's behavior and our brain's interpretations become even more complex. A good example of that is our perception of stressed and unstressed syllables, including schwas. I will claim that vowel reduction manifests itself as contextual assimilation, rather than as a form of centralization, which again has implications for our phone and word models in ASR and for our coarticulation rules in a synthesizer.

Key words: Psycho-acoustics, speech perception, reduction, coarticulation, context, schwa.

1. Introduction

The words 'speech' and 'spoken language' are generally considered to be synonyms. The last one emphasizes more the fact that speech is a linguistic message, generally having a meaning and a communicative function. But speech is also a signal in the acoustic domain, while sometimes even other domains are involved (gestures). As an acoustic signal it has a number of (objective) signal characteristics that can all vary in time and can lead to one- or more-dimensional subjective impressions, see Table 1.

Table 1. Acoustic signal characteristics

subjective	dimensionality	objective
pitch	1	F0
loudness	1	amplitude
timbre	multi	spectrum
direction	3	$\Delta t, \Delta I$

In speech technology it is important to analyze and describe the speech signal with sufficient detail without being overly sensitive, human performance might be an appropriate guideline for that. In section 2 we will give some data about this human performance from the psycho-acoustics point of view for simple and slightly more complex signals, whereas in section 3 some interesting speech perception phenomena related to reduction and coarticulation are presented. In section 4 we discuss what the implications of these perceptual data might be for speech technology.

2. Psycho-acoustics

Human sensitivity to various acoustic signal characteristics can be expressed in detection thresholds or just noticeable differences (jnd). For spectrally simple and stationary signals (pure tones or noise bursts) these figures are well known, see handbooks like Plomp [11], Yost and Nielsen [21], or Moore [8]. Table 2 summarizes some of these data. The absolute threshold phenomena are generally not of immediate importance for speech technology, because most speech events occur well above threshold. However, they do give some indication of the required precision for pitch, duration, intensity, and spectral sensitivity.

Table 2. Detection thresholds and just noticeable differences

phenomenon	threshold/jnd	remarks
threshold of hearing	0 dB at 1000 Hz	frequency dependent
threshold of duration	constant energy at 10 - 300 ms	Energy = Power x Duration
frequency discrimination	1.5 Hz at 1000 Hz	more when < 200 ms
intensity discrimination	0.5-1 dB up to 80 dB SL	
temporal discrimination	≈ 5 ms at 50 ms	duration dependent
masking	psychophysical tuning curve	
pitch complex tones	low pitch	many peculiarities
gap detection	≈ 3 ms for wide-band noise	more at low freq. for narrow-band noise

For slightly more complex, but thus also slightly more speech-like, but still stationary signals, such as complex tones and single-formant periodic signals, the difference limen (DL) are summarized in Table 3. These data are taken from [5] and [7]. Apparently the perceptual precision for stationary formant discrimination is not very high, especially with respect to the bandwidth. Pitch discrimination under these experimental conditions is rather good, but degrades under more realistic conditions.

Table 3. Difference limen for multi-harmonic, single-formant-like periodic signals

feature	difference limen	remarks
formant frequency	3 - 5 %	one formant only < 3 % with more experienced subjects
formant amplitude	≈ 3 dB	F2 in synthetic vowel
overall intensity	≈ 1.5 dB	synthetic vowel, mainly F1
formant bandwidth	20 - 40 %	one-formant vowel
F0 (pitch)	0.3 - 0.5 %	synthetic vowel

The next step is to go from stationary to dynamic speech-like signals. Here the perceptual data become much more sparse. I refer to a recent study of van Wieringen and Pols [19], who studied the difference limen in (formant) transitions for tone sweeps, and single, and complex (containing multiple formants) formant!sweeps, in isolation, or with a stationary vowel-like part. Depending on whether the transition is initial or final, such stimuli sound somewhat like /du/ or /ab/. Trained subjects are asked to make a same-different judgement for stimulus pairs that may differ somewhat in onset/offset transition frequency.

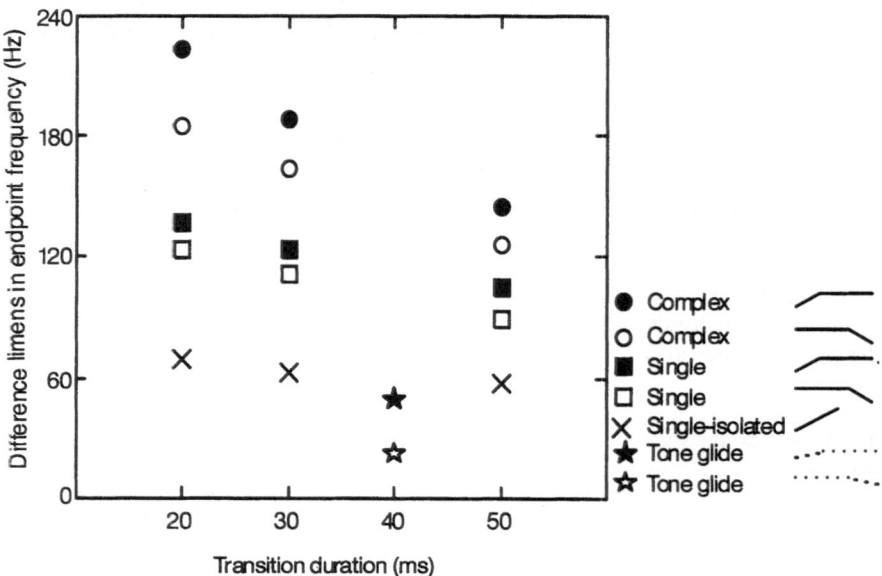

Fig. 1. Difference limen in onset or offset frequency (Hz), for initial or final transitions respectively, of tone sweeps, and of single or complex transitions, in isolation, or with a steady state. DL data are averaged over 4 subjects.

As one can see in Fig. 1, for short (20 ms) complex initial transitions, the DL is as large as 230 ms. For longer transitions the sensitivity becomes better (smaller DL), the same is true for less complex signals, whereas the ear is also more sensitive to final than to initial transitions. Whether this last asymmetry is a universal phenomenon, or at least partly related to the methodology used (a final transition is not masked by anything that might follow, it is also the most recent event before making the decision), is still a matter of debate. This question is also related to the discussion about the possible asymmetry between perceiving initial and final plosives [18]. The better sensitivity for less complex signals is generally understood as a matter of greater attention for single features and lack of masking by neighbouring components in simple signals.

The more speech-like the signal gets, the more speech-like the task generally will (have to) be. For simple stationary stimuli, one can still use psycho-physical tasks such as yes/no detection, or a same/different judgement, however, for (manipulated) speech one has to do things like asking a scale judgement, or doing a phoneme or word identification. By comparing performance over various conditions, one can still derive interesting data about the ear's sensitivity to e.g. temporal and spectral smearing.

Drullman et al. [4] studied the intelligibility of *temporally* degraded speech by low- or high-pass filtering the *signal envelope* in various frequency bands and then using that filtered envelope to amplitude-modulate the original band signal. For LP cutoff frequencies above 4 Hz the phoneme intelligibility for CVC and VCV test words was hardly degraded. The same is true for HP cutoff frequencies lower than 8 Hz. Apparently we are not very sensitive to temporal smearing.

Ter Keurs et al. [6] studied the same for *spectral* envelope smearing. The masked Speech Reception Threshold (SRT), defined as the signal-to-noise ratio in dB at which 50% of the sentences can be reproduced without a single error, was used as an indicator for the effect of smearing. Only when the spectral energy is smeared over a bandwidth wider than 1/3 octave, the masked SRT starts to degrade. This indicates that the intelligibility primarily depends on the global shape of the spectral envelope and not so much on the fine details.

3. Speech Perception

In this section I would like to present some observations on phonetic phenomena like reduction and coarticulation that also might have implications for speech technology, for instance with respect to the choice of the concatenative units in speech synthesis, or the models for the phone-like units in speech recognition.

3.1 Vowel Reduction and Schwa

A distinction can be made between *lexical* vowel reduction (vowel produced as a schwa according to the pronunciation dictionary) and *acoustic* vowel reduction (continuum between full vowel and schwa, actual realization depends on many different factors). Take for instance the vowel in the unstressed syllable 'ba' of the Dutch word 'banaan' (banana) that can be pronounced as either /ə/, /ɑ/, or /a/. There still might be some flavor of the /ɑ/ in that schwa realization. Others say that all such schwas tend towards the same centralized vowel target. Finally there is some recent evidence that it is the local context that fully determines the actual realization of such schwas.

Van Bergem [1] systematically studied both the production and the perception of Dutch reduced vowels. He asked for instance 20 male speakers to pronounce 15 word pairs (high frequent vs. low frequent words, like 'banaan' (banana) vs. 'banier' (banner)), both in isolation (spoken as a list words), and in short phrases. Then 20 listeners had to indicate whether the underlined vowel in each test word (e.g. ba̲naan) was spoken as a full vowel or as a schwa. Table 4 summarizes the results.

We see that more vowels are perceived as a schwa in the sentence condition than in the isolated word condition, and that the low frequent words are more clearly pro-

Table 4. Percentage schwa responses (averaged over 20 listeners), for 15 high frequent (HF) and 15 similar but low frequent (LF) words, spoken in Dutch by 20 males, either in isolation (W), or as part of a sentence (S).

HF		LF	
condition W	condition S	condition W	condition S
27	58	8	17

nounced (less schwa responses) than the high frequent words. In subsequent analyses it became clear that not necessarily the centrally located vowels in F1/F2 get most schwa responses, but rather those that are most adapted to their left and/or right context. Two extreme examples to illustrate this point are the words 'miljoen' (million) and 'bioscoop' (cinema). The /I/ in /mIl'jun/ gets most schwa responses when realized as an /ɔ/-like vowel, thus adapting optimally to the /m-l/ environment, whereas the /ɔ/ in /bijɔs'koːp/ gets most schwa-responses when realized as an /I/-like vowel, according to the /j-s/ environment. This is a strong indication that vowel reduction is much more a process of contextual assimilation than of centralization. By systematically analyzing the formant contours in 'VC$_1$/ə/C$_2$- and C$_1$/ə/'C$_2$V- nonsense words, van Bergem was able to model the coarticulatory effects (of C$_1$, C$_2$, and V) on the schwa with rather high precision. At the center of the F2-contour 92% of the variance could be explained.

3.2 Spectro-temporal Dynamics of Formant Transitions

Formant transitions are not just informative to study the amount of vowel reduction, but influence for instance also consonant identification. One could make a continuum of more and more gradual formant transitions from /p/, to /b/, to /w/, and /u/ to a following vowel. Furthermore, such transitions vary as a function of word position (sentence final lengthening), speaking style (overarticulated vs. sloppy, hyper vs. hypo) and speaking rate. This raises the question how formant transitions can best be described, and how informative they are for perception. Is there a normalization of formant transitions vs. speaking rate, is there a duration-dependent spectral undershoot, or is there perhaps an active adaptation to speaking style? The detailed dynamic analyses of some 550 vowel realizations in normal- and fast-rate speech (by using either 16 equidistant points or Legendre polynomials per vowel segment) of one trained male speaker showed strong evidence for the active adaptation model (Van Son and Pols [15]).

3.3 Consonant Reduction

Van Son and Pols [16] also started a systematic acoustic and perceptual study on consonant reduction. The material consisted of 791 intervocalic Dutch consonants (308 stressed, 483 unstressed) either taken from spontaneous speech, or from the same text read aloud by one male speaker. These VCV-segments were presented to 22 Dutch subjects for *consonant identification*, see Table 5 for some cumulative results. Intervocalic consonants in stressed syllables of read speech are definitely best identified.

Table 5. Percentage error (averaged over 22 listeners) in identifying intervocalic consonants

	Stressed	Unstressed	Total
Spontaneous	21.7	30.3	27.0
Read	14.2	18.1	16.6
Total	18.0	24.2	21.8

Also the amount of *acoustic consonant reduction* was measured between read and spontaneous speech. Unlike for vowels, where stationary or dynamic formant values are good descriptors of reduction, for consonants no such global measures are generally available. We used:

- segment duration
- spectral center of gravity (i.e., the 'mean' frequency, weighted by spectral power)
- intervocalic sound energy difference between consonant and neighbouring vowels
- mean formant distance (in F1/F2 in semitones) between the tautosyllabic vowel of the VCV and the center of vowel reduction for this speaker
- difference between F2 slope at CV and VC consonant borders.

Each of these acoustic measures shows a considerable difference between read and spontaneous speech and between stressed and unstressed realizations. The differences indeed point towards consonant reduction in spontaneous speech and unstressed segments, e.g., shorter durations, lower center of gravity, smaller intervocalic sound energy differences, shorter formant distances of the adjacent vowels, and larger differences of F2 slopes. This change in F2 slopes measures how well articulation speed can keep up with changes in duration. Of course we like to know to what extent a change in the value of each of these acoustic markers for consonant reduction is predictive for a change in identification errors as given in Table 5. Because of the large variation in the data, only for segmental duration and spectral center of gravity there was a statistically significant correlation to reduced identification.

4. Discussion

From the rather sketchy data presentation above, together with the much more detailed publications underlying these data, we believe that we can conclude that:

- *schwa:* the (Dutch) schwa is a vowel without an articulatory target, completely assimilated with its environment
- *vowel reduction:* vowel reduction implies increased contextual assimilation

- *formant transitions:* for the trained speaker that we systematically analyzed, the spectro-temporal dynamics of formant transitions are the result of an active process of speaking style compensation
- *consonant reduction:* consonant reduction is manifest in various acoustic measures, but so far only for segmental duration and spectral center of gravity a relation with reduced consonant identification was statistically significant.

We have the strong impression that not just careful pronunciation, but also appropriate reduction is essential and functional in speech communication. Most of these reduction and coarticulation phenomena are only *implicitly* modeled in speech technology. For instance in diphone synthesis the concatenative units are deliberately chosen to include the coarticulation between neighbouring phonemes, however, how these dynamic aspects should be modified at various speaking rates is not well known. One does see already systems with 2 diphone sets, one for stressed, one for unstressed syllables. Actually what one would like to see is full gradual control over coarticulation and reduction [17]. Similarly, in HMM speech recognition most of the time the monophone or multiphone models are trained with lots of data in which various sources of variability (context, speakers, gender, stress, style, rate, background noise) are supposed to be sufficiently represented. However, for instance the systematicity in the various occurrences of the highly variable schwa is generally neglected. The question may be asked how much random variability actually exists [10], I tend to believe that for most of that variability one or more sources could be identified.

As a phonetician I am pleased to see that in speech recognition more and more specific knowledge is being collected and that one tries to find ways to include this knowledge in the recognition system. Several contributions to this NATO-ASI exemplify this [e.g., 3, 9, 20]. The results, in terms of improved performance, are not always immediately that promising [2, 13], but the approach is certainly worth further exploration. This is also true for a number of other speech perception phenomena, let me just mention scale spacing, redundancy, and variability. In so-called scale spacing the level of precision is selected according to demands, global as long as possible, specific if required [e.g., 14]. Trading one feature against another is a common procedure for robust speech perception, so far this approach is hardly used in ASR.

References

[1]   Bergem, D.R. van (1995), *Acoustic and lexical vowel reduction*, Studies in Language and Language Use 16, Ph.D. thesis, Univ. of Amsterdam.
[2]   Bourlard, H., Hermansky, H. & Morgan, N. (1996), "Towards increasing speech recognition error rates", *Speech Communication* 18, 205-231.
[3]   Deng, L. (forthcoming), "Articulatory features and associated production models in statistical speech recognition", In *Computational Models of Speech Pattern Processing*, (this volume) in *NATO ASI Series F*, 214-224, Springer-Verlag, Berlin, 1998.
[4]   Drullman, R., Festen, J.M. & Plomp, R. (1994), "Effect of reducing slow temporal modulations on speech perception, *J. Acoust. Soc. Am.* 95, 2670-2680.

[5]   Flanagan, J.L. (1972), *Speech analysis synthesis and perception*, Springer Verlag, Berlin, 2nd edition.

[6]   Keurs, M. ter, Festen, J.M. & Plomp, R. (1993), "Effect of spectral envelope smearing on speech reception. II", *J. Acoust. Soc. Am.* 93, 1547-1552.

[7]   Kewley-Port, Li, X., Zheng, Y. & Neel, A. (1996), "Fundamental frequency effects on thresholds for vowel formant discrimination", *J. Acoust. Soc. Am.* 100, 2462-2470.

[8]   Moore, B.C.J. (1989), *An introduction to the psychology of hearing*, Academic Press, London, 3rd edition.

[9]   Ostendorf, M., Kannan, A. & Ronen, O. (forthcoming), "Tree-based Dependence Models for Speech Recognition", In *Computational Models of Speech Pattern Processing*, (this volume) in *NATO ASI Series F*, 40–53, Springer–Verlag, Berlin, 1998.

[10]   Perkell, J.S. & Klatt, D.H. (Eds.) (1986), *Invariance and variability in speech processes*, Lawrence Erlbaum Ass., Publ., Hillsdale, NJ.

[11]   Plomp, R. (1996), *Aspects of tone sensation. A psychophysical study*, Academic Press, London.

[12]   Pols, L.C.W. & Son, R.J.J.H. van (1996), "Acoustics and perception of dynamic vowel segments", *Speech Communication* 13, 135-147.

[13]   Pols, L.C.W., Wang, X. & Bosch, L.F.M. ten (1996), "Modelling of phone duration (using the TIMIT database) and its potential benefit for ASR", *Speech Communication* 19, 161-176.

[14]   Smits, R.L.H.M. (1995), *Detailed versus gross spectro-temporal cues for the perception of stop consonants*, Ph.D. thesis, Univ. of Eindhoven.

[15]   Son, R.J.J.H. van & Pols, L.C.W. (1992), "Formant movements of Dutch vowels in a text, read at normal and fast rate", *J. Acoust. Soc. Am.* 92, 121-127.

[16]   Son, R.J.J.H. van & Pols, L.C.W. (1997), "The correlation between consonant identification and the amount of acoustic consonant reduction", *Proc. Eurospeech'97*, Vol. 4, 2135-2138, Rhodes.

[17]   Sproat, R. (Ed.) (1998), *Multilingual Text-to-Speech Synthesis: The Bell Labs Approach*, Kluwer Acad. Publ., Dordrecht.

[18]   Wieringen, A. van (1995), *Perceiving dynamic speechlike sounds: psycho-acoustics and speech perception*, Ph.D. thesis, Univ. of Amsterdam.

[19]   Wieringen, A. van & Pols, L.C.W. (1995), "Discrimination of single and complex consonant-vowel and vowel-consonant-like formant transitions", *J. Acoust. Soc. Am.* 98, 1304-1312.

[20]   Young, S. (forthcoming), "Acoustic modelling for large vocabulary continuous speech recognition", In *Computational Models of Speech Pattern Processing*, (this volume) in *NATO ASI Series F*, 18–39, Springer–Verlag, Berlin, 1998.

[21]   Yost, W.A. & Nielsen, D.W. (1987), *Fundamentals of hearing. An introduction*, Holt, Rinehart and Winston, New York.

Acoustic Modelling for Large Vocabulary Continuous Speech Recognition

Steve Young

Engineering Dept., Cambridge University
Trumpington Street, Cambridge, CB2 1PZ, UK
email: sjy@eng.cam.ac.uk

Summary. This chapter describes acoustic modelling in modern HMM-based LVCSR systems. The presentation emphasises the need to carefully balance model complexity with available training data, and the methods of state-tying and mixture-splitting are described as examples of how this can be done. Iterative parameter re-estimation using the forward-backward algorithm is then reviewed and the importance of the component occupation probabilities is emphasised. Using this as a basis, two powerful methods are presented for dealing with the inevitable mis-match between training and test data. Firstly, MLLR adaptation allows a set of HMM parameter transforms to be robustly estimated using small amounts of adaptation data. Secondly, MMI training based on lattices can be used to increase the inherent discrimination of the HMMs.

1. Introduction

The rôle of a Large Vocabulary Continuous Speech Recognition (LVCSR) System is to transcribe input speech into an orthographic transcription. Modern LVCSR systems have vocabularies of 5000 to 100000 distinct words and they were developed initially for transcribing carefully spoken dictated speech. Today, however, they are being applied to much more general problems such as the transcription of broadcast news programmes [18, 20] where a variety of speakers, speaking styles, acoustic channels and background noise conditions must be handled.

This chapter describes current approaches to acoustic modelling for LVCSR. Following a brief overview of LVCSR system architecture, HMM-based phone modelling is described followed by an introduction to acoustic adaptation techniques. Finally, some recent research on MMI-based discriminative training for LVCSR is presented as an illustration of possible future developments.

All of the techniques described have been implemented by the author and his colleagues at Cambridge within the HTK LVCSR system [22, 21]. This is a modern design giving state-of-the-art performance and it is typical of the current generation of recognition systems.

2. Overview of LVCSR Architecture

The basic components of an LVCSR system are shown in Fig. 1. The input speech is assumed to consist of a sequence of words and the probability of any specific word sequence can be determined from a *language model*. This is typically a statistical

N-gram model in which the probability of each individual word is conditional only on the identity of the $N - 1$ preceding words.

Each word is assumed to consist of a sequence of basic sounds called *phones*. The sequence of phones constituting each word is determined by a pronouncing dictionary and each phone is represented by a hidden Markov Model (HMM). A HMM is a statistical model which allows the distribution of a sequence of vectors to be represented. Given speech parameterised into a sequence of spectral vectors, each phone model determines the probability that any particular segment was generated by that phone.

Thus, for any spoken input to the recogniser, the overall probability of any hypothesised word sequence can be determined by combining the probability of each word as determined by the HMM phone models and the probability of the word sequence as determined by the language model. It is the job of the *decoder* to efficiently explore all the possible word sequences and find the particular word sequence which has the highest probability. This word sequence then constitutes the recogniser output.

A final step in modern systems is to use the recognised input speech to adapt the acoustic phone models in order to make them better matched to the speaker and environment. This is indicated in Fig. 1 by the broken arrow leading from the decoder back to the phone models.

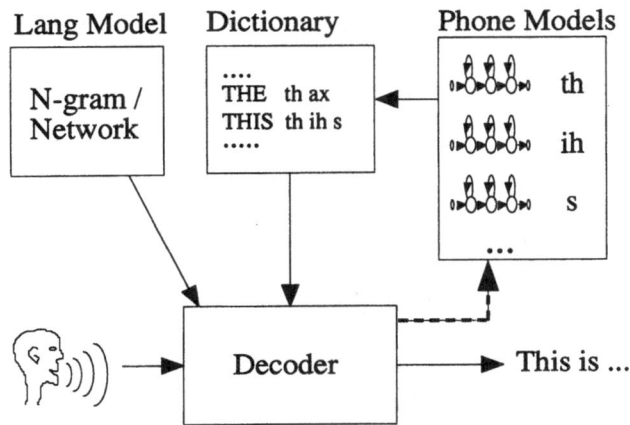

Fig. 1. The main components of an LVCSR system

The mathematical model underlying the above system design was established by Baker, Jelinek and their colleagues from IBM in the 1970's [3, 13]. Figure 2 shows in more detail the way that the probability $P(W|Y)$ of a hypothesised word sequence W can be computed given the parameterised acoustic signal Y.

The unknown speech waveform is converted by the front-end signal processor into a sequence of acoustic vectors, $Y = y_1, y_2, \ldots, y_T$. Each of these vectors is a compact representation of the short-time speech spectrum covering a period of typically 10 msecs. If the utterance consists of a sequence of words W, Bayes' rule

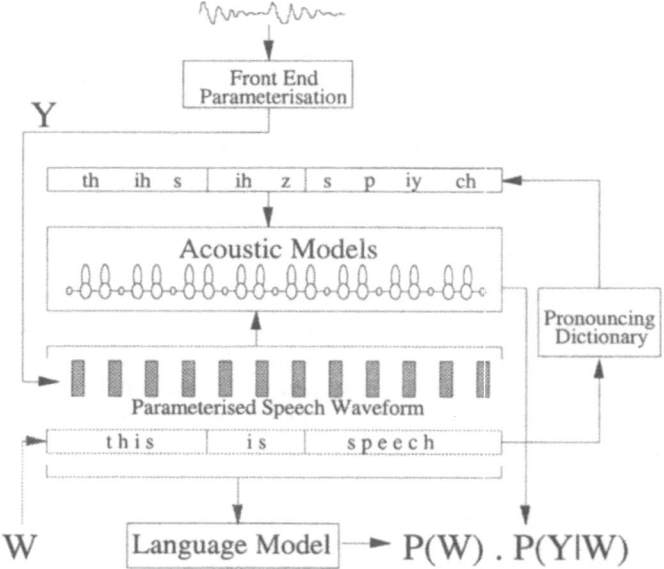

Fig. 2. The LVCSR computational model

can be used to decompose the required probability $P(W|Y)$ into two components, that is,

$$\hat{W} = \underset{W}{\operatorname{argmax}}\, P(W|Y) = \underset{W}{\operatorname{argmax}}\, \frac{P(W)P(Y|W)}{P(Y)}$$

This equation indicates that to find the most likely word sequence W, the word sequence which maximises the product of $P(W)$ and $P(Y|W)$ must be found.

Figure 2 shows how these relationships might be computed. A word sequence $W =$"This is speech" is postulated and the language model computes its probability $P(W)$. Each word is then converted into a sequence of phones using the pronouncing dictionary. The corresponding HMMs needed to represent the postulated utterance are then concatenated to form a single composite model and the probability of that model generating the observed sequence Y is calculated. This is the required probability $P(Y|W)$. In principle, this process can be repeated for all possible word sequences and the most likely sequence selected as the recogniser output[1].

The recognition accuracy of an LVCSR system depends on a wide variety of factors. However, the most crucial system components are the HMM phone models. These must be designed to accurately represent the distributions of each sound in each of the many contexts in which it may occur. The parameters of these models must be estimated from data and since it will never be possible to obtain sufficient data to cover all possible contexts, techniques must be developed which can balance model complexity with available data. Also, the HMM parameters must often

---

[1] In practice, of course, a more sophisticated search strategy is required. For example, LVCSR decoders typically explore word sequences in parallel, discarding hypotheses as soon as they become improbable.

track changing speakers and environmental conditions. This requires the ability to robustly adapt the HMM parameters from small amounts of acoustic data and potentially errorful transcriptions. These are the topics at the heart of acoustic modelling for LVCSR systems and they provide the focus for the rest of this chapter.

3. Front End Processing

As explained in the previous section, the input speech waveform must be parameterised into a discrete sequence of vectors in order to represent its characteristics using a HMM. The main features of this parameterisation process are shown in Fig. 3.

The basic premise is that the speech signal can be regarded as stationary (i.e. the spectral characteristics are relatively constant) over an interval of a few milliseconds. Hence, the input speech is divided into blocks and from each block a smoothed spectral estimate is derived. The spacing between blocks is typically 10 msecs and blocks are normally overlapped to give a longer analysis window, typically 25 msecs. As with all processing of this type, it is usual to apply a tapered window function (e.g. Hamming) to each block. Also the speech signal is often pre-emphasised by applying high frequency amplification to compensate for the attenuation caused by the radiation from the lips.

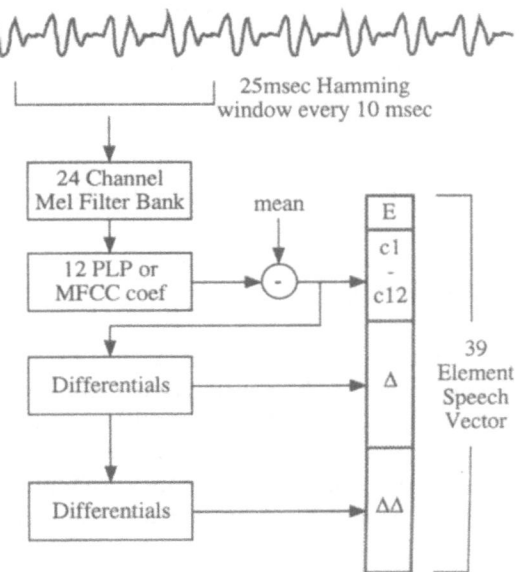

Fig. 3. Front end signal processing

Compared to using a simple linear spectral estimate, performance is improved by using a non-linear Mel-filterbank followed by a Discrete Cosine Transform

(DCT) to form so-called Mel-Frequency Cepstral Coefficients (MFCCs) [6]. The Mel-scale is designed to approximate the frequency resolution of the human ear being linear up to 1000Hz and logarithmic thereafter. The DCT is computed using

$$c_i = \sqrt{\frac{2}{N}} \sum_{j=1}^{N} m_j \cos\left(\frac{\pi i}{N}(j - 0.5)\right)$$

where m_j is the log energy in each Mel-filter band and c_i is the required cepstral coefficient. The DCT compresses the spectral information into the lower order coefficients and it also has the effect of decorrelating the signal thereby improving assumptions of statistical independence. The MFCC coefficients are often normalised by subtracting the mean. This has the effect of removing any long term spectral bias on the input signal.

The static MFCC coefficients are usually augmented by appending time derivatives

$$\Delta_t = \frac{\sum_{\tau=1}^{D} \tau(c_{t+\tau} - c_{t-\tau})}{2\sum_{\tau=1}^{D} \tau^2}$$

The same regression formula can then be applied to the Δ coefficients to give $\Delta\Delta$ (or acceleration) coefficients. These differentials compensate for the rather poor assumption made by the HMMs that successive speech vectors are independent.

MFCC coefficients are widely used in LVCSR systems and give good results. Similar performance can also be achieved by using LP coefficients to derive a smoothed spectrum which is then perceptually weighted to give Perceptually weighted Linear Prediction (PLP) coefficients[10].

An important point to emphasise is the degree to which the design of the front-end has evolved to optimise the subsequent pattern-matching. For example, in the above, the log compression, DCT transform and delta coefficients are all introduced primarily to satisfy the assumptions made by the acoustic modelling component.

4. Basic Phone Modelling

Each basic sound in an LVCSR system is represented by a HMM which can be regarded as a random generator of acoustic vectors (see Fig. 4). It consists of a sequence of states connected by probabilistic transitions. It changes to a new (possibly the same) state each time period generating a new acoustic vector according to the output distribution of that state. The transition probabilities therefore model the durational variability in real speech and the output probabilities model the spectral variability.

4.1 HMM Phone Models

HMM phone models typically have three *emitting* states and a simple left-right topology as illustrated by Fig 4. The entry and exit states are provided to make

it easy to join models together. The exit state of one phone model can be merged
with the entry state of another to form a composite HMM. This allows phone models to be joined together to form words and words to be joined together to cover
complete utterances.

More formally, a HMM phone model consists of

1. Non-emitting entry and exit states
2. A set of internal states x_j, each with output probability $b_j(y_t)$
3. A transition matrix $\{a_{ij}\}$ defining the probability of moving from state x_i to
   x_j[2]

For high accuracy, modern systems uses continuous density mixture Gaussians to
model the output probability distributions, i.e.

$$b_j(y_t) = \sum_{m=1}^{M} c_{jm}\mathcal{N}(y_t; \mu_{jm}, \Sigma_{jm})$$

where $\mathcal{N}(y; \mu, \Sigma)$ is the normal distribution with mean μ and (diagonal) covariance
Σ.

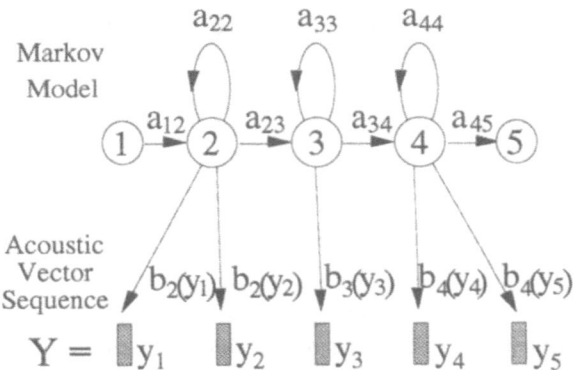

Fig. 4. A HMM phone model

The joint probability of a vector sequence Y and state sequence X given some
model M is calculated simply as the product of the transition probabilities and the
output probabilities. So for the state sequence X in Figure 4

$$P(Y, X|M) = a_{12}b_2(y_1)a_{22}b_2(y_2)a_{23}b_3(y_3)\cdots$$

More formally, the joint probability of an acoustic vector sequence Y and some
state sequence $X = x(1), x(2), x(3), \ldots, x(T)$ is

---

[2] In practice, the transition matrix parameters have little effect on recognition performance
compared to the output distributions. Hence, their estimation is not considered in this
chapter.

$$P(\boldsymbol{Y}, X|M) = a_{x(0)x(1)} \prod_{t=1}^{T} b_{x(t)}(\boldsymbol{y}_t) a_{x(t)x(t+1)} \tag{1}$$

where $x(0)$ is constrained to be the model entry state and $x(T+1)$ is constrained to be the model exit state.

In practice, of course, only the observation sequence $\boldsymbol{Y}$ is known and the underlying state sequence X is hidden. This is why it is called a *Hidden Markov Model*. For recognition, $P(\boldsymbol{Y}|M)$ can be approximated by finding the state sequence which maximises equation 1. A simple algorithm exists for computing this efficiently called the *Viterbi* algorithm and it is the basis of many decoder designs where determination of the most likely state sequence is the key to recognising the unknown word sequence[17].

4.2 HMM Parameter Estimation

In this chapter, the main interest is in designing accurate HMM phone models and estimating their parameters. For the moment, assume that there is a single HMM for each distinct phone and that there is a single spoken example available to estimate its parameters. Consider first the case where each HMM has a single state and each state has only a single Gaussian component. In this case, the state mean and covariance would be given by simple averages

$$\mu_i = \frac{1}{T} \sum_{t=1}^{T} \boldsymbol{y}_t$$

$$\Sigma_i = \frac{1}{T} \sum_{t=1}^{T} (\boldsymbol{y}_t - \mu_i)(\boldsymbol{y}_t - \mu_i)'$$

This can be extended to the case of a real HMM with multiple states and multiple Gaussian components per state, by using weighted averages as follows

$$\mu_{jm} = \frac{\sum_{t=1}^{T} \gamma_{jm}(t) \boldsymbol{y}_t}{\sum_{t=1}^{T} \gamma_{jm}(t)} \tag{2}$$

$$\Sigma_{jm} = \frac{\sum_{t=1}^{T} \gamma_{jm}(t)(\boldsymbol{y}_t - \mu_i)(\boldsymbol{y}_t - \mu_i)'}{\sum_{t=1}^{T} \gamma_{jm}(t)} \tag{3}$$

where $\gamma_{jm}(t)$ is the so-called *component occupation probability*. The key idea here is that each training vector is distributed amongst the HMM Gaussian components according to the probability that it was generated by that component. Since $\gamma_{jm}(t)$ depends on the existing HMM parameters, an iterative procedure is suggested

1. choose initial values for the HMM parameters
2. compute the component occupation probabilities in terms of the existing HMM parameters

3. update the HMM parameters using equations 2 and 3

The component occupation probabilities can be computed efficiently using a recursive procedure known as the *Forward-Backward* algorithm. Firstly, define the *forward probability* $\alpha_j(t) = P(\boldsymbol{y}_1 \ldots \boldsymbol{y}_t, x_t = j)$. As illustrated by Fig. 5, this can be computed recursively by

$$\alpha_j(t) = \left\{ \sum_{i=1}^{N} \alpha_i(t-1)a_{ij} \right\} b_j(\boldsymbol{y}_t)$$

Similarly, the *backward probability* is defined as $\beta_j(t) = P(\boldsymbol{y}_{t+1} \ldots \boldsymbol{y}_T | x_t = j)$, this can also be computed recursively by

$$\beta_i(t) = \sum_{j=1}^{N} a_{ij} b_j(\boldsymbol{y}_{t+1}) \beta_j(t+1)$$

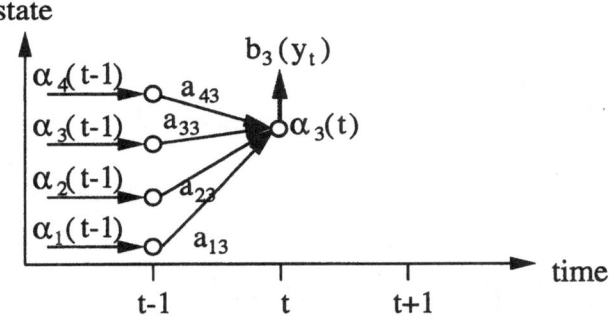

Fig. 5. The forward probability calculation

Given the forward and backward probabilities, the state occupation probability is simply

$$\gamma_j(t) = \frac{1}{P} \alpha_j(t) \beta_j(t)$$

where $P = P(\boldsymbol{Y}|M) = \alpha_N(T)$, and the component occupation probability is

$$\gamma_{jm}(t) = \frac{1}{P} \sum_{i=1}^{N} \alpha_i(t-1) a_{ij} c_{jm} \mathcal{N}(\boldsymbol{y}_t; \boldsymbol{\mu}_{jm}, \boldsymbol{\Sigma}_{jm}) \beta_j(t)$$

The estimation of HMM parameters using the above procedure is an example of the Expectation-Maximisation (EM) algorithm and it converges such that the likelihood of the training data given the HMM i.e. $P(\boldsymbol{Y}|M)$ achieves a local maximum [4, 7].

Although the above is now established text-book material, it is not usually presented in terms of simple weighted averages. This is a pity since even though it lacks mathematical rigour, it offers considerable insight into the reestimation process. For example, it is easy to see that when multiple training instances are provided, the same basic equations 2 and 3 still apply. The sums required to compute the numerators and denominators of these equations are first accumulated over all of the data, and then the parameters are updated.

To complete the presentation of basic HMM phone model estimation, one final unrealistic assumption must be removed. In practice, there is no access to individual speech segments corresponding to a single phone model. Instead, the training data consists of naturally spoken utterances annotated at the word level. Rather than attempting to segment this data, it can be used directly for parameter estimation by adopting an *embedded training* paradigm as illustrated in Fig. 6. The phone sequence corresponding to each training utterance is determined from a dictionary. Then a composite HMM is constructed by concatenating all of the phone models and the numerator and denominator statistics needed for equations 2 and 3 are accumulated for all of the phones in the sequence. This is repeated for all of the training data and finally, all of the phone model parameters are re-estimated in parallel.

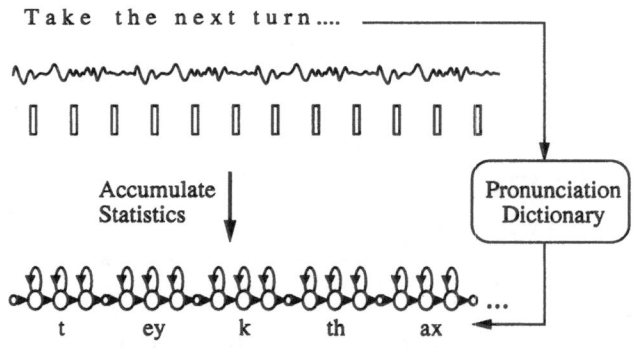

Fig. 6. Embedded HMM training

4.3 Context-Dependent Phone Models

So far there has been an implicit assumption that only one HMM is required per phone, and since approximately 45 phones are needed for English, it may be thought that only 45 phone HMMs need be trained. In practice, however, contextual effects cause large variations in the way that different sounds are produced. Hence, to achieve good phonetic discrimination, different HMMs have to be trained for each different context. The simplest and most common approach is to use *triphones* whereby every phone has a distinct HMM model for every unique pair of left and right neighbours. For example, suppose that the notation x–y+z represents the phone y occurring after phone x and before phone z. The phrase, "Beat it!" would be represented by the phone sequence sil b iy t ih t sil, and if triphone HMMs were used the sequence would be modelled as

```
sil sil-b+iy b-iy+t iy-t+ih t-ih+t ih-t+sil sil
```

Notice that the triphone contexts span word boundaries and the two instances of the phone t are represented by different HMMs because their contexts are different. This use of so-called *cross-word triphones* gives the best modelling accuracy but leads to complications in the decoder. Simpler systems result from the use of *word-internal triphones* where the above example would become

```
sil b+iy b-iy+t iy-t ih+t ih-t sil
```

Here far fewer distinct models are needed simplifying both the parameter estimation problem and decoder design. However, the cost is an inability to model contextual effects at word boundaries and in fluent speech these are considerable.

The use of Gaussian mixture output distributions allows each state distribution to be modelled very accurately. However, when triphones are used they result in a system which has too many parameters to train. For example, a large vocabulary cross-word triphone system will typically need around 60,000 triphones[3]. In practice, around 10 mixture components per state are needed for reasonable performance. Assuming that the covariances are all diagonal, then a recogniser with 39 element acoustic vectors would require around 790 parameters per state. Hence, 60,000 3-state triphones would have a total of 142 million parameters!

The problem of too many parameters and too little training data is absolutely crucial in the design of a statistical speech recogniser. Early systems dealt with the problem by tying all Gaussian components together to form a pool which was then shared amongst all HMM states. In these so-called tied-mixture systems, only the mixture component weights were state-specific and these could be smoothed by interpolating with context independent models[11, 5]. Modern systems, however, commonly use a technique called *state-tying* [12, 24]. in which states which are acoustically indistinguishable are tied together. This allows all the data associated with each individual state to be pooled and thereby gives more robust estimates for the parameters of the tied-state.

State-tying is illustrated in Fig 7. At the top of the figure, each triphone has its own private output distribution. After clustering similar states together and tying, several states share distributions. This figure also illustrates an important practical advantage of using Gaussian mixture distributions in that it is very simple to increase the number of mixture components in a system by so-called *mixture splitting*. In mixture-splitting, the more dominant Gaussian components in each state are cloned and then the means are perturbed by a small fraction of the standard deviation. The resulting HMMs are then re-estimated using the forward-backward algorithm. This process can be repeated so that a single Gaussian system can be converted to the required multiple mixture component system in just a few iterations.

Mixture-splitting allows a tied-state system to be built using single Gaussians and then converted to a multiple component system after the states have been tied. This avoids the problem of having too little data to train untied mixture Gaussians

---

[3] With 45 phones, there are $45^3 = 91125$ possible triphones but not all can occur due to the phonotactic constraints of the language

Conventional triphones

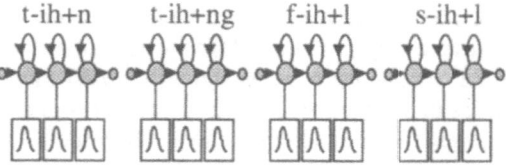

State Clustered single Gaussian Triphones

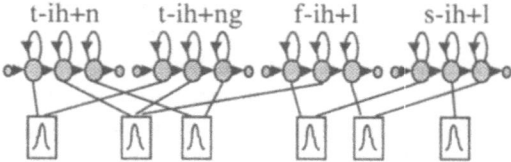

State Clustered mixture Gaussian Triphones

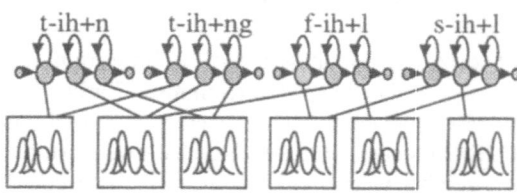

Fig. 7. Tied-state triphone construction

and it simplifies the clustering process since it is much easier to compute the similarity between single Gaussian distributions.

Although almost any clustering technique could be used to decide which states to tie, in practice, the use of *phonetic decision trees*[2, 14, 23] is preferred. In decision tree-based clustering, a binary tree is built for each phone and state position. Each tree has a yes/no phonetic question such as "Is the left context a nasal?" at each node. Initially all states for a given phone state position are placed at the root node of a tree. Depending on each answer, the pool of states is successively split and this continues until the states have trickled down to leaf-nodes. All states in the same leaf node are then tied. For example, Fig 8 illustrates the case of tying the centre states of all triphones of the phone /aw/ (as in "out"). All of the states trickle down the tree and depending on the answer to the questions, they end up at one of the shaded terminal nodes. For example, in the illustrated case, the centre state of s-aw+n would join the second leaf node from the right since its right context is a central consonant, and its right context is a nasal but its left context is not a central stop.

The questions at each node are chosen from a large predefined set of possible contextual effects in order to maximise the likelihood of the training data given the

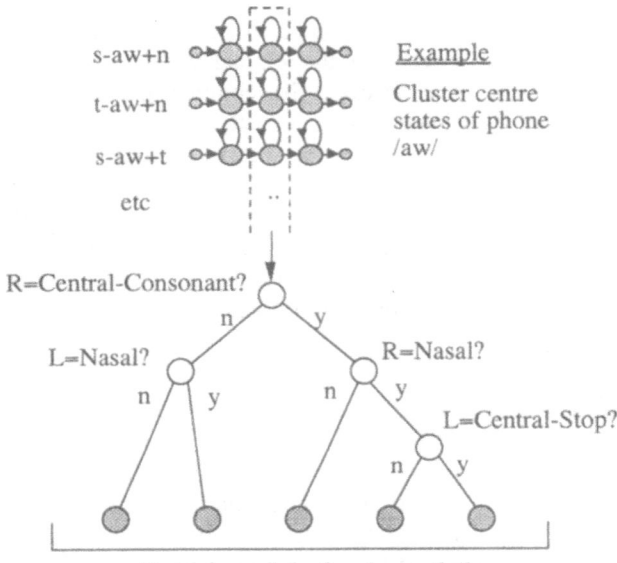

s-aw+n

t-aw+n

s-aw+t

etc

Example

Cluster centre
states of phone
/aw/

R=Central-Consonant?

n   y

L=Nasal?   R=Nasal?

n   y   n   y

L=Central-Stop?

n   y

States in each leaf node are tied

Fig. 8. Phonetic decision tree-based clustering

final set of state tyings. The tree is grown starting at the root node which represents all states as a single cluster. Each state s_i has an associated set of observations $Y = \{y_{i,1}, \ldots, y_{i,N_i}\}$. If $S = \{s_1, s_2, \ldots, s_k\}$ defines a pool of states, then the log likelihood of the data associated with this pool is defined as

$$L(S) = \sum_{i=1}^{K} log P(Y_i | \mu_S, \Sigma_S)$$

This is the likelihood of the data if all of the associated states are merged to form a single Gaussian with mean μ_S and variance Σ_S.

This pool of states S is now split into two partitions by asking a question based on the phonetic context. Since the likelihood of each partition is computed using the overall mean and variance for that partition, the total likelihood of the partitioned data will increase by an amount

$$\Delta = L(S_y) + L(S_n) - L(S)$$

Δ is therefore computed for all possible questions and the question q* which maximises it is selected. The process then repeats by splitting each of the two newly formed nodes. It is terminated when either Δ falls below a predefined threshold or when the amount of data associated with one of the split nodes would fall below a threshold.

Note that provided the state occupancy counts γ_j are retained from the reestimation of the original untied single Gaussian system, all of the likelihoods needed for the above tree growing procedure can be computed directly from the model parameters and no reference is needed to the original data.

In practice, phonetic decision trees give compact good-quality state clusters which have sufficient associated data to robustly estimate mixture Gaussian output probability functions. Furthermore, they can be used to synthesise a HMM for any possible context whether it appears in the training data or not, simply by descending the trees and using the state distributions associated with the terminating leaf nodes. Finally, phonetic decision trees can be used to include more than simple triphone contexts. For example, questions spanning ±2 phones can be included and they can also take account of the presence of word boundaries.

5. Adaptation for LVCSR

Large vocabulary speech recognisers require very large databases of acoustic data to train them. These databases usually contain many speakers recorded under controlled conditions, typically noise-free and wide-band. The resulting HMMs are therefore speaker independent (SI) and optimised for a specific microphone and environment.

For practical applications, an LVCSR system trained in this way results in a number of limitations

- SI performance is inferior to speaker dependent (SD) performance
- many speakers are outliers with respect to the original training population and will therefore be poorly recognised
- channel conditions will vary with different microphones and recording conditions
- background noise is common

Hence, there is often a *mis-match* between the training and testing conditions and it is important to reduce this mis-match as much as possible by using the test data itself to adapt the HMM parameters to be more suited to the current speaker, channel and environmental conditions.

There are a number of distinct modes of adaptation

- *Supervised* – an exact transcription of all the adaptation data is available
- *Unsupervised* – the recogniser output is used to transcribe the adaptation data
- *Enrolment Mode* – the adaptation data is applied off-line prior to recognition
- *Incremental Mode* – each new recogniser output is used to augment the adaptation data.
- *Transcription Mode* – non-causal, all recognised speech is saved, used for adaptation, then all speech is re-recognised

Clearly the choice and combination of modes depends on the application and ergonomic considerations. For example, a personal desk-top dictation system will typically use supervised enrolment, whereas an off-line broadcast news transcription service will use unsupervised transcription mode.

5.1 Maximum Likelihood Linear Regression

There are many different approaches to adaptation, but one of the most versatile is Maximum Likelihood Linear Regression (MLLR) [15, 9]. MLLR seeks to find an affine transform of the Gaussian means which maximises the likelihood of the adaptation data, i.e.

$$\hat{\mu}_r = A_m \mu_r + b_m = W_m \eta_r$$

where $W_m = [b_m \ A_m]$ and $\eta_r = \begin{bmatrix} 1 \ \mu_r^T \end{bmatrix}^T$.

The key to the power of this adaptation approach is that a single transformation W_m can be shared across a set of Gaussian mixture components. When the amount of adaptation data is limited, a single transform can be shared across all Gaussians in the system. As the amount of data increases, the HMM state components can be grouped into classes with each class having its own transform. As the amount of data increases further, the number of classes and therefore transforms increases correspondingly leading to better and better adaptation.

The number of transforms is usually determined automatically using a *regression class tree* as illustrated in Fig. 9. Each node represents a regression class i.e. a set of Gaussian components which will share a single transform. For a given adaptation set, the tree is descended and the most specific set of nodes is selected for which there is sufficient data (for example, the filled-in nodes in the figure). The regression class tree itself can be built using similar techniques to those described in the previous section for state-clustering [8].

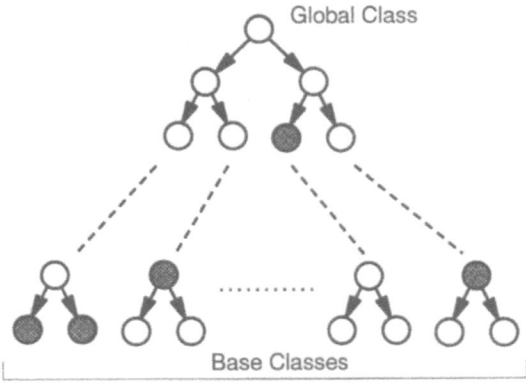

Fig. 9. An MLLR regression tree

5.2 Estimating the MLLR Transforms

As its name suggests, the parameters of the transforms W_m are estimated so as to maximise the likelihood of the adaptation data with respect to the transformed HMM parameters. This log likelihood $\mathcal{L}$ is given by

$$\mathcal{L} = \sum_{r=1}^{R} \sum_{t=1}^{T} \gamma_r(t) log \left\{ K_r exp(-\frac{1}{2}(\boldsymbol{y}(t) - \boldsymbol{W}_m \boldsymbol{\eta}_r)' \Sigma_r^{-1} (\boldsymbol{y}(t) - \boldsymbol{W}_m \boldsymbol{\eta}_r)) \right\}$$

where r ranges over the R Gaussian components belonging to the regression class associated with transform $\boldsymbol{W}_m$ and K_r are normalising constants. Differentiating wrt to $\boldsymbol{W}_m$ and setting the result equal to zero gives

$$\sum_{r=1}^{R} \sum_{t=1}^{T} \gamma_r(t) \Sigma_r^{-1} \boldsymbol{y}(t) \boldsymbol{\eta}_r' = \sum_{r=1}^{R} \sum_{t=1}^{T} \gamma_r(t) \Sigma_r^{-1} \boldsymbol{W}_m \boldsymbol{\eta}_r \boldsymbol{\eta}_r'$$

which can be written in matrix form as

$$\boldsymbol{Z} = \sum_{r=1}^{R} \boldsymbol{V}^r \boldsymbol{W}_m \boldsymbol{D}^r$$

There is no computationally efficient solution for this in the full covariance case. However, for diagonal covariance, the i'th row of $\boldsymbol{W}_m$ is given by

$$\boldsymbol{z}_i' = \boldsymbol{w}_i' \sum_{r=1}^{R} v_{ii}^r \boldsymbol{D}^r$$

which can be solved by inverting the matrix $\boldsymbol{D}^r$.

In addition to mean adaptation, variance adaptation is also possible. A particularly simple form of transform to use for this is $\boldsymbol{H}_m$ where

$$\hat{\Sigma}_r^{-1} = C_r \boldsymbol{H}_m^{-1} C_r'$$

and where C_r is the Choleski factor of Σ_r^{-1}. $\boldsymbol{H}_m$ is easy to estimate, because rewriting the quadratic in the exponent of the Gaussian as

$$\frac{1}{2} \left((C_r' \boldsymbol{y}(t) - C_r' \boldsymbol{\mu}_r)' \boldsymbol{H}_m^{-1} (C_r' \boldsymbol{y}(t) - C_r' \boldsymbol{\mu}_r) \right)$$

it can be seen that the form is the same as for the re-estimation of the HMM variances using equation 3, i.e.

$$\hat{\boldsymbol{H}}_m = \frac{C_m' \left[\sum_{t=1}^{T} \gamma_m(t)(\boldsymbol{y}(t) - \boldsymbol{\mu}_m)(\boldsymbol{y}(t) - \boldsymbol{\mu}_m)' \right] C_m}{\sum_{t=1}^{T} \gamma_m(t)}$$

Instead of having a separate transform for the means and variances, a single *constrained* transform can be applied to both, i.e.

$$\hat{\boldsymbol{\mu}}_r = \boldsymbol{A}_m \boldsymbol{\mu}_r + \boldsymbol{b}_m$$
$$\hat{\boldsymbol{\Sigma}}_r = \boldsymbol{A}_m \boldsymbol{\Sigma}_r \boldsymbol{A}_m'$$

This has no closed-form solution but an iterative solution is possible [9]. A key advantage of this form of adaptation is that the likelihoods can be calculated as

$$\mathcal{L}(\boldsymbol{y}(t); \boldsymbol{\mu}, \boldsymbol{\Sigma}, \boldsymbol{A}, \boldsymbol{b}) = \mathcal{N}(\boldsymbol{A}\boldsymbol{y}(t) + \boldsymbol{b}; \boldsymbol{\mu}, \boldsymbol{\Sigma}) + \log(|\boldsymbol{A}|)$$

This means that the transform can be applied to the data rather than the HMM parameters which may be more convenient for some applications. When using incremental adaptation, this transform can also be more efficient to compute since although it is iterative, only one iteration is needed for each new increment of adaptation data and, unlike the unconstrained case, it does not require any expensive matrix inversions.

Finally, it should be noted that for unsupervised adaptation, the quality of the transforms depends on the accuracy of the recogniser output. One obvious way to improve this is to iterate the recognition and adaptation cycle.

6. Progress in LVCSR

Progress in LVCSR over the last decade has been tracked by the US National Institute of Standards and Technology (NIST) in the form of annual speech recognition evaluations. These have evolved over the years but the basic style is that participating organisations are provided with the necessary training data and some development test data at the start of the year. Towards the end of the year, NIST then distribute *unseen* evaluation test data and each organisation then recognises this data and sends the output back to NIST for scoring. Initially, the participating organisations were all US funded research groups, but since 1992, the evaluations have been open to non-US groups.

Table 1 lists the different evaluation tasks along with their main characteristics. In this table, the test mode indicates whether or not the evaluation data has a closed or open vocabulary. If the vocabulary is open, then the test data will contain so-called Out-of-Vocabulary (OOV) words which contribute to the error rate. PP denotes perplexity which is similar to the *average branching factor* and indicates the degree of uncertainty as each new word is encountered. The % word error (WER) rates indicate the approximate performance of the best systems at the time they were tested.

RM denotes the *Naval Resource Management Task* which is an artificial task based on spoken access to a database of naval information. WSJ (Wall Street Journal) and NAB (North American Business news) are large vocabulary dictation tasks in which the source material is taken from either the WSJ or more generally, a range of US newspapers (NAB). Finally, the current BN (Broadcast News) task involves the transcription of arbitrary broadcast news material. This challenging task introduces many new problems including the need to segment and classify a continuous audio stream, handle a range of speakers and channels, and cope with a wide variety of interfering signals including noise, music and other speakers. Note that all of these tasks involve speaker independent recognition of continuous speech.

As can be seen from the table, the state of the art on clean speech dictation within a limited domain such as business news is around 7%WER. The LVCSR systems which can achieve this are typically of the sort described in this chapter i.e. tied-state mixture Gaussian HMM based with cross-word triphones, N-gram language models and incremental unsupervised MLLR. The error rates for broadcast news transcription are much higher reflecting the many additional problems that it poses. However, this is an active area of research and the error rates will fall quickly.

Table 1. ARPA evaluation tasks and performance of best systems at the time

When	Task	Train Data	Vocab Size	Test Mode	PP	WER %
87-92	RM	4 Hrs	1k	Closed	60	4
92-94	WSJ	12 Hrs	5k	Closed	50	5
92-94	WSJ	66 Hrs	20k	Open	150	10
94-95	NAB	66 Hrs	65k	Open	150	7
95-96	BN	50 Hrs	65k	Open	200	30

7. Discriminative Training for LVCSR

All of the methods described in the preceding sections are so-called *Maximum Likelihood* (ML) methods. They are based on the simple premise that the parameters of an LVCSR system should be designed to give the closest possible fit to the training data, and where appropriate the adaptation data. Unfortunately, as noted already, there is often a mis-match between the training and test data so that maximising the fit to the training data does not necessarily mean that the ultimate recognition performance will be optimised.

All this has been well-known for many years and several alternative parameter estimation schemes have been developed. In particular, a maximum mutual information (MMI) criterion can be used [1] which seeks to increase the *a posteriori* probability of the model sequence corresponding to the training data given the training data.

More formally, for R training observations $\{Y_1, \ldots, Y_r, \ldots Y_R\}$ with corresponding transcriptions $\{w_r\}$, the MMI objective function is given by

$$\mathcal{F}(\lambda) = \sum_{r=1}^{R} \log \frac{P_\lambda(Y_r | \mathcal{M}_{w_r}) P(w_r)}{\sum_{\hat{w}} P_\lambda(Y_r | \mathcal{M}_{\hat{w}}) P(\hat{w})}$$

where $\mathcal{M}_w$ is the composite model corresponding to the word sequence w and $P(w)$ is the probability of this sequence as determined by the language model.

The numerator of $\mathcal{F}(\lambda)$ corresponds to the likelihood of the training data given the correct model sequence, whereas the denominator corresponds to its likelihood given all the other possible sequences. Maximising the numerator whilst simultaneously minimising the denominator gives HMMs trained using the MMI criterion improved discrimination compared to ML.

The problem with using MMI in practice is that the denominator is impossible to compute for anything other than simple isolated word systems which have a finite number of possible model sequences to consider. Modern LVCSR systems, however, are capable of generating lattices of alternative recognition hypotheses. This last section on acoustic modelling explains how these lattices can be used to discriminatively train the HMMs of an LVCSR system using the MMI criterion [19].

To make the evaluation of $\mathcal{F}(\lambda)$ tractable, the denominator can be approximated by

$$\sum_{\hat{w}} P_\lambda(\boldsymbol{Y}_r|\mathcal{M}_{\hat{w}})P(\hat{w}) \Rightarrow P_\lambda(\boldsymbol{Y}_r|\mathcal{M}_{rec})$$

where $\mathcal{M}_{rec}$ is a model constructed such that for all paths in every $\mathcal{M}_{\hat{w}}$ there is a corresponding path of equal probability in $\mathcal{M}_{rec}$ i.e. $\mathcal{M}_{rec}$ is the model used for recognition. Thus, the MMI objective function now becomes

$$\mathcal{F}(\lambda) = \sum_{r=1}^{R} \log \frac{P_\lambda(\boldsymbol{Y}_r|\mathcal{M}_{cor})}{P_\lambda(\boldsymbol{Y}_r|\mathcal{M}_{rec})}$$

Unlike the ML case, it is not possible to derive provably convergent re-estimation formula. However, Normandin has derived the following formulae which work well in practice [16]

$$\hat{\mu}_{j,m} = \frac{\{\theta_{j,m}^{cor}(\boldsymbol{Y}) - \theta_{j,m}^{rec}(\boldsymbol{Y})\} + D\mu_{j,m}}{\{\gamma_{j,m}^{cor} - \gamma_{j,m}^{rec}\} + D} \tag{4}$$

$$\hat{\sigma}_{j,m}^2 = \frac{\{\theta_{j,m}^{cor}(\boldsymbol{Y}^2) - \theta_{j,m}^{rec}(\boldsymbol{Y}^2)\} + D(\sigma_{j,m}^2 + \mu_{j,m}^2)}{\{\gamma_{j,m}^{cor} - \gamma_{j,m}^{rec}\} + D} - \hat{\mu}_{j,m}^2 \tag{5}$$

where

$$\theta_{j,m}(x) = \sum_{r=1}^{R}\sum_{t=1}^{T_r} x^r(t)\,\gamma_{j,m}^r(t)$$

and

$$\gamma_{j,m} = \sum_{r=1}^{R}\sum_{t=1}^{T_r} \gamma_{j,m}^r(t)$$

In these equations, D is a constant which determines the rate of convergence of the re-estimation formula. If D is too big then convergence is too slow, if it is too small then instability can occur. In practice, D should be set to ensure that all variances remain positive. It is also beneficial to compute separate values of D for each phone model.

As with ML-based parameter estimation, the crucial quantities to compute are the component occupation probabilities $\gamma_{j,m}^{cor}$ and $\gamma_{j,m}^{rec}$. The former is straightforward but the latter requires all possible word sequences to be considered. As noted earlier, however, lattices provide a tractable way of approximating this. A lattice is a directed graph in which each arc represents a hypothesised word. Within any given lattice, it is simple to compute the probability of being at any node using the forward-backward algorithm. For node l in the lattice and preceding words $w_{k,l}$ spanning nodes k to l, the forward probability is given by

$$\bar{\alpha}_l = \sum_k \bar{\alpha}_k P_{acoust}(w_{k,l}) P_{lang}(w_{k,l})$$

where P_{acoust} is the likelihood of word $w_{k,l}$ hypothesised between the time instances corresponding to nodes k and l, and P_{lang} is the language model probability of $w_{k,l}$. The backward probabilities $\bar{\beta}_k$ are computed in a similar fashion starting from the end of the lattice. For each pair of nodes k and l, the corresponding $\bar{\alpha}_k$ and $\bar{\beta}_l$ can be used to compute the required occupation probabilities within the word hence the quantities needed to compute the reestimation equations 4 and 5 can be calculated.

The overall framework of MMI training using lattices is illustrated in Fig. 10. First a pair of lattices is generated for each sentence in the training database: one for the numerator using the recogniser constrained by the correct word sequence, and the other using the unconstrained recogniser. The re-estimation process then consists of rescoring the lattices with the current model set, computing the occupation probabilities and finally, updating the parameters. Note that strictly the lattices should be recomputed at every reestimation cycle but this would be computationally very expensive and probably unnecessary since the set of confusable word sequences will change very little.

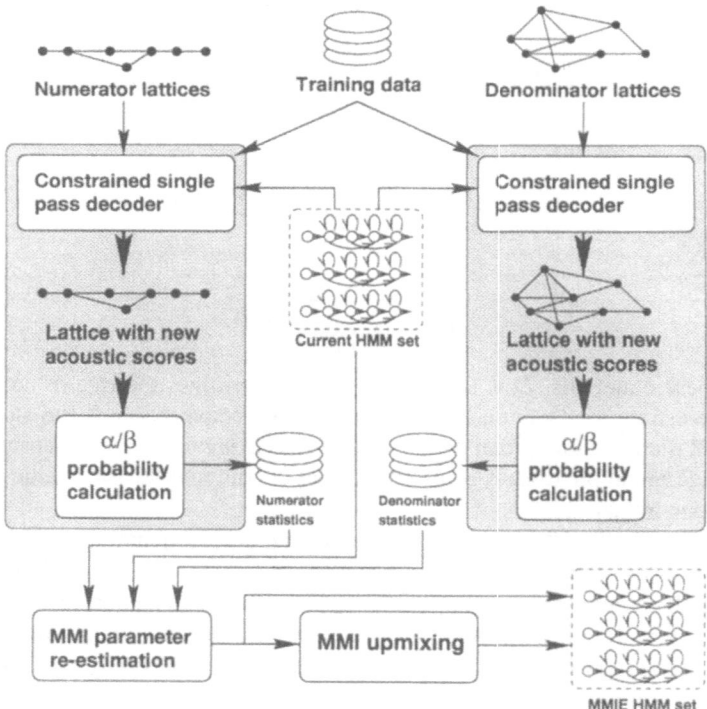

Fig. 10. Lattice-based framework for MMI training of an LVCSR system

The effectiveness of the MMI training procedure is illustrated in Fig. 11 which shows the training of a simple single Gaussian WSJ system using 60 hours of training data. The diagram on the left shows the way the MMI objective function increases at each iteration. The diagram on the right plots the % WER on both the training data and an evaluation test set. As can be seen, the errors on the training set are substantially reduced whereas much more modest improvements on the test set are obtained. More formal testing of the lattice-based MMI training procedure on a full WSJ system has shown that between 5% and 15% relative reductions in error rate can be achieved [19]. More importantly, perhaps, it appears that MMI is most effective with smaller less complex systems (i.e. systems with relatively few mixture components per state). Thus, MMI training may be particularly useful for making small compact LVCSR systems without sacrificing accuracy.

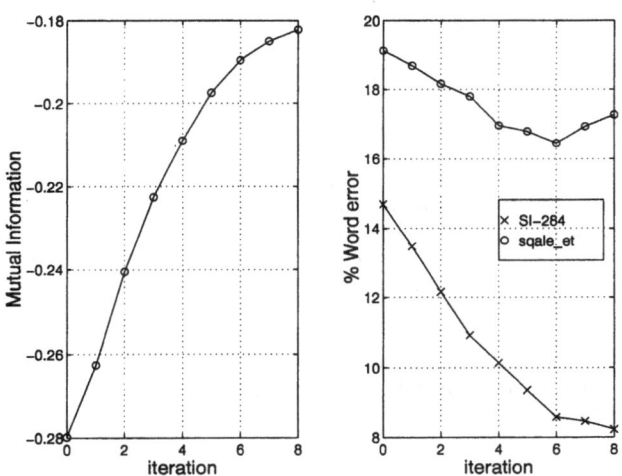

Fig. 11. MMI training performance

8. Conclusions

This chapter has described acoustic modelling in modern HMM-based LVCSR systems. The presentation has emphasised the need to carefully balance model complexity with available training data. The methods of state-tying and mixture-splitting allow this to be achieved in a simple and straightforward way. Iterative parameter re-estimation using the forward-backward algorithm has been described and the importance of the component occupation probabilities has been emphasised. Using this as a basis, two powerful methods have been presented for dealing with the inevitable mis-match between training and test data. Firstly, MLLR adaptation allows a set of HMM parameter transforms to be robustly estimated using small amounts of adaptation data. Secondly, MMI training based on lattices can be used to increase the inherent discrimination of the HMMs.

Taken together, the methods described allow speaker independent LVCSR systems to be built with average error rates well below 10%. Future developments will aim to reduce this figure further. They will also focus on more general transcription tasks such as the transcription of broadcast news material making the deployment of LVCSR technology feasible across a wide range of IT applications.

References

[1] L. Bahl, P. Brown, P. de Souza, and R. Mercer. Maximum Mutual Information Estimation of Hidden Markov Model Parameters for Speech Recognition. In *Proc ICASSP*, pages 49–52, Tokyo, 1986.

[2] L. Bahl, P. de Souza, P. Gopalakrishnan, D. Nahamoo, and M. Picheny. Context Dependent Modeling of Phones in Continuous Speech Using Decision Trees. In *Proc DARPA Speech and Natural Language Processing Workshop*, pages 264–270, Pacific Grove, Calif, Feb. 1991.

[3] J. Baker. The Dragon System - an Overview. *IEEE Trans ASSP*, 23(1):24–29, 1975.

[4] L. Baum. An Inequality and Associated Maximisation Technique in Statistical Estimation for Probabilistic Functions of Markov Processes. *Inequalities*, 3:1–8, 1972.

[5] J. Bellegarda and D. Nahamoo. Tied Mixture Continuous Parameter Modeling for Speech Recognition. *IEEE Trans ASSP*, 38(12):2033–2045, 1990.

[6] S. Davis and P. Mermelstein. Comparison of Parametric Representations for Monosyllabic Word Recognition in Continuously Spoken Sentences. *IEEE Trans ASSP*, 28(4):357–366, 1980.

[7] A. Dempster, N. Laird, and D. Rubin. Maximum likelihood from incomplete data via the EM algorithm. *J Royal Statistical Society Series B*, 39:1–38, 1977.

[8] M. Gales. The Generation and Use of Regression Class Trees for MLLR adaptation. Technical Report CUED/F-INFENG/TR.263, Cambridge University Engineering Department, 1996.

[9] M. Gales. Maximum Likelihood Linear Transformations for HMM-Based Speech Recognition. Technical Report CUED/F-INFENG/TR.291, Cambridge University Engineering Department, 1997.

[10] H. Hermansky. Perceptual Linear Predictive (PLP) Analysis of Speech. *J Acoustical Soc America*, 87(4):1738–1752, 1990.

[11] X. Huang and M. Jack. Semi-continuous hidden Markov models for Speech Signals. *Computer Speech and Language*, 3(3):239–252, 1989.

[12] M.-Y. Hwang and X. Huang. Shared Distribution Hidden Markov Models for Speech Recognition. *IEEE Trans Speech and Audio Processing*, 1(4):414–420, 1993.

[13] F. Jelinek. Continuous Speech Recognition by Statistical Methods. *Proc IEEE*, 64(4):532–556, 1976.

[14] A. Kannan, M. Ostendorf, and J. Rohlicek. Maximum Likelihood Clustering of Gaussians for Speech Recognition. *IEEE Trans on Speech and Audio Processing*, 2(3):453–455, 1994.

[15] C. Leggetter and P. Woodland. Maximum Likelihood Linear Regression for Speaker Adaptation of Continuous Density Hidden Markov Models. *Computer Speech and Language*, 9(2):171–185, 1995.

[16] Y. Normandin. *Hidden Markov Models, Maximum Mutual Information Estimation, and the Speech Recognition Problem*. PhD thesis, Dept of Elect Eng McGill University, Mar. 1991.

[17] J. Odell, V. Valtchev, P. Woodland, and S. Young. A One-Pass Decoder Design for Large Vocabulary Recognition. In *Proc Human Language Technology Workshop*, pages 405–410, Plainsboro NJ, Morgan Kaufman Publishers Inc, Mar. 1994.

[18] D. Pallett, J. Fiscus, and Przybocki. 1996 Preliminary Broadcast News Benchmark Tests. In *Proc DARPA Speech Recognition Workshop*, pages 22–46, Chantilly, Virginia, Feb. 1997. Morgan Kaufmann.

[19] V. Valtchev, P. Woodland, and S. Young. Lattice-based Discriminative Training for Large Vocabulary Speech Recognition. In *Proc ICASSP*, volume 2, pages 605–608, Atlanta, May 1996.

[20] P. Woodland, M. Gales, D. Pye, and S. Young. Broadcast News Transcription using HTK. In *Proc ICASSP*, volume 2, pages 719–722, Munich, Germany, 1997.

[21] P. Woodland, M. Gales, D. Pye, and S. Young. The Development of the 1996 HTK Broadcast News Transcription System. In *Proc DARPA Speech Recognition Workshop*, pages 73–78, Chantilly, Virginia, Feb. 1997. Morgan Kaufmann.

[22] P. Woodland, C. Leggetter, J. Odell, V. Valtchev, and S. Young. The 1994 HTK Large Vocabulary Speech Recognition System. In *Proc ICASSP*, volume 1, pages 73–76, Detroit, 1995.

[23] S. Young, J. Odell, and P. Woodland. Tree-Based State Tying for High Accuracy Acoustic Modelling. In *Proc Human Language Technology Workshop*, pages 307–312, Plainsboro NJ, Morgan Kaufman Publishers Inc, Mar. 1994.

[24] S. Young and P. Woodland. State Clustering in HMM-based Continuous Speech Recognition. *Computer Speech and Language*, 8(4):369–384, 1994.

Tree-based Dependence Models for Speech Recognition

Mari Ostendorf, Ashvin Kannan and Orith Ronen

Electrical and Computer Engineering Department, Boston University
8 St. Mary's St., Boston, MA 02215 USA
email: mo@bu.edu

Summary. The independence assumptions typically used to make speech recognition practical ignore the fact that different sounds in speech are highly correlated. Tree-structured dependence models make it possible to represent cross-class acoustic dependence in recognition when used in conjunction with hidden Markov or other such models. These models have Markov-like assumptions on the branches of a tree, which lead to efficient recursive algorithms for state estimation. This paper will describe general approaches to topology design and parameter estimation of tree-based models and outline more specific solutions for two examples: discrete-state hidden dependence trees and continuous-state multiscale models, drawing analogies to results for time series models. Initial results for both cases will be described, followed by a discussion of questions raised by the experiments.

1. Introduction

In speech recognition, independence assumptions are typically made to reduce the complexity of automatic training and the recognition search. In particular, a standard assumption used in virtually all recognition systems is that each vector or segment is generated independently given an underlying state or phone sequence. In other words, in a speaker-independent system, there is no notion that an /aa/ and an /ah/ (or even another /aa/) in the same utterance or speaker session have something in common because they came from the same vocal tract. The assumption effectively allows two phones at different times in an utterance to come from different speakers. Vocal tract length (VTL) normalization (e.g. [1]) compensates for this problem to some extent, but it is clear that VTL normalization does not account for all speaker-dependent effects because gains are additive when it is used in combination with acoustic model adaptation. In addition, sounds can be correlated for other reasons, such as the recording environment or dialect-related pronunciation patterns.

Acoustic model adaptation is used to overcome this problem, but in large vocabulary recognition one often has very little data from the target speaker with which to adapt possibly millions of parameters. Therefore, most current adaptation techniques assume classes of models over which the adaptation transformation is tied (e.g. [2, 3]), or they may approximate the joint dependence of different speech sounds by defining regions of local dependence ([3, 4]). However, such approaches do not take full advantage of the predictive power that observations from one phone have for another.

Ignoring the speech recognition application for the moment, our problem is to estimate a probability distribution that represents the joint dependence of variables in a very high dimensional space and to use this distribution to make inferences about missing variables. We refer to such a probability distribution as a dependence model. For such a model to be practical, Markov-like assumptions are required; some examples include Markov random fields and Bayesian networks. In this work, we focus on a particular class of Markov dependence models that are tree-based, because of the additional efficiency of estimation and prediction algorithms and because topology design is simpler and arguably more robust for trees.

The remainder of the chapter is organized as follows. In the next section, we introduce a general hidden tree framework that handles the type of variable-length observations encountered in speech applications. Then, we describe two specific examples of dependence models and how they can be used in speech processing applications. We conclude with a brief summary and discussion of open questions.

2. Hidden Tree Framework

The problem of acoustic modeling of speech sounds includes the important issue of characterizing variable-length observations. A speech sound may occur a different number of times (or not at all) and each instance has a randomly varying length. We handle this problem in the dependence model by defining a fixed-dimension hidden state $X = [x_1, \ldots, x_N]$ that represents the joint dependence of speech sounds x_i, which are associated with observations $Y = \{Y_i; i = 1, \ldots, N\}$. For a tree-based model, x_i corresponds to a node in the tree, and N is equal to the number of nodes in the tree. Each observation $Y_i = [y_{i,1}, \ldots, y_{i,L_i}]$ is a concatenation of the L_i different instances of sound i (ignoring time order), and can be thought of as a random process with characteristics depending on the hidden state. Figure 1 illustrates a tree with a variable number of observations associated with each node. Depending on the underlying model of speech, $y_{i,j}$ may correspond to a frame of speech (i.e. a vector of cepstral coefficients, the "observation" associated with an HMM state) or a segment trajectory (i.e. the vector of coefficients describing a trajectory of features, the sufficient statistics for the observation associated with a polynomial segment model [19]). The complete set of observations Y may correspond to a single utterance or a collection of utterances.

The probabilistic model is specified by: a function that defines the tree topology, e.g. $\pi(i)$ is the parent of node i; Markov state distributions associated with branches in the tree, $p(x_i|x_{\pi(i)})$; and observation distributions $p(Y_i|x_i)$, assuming that observations are conditionally independent given the state. The conditional independence assumptions, like those in a hidden Markov model (HMM), make implementation practical. The differences with respect to an HMM are that dependence is between un-ordered sound classes rather than sequential states in time, and that the state sequence is a fixed-length vector rather than a random process.

As with an HMM, there are three problems to solve in applying the dependence model: optimal state estimation, efficient computation of the likelihood $p(Y)$, and model design (i.e. topology design and distribution parameter estimation). These problems are solved using recursive algorithms that take advantage of assumptions of conditional independence of nodes and subtrees given the value of an intermediate node. The solutions are analogous to the corresponding HMM algorithms but

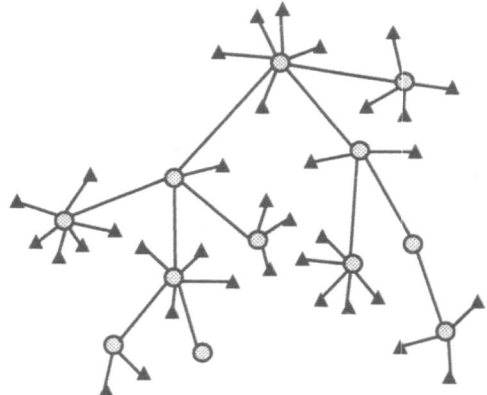

Fig. 1. Illustration of a tree-structured model with a hidden state: open circles indicate the nodes of the tree that form the hidden state X; filled triangles denote observations associated with a node Y

differ in that the updates follow the tree structure rather than a linear time sequence. Details of the algorithms depend on the particular state and observation distribution assumptions, but there are some issues that apply in general, as described below.

There are two main types of tree topologies that could describe a collection of sound classes, and the relationship between the number of nodes in the tree N and the number of sound classes M depends on the particular type of topology. At one extreme is the graph of connections between sound classes, where every node in the graph corresponds to one of a disjoint collection of classes (one x_i per class). In this case, $N = M$, and topology design involves finding the best graph that connects the classes. At the other extreme is a hierarchically organized tree, where the target sound classes comprise the leaves of the tree and sub-classes representing different levels of granularity are introduced at internal nodes. In this case, topology design involves clustering the M classes. If sub-classes are defined using a binary tree, then $N = 2M - 1$. Hybrid versions can be envisioned and probably will be the most effective solution, in part because the introduction of sub-classes is a useful tool for robust topology design when M is large.

Parameter estimation for dependence models with a hidden tree structure must address the problem of unobserved variables in estimation, which is generally solved using the Expectation-Maximization (EM) algorithm [5]. The algorithm involves two steps: 1) finding the expected joint likelihood of the hidden state and observations given the current parameter estimates, and 2) computing the maximum likelihood estimate of the parameters in terms of the statistics found in step (1). Making Markov assumptions on the tree, the first step can be implemented efficiently with an algorithm analogous to the forward-backward algorithm used in HMM parameter estimation, except that it runs upward and downward on the tree rather than forward and backward in time.

In the two sections that follow, we introduce two different examples of tree-based dependence models: the hidden dependence tree, which represents a discrete-valued hidden state using the disjoint-class topology; and the multiscale model, which represents a continuous-valued state using a hierarchical topology. Each section first describes the mathematical framework of the model and its application to speech recognition, followed by the algorithms for topology design and parameter estimation, and finally presents some experimental results.

3. Hidden Dependence Trees

Hidden dependence trees are an extension of the discrete dependence trees introduced by Chow and Liu [6] to efficiently model the dependence among a set of random variables. The dependence tree represents a discrete underlying state, and the extension allows for variable-length and continuous-valued observations.

3.1 The Mathematical Framework

In the discrete-state case, the joint probability function $P(X)$ is modeled using a dependence tree [6]. A component x_i is assigned to node i in the tree, and each edge in the tree is associated with the conditional probability function of the two variables connected by the edge, i.e. $P(x_i|x_j)$ for the edge connecting the node of x_i to its parent x_j. The parent of node i, denoted by $\pi(i)$, is the first node on the path connecting node i to the root. The root of the tree is associated with the component x_0, which is introduced for notational purposes and is not actually a component of X. The nodes x_i connected to the root have $\pi(i) = 0$ as their parent, and edges $P(x_i|x_0)$ are defined to be $P(x_i)$. In other words, the x_i with $\pi(i) = 0$ are independent and hence so are the respective subtrees associated with those nodes. The dependence tree state distribution model is

$$P(X) = \prod_{i=1}^{N} P(x_i|x_{\pi(i)}),\tag{1}$$

and the joint observation-state distribution is

$$p(Y, X) = \prod_{i=1}^{N} p(Y_i|x_i)P(x_i|x_{\pi(i)}) = \prod_{i=1}^{N} \left[\prod_{j=1}^{L_i} p(y_{i,j}|x_i) \right] P(x_i|x_{\pi(i)}),\tag{2}$$

assuming that $\{y_{i,j}\}$ are conditionally independent and identically distributed given the state. (Note that upper case P is used to denote a probability mass function, and lower case p a density function.) As in an HMM, the probability of a set of observations is computed by summing over the possible state vectors

$$p(Y) = \sum_{X} \prod_{i=1}^{N} p(Y_i|x_i)P(x_i|x_{\pi(i)}).\tag{3}$$

The sum can be computed efficiently using a recursive algorithm that incorporates probabilities from the leaves upward to the root of the tree, analogous to the forward algorithm for HMMs.

3.2 Application to Speech

The dependence tree state is hidden in the same sense that the mode of a Gaussian mixture distribution is hidden; observations are continuous-valued cepstral features described by Gaussian distributions conditioned on the hidden state. The difference is that the dependence tree state is vector-valued, unlike the scalar mode of a Gaussian mixture distribution. An HMM that uses Gaussian mixture observation distributions also has a multi-dimensional state, but there are important differences with respect to dependence trees. The HMM state sequence is variable-length and time-ordered. In the hidden dependence tree, on the other hand, the state dimension and order is fixed, and there is no notion of time. The state probability distributions in an HMM ($a_{ij} = P(s_t = i | s_{t-1} = j)$) describe sequence length and ordering, while the state probability distributions in a dependence tree ($a_{i,jk} = P(x_i = j | x_{\pi(i)} = k)$) describe the relationship between the values of states in a fixed order.

The analogy of the hidden dependence tree to a Gaussian mixture and the differences with respect to an HMM suggest an application of the dependence tree in acoustic modeling. Consider an HMM that uses Gaussian mixture distributions. Let $x_i = j$ in the dependence tree indicate that the mixture mode of HMM state i is j. Then the hidden dependence tree provides a model for correlation of the mixture modes across sound classes. With this interpretation, one can envision different applications of the dependence model used in conjunction with an HMM. Assume that an HMM is first used to provide a "transcription" and segmentation of an utterance in terms of the N sound classes in the dependence tree. The transcription is used to group the observations into subsets Y_i, and the hidden dependence tree model is then used to compute $p(Y)$. This probability can be used in a likelihood ratio test of whether two segments of speech came from one vs. two speakers (or in other text-independent speaker/language identification problems), or as an additional "consistency score" in N-best rescoring of hypotheses for word recognition. Alternatively, the observations can be used to re-estimate mixture weights, i.e. $\hat{\lambda}_{ij} = P(x_i = j | Y)$ for the j-th mixture weight associated with state i, for use in a subsequent decoding pass. This probability is computed using the upward-downward algorithm used for state estimation in model design, described next.

3.3 Topology Design and Parameter Estimation

In this discussion, we will assume a non-hierarchical topology for the dependence tree structure; that is, the classes represented by the tree are disjoint. For the case where X is discrete-valued and fully observable, Chow and Liu [6] describe an algorithm for estimating both dependence tree structure and its parameters. In our case, where both the tree and subsets of the observations Y_i are unobserved, we divide topology design and parameter estimation into two steps. However, we build on the Chow-Liu algorithm by defining an intermediate, partially observable discrete state, as described below.

Class Definition and Topology. Topology design requires finding $\pi(i)$ for all nodes $i = 1, \ldots, N$. The Chow and Liu algorithm finds the tree topology that minimizes the difference of the information contained in the true probability function and that contained in its approximation by a dependence tree. This minimization criterion is equivalent to maximizing the total weight on the edges of the tree, where the weight of the edge connecting nodes x_i and x_j is the mutual information $I(x_i; x_j)$ based on relative frequency estimates of their joint probability distribution. Given all possible $I(x_i; x_j)$, topology design is a minimum spanning tree search problem. The Chow and Liu algorithm works well when the samples of the vector X are complete, meaning that all the components of samples are observed, and when the number of samples of the vector X is large relative to the number of values an x_i can take on. When there are a small number of samples for a pair of variables, the mutual information estimate is biased above the true value, so (x_i, x_j) pairs that are infrequently observed may be incorrectly assigned links in the tree.

In order to use the Chow-Liu algorithm, we estimate a discrete state vector X for each training sample by setting x_i equal to the index of the vector quantization (VQ) codeword that minimizes the total distortion of the observations $y_j \in Y_i$. In order to keep the number of values for x_i small and still have a reasonable sampling of the vector observation space, node-dependent codebooks are designed. Assuming the tree nodes correspond to phonetic or sub-phonetic units, there will be some missing elements of the estimated state vectors, because of the wide variation in frequencies of occurrence of different phonemes. This imbalance of phone-pair occurrence rates can lead to bad estimates of the mutual information and poor tree topologies. To obtain robust trees, we modified the Chow-Liu algorithm to include a threshold on the number of co-occurrences of every phone pair for allowing a link between the pair and a limit on the number of connections in the tree. In addition, we used random sampling to obtain speaker-level state vectors to reduce the number of missing elements relative to utterance-level vectors.

Parameter Estimation. Two sets of distribution parameters are needed to characterize the hidden dependence tree: the tree edge distributions $P(x_i|x_{\pi(i)})$ and the observation distributions $p(y|x_i = j) \sim \mathcal{N}(\mu_{ij}, \Sigma_{ij})$, where $\mathcal{N}(\mu, \Sigma)$ denotes a Gaussian with mean μ and covariance Σ. The above topology design gives an initial estimate for the edge distributions. Initial estimates for μ_{ij} and Σ_{ij} are given by the VQ codewords and associated error covariances. Given an initial estimate, the parameters can be refined based on the actual observations using the iterative EM algorithm. As mentioned earlier, the expectation step involves a recursive upward-downward algorithm that is analogous to the forward-backward algorithm used in HMM parameter estimation. If the VQ-estimated observations of X are available, then the tree edge distributions can be estimated using the upward-downward algorithm for discrete dependence trees [7], which is an extension of Pearl's algorithm for belief propagation in causal trees [8] and a special case of the more general algorithm for Bayesian networks described by Lucke [9]. To estimate both the edge distributions and the observation distributions, the upward-downward algorithm is extended to use the observations Y_i in the upward step during the update of node i, and parameter re-estimation for the node-dependent Gaussians is added to the maximization step [10]. The complete parameter estimation algorithm is similar to that used for HMMs with Gaussian mixture distributions, with the difference being added complexity due to tree-structured rather than time-based dependence.

3.4 Experiments

Experiments assessing various methods for topology training and the usefulness of the hidden dependence model were conducted on two large vocabulary continuous speech recognition tasks using the Wall Street Journal [11] and the Switchboard [12] corpora, in both cases training on roughly 120 hours of speech. The WSJ corpus is based on read business news, and the Switchboard corpus comprises telephone-quality conversational speech on a variety of topics. The feature vectors included 14 mel-warped cepstra (no derivatives), computed at a 10ms frame rate using cepstral mean subtraction. In the experiments on Switchboard, the features were also normalized to compensate for vocal tract length [1]. The X vector associated each node with a phone, so the dimension was 50-60 (i.e. context-independent models) and therefore the dependence tree models used only the frames in the center of the phone segment to minimize coarticulation influences.

In development of the topology design approach, we evaluated the performance of the dependence tree models by computing the likelihood of an independent test set. The results showed that the dependence tree performed better than an independent-phone model, and that constraints on the topology of the tree generally improved performance. In addition, the automatically designed dependence tree outperformed a tree that had been specified by hand according to manner of articulation and other differences in articulatory features. As an example, the tree topology designed on the Switchboard corpus is given in Figure 2, illustrating learned dependence that reflects articulation manner (e.g. among fricatives and nasals) but also some connections that were probably dominated by co-articulation effects (e.g. /aa/-/er/ since /aa/ is often followed by /r/).

We evaluated the performance of the hidden dependence tree model by using the likelihood $p(Y)$ as an additional score in N-best rescoring experiments. In these experiments an HMM-based recognition system from BBN [13] provided an N-best list of hypotheses ($N = 100$) for all the utterances in the test set, along with an HMM acoustic score and a trigram language model score for each hypothesis. These hypotheses were rescored by the hidden dependence tree model. A linear combination of these scores plus word and phone counts (insertion penalties) was used for re-ranking the list of hypotheses and producing the final recognized output. The weights of the different scores were optimized on a development test set. The dependence tree model used in this experiment was a gender-dependent model with 10 node-dependent codewords per phone and a constrained tree. Table 1 shows the results of these experiments. There is a slight improvement when combining the likelihood score obtained from the dependence tree model with the HMM score. Further gains should be obtained by using more detailed sound classes, such as triphone states, but the resulting dependence tree would be large and likely require a hybrid hierarchical and Chow-Liu topology design strategy.

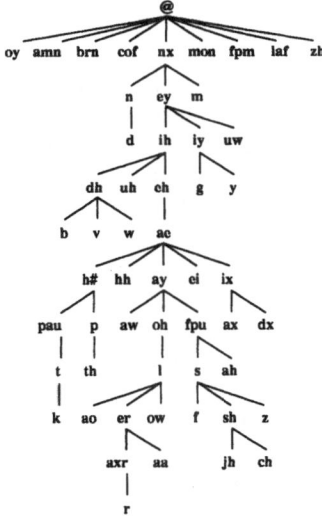

Fig. 2. Discrete dependence tree designed on the Switchboard corpus, where subtrees connected to the root node (indicated by "@") are independent

Table 1. N-best rescoring results (word error rates) on the 1993 WSJ evaluation test and the 1996 SWBD evaluation test. The knowledge sources are the HMM acoustic score, the dependence tree score (DT), and a trigram language model (LM).

Knowledge Sources	WSJ Eval93	SWBD Eval96
HMM, LM	15.7	44.6
HMM, DT, LM	15.5	44.5

4. Multiscale Tree Processes

Multiscale stochastic processes represent an important class of models, of which a particularly useful subclass is based on scale-recursive dynamics on trees [14, 15]. These models allow efficient algorithms for both estimation and likelihood calculation resulting in a variety of applications. In this section, we describe the general framework and application to acoustic model adaptation in speech recognition.

4.1 The Mathematical Framework

Denoting a node in a tree by t with parent[1] $t\bar{\gamma}$, a state-space model for the evolution in the tree of the Gaussian process X and its noisy observation Y is given by

---

[1] The notation $t\bar{\gamma}$ represents the same information as $\pi(i)$ for the hidden dependence tree; the two notations are used to be consistent with other literature in the respective areas.

$$x_t = A_t x_{t\bar{\tau}} + w_t \tag{4}$$

$$y_{t,i} = C_t x_t + v_{t,i} \tag{5}$$

where x_t is the vector state of the process at node t. The root node state x_0 has distribution $\mathcal{N}(0, \Sigma_0)$. The process noise w_t is white, independent of x_0, and has distribution $\mathcal{N}(0, Q_t)$. The state x_t is observed via a noisy measurement $y_{t,i}$, where the measurement noise $v_{t,i}$ is white, independent of x_0 and w_t, and has distribution $\mathcal{N}(0, R_t)$. Thus, $\Theta_X = (\Sigma_0, \{A_t, Q_t\})$ are the parameters of $p(X)$, and $\Theta_{Y|X} = (\{C_t, R_t\})$ are the parameters of $p(Y|X)$. The zero-mean assumptions of the root node x_0 and the noise terms are not a requirement of the model, but result in simpler estimation equations.

A degenerate tree with only one leaf node (parent nodes have only one child) can be interpreted as a standard linear dynamical system, i.e. having a time-like index. As an acoustic model for speech recognition, the standard dynamical system is a continuous-state alternative to the discrete-state HMM, where likelihood is computed using Kalman filtering recursions to obtain innovations and associated distribution parameters [16]. A similar approach can be used for the multiscale model extending the Kalman recursions on the tree [17].

For the adaptation application, state estimation is more important than likelihood computation. Given Y, the set of all available observations, the smoothed estimate[2] of the state $\hat{x}_t = E\{x_t|Y\}$ and the associated error covariance $P_{t|Y} = E\{[x_t - \hat{x}_t][x_t - \hat{x}_t]^T\}$ is computed using a generalization of the Rauch-Tung-Striebel (RTS) algorithm [14]. Smoothing is done in two sweeps: an upward sweep from the leaves to the root, followed by a downward one from the root to the leaves. The complexity of the tree RTS smoother is $O(d^3 N)$ where d is the dimensionality of the state (the d^3 is due to matrix inversions), and N is the number of nodes in the tree.

4.2 Application to Speech

The multiscale model can be used for the adaptation of means of acoustic models to a new speaker or new environmental conditions. For example, let each leaf τ of the multiscale tree be associated with a set of Gaussians $\mathcal{G}_\tau$, and adapt the means of all Gaussians in class τ by a common shared shift x_τ:

$$\mu_i^a = \mu_i + x_\tau, \ \forall \, i \in \mathcal{G}_\tau, \tag{6}$$

where μ_i^a denotes the mean μ_i after adaptation. Such a shared shift approach has been used for Gaussians in hidden Markov models (HMMs) [3] and the stochastic segment model (SSM) [18], and for polynomial segment models (PSMs) [19]. The observations $y_{\tau,i} \in Y_\tau$ associated with node τ are differences between the speaker-independent means μ_i and the average of feature vectors observed for sound $i \in \mathcal{G}_\tau$ in an utterance.

---

[2] The term "smoothed estimate" refers to the linear least squares (or for Gaussians, the minimum mean squares) estimate. It also corresponds to the maximum *a posteriori* estimate.

Initial independent estimates for the shift $\bar{x}_\tau$ and associated error covariance P_τ can be obtained from adaptation data for each class l by averaging the observed shifts for that class ($y_{\tau,i} \in Y_\tau$) and computing the equivalent covariance of the averaged variable [19]. Let us define a Gaussian tree-based shift process (Equation 4) with M leaves, and associate the leaf node states with the shifts of the M classes we wish to model dependence between. Given $\bar{x}_\tau$ and P_τ at a subset of leaves, adaptation involves estimating the hidden states $\hat{x}_\tau$ for all leaves using the tree RTS smoother and then shifting the models within the respective classes. Due to the Bayesian nature of the estimate, the smoothed shift approaches the unsmoothed shift and converges to the standard ML speaker-dependent estimate as the amount of adaptation data increases.

A similar tree-based model for adaptation is described in [20], but the upward-downward propagation of mean shifts is based on a heuristic that does not account for degree of correlation or variance differences. Another Markovian model used in adaptation is based on Markov random fields [21]. Multiscale models offer a number of advantages over Markov random fields including a constant per-node complexity, the availability of an error covariance associated with smoothing, and the fact that state estimation algorithms are efficient, non-iterative, recursive and parallelizable.

4.3 Topology Design and Parameter Estimation

To use the multiscale model of dependence in adaptation, we need to define the adaptation classes and tree topology, as well as estimate parameters for the process.

Class Definition and Topology. In continuous speech recognition, context-dependent models are frequently clustered in the form of a tree for each region (or state) of a phone using ML clustering of Gaussians [22, 23]. Figure 3(a) illustrates the tree for one region. Each node of the tree represents an equivalence class of triphones. Nodes at a certain "cut" through the tree, the boxes in Figure 3(a), define terminal adaptation classes to share shifts (Equation 6). One popular, but ad-hoc, option for adaptation is the "back-off" strategy, where the shift is computed at the most detailed node which has more than some threshold of adaptation frames and copied to all child terminal adaptation classes as shown in Figure 3(b). The topology of the clustering tree can also be used for multiscale smoothing. Class-dependent shifts $\bar{x}_t$ are computed at the terminal adaptation nodes and then smoothed using the multiscale model (Figure 3(c)) to get shift estimates $\hat{x}_t$ at *all* nodes.

Parameter Estimation. Maximum-likelihood estimates of the parameters of the tree process ($\Sigma_0, A_t, Q_t, C_t, R_t$) can be obtained by applying the RTS and EM algorithms to multiple independent sample vectors Y [24], where each conversation side contributes one sample. The general approach follows that described in [16] for a time-ordered dynamical system, which involves iteratively finding expected sufficient statistics of the hidden state (E-step), and then using multivariate regression to compute new process parameters (M-step). The main difference is in the recursions used in the E-step, which build on the tree algorithms developed in [14, 25]. Here, we assume $C_t = I$ and R_t is effectively given by the sample variance of the observations, so there is no need to estimate C_t and R_t. In the experiments described

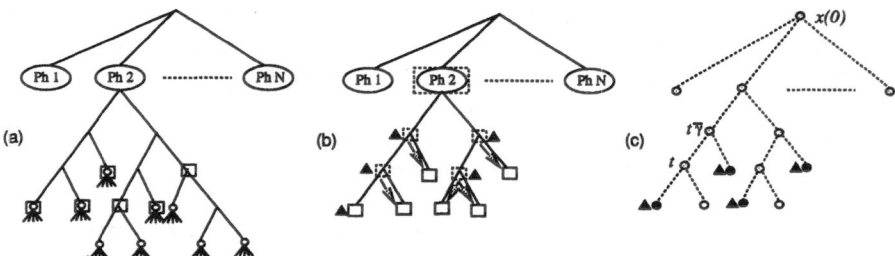

Fig. 3. Trees used for adaptation: (a) shows the clustering tree with squares indicating terminal adaptation classes, (b) illustrates the "back-off" method of adaptation with dashed squares indicating back-off classes, and (c) shows the corresponding multiscale smoothing approach. In both (b) and (c), triangles indicate observations.

next, the A and Q parameters are shared among all nodes of a phone; i.e. for K phones, there are K trees each with an (A, Q) pair. To start the EM iterations, we need initial estimates of $\Sigma(0)$, the A's and the Q's. For each speaker in the training set we compute covariances of ML (unsmoothed) shifts at each terminal shift node. A frequency-weighted average of these covariances across all speakers is used for initializing $\Sigma(0)$ and all Q_t, and initial $A_t = I$ for all t.

4.4 Experiments

Experiments were conducted on the Switchboard corpus. The feature vectors were the same as for the hidden dependence tree experiments except that energy and feature derivatives are used. N-best rescoring is also used here, with the segment-model acoustic score substituted for the HMM score. Most experiments use sixty hours of speech for training the acoustic and multiscale (MS) models; 123 hours are used in the guided adaptation experiments.

The PSM systems used a 2-region model, with each region modeled by a linear trajectory Gaussian process with a single full covariance. The SSM systems used a 5-region model, with each region represented by a full covariance Gaussian. Both cases used gender-dependent models and ML clustered triphones. The PSM and SSM adaptation systems had 300 and 150 terminal adaptation classes/region, respectively. For a fair comparison of MS smoothing vs. back-off approaches, the same topology is used for both types of adaptation.

In batch-mode adaptation, the first half of each conversation is used as adaptation data and the second half for testing. The results in Table 2 for supervised adaptation indicate the MS-smoothing is better than the back-off approach. However, performance for both algorithms degrades relative to the baseline in unsupervised adaptation, indicating a sensitivity to the high error rate in the Switchboard tasks. Consequently, we use guided adaptation in further unsupervised experiments.

In unsupervised transcription-mode adaptation two passes are made over the speech: the first to collect statistics for adaptation, and the second to perform recognition with the adapted models. Adaptation is *guided* in the sense that we adapt only

Table 2. Supervised batch recognition with 2-region PSMs. Error rates are on the second half of conversations in the Dev96 test set.

SI baseline	back-off	multiscale
44.5%	44.1%	43.9%

with data from the subset of words in the top first-pass hypothesis with confidence over a specified threshold. This serves to lower the error-rate for the speech used in adaptation, which benefits both ML and MS adaptation. Guided adaptation also tests the generalization capability of the dependence model for unseen classes (in the "incorrect" parts of the speech). In experiments on the Dev97 test set with a 5-region SSM, we found that the ML back-off approach improved a 40.9% WER baseline[3] to 40.4% and that further improvement to 40.0% was obtained with MS smoothing. Table 3 shows that MS adaptation gives about 1% absolute improvement in performance. A much greater gain is expected from using lattice decoding rather than N-best rescoring, based on BBN adaptation experiments [26].

Table 3. Guided unsupervised transcription mode adaptation with a 5-region SSM system

	Dev97	Eval97
Baseline	40.9	52.5
MS adapt	40.0	51.6

5. Discussion

In summary, tree-based models of dependence provide an efficient framework for representing correlation across phones in speech (or sub-phonetic units represented by HMM states), for use in adaptation as well as other applications. Dependence models are a supplement to and not a replacement for existing techniques, such as HMMs, in that they model correlation across classes but not time. Markov-like assumptions combined with a tree structure make for efficient algorithms for computing the expected state given a set of observations. Dependence model design involves first finding the tree topology, which can be an direct connection of classes or a hierarchy of sub-classes, and then EM parameter estimation using an upward-downward algorithm for handling the hidden state. Two important examples are described: the hidden dependence tree, which has a discrete hidden state and can be thought of as a mechanism for loosely coupling Gaussian mixture modes of different models; and the multiscale model, which has a continuous hidden state and relates models via a hierarchy with different levels of granularity. The two approaches differ primarily in the discrete vs. continuous hidden state, but they also

---

[3] The lower baseline error rate is due to differences in the test set, language model, signal processing parameters, and training on 123 hours of speech.

illustrate two extremes of topology design. It is an open question as to which of the two models is more useful: the hidden dependence tree is better at capturing non-linear dependence between classes, but the mixture mode dependence may be too weak a coupling. Initial results for both models are promising, but much work remains to explore their full potential. For example, the speaker adaptation experiments did not take advantage of several variations known to improve results, such as speaker-adaptive training [27] and iterative adaptation and decoding.

Several questions are raised by the initial experimental results, particularly related to topology design and parameter tying. How can we best integrate the mutual information clustering technique, which is more general but not very robust, with hierarchical clustering techniques? What is the right number of classes to represent with the tree? For adaptation, theoretically it is better to use a large number of classes, but in practice we do not find this to be the case, probably because of inaccuracies in the model exacerbated by parameter tying assumptions. In the multiscale model experiments, we assumed that all branches of the tree for a particular phone shared the same transition matrix and process noise covariance. Is it possible to learn finer grained parameter sharing automatically? Of course, there is also the question of whether better results can be obtained by relaxing the tree-structure assumption and using less restrictive models such as Bayesian networks. However, it is likely that the computational efficiency of the tree structure will make the tree-based dependence models more attractive in the near term.

Acknowledgement. This work was supported by the United States DoD, grant ONR-N00014-92-J-1778.

References

[1] E. Eide and H. Gish, "A parametric approach to vocal tract length normalization," *Proc. Inter. Conf. on Acoust., Speech and Signal Proc.*, vol. 1, pp. 346-348, May 1996.

[2] C. J. Leggetter and P.C. Woodland, "Flexible Speaker Adaptation Using Maximum Likelihood Linear Regression," *Proc. ARPA Workshop on Spoken Language Technology*, pp. 110-115, January 1995.

[3] G. Zavaliagkos, R. Schwartz, J. McDonough, and J. Makhoul, "Adaptation algorithms for large scale HMM recognizers," *Proc. European Conference on Speech Comm. and Tech.*, vol. 2, pp. 1131-1134, September 1995.

[4] Q. Huo and C.-H. Lee, "On-line adaptive learning of the correlated continuous density hidden Markov models for speech recognition," *Proc. Inter. Conf. on Acoust., Speech and Signal Proc.*, vol. 2, pp. 705-708, May 1996.

[5] A.P. Dempster, N.M. Laird, and D.B. Rubin, "Maximum likelihood estimation from incomplete data," *Journal of the Royal Statistical Society (B)*, vol. 39, no. 1, pp. 1-38, 1977.

[6] C.K. Chow and C.N. Liu, "Approximating discrete probability distributions with dependence trees," *IEEE Trans. Information Theory*, vol. IT-14, no. 3, pp. 462-467, May 1968.

[7] O. Ronen, J.R. Rohlicek, and M. Ostendorf, "Parameter estimation of dependence tree models using the EM algorithm," *IEEE Signal Processing Letters*, vol. 2, no. 8, pp. 157-159, 1995.

[8] J. Pearl, *Probabilistic Reasoning in Intelligent Systems: Networks of Plausible Inference*, Morgan Kaufmann, San Mateo, CA, 1988.

[9] H. Lucke, "Which stochastic models allow Baum-Welch training?" *IEEE Trans. Signal Proc.*, vol. 44, no. 11, pp. 2746-2756, 1996.

[10] O. Ronen, *Dependence tree models of intra-utterance phone dependence*, Boston University Ph.D. Thesis, 1997.

[11] F. Kubala *et al.*, "The hub and spoke paradigm for CSR evaluation," *Proc. of the ARPA Human Language Technology Workshop*, pp. 37-42, March 1994.

[12] J.J. Godfrey, E.C. Holliman, and J. McDaniel, "SWITCHBOARD: Telephone speech corpus for research and development," *Proc. Inter. Conf. Acoust., Speech, and Signal Proc.*, vol. 1, pp. 517-520, March 1992.

[13] L. Nguyen *et al.*, "The 1994 BBN/BYBLOS speech recognition system," *Proc. of the ARPA Spoken Language Systems Technology Workshop*, pp. 77-81, January 1995.

[14] K. C. Chou, A. S. Willsky, and A. Benveniste, "Multiscale recursive estimation, data fusion, and regularization," *IEEE Trans. Automatic Control*, vol. 39, no. 3, pp. 464-478, 1994.

[15] M. R. Luettgen, W. C. Karl, A. S. Willsky, and R. R. Tenney, "Multiscale representations of Markov random fields," *IEEE Trans. Signal Proc.*, vol. 41, no. 12, pp. 3377-3396, 1993.

[16] V. Digalakis, J.R. Rohlicek, and M. Ostendorf, "ML estimation of a stochastic linear system with the EM algorithm and its application to speech recognition," *IEEE Trans. Speech and Audio Proc.*, vol. 1, no. 4, pp. 431-442, 1993.

[17] M. R. Luettgen and A. S. Willsky, "Likelihood calculation for a class of multiscale stochastic models, with application to texture discrimination," *IEEE Trans. Image Proc.*, vol. 4, no. 2, pp. 194-207, 1995.

[18] A. Kannan and M. Ostendorf, "Modeling Dependency in Adaptation of Acoustic Models using Multiscale Tree Processes," *Proc. Eurospeech*, vol. 4, pp. 1863-1866, 1997.

[19] A. Kannan and M. Ostendorf, "Adaptation of polynomial trajectory segment models for large vocabulary speech recognition," *Proc. Inter. Conf. Acoust., Speech and Signal Proc.*, vol. 2, pp. 1411-1414, April 1997.

[20] D. Paul, "Extensions to phone-state decision-tree clustering single tree and tagged clustering," *Proc. Inter. Conf. Acoust., Speech and Signal Proc.*, vol. 2, pp. 1487-1490, April 1997.

[21] B. M. Shahshahani, "A Markov random field approach to Bayesian speaker adaptation," *IEEE Trans. Speech and Audio Proc.*, vol. 5, no. 2, pp. 183-191, 1997.

[22] A. Kannan, M. Ostendorf and J. R. Rohlicek, "Maximum likelihood clustering of Gaussians for speech recognition," *IEEE Trans. Speech and Audio Proc.*, vol. 2, no. 3, pp. 453-455, 1994.

[23] S. J. Young, J. J. Odell and P. C. Woodland, "Tree-based state tying for high accuracy acoustic modeling," *Proc. ARPA Workshop on Human Language Technology*, pp. 307-312, March 1994.

[24] A. Kannan, M. Ostendorf, D. A. Castañon, and W. C. Karl, "ML parameter estimation of a multiscale tree process using the EM algorithm," Technical Report ECE-96-009, Boston University, November 1996. Available from `ftp://raven.bu.edu/pub/reports`.

[25] M. R. Luettgen and A. S. Willsky, "Multiscale smoothing error models," *IEEE Trans. Automatic Control*, vol. 40, no. 1, pp. 173-175, 1995.

[26] G. Zavaliagkos, personal communication.

[27] T. Anastasakos, J. McDonough, R. Schwartz, and J. Makhoul, "A compact model for speaker-adaptive training," *Proc. of the Inter. Conf. on Spoken Language Processing*, vol. 2, pp. 1137–1140, October 1996.

Connectionist and Hybrid Models for Automatic Speech Recognition

Jean-Paul Haton

LORIA/Université Henri Poincaré, Nancy 1 BP 239, 54506 Vandœuvre-lès-Nancy, France
email: jph@loria.fr

Summary. Automatic speech recognition (ASR) has now reached the point where practical applications can be envisaged. However, the models that are presently used still have to be enhanced, especially to improve the robustness of recognition in real conditions. Most of present systems are based on stochastic models, especially hidden Markov models (HMMs). In the past few years, a quite large number of projects have been directed toward the development of a new class of models: the connectionist artificial neural networks (ANNs).

The present chapter proposes an overview of ANNs and their various uses in the field of speech signal analysis, ASR, and speaker verification. After a brief review of the basic principles of these models and of their properties, the different models used in ASR are presented and compared: multi-layer perceptrons, self-organizing maps, recurrent ANNs, etc. The modifications made to these models in order to take into account the dynamic aspects of speech are then discussed. Various types of hybrid models are finally presented. Such models combine in different ways connectionist models with other models, especially stochastic models.

Key words: Speech recognition, speaker verification, neural networks, discriminant learning, multi-layer perceptrons, self-organizing maps, recurrent networks, time delays, hybrid connectionist-stochastic network

1. Introduction

Automatic speech recognition (ASR) is a topic which has attracted particular attention and research effort in this area. In spite of this and notwithstanding important progress and cumulated improvements, the performance of ASR systems, under natural and realistic conditions, are still far from the level attained by human abilities under similar conditions. Thus, ASR is still one of the most challenging area of artificial intelligence and pattern recognition.The search for solutions to the problem of automatic speech recognition has led to the development of a large number of techniques and models [1]. Artificial Neural Networks (ANNs) have been applied with some success over the past decade. This is not really surprising, since, for a long time, ANNs have been successfully used to solve complex problems of pattern classification and recognition [2].

This chapter presents an overview of the different attempts to apply ANN models to speech and speaker recognition. There are several good reasons in favour of the use of ANNs in ASR, besides the fallacious reason that the human cortex is able to recognize speech and that ANNs represent crude models of the cortex:

- ANNs can learn from examples, a feature particularly crucial for ASR,
- ANNs, unlike HMMs, do not require strong assumptions about the statistical properties of the input data,

- the ANN formalism is in general intuitively accessible,
- ANNs are capable of producing highly non-linear functions of the inputs [3],
- ANNs have highly parallel and regular structures which make them particularly suitable for high performance hardware implementations.

The organization of the chapter is as follows. Section 2 proposes a brief introduction to ANNs and to the models that are most commonly used in the speech processing area. Section 3 focuses on the initial step of signal processing and feature extraction, and considers the role of ANN at this level. Section 4 addresses the classical problem of static speech pattern identification by neural networks. Section 5 addresses the dynamic aspects of speech recognition. Section 6 presents and discusses hybrid models the design of which is an important topic of ongoing research. Such hybrid models combine ANN with stochastic models, especially with hidden Markov models (HMMs). In the concluding section, we outline some future research perspectives.

2. A Brief Overview of Neural Networks

2.1 Basic Principles

ANN research started in the 1940s with the pioneering work of McCulloch and Pitts on the formal neuron [4] (see below). A second important milestone was the perceptron of Rosenblatt in the early 1960s [5]. There was then a dramatic decrease in research effort, due to a book by Minsky and Papert showing the limitations of the perceptron [6]. ANNs have finally received important renewed interest in the 1980s with Hopfield's approach [7] and the error backpropagation learning algorithm [8] [9].

The basic idea underlying ANN research is to take limited inspiration from neurobiology data about the human cortex in order to develop automatic models that feature some "intelligent" behaviour. The human cortex is made up of a very large number of neurons. A neuron is a biological cell capable of processing information which is massively connected with other neurons. Information processing in the cortex is a very complex phenomenon that involves electrical and chemical processes. In fact, the properties of the cortex do not come from the way neurons communicate, but from the complex interconnections between them. McCulloch and Pitts proposed a formal model of a neuron under the form of a binary threshold unit. This "neuron" calculates a weighted sum of its inputs, and produces a binary output 1 if this sum exceeds a predetermined threshold, as illustrated in figure 1.

The neuron model proposed by McCulloch and Pitts is a rough approximation of the biological neuron. It contains a number of simplifications, so that it does not reflect the actual behaviour of cortical neurons. However, most of today's ANNs are based on this initial model with various generalizations. The most common one is to replace the threshold decision function by a more complex one, usually a sigmoid function, or a Gaussian function. An ANN is made up of a large number of such interconnected neurons. The topology of the network is an essential characteristic of an ANN. Two main categories can be distinguished:

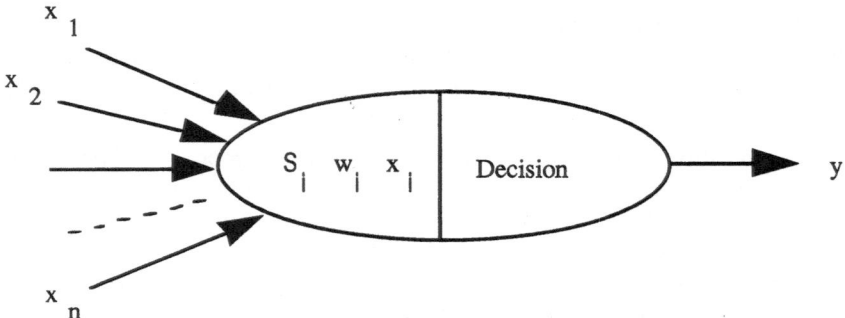

Fig. 1. Principle of McCulloch and Pitts neuron

- feedforward networks, in which no loops exist in the connections between neurons,
- recurrent networks, which contain loops between neurons corresponding to feedback connections.

In summary, ANNs present characteristics which make their use interesting in ASR especially because:

- they have a classification ability that comes from their inherently discriminant learning algorithms. Besides, the discriminant functions that are learned can be as complex as necessary. Once the learning phase is over, ANNs also have a capacity of generalization for input stimuli which are unknown or degraded;
- they have regular, parallel structures that make them implementable on VLSI circuits (even though most of the present ANN systems are simulated on traditional computers).

2.2 Main Models for ASR

Most of the numerous ANN systems that have been developed so far belong to four main categories, i.e., multi-layer perceptrons, self-organizing maps, probabilistic networks, and recurrent networks.

Multi-layer perceptrons (MLPs) are feedforward networks in which the outputs of layer i constitute the entries of layer i+1. MLPs are direct generalization of Rosenblatt's single layer perceptron. They can learn arbitrarily complex decision boundary functions from sample examples in a supervised manner. The development of the error gradient backpropagation learning algorithm has made these networks the most popular, especially for classification problems. MLPs generally use sigmoids as decision function.

Self-organizing maps (SOM) [10] are based on the property of topology preservation. This property, found in the cortex of humans and of highly developed animals, consists of mapping similar input stimuli on nearby neurons of the cortex. A Kohonen's SOM is a special kind of competitive network that computes a neighbourhood for each output neuron from learning data in an unsupervised manner.

It is basically made up of two layers of neurons, i.e., an input layer and a two-dimensional array of neurons, each neuron of this array connected to all input neurons and to neighbouring neurons through a lateral inhibition mechanism. SOM have been successfully used for various tasks in ASR, including feature extraction, vector quantization, density approximation, etc (cf. section 4.2).

Recurrent networks are networks in which connections are allowed both ways between a pair of neurons, or even from a neuron to itself. Such networks exhibit dynamic behaviour, and they possess modifiable feedback pathways which result in a remaining activity after the removal of the input stimuli. Examples are the Hopfield model and the Boltzmann machine. Partially recurrent models have been extensively studied in ASR. In these models, the connections between neurons are mainly feedforward, and include a set of particular feedback connections.

3. Signal Processing and Feature Extraction using ANNs

We first consider the use of ANNs in the front-end processing tasks of a speech or a speaker recognizer involving signal processing and feature extraction. The current literature reports interesting applications in this area, even though it is by no means the area where ANNs are most applied.

Conventional signal processing primarily resorts to linear methods. ANNs offer the potential of non-linear processing. A multi-layer perceptron has for instance been used to derive an efficient solution for the least mean-square algorithm used in adaptive filtering for correction estimation [11]. The authors reported that the MLP filter produces consistently lower bit errors than a linear transversal filter.

ANNs are also used to reduce the level of noise in a corrupted speech signal. The idea is to view speech enhancement as the process of transforming noisy speech into clean speech by some kind of mapping into a parametric space. ANNs can learn arbitrarily complex space mappings. For instance, an MLP-based system has been tested successfully in an auditory preference test with human listeners [12]. Since then, some improvements have been brought to the method. These improvements are based on the observation that it is easier to separate speech from noise at the output of a hidden layer of the network than to do so in the initial physical space [13].

ANNs can play a very useful role in the extraction from the speech wave of efficient features for recognition. This can be done in several ways. In the first place, the different hidden layers of an MLP can be seen as producing new and more suitable internal representations of an input signal for its subsequent classification. This is a trainable feature extraction process extensively investigated in [14]. Classically, feature extraction in pattern recognition proceeds by data compression using algorithms akin to data analysis like the Karhunen-Loeve expansion. It was shown that auto-associative MLPs with linear output units provide an efficient way of performing this transform [15].

In the second place, self-organizing ANNs and those endowed with the capability of unsupervised learning can also be used for automatic extraction of relevant features from raw data. A classical example is the Kohonen feature map network [16] which is basically a two-layer, fully connected, feedforward network with lateral inhibition. This model can further be used for phone or word recognition as will be shown later.

4. Neural Networks as Static Pattern Classifiers

4.1 Speech Pattern Classification with Perceptrons

ANNs have been widely applied in the classification of static speech patterns which are obtained through a kind of spectral analysis of isolated or pre-segmented words or phonemes [17]. In such cases, the input data are considered as a single time-frequency pattern. MLPs are the most commonly used model in these experiments [18], [19], [20], [21]. The results obtained for small vocabularies involving about ten words compare favourably with the results of advanced HMM-based recognizers.

As in the case of the previous systems, the decision part of a neuron in a MLP is commonly based on a sigmoid function. This is the case for the systems just presented. The use of Gaussian functions has also been proposed [22], the basic idea being that of the so-called Radial Basis Functions (RBF) [23]. In conventional MLPs, complex non-linear decision surfaces are approximated by combining hyperplanes, whereas RBF use more complex functions such as Gaussian functions or higher order polynomials. Like MLPs, RBF networks can be used for classification and approximation tasks. Their main advantage is that they keep the pattern classification performances of other ANNs with a much lower training computation cost (the convergence to a solution can be up to several orders of magnitude faster than the MLP). These networks also feature good generalization capabilities. However, they require more training data than MLPs for achieving the same level of accuracy. The two approaches have so far produced comparable results in speech and speaker recognition experiments [24].

4.2 Feature Maps

Other ANN models have also been used for speech pattern classification, in particular the Kohonen feature map. In this model, the learning algorithm is unsupervised, and the development of the cell responses takes place in a self-organizing manner [25]. The model is made up of two layers of cells (cf. section 2.2). The effect of the lateral connectivity during learning is a winner-take-all type of activity. For instance, a phoneme or a short word will correspond to the excitation of a small region of the map centered around a particular cell. This model was successfully used to recognize phonemes [26].

It is also possible to change the learning based on self-organizing feature map into a supervised learning algorithm by defining "target" cells in the map. The resulting Learning Vector Quantization (LVQ) [27] is often used in speech recognition.

Other feature map classifiers, like the heirarchical feature map classifier which is similar to LVQ, have been tested for the recognition of vowel sounds [20]. In the hierarchical model, hidden nodes are used to compute kernel functions related to the Euclidean distance between input data and the clusters represented by these nodes. As in LVQ, the combination of supervised and unsupervised learning drastically reduces the amount of training data needed.

5. Dynamic Aspects

5.1 Position of the Problem

Speech is essentially a time-varying phenomenon. Since the basic connectionist formalism is not tailored for time sequential input pattern processing, it was found necessary to develop solutions (at least partially satisfactory) to deal properly with the dynamic nature of speech. The use of time delays has been advocated as a first solution to this problem (cf. section 5.2). The adjunction of dynamic time warping techniques has also been proposed (cf. section 5.3). Recurrent networks constitute another attractive model for dealing with time (cf. section 5.4).

5.2 Time Delays

The specific problem of handling sequences of speech input vectors has been considered from different angles. A first solution uses the Time-Delay Neural Network (TDNN) [28] [29]. A TDNN is an MLP with fixed time delays. Each cell of a TDNN weights the current input feature vector f(t) as well as the N preceding vectors f(t-n).

Phoneme recognition and syllable spotting in continuous speech [30] [31] are examples of ASR tasks in which TDNNs have successfully been applied. Figure 2 shows the architecture of a TDNN designed for the identification of the English plosive consonants /b,d,g/.

5.3 Dynamic Classifiers

ANN based dynamic classifiers can be obtained by combining a dynamic time warping (DTW) algorithm with static ANN models of the MLP type. In such systems, the outputs of the ANN are used as the local distance measure by the DTW. Results obtained with such a configuration show substantial improvements when compared to those obtained with classical DTW systems.

Early examples of the preceding type of systems for continuous speech recognition involving a 918 word vocabulary can be found in [32], and examples for the E-set recognition in [33]. A further example is provided by the Dynamic programming Neural Network (DNN) [34]. The DNN model is basically a time-delay MLP with back-propagation learning in which the learning phase requires a pre-segmentation of the speech data. The TDNN model has also been combined with DTW to yield the so-called multi-state TDNN (MS-TDNN) [35]. The principle of MS-TDNN is best described as a two-phased process. The first transforms the input signal using a time-delay network while the second performs dynamic time alignment. Once again, learning requires segmented data.

Output
Layer

Second Hidden
Layer

First Hidden
Layer

Input
Layer

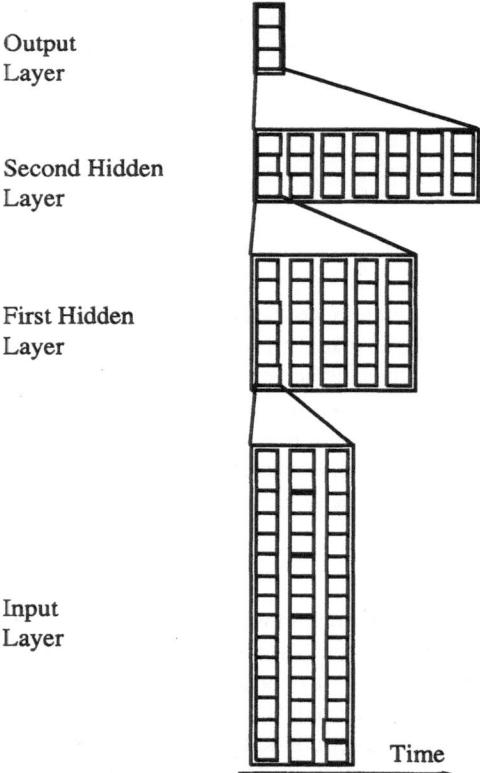

Time

Fig. 2. A TDNN for recognition of /b,d,g/ (after [30])

5.4 Recurrent NNs

Temporal sequences may also be handled using recurrent networks. As illustrated in figure 3, such networks operate on the principle that a recurrent internal state is a function of both the current input and the previous internal state.

A variety of recurrent networks have been designed and tried in ASR [36]. Their design often comprises a feedforward MLP with a time-delayed loop on the output and/or hidden units. For input sequences of sufficiently short duration, it is possible to solve the learning problem by replacing the recurrent net with an equivalent feedforward net obtained by unfolding over time [37]. This is the solution implemented in [38] for instance . Recurrent nets have also been implemented using ANN models other than MLPs. The Boltzmann machine with "carry units" [39] is an example of this type of implementation. The learning algorithm of this machine is based on a simulated annealing algorithm which turns out to be exceedingly time consuming for learning and recognition to be of practical use.

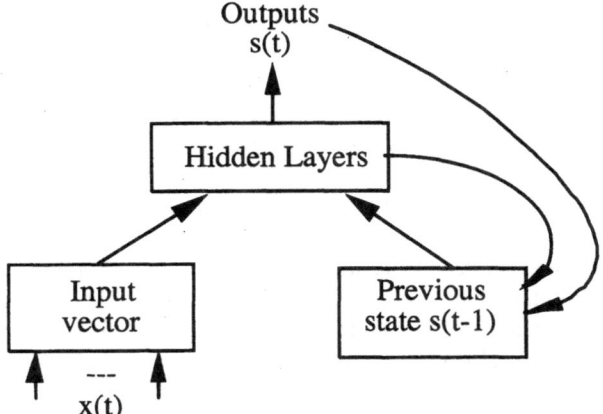

Fig. 3. Principle of a recurrent network

6. Hybrid Models

6.1 Position of the Problem

As stated previously, HMMs are now widely used in ASR to recognize both isolated words and continuous sentences. These models produce very good results but suffer from a number of limitations which are mainly related to the strong hypotheses on which their optimization algorithms are based. ANNs have also proved useful in the classification of speech patterns but require improvement in their ability to deal with the temporal and sequential nature of speech.

From what precedes, combining the time-domain modelling capability of HMMs with the discriminative power of ANNs can constitute an attractive solution. It is however not so easy to devise an adequate interface with which the respective qualities of the two models can be fully exploited. The next section presents a number of solutions found in the current literature.

6.2 Proposed Solutions

ANNs and HMMs can be made to co-operate in a variety of ways in hybrid systems. A large number of studies have concentrated on the use of ANNs as front-end for HMMs. It was proved (see for instance [40], [36], [41], or [42]) that a properly trained MLP for pattern classification is asymptotically equivalent to an estimator of *a posteriori* class probability. Several experiments have confirmed the quality of this estimation. So, from a practical point of view, the outputs of an ANN in a perfectly mastered framework are reliable.

Estimators other than MLPs, especially recurrent nets and RBF nets have also been used in such hybrid systems[43], [44]. The use of TDNNs has also been investigated in [45]. In the proposed system, a TDNN is incorporated into a Viterbi framework to perform time alignment. The outputs of the TDNN can then be normalized so as to deliver *a posteriori* probabilities which are then fed into an HMM.

Hybrid systems like those presented above are also reported to somewhat enhance the recognition performances of HMM systems (see for instance [46] for a test on the ARPA Resource Management (RM) task). However, the implementation and use of these systems are not straightforward, owing to the large number of parameters involved and to the cost of computation. Moreover, the network must be trained with a sufficient amount of data to ensure convergence to a global minimum. A variety of solutions have been proposed to tackle this problem. A solution uses the ANN to compute an additional set of symbols as transformed observations for the HMM [47]. A further improvement of this method is achieved through a global optimization of both the ANN and HMM [48]. This approach uses the gradient of the HMM optimization criterion with respect to the transformed observations to estimate the weights of the ANN connections. In another solution, the ANN is used not as a probability estimator, but as a labeller for a discrete parameter HMM [49]. The authors view an MLP as playing the role of assigning a phonetic label to each speech frame which label is then passed on to the HMM. This amounts to vector quantization involving phonetic information. A similar idea is described in [50], but the ANN is trained using an information theory based unsupervised training algorithm.

A further approach to hybrid system design considers ANNs as HMM postprocessors. This allows efficient combination of the time-alignment capability of HMMs with the discriminative power of ANNs, a feature which is of particular relevance to continuous speech recognition. Some systems feed sentences recognized by the HMM into the ANN [51], [52]. This solution works only for applications requiring a limited vocabulary. Another approach is to limit the number of hypotheses sent to the ANN as candidates elected by an N-best search algorithm [53], [54], [55].

The system described in [62] for instance, uses a so-called Segmental Neural Network which is an ANN accepting the acoustic frames of a phonetic segment as input and produces an estimation of the probability of a phoneme corresponding to this segment. To be efficient, this network requires a reliable phone segmentation procedure.

A similar system was used in [56] to recognize connected spelled letters. This system is made up of two parts, the first of which is a second order HMM [57] trained to recognize letters of the alphabet. The second is a selectively trained neural net (STNN) [58], basically an MLP. The latter incorporates acoustic-phonetic knowledge to help separate confusable letter pairs like /p/ and /t/ on the one hand, and /m/ and /n/ on the other hand. The STNN focuses on the discriminative part of words, i.e. the area where the distinct acoustic information is localized. For instance, in the case of the plosive consonants /p/ and /t/, the discriminative part is made up of the shape of the burst and the vowel transitions. The HMM provides an N-best response string with indications of word boundaries. These boundaries are then used by the STNN to localize the discriminative parts of words. It was shown that the combination of HMM and STNN substantially improves the identification of confusable letters both in English and in French.

Finally, another approach to ANN-HMM hybrid systems considers a unified paradigm for the two components. It was for instance proposed that the forward pass (hence the name of the method) of the Baum-Welch learning algorithm of HMMs in alphanets [59] can viewed as a particular type of recurrent network. Other frameworks are also proposed in [60], [61] and [62].

7. Conclusion

The various applications developed until now have proven the potential of ANN models for automatic speech processing, even though most of these applications are still limited. It is certainly too early to conclude on the long term impact of ANNs on ASR. Nevertheless, significant results are already available and new improvements could be expected in the near future.

This chapter has mainly addressed the aspects of speech analysis and of speech and speaker recognition. The ANN technology is equally relevant to other aspects of automatic speech processing such as speech synthesis [63], speaker verification [64] or else language acquisition [65].

Existing architectures and associated learning procedures are mainly devoted to static pattern classification. They are not yet able to handle the temporal structure of speech in a fully satisfactorily way. Research work is still needed in order to efficiently take into account the inherent dynamic nature of speech.

Besides the study of the mathematical properties of ANNs, a promising research avenue consists of designing efficient combinations of ANNs inspired by the structure of the human cortex. New architectures of interconnected sub-networks should provide effective ways of introducing diverse levels of knowledge (including language models) into the speech understanding process. Similarly, neurosciences could contribute in a decisive way to the design of innovative ANN models.

The development of more efficient, integrated hybrid systems which combine the ANNs formalism with other formalisms, notably that of stochastic models, is also a topic that is worth investigating.

References

[1]   L.R. Rabiner and B.H. Juang, *Fundamentals of Speech Recognition*, Prentice-Hall, Englewood Cliffs, 1993.

[2]   J.M. Mendel and K.S. Fu, eds, *Adaptive Learning and Pattern Recognition Systems*, Academic Press, New-York, 1970.

[3]   R.P. Lippmann, "An Introduction to Computing with Neural Nets", *IEEE ASSP Magazine*, 4, no. 2, pp. 4-22, 1987.

[4]   W. McCullough and W. Pitts , "A Logical Calculus of Ideas Immanent in Nervous Activity", *Bull. Math. Biophysics*, 5, pp. 115-133, 1943.

[5]   F. Rosenblatt, *Principles of Neurodynamics*, Spartan Books, New York, 1962.

[6]   M. Minsky and S. Papert, *Perceptrons*, MIT Press, Cambridge, Mass, 1969.

[7]   J.J. Hopfield, "Neural Networks and Physical Systems with Emergent Collective Computational Abilities" , Proc. Nat. Acad. Sci. USA, vol. 79, pp. 2554-2558, 1982.

[8]   Y. Le Cun, "Une procédure d'apprentissage pour réseau à seuil asymétrique", Proc. Cognitiva, pp.599-604, 1985.

[9]   D.E. Rumelhart, G.E. Hinton and R.J. Williams, "Learning Representations by back-propagating errors", Nature, vol. 332, pp. 533-536, 1986.

[10] T. Kohonen, "Self-organized Formation of Topologically Correct Feature Maps", Biological Cybernetics, vol. 43, pp. 59-69, 1982.

[11] G.J. Gibson et al., "Multilayer Perceptron Structures Applied to Adaptive Equalisers for Data Communications", Proc. IEEE ICASSP, Glasgow, pp. 1183-1186, 1989.

[12] S. Tamura and A. Waibel, "Noise Reduction using Connectionist Models", Proc. IEEE ICASSP, New York, pp. 553-556, 1988.

[13] S. Tamura and M. Nakamura, "Improvements to the Noise Reduction Neural Network", Proc. IEEE ICASSP, Albuquerque, pp. 825-828, 1990.

[14] J. Elman and D. Zipser, "Learning the Hidden Structure of Speech", JASA, vol. 83, pp. 615-626, 1988.

[15] H. Bourlard and Y. Kamp, "Auto-association by Multilayer Perceptrons and Singular Value Decomposition", Biological Cybernetics, vol. 59, pp. 291-294, 1988.

[16] T. Kohonen et al., "Phonotopic Maps: Insightful Representation of Phonological Features for Speech Recognition", Proc. ICPR, Montréal, pp. 182-185, 1984.

[17] R.P. Lippmann, "Review of Neural Networks for Speech Recognition", Neural Computation, vol. 1, pp. 1-38, 1989.

[18] R.P. Lippmann and B. Gold, "Neural Classifiers Useful for Speech Recognition", 1st International Conference on Neural Networks, IV-417, 1987.

[19] S.M. Peeling and R.K. Moore, "Experiments in isolated Digit Recognition Using the Multi-layer Perceptron", Technical Report no. 4073, Royal Signals and Radar Establishement, Malvern, Worcester, GB, 1987.

[20] W.M. Huang and R.P. Lippmann, "Neural Net and Traditional Classifiers", in Neural Information Processing Systems, D. Anderson, editor, pp. 387-396, American Institute of Physics, New York, 1988.

[21] F. Yang et al., "Reconnaisance de mots isolés en utilisant un réseau de neurones", Proc. XVIIe Journées d'Etude sur la Parole, Nancy, pp. 127-134, 1988.

[22] T. Poggio and F. Gorosi, "A Theory of Network for Approximation and Learning", Report no. AI-1140, MIT Artificial Intelligence Laboratory, Cambridge, 1989.

[23] J. Moody, "Fast Learning in Multi-resolution Hierarchies", in Advances in Neural Information Processing Systems, D.S. Touretsky, editor, Morgan Kaufman, 1989.

[24] J.P. Haton, "Probabilistic Neural Networks and Hybrid Connectionist/Stochastic Networks" in Handbook of Neural Networks for Speech Processing, S. Katagiri, editor, Artech House, Boston, USA., (in press).

[25] T. Kohonen, "Clustering, Taxonomy and Topological Maps of Patterns", Proc. 6th International Conference on Pattern Recognition, pp. 114-128, 1982.

[26] T. Kohonen et al., "Phonotopic Maps - Insightful Representation of Phonological Features for Speech Recognition", Proc. 7th International Conference on Pattern Recognition, pp. 182-185, 1984.

[27] T. Kohonen, "Learning Vector Quantization", 1st Annual International Neural Network Society Meeting, p. 303, 1988.

[28] A. Waibel et al., "Phoneme Recognition: Neural Networks vs Hidden Markov Models", Proc. IEEE ICASSP, New-York, pp. 107-110, 1988.

[29] K.J. Lang and G.E. Hinton, "The Development of the Time-delay Neural Network Architecture for Speech Recognition", Technical Report CMU-CS-88-152, Carnegie-Mellon University, 1988.

[30] A. Waibel et al., "Phoneme Recognition Using Time-delay Neural Networks", IEEE Trans. ASSP, vol. 37, no. 3, pp. 328-339, 1989.

[31] H. Sawai *et al.*, "Spotting Japanese CV - Syllables and Phonemes Using Time-Delay Neural Networks", *Proc. IEEE ICASSP*, Glasgow, pp. 25-28, 1989.

[32] H. Bourlard and C.J. Wellekens, "Speech Pattern Discrimination and Multilayer Perceptrons", *Computer Speech and Language*, vol. 3, pp. 1-19, 1989.

[33] D.J. Burr, "Experiments on Neural Net Recognition of Spoken and Written Text", *IEEE Transactions on ASSP*, vol. 36, pp. 1162-1168, 1988.

[34] H. Sakoe *et al.*, "Speaker-independent Word Recognition Using Dynamic Programming Neural Networks", *Proc. IEEE ICASSP*, Glasgow, pp. 29-32, 1989.

[35] P. Haffner *et al.*, "Integrating Time Alignment and Neural Networks for High Performance Continuous Speech Recognition", *Proc. IEEE ICASSP*, Toronto, pp. 105-108, 1991.

[36] H. Bourlard and C.J. Wellekens, "Links between Markov Models and Multilayer Perceptrons", *IEEE Trans. PAMI.*, vol. 12, pp 1167-1178, 1990.

[37] D.E. Rumelhart *et al.*, "Learning Internal Representations by Error Propagation", in *Parallel Distributed Processing. Exploration of the Microstructure of Cognition, vol. 1: Foundations*, D.E. Rumelhart and J.M. McClelland, editors, MIT Press, 1986.

[38] R.I. Watrous and L. Shastri, "Learning Phonetic Features Using Connectionist Networks: An Experiment in Speech Recognition", *1st International Conference on Neural Networks*, vol. 2, San Francisco, pp. 619-627, 1987.

[39] R.W. Prager *et al.*, "Boltzmann Machines for Speech Recognition", *Computer Speech and Language*, vol. 1, pp. 2-27, 1986.

[40] J.J. Hopfield, "Learning Algorithms and Probability Distributions in Feed-forward Networks", *Proc. National Academy of Sciences*, pp. 8429-8433, 1987.

[41] H. Gish, "A Probabilistic Approach to the Understanding and Training of Neural Network Classifiers", *Proc. IEEE ICASSP*, Albuquerque, pp. 1361-1364, 1990.

[42] M.D. Richard and R.P. Lippmann, "Neural Network Classifiers Estimate Bayesian *a posteriori* Probabilities", *Neural Computation*, vol. 3, pp. 461-483, 1991.

[43] S. Renals *et al.*, "Connectionist Probability Estimation in the Decipher Speech Recognition System", *Proc. IEEE ICASSP*, San Francisco, vol. 1, pp. 601-604, 1992.

[44] E. Singer and R. Lippmann, "A Speech Recognizer using Radial Basis Function Neural Networks in a HMM Framework", *Proc. IEEE ICASSP*, San Francisco, vol. 1, pp. 629-632, 1992.

[45] C. Dugast *et al.*, "Combining TDNN and HMM on a Hybrid System for Improved Continuous-speech Recognition", *IEEE Transactions on Speech and Audio Processing*, vol. 2, no. 1, pp. 217-223, 1994.

[46] S. Renals *et al.*, "Connectionist Probability Estimators in HMM Speech Recognition", *IEEE Transactions on Speech and Audio Processing*, vol. 2, no. 1, pp. 161-174, 1994.

[47] Y. Bengio *et al.*, "A Hybrid Coder for Hidden Markov Models using a Recurrent Neural Network", *Proc. IEEE ICASSP*, Albuquerque, pp. 537-540, 1990.

[48] Y. Bengio *et al.*, "Global Optimization of a Neural Network-hidden Markov Model Hybrid", *IEEE Transactions on Neural Networks*, vol. 3, no. 2, pp. 252-259.

[49] P. Le Cerf *et al.*, "Multilayer Perceptrons as Labelers for Hidden Markov Models", *IEEE Transactions on Speech and Audio Processing*, vol. 2, no. 1, 1994.

[50] G. Rigoll, "Maximum Mutual Information Neural Networks for Hybrid Connectionist-HMM Speech Recognition Systems", *IEEE Transactions on Speech and Audio Processing*, vol. 2, no. 1, pp. 175-184, 1994.

[51] D.N.L. Howell, "The Multi-layer Perceptron as a Discriminating post Processor for Hidden Markov Networks", *Proc. of Speech'88 7th FASE Symposium*, 1988.

[52] W.Y. Huang and R.P. Lippmann, "HMM Speech recognition with Neural Net Discrimination" in *Advances in Neural Information Processing System 2*, Morgan Kaufmann, San Mateo, pp. 194-202, 1990.

[53] M. Ostendorf *et al.*, "Integration of Diverse Recognition Methlogies through Reevaluation of the N-best Sentence Hypotheses", *Proc. of DARPA Speech and Natural Language Workshop*, Morgan Kaufmann, San Mateo, 1991.

[54] S. Austin *et al.*, "Speech Recognition using Segmental Neural Nets", *Proc. IEEE ICASSP*, San Francisco, pp. 625-628, 1992.

[55] G. Zavaliagkos *et al.*, "A Hybrid Segmental Neural Net/Hidden Markov Model System for Continuous Speech Recognition", *IEEE Transactions on Speech and Audio Processing*, vol. 2, no. 1, pp. 151-160, 1994.

[56] J.F. Mari, D. Fohr, Y. Anglade, and J.C. Junqua,"Hidden Markov Models and Selectively Trained Neural Networks for Connected Confusable Word Recognition", *Proc. International Conference on Spoken Language Processing*, Yokohama, vol. 3, pp. 1519-1522, 1994.

[57] J.F. Mari and J.P. Haton, "Automatic Word Recognition Based On Second-order Hidden Markov Models", *Proc. International Conference on Spoken Language Processing*, Yokohama, vol. 1, pp 247-250, 1994.

[58] Y. Anglade *et al.*, "Speech Discrimination in Adverse Conditions using Acoustic Knowledge and Selectively Trained Neural Networks", *Proc. IEEE ICASSP*, Minneapolis, pp. 279-282, 1993.

[59] J.S. Bridle, "Training Stochastic Model Recognition Algorithms as Networks can lead to Maximum Mutual Information Estimation of Parameters", in *Advances in Neural Information Processing Systems 2*, D.S. Touretzky, editor, Los Altos, CA, Morgan Kaufman, pp. 211-217, 1990.

[60] L.T. Niles and H.F. Silverman, "Combining Hidden Markov Models and Neural Network Classifiers", *Proc. IEEE ICASSP*, Albuquerque, pp. 417-420, 1990.

[61] S.J. Young, "Competitive Training in Hidden Markov Models", *Proc. IEEE ICASSP*, Albuquerque, pp. 681-684, 1990.

[62] J.N. Hwang *et al.*, "A Systolic Neural Network Architecture for Hidden Markov Models", *Proc. IEEE ICASSP*, Glasgow, pp. 1967-1979, 1989.

[63] T.J. Sejnowski and C.R. Rosenberg, "Parallel Networks that Learn to Pronounce English Text", *Complex Systems*, vol. 1, pp. 145-168, 1987.

[64] Y. Bennani and P. Gallinari, "Connectionist Approaches for Automatic Speaker Recognition", *Proc. ESCA Workshop on Automatic Speaker Recognition, Identification, Verification*, Martigny, pp. 95-102, 1994.

[65] A.L. Gorin *et al.*, "An Experiment in Spoken Language Acquisition", *IEEE Transactions on Speech and Audio Processing*, vol. 2, no. 1, part II, pp. 224-240, 1994.

Computational Models for Auditory Speech Processing

Li Deng

Department of Electrical and Computer Engineering
University of Waterloo, Waterloo, Ontario, Canada N2L 3G1
email: deng@crg6.uwaterloo.ca

Summary. Auditory processing of speech is an important stage in the closed-loop human speech communication system. A computational auditory model for temporal processing of speech is described with details of numerical solution and of the temporal information extraction method given. The model is used to process fluent speech utterances and is applied to phonetic classification using both clean and noisy speech materials. The need for integrating auditory speech processing and phonetic modeling components in machine speech recognizer design is discussed within a proposed computational framework of speech recognition motivated by the closed-loop speech chain model for integrated human speech production and perception behaviors.

1. Introduction

Auditory speech processing is an important component in the closed-loop speech chain underlying human speech communication. The roles of this component are to receive and to subsequently transform the raw speech signal, which is often severely distorted and significantly modified from that generated by the human speech production system, into suitable forms that can be effectively used by the linguistic decoder or "interpreter" based on its internal "generative" model for optimal decoding of the phonologically-coded messages. The computational approach to auditory speech processing to be described in this paper has been developed from a detailed biomechanical model of the peripheral auditory system up to the level of auditory nerve (AN) [5, 2, 7]. The processing stages in the auditory pathway beyond the AN level will not be covered here and interested readers are referred to a few recent, excellent review articles (e.g. [1, 9]) and to some preliminary work published in [8].

   The component modeling approach to auditory speech processing described in this paper appears to be a rightfully viable one at the present stage of the auditory-model development. This contrasts the development of speech production models where global modeling has been the main focus [4]. Development of appropriate statistical structures in global auditory models in the future will rely on considerable further efforts in the development of component models.

2. A nonlinear computational model for basilar membrane wave motions

The computational model of the basilar membrane (BM) used for speech processing is of a nonlinear, transmission-line type, which has been motivated by a number of key biophysical mechanisms known to be operative in actual ears [5, 2]. The final mathematical expression which succinctly summarizes the model is the following nonlinear partial differential equation (wave equation):

$$\frac{\partial^2}{\partial x^2} \left(m \frac{\partial^2 u}{\partial t^2} + r(x, u) \frac{\partial u}{\partial t} + s(x)u - K(x) \frac{\partial^2 u}{\partial x^2} \right) - \frac{2\rho\beta}{A} \frac{\partial^2 u}{\partial t^2} = 0, \qquad (1)$$

where $u(x, t)$ is BM displacement function of time along longitudinal dimension x; $m, s(x)$, and $r(x, u)$ are model parameters for BM unit mass (constant), stiffness (space dependent), and damping (space and output dependent), respectively, and $K(x)$ is BM lateral stiffness coupling coefficient. Nonlinearity of the model comes from output-dependent damping parameter $r(x, u)$, whose biophysical mechanisms and functional significance in speech processing have been discussed in detail in [5, 2, 7]. Input speech waveforms or other arbitrary acoustic inputs to the model enter into the partial differential equation (1) via the boundary condition at $x = 0$ (stapes).

The derivation of the above model is based on 1) Newton's second law; 2) fluid mass conservation law; 3) mechanical mass-spring-damping properties of the basilar membrane; and 4) outer hair-cell motility properties (which produce nonlinear damping $r(x, u)$). The model's output, $u(x, t)$, can be viewed as nonlinear travelling waves along the longitudinal dimension of the BM, or as a highly-coupled bank of nonlinear filter outputs. Both the derivation and the wave properties of this BM model are very similar to those of the partial differential equation governing vocal tract acoustic wave propagation (except the latter typically gives linear wave propagation). [1]

3. Frequency-domain and time-domain computational solutions to the BM model

The nonlinear partial differential equation (1) does not have analytic solution for arbitrary acoustic input signals. The only viable approach to obtaining model outputs appears to be computational means by numerical solution. Two methods of numerical solution, frequency-domain and time-domain methods based on the finite-difference scheme , will be described with their respective strengths and weaknesses discussed.

The frequency-domain method is significantly faster than the time-domain counterpart, but requires batch processing (non real-time) and linearization of the BM model. Linearization of the BM model results in some degrees of loss in the model solution's accuracy. This, however, can be somewhat but not fully mitigated by using adaptive linearization [2].

---

[1] In this parallel, the mechanical property of the BM which consists of a damped mass-spring system causing BM vibration is analogous to the vocal tract wall vibration arising also from a damped mass-spring system . The same Newton's second law and mass conservation law lead to wave properties of the BM travelling wave and of the vocal tract acoustic wave.

When Eqn.(1) is linearized by eliminating output-dependence of the damping term $r(x, u)$, frequency-domain solution of the model can be obtained using Fourier transforms:

$$u(x, t) \longleftrightarrow u(x, j\omega),$$

$$\frac{\partial u(x, t)}{\partial t} \longleftrightarrow jwu(x, j\omega),$$

$$\frac{\partial^2 u(x, t)}{\partial t^2} \longleftrightarrow -\omega^2 u(x, j\omega).$$

This turns Eqn. (1) into an ordinary differential equation:

$$\frac{d^2}{dx^2}\left\{\left(-m\omega^2 + s(x) + j\omega r(x)\right)u - k(x)\frac{d^2 u}{dx^2}\right\} + \frac{2\rho\beta}{A}\omega^2 u = 0. \qquad (2)$$

Numerical solution of the above frequency-domain model by the finite-difference method requires that the spatial dimension be represented by a finite number of discrete points. The solution is obtained for the displacement of the BM, $u(x, j\omega)$, as a function of the distance from the stapes, x, for selected input frequencies, ω. To discretize the frequency-domain model, the derivatives in Eqn.(2) are approximated by the conventional central differences:

$$\frac{du}{dx} = \frac{u_{i+1} - u_{i-1}}{2\Delta x},$$

$$\frac{d^2 u}{dx^2} = \frac{u_{i+1} - 2u_i + u_{i-1}}{(\Delta x)^2},$$

$$\frac{d^4 u}{dx^4} = \frac{u_{i+2} - 4u_{i+1} + 6u_i - 4u_{i-1} + u_{i-2}}{(\Delta x)^4}.$$

This then turns ordinary differential equation (2) into a linear algebraic equation, which can be solved by straightforward matrix inversion to give $u(x, j\omega)$. The time-domain output is finally obtained by taking inverse Fourier transform of $u(x, j\omega)$, one for each discrete point along the x dimension.

The time-domain numeric solution allows on-line processing, and solve arbitrarily complex nonlinear BM model without performing model linearization. But the computational load is significantly greater than the frequency-domain method since one matrix inversion is required for each sample of speech. The reason for the computational load is that we can no longer use Fourier transform due to nonlinear element(s) in the model. Hence, both time and space variables need to be discretized. After the discretization, we use the following finite difference approximation to all partial derivatives, from order one to order four, in Eqn. (1):

$$\frac{\partial u}{\partial t} = \frac{u_i^{n+1} - u_i^n}{\Delta t},$$

$$\frac{\partial^2 u}{\partial t^2} = \frac{u_i^{n+1} - 2u_i^n + u_i^{n-1}}{(\Delta t)^2},$$

$$\frac{\partial u}{\partial x} = \frac{u_{i+1}^n - u_i^n}{\Delta x},$$

$$\frac{\partial^2 u}{\partial x^2} = \frac{u_{i+1}^n - 2u_i^n + u_{i-1}^n}{(\Delta x)^2},$$

$$\frac{\partial^4 u}{\partial x^4} = \frac{u_{i+2}^n - 4u_{i+1}^n + 6u_i^n - 4u_{i-1}^n + u_{i-2}^n}{(\Delta x)^4},$$

$$\frac{\partial^3 u}{\partial t \partial x^2} = \frac{u_{i+1}^{n+1} - 2u_i^{n+1} + u_{i-1}^{n+1} - u_{i+1}^n + 2u_i^n - u_{i-1}^n}{\Delta t \, (\Delta x)^2},$$

$$\frac{\partial^4 u}{\partial t^2 \partial x^2} = \frac{u_{i+1}^{n+1} - 2u_i^{n+1} + u_{i-1}^{n+1} - 2u_{i+1}^n}{(\Delta t)^2 \, (\Delta x)^2}$$
$$\frac{+4u_i^n - 2u_{i-1}^n + u_{i+1}^{n-1} - 2u_i^{n-1} + u_{i-1}^{n-1}}{(\Delta t)^2 \, (\Delta x)^2}.$$

This turns the partial differential equation into a large algebraic equation with the solution variable $u(x, t)$ indexed by both time t and space x. The numerical procedure proceeds by first fixing each time t index and finding the solution for u as a function of space index x via matrix inversion . Then, by advancing time one sample after another, the entire solution for $u(x, t)$ is obtained.

The above solution has been used to process a large amount of speech data (rf. [7, 8]). Theoretical work on stability analysis of the model solution, which is essential to guarantee successful use of the model for automatic processing large-sized data, has been carefully carried out in the work reported in [6].

4. Interval analysis of auditory model's outputs for temporal information extraction

The BM model's output obtained by the finite-difference method described in the preceding section is used as the input to the inner hair cell model, which consists of hyperbolic tangent compression followed by low-pass filtering. The final stage of the auditory model is for the AN-synapse, which receives the input as the inner hair cell model's output. The AN-synapse consists of pools of neurotransmitters, separated by membranes of varying permeability, which simulate the temporal adaptation phenomenon experimentally observed in the AN.

The above composite auditory model's output is an array of temporally varying AN firing probabilities in response to input speech sounds to the BM model. This output is subject to an interval analysis for temporal information extraction. The analysis is based on construction of the Inter-Peak-Interval Histogram (IPIH) of the dominant intervals measured from autocorrelation of 10-ms segments of the auditory model's output. In the IPIH construction, increment of each bin in the histogram is multiplied by the amplitude of the peak at the start of the corresponding interval.
[2] Further, in the IPIH construction, a fixed number of intervals in the autocorrela-

---

[2] This permits the IPIH to code the firing rate information in addition to the otherwise temporal information only.

tion function are counted which are common across all AN output channels. This gives rise to approximately exponential temporal analysis windows, with the low-frequency channels occupying longer windows than the high-frequency channels. Finally, to reduce the data rate, the IPIHs constructed for all AN output channels are amalgamated, resulting in a single histogram per time frame. [3] Figure 1 shows an example of the process in the IPIH construction described above.

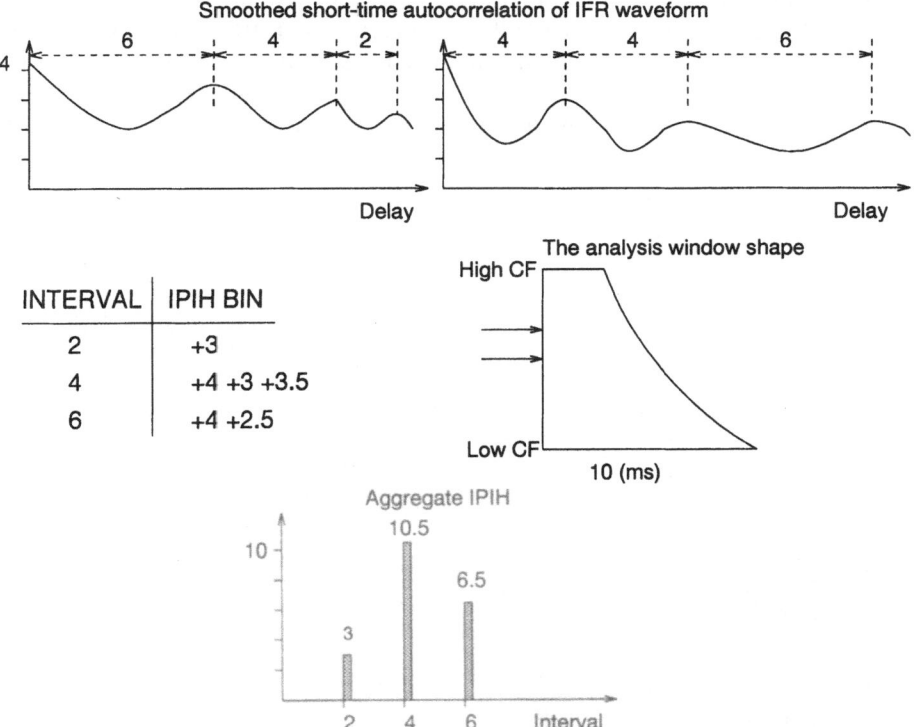

Fig. 1. Construction of IPIH from the autocorrelation of the modeled AN instantaneous firing rate function

5. IPIH representation of clean and noisy speech sounds

We have run the auditory model and carried out the consequent IPIH analysis on a number of utterances in the TIMIT database which cover a wide range of acoustic phonetic classes in American English. The model has been run for both clean speech and speech embedded in additive noise. A few examples are provided here

---

[3] Note that the length of the time frame is frequency dependent (i.e. conditioned on the AN channel' center frequency).

to illustrate how various classes of speech sounds are represented in the form of
IPIH constructed from the time-domain output of the auditory model as a temporal-
nonplace code, and to show robustness of the representation to noise degradation.

Plotted in Figure 2 are the IPIHs for clean utterance *heels* (a) and *semi* (b),
respectively, both presented to the auditory model at 69 dB SPL. The prominent
acoustic characteristics of these utterances are the wide range of the formant transitions in the vocalic segments. For [iy] in *heels*, F2 moves drastically down from near
2100 Hz toward near 1300 Hz (F2 of the postvocalic [l]); this acoustic transition
is reflected in the corresponding peak movement in the IPIH from about 0.48-ms
inter-peak interval (starting at 60 ms) to the interval of 0.75 ms (ending at around
200 ms). Similarly, the slow rising F1 transition in acoustics is represented as the
slow falling IPIH peaks. For [ay] in *semi*, the rising F2 from about 1200 Hz to 2000
Hz is reflected in the falling IPIH peak from around 0.85-ms to 0.5-ms.

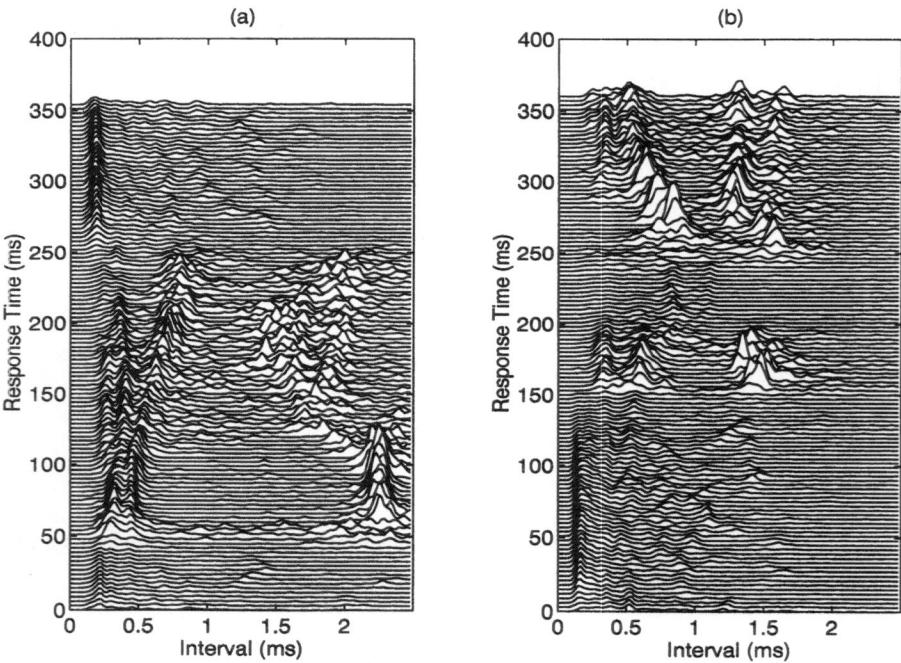

Fig. 2. Modeled IPIHs for words (a) *heels* (b) *semi*

We have produced and analyzed the IPIHs for the words from several TIMIT
sentences in much the same qualitative way as described above. ¿From the analysis
we find that all the significant acoustic properties of all classes of American English
sounds that can be identified from spectrograms can also be identified, albeit to a
varying degrees of modification, from the corresponding IPIH.

To evaluate noise robustness of the speech representation in terms of the interval
statistics collected from the auditory-nerve population, we performed the identical

IPIH analysis for the speech sounds identical to the ones described above except adding white Gaussian noise with 10-dB signal-to-noise-ratio (SNR) into the speech stimuli before running the auditory model. The resulting IPIHs for noisy versions of the utterances, *heels* and *semi* of Figure 2, are shown in Figure 3. A comparison between the IPIHs in Figures 2 and 3 shows that aside from some relatively minor distortions in the nasal murmur and in the aspiration, the major characteristics in the IPIH representation for the clean speech have been well preserved. In contrast to the above IPIH-based temporal representation in the auditory domain, the differences in the acoustic (spectral) domain between the clean and noisy versions of the speech utterances are found to be vast (not shown here).

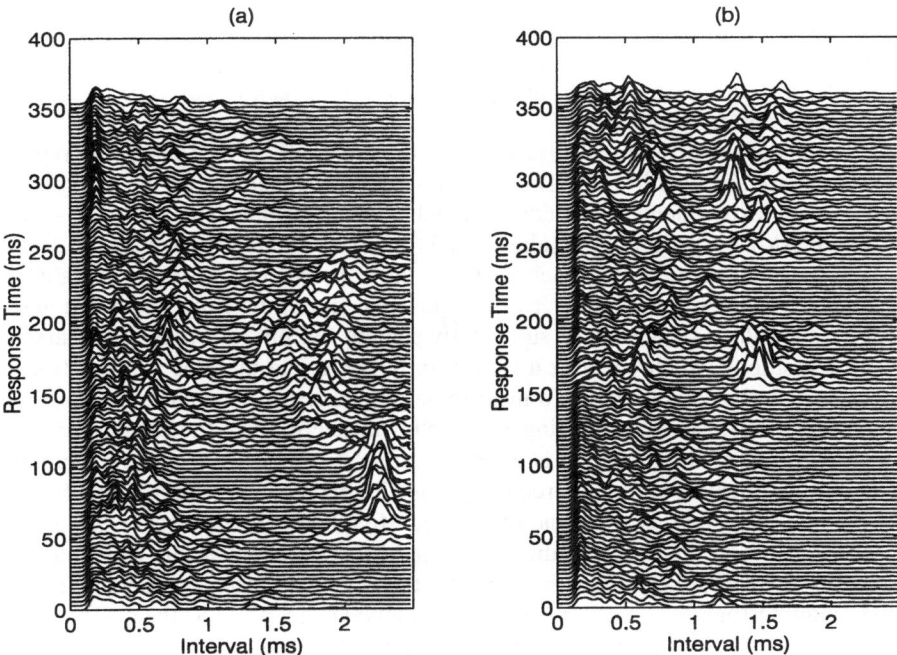

Fig. 3. Modeled IPIH for words (a) *heels* (b) *semi* embedded in white noise with 10-dB SNR

6. Speech recognition experiments

The IPIH speech analysis results we have obtained demonstrated that the IPIH-based temporal representation preserves major acoustic properties of the speech utterances for all classes of English sounds in the magnitude-spectral domain, and that such a representation is robust to additive noise. One additional advantage of such a temporal representation over the conventional spectral representation in speech analysis is that the frequency resolution and time resolution can be controlled independently, rather than being constrained by an inverse, trade-off relationship. In our IPIH analysis, the time resolution is controlled by the frame size and by the overlap between adjacent frames, while the frequency resolution is independently determined by the number of cochlear channels set up in the model and by the bin width used to construct the IPIH. In principle, both the time and frequency resolutions can be increased simultaneously with no limits.

Despite these advantages, the IPIH-based temporal representation contains a much greater data dimensionality than that from the conventional magnitude-spectral analysis. Unfortunately, the current speech modeling methodology has not been advanced to the extent that the large data dimensionality required by the auditory temporal representation can be adequately accommodated and the data complexity associated with the large dimensionality be faithfully modeled. As such, heuristics-driven data dimensionality and complexity reduction methods have to be devised in order to interface the temporal representation of speech to any type of speech recognizer currently available.

Details of the experiments designed to evaluate the IPIH-based auditory representation are reported in [10]. The speech model embedded within the recognizer used in the experiments is the conventional, context-independent, stationary-state mixture HMM. This model requires that 1) the data inputs be organized to form a vector-valued sequence; 2) each vector in the sequence (i.e. a frame) contain an identical, relatively small number of components; and 3) the temporal variation of the vector-valued sequences be sufficiently smooth (except for occasional Markov state transitions which occur at a significantly lower rate than the frame rate but greater than the sample rate). To meet these requirements, we transform the IPIH representation of speech according to the following steps. First, the IPIH associated with each 10-ms time window is divided into a set of interval bands corresponding to the critical bands in the frequency domain. Each band contains a number of histogram bins, ranging from one for the high-frequency IPIH points to 15 for the low-frequency points. Second, the maximum histogram count within each interval band of the IPIH is kept while throwing out the remaining histogram counts. These maximum histogram counts, one from each interval band, preserve the overall IPIH profile while drastically reducing the data complexity. Third, this simplified IPIH is subject to further data complexity reduction via a standard cosine transform.

In the evaluation experiments, the speech data consist of eight vowels ([aa], [ae], [ah], [ao], [eh], [ey], [ih], [iy]) extracted from the speaker-independent TIMIT corpus. Tokens of the eight vowels (clean speech) from 40 male and female speakers (a total of 2000 vowel tokens) are used for training and those from disjoint 24 male and female speakers (a total of 1200 vowel tokens) for testing. Both clean vowel tokens and their noisy version created by adding white Gaussian noise with varying levels of SNR are used as training and test tokens. The performance results, organized as the vowel classification rate as a function of the SNR level and of the two types of the speech preprocessor (IPI-based one with solid line vs. benchmark, MFCC-based one with dashed line), are shown in Figure 4. The results demonstrate that the auditory IPI-based preprocessor consistently outperforms the MFCC-based counterpart over a wide range of the SNR level (0 dB to over 15 dB). Only for near-clean vowels (20-dB SNR level), the two preprocessors become comparable in performance. [4]

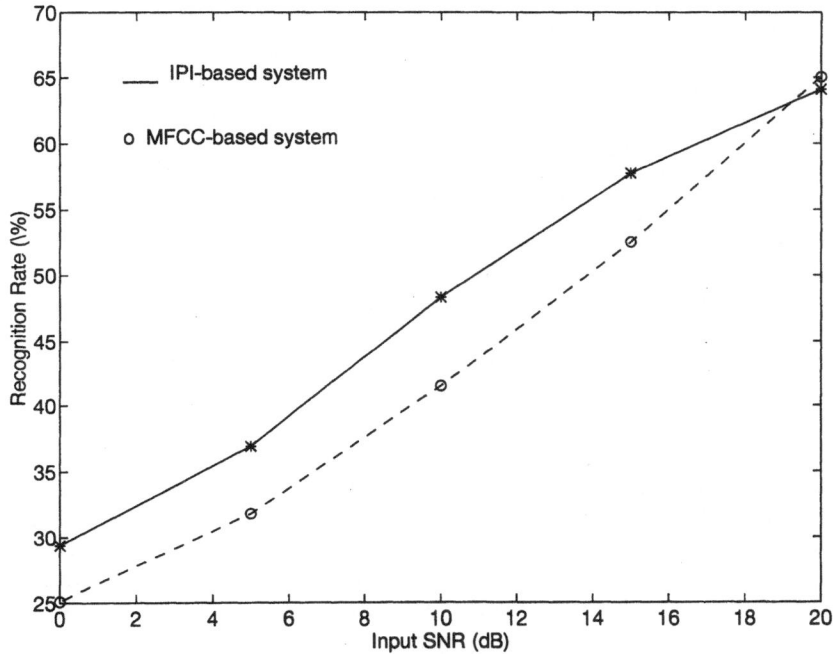

Fig. 4. Comparative average classification rates for TIMIT vowels

7. Summary and discussions

With use of the computational auditory model described in this paper to process the speech utterances contained in the TIMIT database, it has been shown that not only for limited and isolated speech tokens but also for a comprehensive range of manner classes of fluently spoken speech sounds, the auditory temporal representation on the basis of interval statistics collected from AN firing patterns preserves (with modification) the major acoustic properties of the speech utterances that can be identified from spectrograms. The temporal nature of the representation makes it robust to changes in the loudness level of the speech sounds and to the noise effect. The rate-level representation, which is closely related to the conventional spectral analysis, lacks such robustness.

---

[4] For evaluation experiments on other tasks and for details of the benchmark system, see [10].

Although the direction of exploring properties and constraints of the auditory system as a guiding principle for robust speech representation against noise effects in speech recognizer design appears to be promising, most experimental results (including ours and many other research groups' (too long to be listed here)) on recognition of noise-free speech have not been as successful as those for noisy speech compared with the conventional MFCC-based representation based more on traditional signal processing than on auditory properties. This is apparently caused by two competing factors working against each other. On the one hand, the independent specification of the time and frequency resolutions in speech preprocessing offered by the auditory interval-based representation allows potentially unlimited analysis resolutions for both time and frequency. On the other hand, however, the simultaneously greater resolutions enabled by the auditory representation are necessarily linked to a greater data dimensionality, causing problems for the speech modeling component of any current recognizer which requires relatively smoothed and redundancy-free patterns produced from the pre-processor. These two competing factors cannot be reconciled within the current HMM-based speech recognition framework. Any success in incorporating hearing science into speech recognition technology must come from *integrated* investigation of faithful auditory representation of speech and of the modeling component of the overall recognition system capable of taking full advantages of the information contained in the auditory representation. This integrated nature of the engineering system design can be closely paralleled with the biological counterpart of the closed-loop human speech communication system, where the auditorily received and transformed speech information must be fully compatible with what is expected from the listener's internal "generative" model approximating the speaker's linguistic behavior (and acting as an optimal decoder on the listener's part). Following this parallel, the integration of auditory representation and speech modeling components discussed here can be gratefully accomplished in the speech recognition architecture described in [3] which has been motivated by the global structure of the human closed-loop speech chain. Within this architecture, the role of computational auditory models will be to provide proper levels of auditory representation of the speech acoustics which will facilitate construction and learning of the nonlinear mapping between such representation and the internal production-affiliated variables. When this mapping is modeled within a global dynamic neural network system [4], then how to choose the output variables of the network to make model learning effective will place a strongest demand on the level of details of auditory modeling which becomes a critical component of the integrated speech recognition architecture.

References

[1] Delgutte B. (1997) "Auditory neural processing of speech," in *The Handbook of Phonetic Sciences*, W. J. Handcastle and J. Lavar (eds.), Blackwell, Cambridge, pp. 507-538.
[2] Deng L. (1992) "Processing of acoustic signals in a cochlear model incorporating laterally coupled suppressive elements," *Neural Networks*, Vol.5, No.1, pp.19–34.

[3] Deng L. (1998) "Articulatory features and associated production models in statistical speech recognition," .

[4] Deng L. (1998) "Computational models for speech production," In *Computational Models of Speech Pattern Processing*, (this volume) in *NATO ASI Series F*, pp. 214–224, Springer–Verlag, Berlin, 1998.

[5] Deng L. and C.D. Geisler D. (1987) "A composite auditory model for processing speech sounds," *J. Acoust. Soc. Am.*, Vol. 82, No. 6, pp. 2001–2012.

[6] Deng L. and Kheirallah I. (1993) "Numerical property and efficient solution of a nonlinear transmission-line model for basilar-membrane wave motions," *Signal Processing*, Vol. 33, No. 3, pp. 269–286.

[7] Deng L. and Kheirallah I. (1993) "Dynamic formant tracking of noisy speech using temporal analysis on outputs from a nonlinear cochlear model," *IEEE Transactions on Biomedical Engineering*, Vol. 40, No. 5, pp. 456–467.

[8] Deng L and Sheikhzadeh H. (1996) "Temporal and rate aspects of speech encoding in the auditory system: Simulation results on TIMIT data using a layered neural network interfaced with a cochlear model," *Proc. European Speech Communication Association Tutorial and Research Workshop on the Auditory Basis of Speech Perception*, Keele Univ., U.K., pp. 75-78.

[9] Greenberg S. (1995) "Auditory processing of speech," in *Principles of Experimental Phonetics*, Ed. N. Lass, Mosby: London, pp. 362-407.

[10] Sheikhzadeh H. and Deng L. (1997) "Speech analysis and recognition using interval statistics generated from a composite auditory model," *IEEE Trans. Speech Audio Processing*, to appear.

Speaker Adaptation of CDHMMs Using Bayesian Learning

Claudio Vair and Luciano Fissore

CSELT - Centro Studi E Laboratori Telecomunicazioni
Via G. Reiss Romoli 274 - 10148 Torino, Italy
email: {vair, fissore}@cselt.it

Summary. We investigate the Bayesian Learning approach (also known as *Maximum A Posteriori* - MAP) to the speaker adaptation of Continuous Density Hidden Markov Models (CDHMMs). The parameters of the Gaussian mixture output densities are adapted using the exponential forgetting mechanism and performing the *a priori* parameter estimation in a model based outline. Moreover a channel adaptation is carried out by means of the cepstral mean normalization method (CMN).

1. Introduction

The model adaptation techniques are a good compromise between the comprehensive modeling of all speech phenomena in a speaker independent recognizer and the specific modeling of a better performing speaker dependent recognizer. Since MAP gives an ideal framework to combine existing SI models with new SD training data, it can be successfully used in speaker adaptation. Two key issues for the on-line use of the Bayesian Learning concern the a priori parameters estimation and the speed of adaptation. In this work we estimate the a-priori parameters directly from the SI models. This allows an easy integration of MAP in the recognition environment and the incremental updating of both models and priors. To speed up the adaptation, we use the forgetting mechanism proposed in [3].

During the tests we adapted sub-word units CDHMM, trained on SI data and coming from PSTN to a particular speaker, recorded through PBX. In order to compensate the variation of the transmission channel, the CMN technique has been adopted.

2. Bayesian Estimation of CDHMMs

We apply the MAP approach to the adaptation of the CDHMM parameters $\theta = (\omega_1 \ldots \omega_K, \mu_1 \ldots \mu_K, r_1 \ldots r_K)$ of the gaussian mixtures. In this case, the mixture weight ω_k, the mean vector μ_k and the precision matrix r_k (inverse of the diagonal covariance matrix Σ_k) are considered random values, realization of a random variable Θ. Given the sample X related to a specific application condition (environment, task, speaker and so on), the maximum a posteriori training problem consists in finding the most likely realization θ_{MAP} of Θ, solution of:

$$\theta_{MAP} = \underset{\theta}{\mathrm{argmax}}\, g(\theta \mid X) \tag{1}$$

where $g(\theta \,|\, \mathbf{X})$ is the posterior distribution of θ. According to the Bayes rule, we can introduce in (1) the prior distribution $g(\theta \,|\, \mathbf{X})$ of θ:

$$\theta_{MAP} = \underset{\theta}{\operatorname{argmax}}\, f(\mathbf{X} \,|\, \theta) g(\theta \,|\, \varPhi) \qquad (2)$$

where $f(\mathbf{X} \,|\, \theta)$ is the likelihood of the sample $\mathbf{X}$ and $\varPhi$ are the parameters of the prior distribution. The prior distribution $g(\theta \,|\, \varPhi)$ constrains the model shape before any observation is made. In this work we use a *segmental* version of MAP [2] for the solution of (2). The algorithm works with two iterative steps: first, the adaptation frames are aligned on the current model, getting the optimal state sequence $\mathbf{S}$; second, the new model parameters are computed on the complete data $(\mathbf{X}, \mathbf{S})$:

$$\theta_{MAP} = \underset{\theta}{\operatorname{argmax}}[\underset{\mathbf{S}}{\max}\, f(\mathbf{X}, \mathbf{S} \,|\, \theta) g(\theta \,|\, \varPhi)] \qquad (3)$$

Equation (3) is a good approximation of (2) with reduced computational load. According to the segmental MAP, the likelihood of the observation $X = (x_1, ..., x_T)$ assigned to a given state by the alignment procedure is:

$$f(\mathbf{X} \,|\, \theta) = \prod_{t=1}^{T} \sum_{k=1}^{K} \omega_k \, \mathcal{N}(x_t \,|\, \mu_k, \Sigma_k) \qquad (4)$$

2.1 Prior Density Definition

Assuming the independence between the mixture weights and the parameters of each mixture component, we can define [2] the joint prior density of (4). Therefore we define the prior p.d.f. as product of a Dirichlet density for the weight and a normal-Wishart density for average vector and precision matrix [2]:

$$g(\omega_k \,|\, \nu_k) \;\propto\; \omega_k^{\nu_k - 1}$$
$$g(\mu_k, \mathbf{r}_k, \,|\, \tau_k, m_k, \alpha_k, \mathbf{U}_k) \;\propto\; |\mathbf{r}_k|^{\frac{1}{2}} \exp[-\tfrac{\tau_k}{2}(\mu_k - m_k)^t \mathbf{r}_k(\mu_k - m_k)]$$
$$\times\; |\mathbf{r}_k|^{\frac{\alpha_k - L - 1}{2}} \exp[-\tfrac{1}{2} tr(\mathbf{U}_k \mathbf{r}_k)] \qquad (5)$$
$$g(\theta \,|\, \varPhi) \;=\; \prod_{k=1}^{K} g(\omega_k) g(\mu_k, \mathbf{r}_k)$$

where $\varPhi = \{\nu_k, \tau_k, m_k, \alpha_k, \mathbf{U}_k\}$ are the prior density parameters (also known as *hyper-parameters*), and L is the features vector dimension.

2.2 Forgetting Mechanism

The definition of the prior density given in (5) allows the use of the EM algorithm for the approximate MAP estimation by repeating the following steps:

Estimation: $R(\theta, \widehat{\theta}) = E\{\log f(\mathbf{X}, \mathbf{S} \,|\, \theta) \,|\, \mathbf{X}, \widehat{\theta}\} + \log g(\theta \,|\, \varPhi)$

Maximization: $\widehat{\theta} = \operatorname{argmax}_\theta R(\theta, \widehat{\theta})$

To balance the importance of the prior information and the adaptation data, we use the exponential forgetting mechanism [3] in the estimation step, introducing a *forgetting factor* $\rho \in [0, 1]$ as a weight associated to the prior density:

Estimation: $R(\theta, \widehat{\theta}) = E\{\log f(\mathbf{X}, \mathbf{S} \mid \theta) \mid \mathbf{X}, \widehat{\theta}\} + \rho \log g(\theta \mid \Phi)$

When $\rho = 1$ we get the usual MAP estimation, without any forgetting; when $\rho = 0$ the prior knowledge is ignored and the definition of $R(\theta, \widehat{\theta})$ become a classical Maximum Likelihood (ML) estimation. The forgetting mechanism is fundamental to achieve good adaptation performance even with a small adaptation corpus.

2.3 Prior Parameter Estimation and MAP Solution

In this work we exploit existing SI models to estimate all the hyper-parameters without any additional data, using a *model based* approach. Let $\Phi = \{\nu_k, \tau_k, m_k, \alpha_k, \mathbf{U}_k\}$ be the unknown prior parameters vector for a given HMM state and $\theta = (\omega_k, \mu_k, \mathbf{r}_k)$ the corresponding model parameters from which we want obtain the estimation. Let's assume, moreover, that the model θ was trained with N data sample. The prior mean m_k and covariance $\mathbf{U}_k$ are estimated directly from the mean μ_k and the covariance matrix $\Sigma_k = \mathbf{r}_k^{-1}$ of the SI model:

$$m_k = \mu_k; \qquad\qquad \mathbf{U}_k = n_k \Sigma_k \qquad (6)$$

Since we use $n_k = \omega_k N$ data sample to estimate the prior m_k and $\mathbf{U}_k$ we chose to fix the precision scaling constant τ_k, the degree of freedom of Wishart distribution α_k and the component of Dirichlet function ν_k as follow [4]:

$$\tau_k = n_k; \qquad \nu_k - 1 \simeq \nu_k = n_k; \qquad \alpha_k - L \simeq \alpha_k = n_k \qquad (7)$$

With the assumptions (6) and (7), and using the EM algorithm with the forgetting mechanism, we obtain the following recursive equations, solution of the MAP estimation:

$$\widetilde{\omega}_k = \frac{\rho n_k + \xi_k}{\sum_{h=1}^{K} \rho n_h + \xi_h}; \qquad \widetilde{\mu}_k = (1 - \lambda_k)\mu_k + \lambda_k \mu_{xk}; \qquad \widetilde{n}_k = \rho n_k + \xi_k$$

$$\widetilde{\Sigma}_k = (1 - \lambda_k)\Sigma_k + \lambda_k \Sigma_{xk} + \lambda_k(1 - \lambda_k)(\mu_k - \mu_{xk})(\mu_k - \mu_{xk})^t \qquad (8)$$

where

$$\gamma_{t,k} = \frac{\widehat{\omega}_k \mathcal{N}(x_t \mid \widehat{\mu}_k, \widehat{\mathbf{r}}_k)}{\sum_{h=1}^{K} \widehat{\omega}_h \mathcal{N}(x_t \mid \widehat{\mu}_h, \widehat{\mathbf{r}}_h)}; \qquad \xi_k = \sum_{t=1}^{T} \gamma_{t,k}; \qquad \lambda_k = \frac{\xi_k}{\xi_k + \rho n_k}$$

$$\mu_{xk} = \frac{\sum_{t=1}^{T} \gamma_{t,k} x_t}{\xi_k}; \qquad\qquad \Sigma_{xk} = \frac{\sum_{t=1}^{T} \gamma_{t,k}(x_t - \mu_{xk})(x_t - \mu_{xk})^t}{\xi_k}$$

A segmentation step, with the current model parameters $\widehat{\theta} = (\widehat{\omega}_k, \widehat{\mu}_k, \widehat{\mathbf{r}}_k)$, must come before re-estimation (8). Then, the current model is updated with the re-estimated values $\widetilde{\theta} = (\widetilde{\omega}_k, \widetilde{\mu}_k, \widetilde{\mathbf{r}}_k)$ and a new EM step may be performed.

In figure 1 the adaptation flow is shown: n refers to the number of incremental adaptation steps while m refers to EM iteration $(1...M)$. In the step ① the general model $\theta^{(n)}$ is used to extract the prior parameters $\Phi^{(n)}$ and as a first estimation of the current model $\widehat{\theta}$; we use multi-style SI model (ML trained) as general model when $n = 0$. The EM iterations are performed by steps ② to ④: first, the adaptation frames are Viterbi aligned in a supervised mode and then the MAP re-estimation is applied. In the step ⑤ the general model is replaced with the adapted one. The last step is necessary only if incremental adaptation is required.

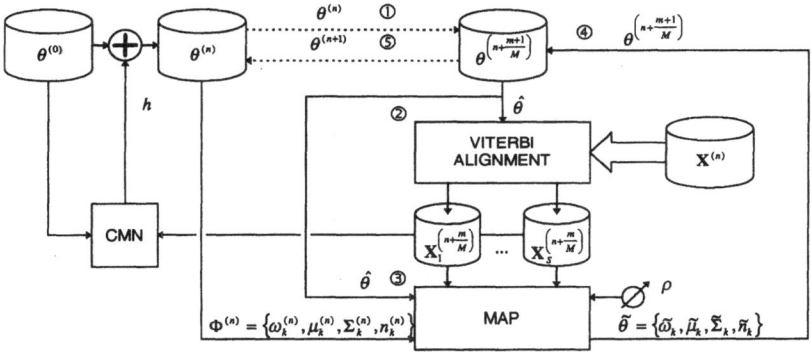

Fig. 1. The adaptation flow

3. Acoustic Normalization

To cope with the variation of the acoustic conditions between the SI and SD environment, we adopt the CMN technique, embedded in the adaptation framework (figure 1). The aim of CMN is to cancel the bias due to the time invariant channel characteristics. We compute the average value of the adaptation frames and the barycentre of the general model, obtaining the compensation vector:

$$h = \frac{1}{N_x} \sum_{t=1}^{N_x} x_t - \sum_{s \in \{S\}} \frac{N_s}{N} \sum_{k=1}^{K} \omega_{k,s} \dot{\mu}_{k,s} \qquad (9)$$

where x_t are the N_x adaptation frames, $\omega_{k,s}$, $\mu_{k,s}$ the weight and average vector of the mixture k of each state s for the general model; each state was trained with N_s data sample and $N = \sum_{s \in \{S\}} N_s$.

4. Tasks, Corpus and System

We tested the effectiveness of adaptation technique using CDHMM with Gaussian mixture output density (max. 32 elements). We adopted two kinds of subword unit: 53 context independent phonemes (53pho) and 391 transitional units (391tr) for Italian language [1]. The 391tr, that model separately the transitional and stationary segment, are more accurate than 53pho, using the same number of states for a given word. The front-end produces 12 Mel Frequency Cepstral Coefficient (MFCC). Δ-cepstral, energy and Δ-energy parameters are added to the features vector.

The SI data base ARV+ARV2 consists of 3399 isolated word and 4532 sentences uttered by 1133 speakers through PSTN; the SD data base BI_MICRO utterances recorded by each one of 4 speakers (mg8, ri8, ma8, dm8) in a quiet room and collected through PBX. A subset, made up of 1400 utterances, is used for the adaptation data. The first 1040 utterances (SD PAN) belong to a phonetic balanced corpus while the remaining 360 contain a vocabulary dependent component. The test set consists of 200 utterances for each speaker, with a vocabulary of 247 words. The average utterance duration is 2.8 seconds.

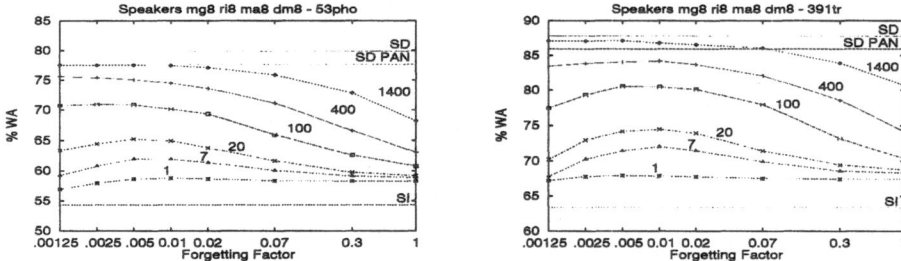

Fig. 2. Forgetting factor effect on word accuracy for different number of adaptation utterances

5. Speaker Adaptation Experiments

The experiments have been carried out according to the flow illustrated in figure 1 but performing only one EM iteration and one incremental adaptation. We trained the SI model using the ARV+ARV2 corpus: this model was used to extract the hyper-parameters and to perform the Viterbi alignment after the application of CMN. Most of the tests were performed varying the number of adaptation utterances; the results are averaged over 5 different adaptation data subset. In the plots we included three reference lines: SD refers to ML training over all the adaptation set (1400 utterances), SD PAN to the ML training over the phonetically balanced adaptation component (1040 utterances) and SI to the use of the ARV+ARV2 model, without any adaptation.

Table 1. Word accuracy comparison as a function of adaptation data

Sent.	53pho ML WA	53pho MAP WA	53pho MAP Rwa	391tr ML WA	391tr MAP WA	391tr MAP Rwa
1	-14.1	58.6	16.7%	-13.5	67.9	18.8%
5	9.7	61.2	26.8%	-21.1	70.5	29.4%
10	32.0	62.6	32.3%	-0.4	72.0	35.4%
20	50.0	65.2	42.7%	27.4	74.2	44.3%
40	57.3	67.4	51.3%	47.2	77.1	56.3%
200	72.1	73.5	74.8%	78.4	82.2	77.1%
1040	77.6	76.0	84.9%	85.9	85.3	89.6%
1400	79.9	77.5	90.4%	87.8	87.1	97.2%
SI:	54.3			63.3		

In figure 2 we show the effect of the forgetting factor on the adaptation performance: when ρ is too small, we lose the constraint given by the a-priori information and the performance falls down to the ML result; when ρ is too close to 1, the contribute of the a-priori is too strong and the adaptation with few data is insignificant. We empirically chose $\rho = 0.005$, based on figure 2, for both 53pho and 391tr, in all the following tests.

In figure 3 (A) and (B) we compare the word accuracy as a function of the adapted HMM parameters. When the adaptation data exceed about 1 minute (20 ut-

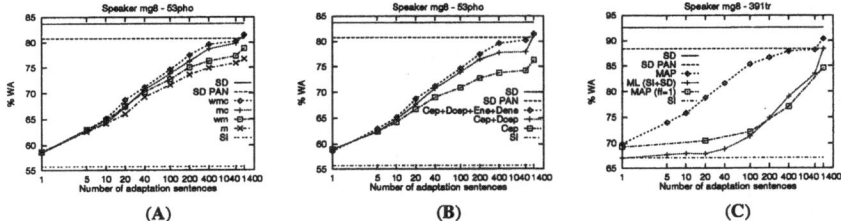

Fig. 3. (A) adaptation of weight (w), average vector (m), and precision matrix (c); (B) adaptation of cepstral (Cep), Δ-cepstral (Dcep), energy (Ene) and Δ-energy (Dene); (C) comparison of ML training (SI+SD), MAP and MAP with no forgetting $\rho = 1$ (ff=1)

terances), sensible improvement are obtained by adapting all the HMM parameters and the Δ features. Figure 3 (C) shows a comparison between ML training, over all the SI and SD data, and the MAP adaptation. Table 1 summarizes the MAP adaptation result; column ML(SD) refers to training ML over the SD data, while column Rwa shows the relative error rate reduction (Rwa=0% for SI result and Rwa=100% for SD performance).

6. Conclusions

A framework for the on-line use of MAP adaptation is proposed and evaluated. The importance of the forgetting mechanism is pointed out by the experiments, in particular with small adaptation data set. The model based approach permits an easy hyper-parameters estimation and updating.

The comparison between ML and MAP with forgetting mechanism, shows that Bayesian models outperform the SI+SD ML ones. Moreover the MAP models are better than SD models when the training data are below 20 minutes of speech.

Further works are required to provide an automatic setting of the forgetting factor and to improve the acoustic normalization.

References

[1] L. Fissore, F. Ravera, and P. Laface. Acoustic-phonetic modeling for flexible vocabulary speech recognition. In *Proc. of EUROSPEECH*, pages I–799–802. Madrid, Spain, 1995.

[2] J.-L. Gauvain and C.-H. Lee. Maximum a posteriori estimation for multivariate gaussian mixture observations of markov chains. *IEEE Trans. on Speech and Audio Processing*, 2(2):291–298, Apr. 1992.

[3] Q. Huo and C.-H. Lee. A study of on-line quasi-bayes adaptation for cdhmm-based speech recognition. In *Proc. of ICASSP*, pages II–705–708. Atlanta, 1996.

[4] Y. Zhao. Self-learning speaker and channel adaptation based on spectral variation source decomposition. *Speech Communication*, 18:65–77, Jan. 1996.

Discriminative Improvement of the Representation Space for Continuous Speech Recognition

Ángel de la Torre, Antonio M. Peinado, Antonio J. Rubio, José C. Segura

Dpto. de Electrónica y Tecnología de Computadores
Universidad de Granada, 18071 GRANADA (Spain)
e-mail: atv@hal.ugr.es

Summary. Signal representation is a very important issue of the design of speech recognizers. An appropriate representation of the speech signal improves the recognizer performance. Recently, the *Discriminative Feature Extraction* (DFE) method has been applied for estimating transformations of the representation space for speech recognizers. In this work, a variant of the DFE method is applied in order to improve the representation space for Continuous Speech Recognition.

1. Introduction

Nowadays, the cepstral-based representations (either based on filter-bank or LPC) are very utilized for speech recognition. Usually, the initial parameters must be transformed. For example, liftering windows are usually applied to the cepstral coefficients in order to enhance the most discriminative ones and to reduce the weight of those more affected by the variability [1].

Recently, a method called *Discriminative Feature Extraction* (DFE) was proposed as a way for improving the representation of the speech signal. The feature extractor, which is usually a linear transformation applied to the original feature space, is discriminatively trained in order to minimize the error-rate. In the transformed representation space, the most discriminative components are enhanced and the recognition accuracy is improved. Biem and Katagiri have applied DFE to compute liftering windows [2] and to design a filter-bank [3]. Torre et al. [4] proposed to obtain the DFE transformation in a pre-training stage by using a very simple classifier in order to obtain a proper algorithm convergence.

In this work we propose an adaptation of the DFE method for Continuous Speech Recognition (CSR) using phoneme-like units (PLUs). The application of the method and the improvement of the recognition results are reported.

2. Discriminative Feature Extraction

The purpose of the DFE method is the estimation of the transformation V by using the *Minimum Classification Error* (MCE) criterion. The elements of the transfor-

This work is funded by CICyT (Spanish Governmental Research Agency) under project TIC96-0956-C04-04.

mation $v_{n,p}$ are iteratively computed (by a gradient descent) in order to minimize a cost function L which represents the classification error,

$$v_{n,p}^k = v_{n,p}^{k-1} - \eta \frac{\partial L}{\partial v_{n,p}} \tag{1}$$

where η is a convergence coefficient. Let $\{X_1, \ldots, X_M\}$ be the set of training sequences and $\{\lambda_1, \ldots, \lambda_I\}$ the set of classes; the cost function is defined as,

$$L = \sum_{m=1}^{M} l_m(X_m); \qquad l_m(X_m) = \frac{1}{1 + e^{-\alpha d_m(X_m)}}; \tag{2a}$$

$$d_m(X_m) = -g_{k(m)} + \frac{1}{\beta} \log \left[\frac{1}{I-1} \sum_{j \neq k(m)} e^{\beta g_j} \right] \tag{2b}$$

where $g_i = g_i(X_m, \lambda_i)$ are the *discriminant functions* (the recognized class is the one whose discriminant function is the largest one). $\lambda_{k(m)}$ is the correct class for the sequence X_m. Thus, $d_m < 0$ (and $l_m \to 0$) if the classification is clearly correct, and $d_m > 0$ (and $l_m \to 1$) if clearly incorrect (l_m is a smooth and derivable classification error function for sequence X_m). In order to compute $\partial L / \partial v_{n,p}$ it is necessary to know the discriminant functions, which are given by the definition of the classifier.

For the DFE estimation of V, we have used a classifier (different than the final one used for recognition) which models every class as a *Single Gaussian pdf*. This way, the obtained transformation improves the representation independently of the classifier utilized for the recognition process, and the training procedures for this recognizer are performed in the improved representation space. Thus, the transformation is computed according to the *Single-Gaussian DFE* (SGDFE) strategy described in [4].

3. SGDFE Algorithm for CSR

The SGDFE algorithm optimizes the transformation from a set of labelled sequences of vectors $\{X_1, \ldots, X_M\}$, where each sequence X_m belongs to a certain class $\lambda_{k(m)}$. In a phoneme-based CSR system, each class roughly corresponds to a phoneme. Thus, in order to perform the SGDFE algorithm, the training data base must be segmented into phonemes and every segment X_m must be labelled. For a given segmentation, a transformation is obtained, and then, the recognizer can be updated to the new representation space.

The updated recognizer leads to a more accurate segmentation of the training data base, and this new segmentation to a more discriminative transformation by the SGDFE algorithm. So, in order to improve the representation for the CSR system, it is necessary to iteratively segment the training data base (using the current recognizer and transformation), update the transformation for the current segmentation and update the recognizer. This segmental procedure is depicted in Figure 1.

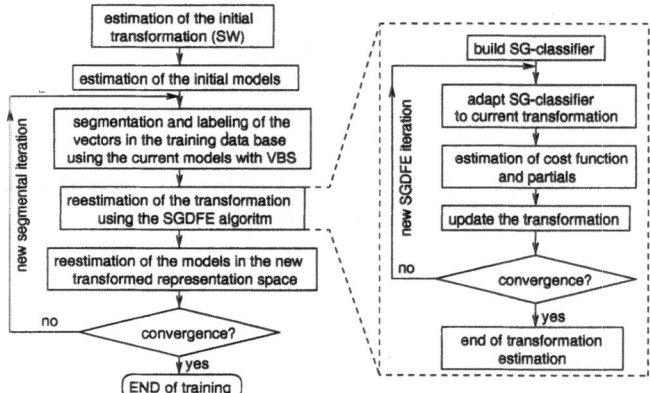

Fig. 1. SGDFE segmental procedure for CSR

The initial transformation consists in a normalization of the standard deviations of the components in the feature vector, this is the *Statistically Weighted* transformation [5]. The segmentation and labelling of the training data base is performed using the *Viterbi Beam Search* algorithm. At every segmental iteration in the algorithm presented in the Figure 1, the current transformation is improved by the iterative minimization of the cost function defined in the equations (2).

4. Experimental Results

The speech signal is sampled (sample frequency f_s=8kHz) and segmented into frames, where the frame length is 30ms and the frame period is 10ms. Every frame is represented by a vector that contains the energy, a cepstral vector containing 14 MFCC coefficients, and the first and second derivatives of these coefficients, which amounts to 45 components.

Three DHMM-based recognition systems have been implemented for the experiments using *Vector Quantization codebooks* with 128, 256 and 512 centroids. 24 Phoneme-Like Units (PLUs), including silence, are modeled using discrete HMM with 3 states and left-to-right topology. The training data base is composed of 804 sentences spoken by 40 speakers. The same data base is used for the estimation of the SGDFE transformation.

In the baseline systems, the components of the feature vector are normalized by applying a *Statistically Weighted transformation*, to compensate the differences in the standard deviation among these parameters. The baseline representation space is improved by the SGDFE method. Due to the small correlation among the different components in the feature vector, the SGDFE transformation is assumed to be diagonal. As criterion for convergence of the SGDFE iterations, (for a given segmental iteration) the loop ends when the reduction of the cost function in the current iteration is smaller than 1/2 times the reduction in the first one. This way, the change

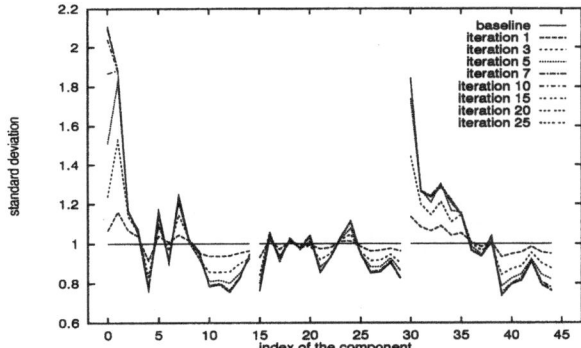

Fig. 2. Evolution of the transformations with the segmental iterations for the 256 centroid system. For the iteration 0 (baseline), the Statistically Weighted transformation is applied, and therefore the standard deviation is 1 for all the components. From the segmental iteration number 10 on, only a small change is observed in the transformations.

in the transformation is small, and the new segmentation is not very different to the last one. This SGDFE convergence criterion makes easier the convergence of the segmental procedure. Figure 2 shows the standard deviation of the transformed components for the transformations obtained at different segmental iterations. We have performed recognition experiments at every segmental iteration to investigate the evolution of the recognition systems.

Three tasks are considered in order to study the behavior of the recognition systems when the SGDFE transformations are applied. The *Task 1* consists on the recognition of continuous speech with a 1104-word vocabulary and a grammar perplexity of 6.9 (using a bigram). In this case the texts of the test sentences were presented in the training data set. The perplexity of *Task 2* is 5.9 and the vocabulary is composed of 203 words. In this case the texts are independent of the ones presented for training. *Task 3* consists in the recognition of the phoneme-like units. For this task, a phoneme grammar was estimated from the training data set. The data base used for this task is the same as that of *Task 2*. The test set used in *Task 1* contains 400 sentences and the one used in *Task 2* and *Task 3* contains 600 sentences. The speakers of the sentences used for test were different than the ones used for training in all cases. Two different *Pruning Thresholds* (PrTh) have been applied in *Task 1* and *Task 2* and no pruning is applied for *Task 3*.

The effect of applying the SGDFE transformation is not only the increment of the recognizer accuracy (see Figure 3). The number of active nodes in the recognition tree is also reduced for a given pruning threshold, because of the increment of accuracy of the HMMs. This reduces the requested memory and the recognition time. Table 1 summarizes the results of the recognition experiments for the different tasks and pruning thresholds. This table includes Word Error Rate (WER), Average Number of active Nodes (ANN) and average Recognition-Time/Duration ratio (RT/D). Due to the fluctuation of the plots (see Figure 3) the results in this table are

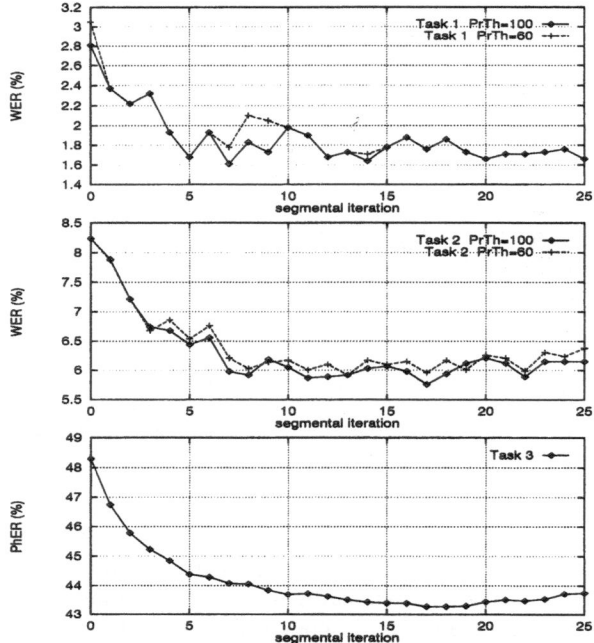

Fig. 3. Evolution of the recognition error-rate for the 256 centroid system for the *Task 1*, the *Task 2* and the *Task 3*. A reduction of the error-rate can be observed. From the segmental iteration number 10 on, only a small variation is observed.

the average from the segmental iteration number 15 to 25. The results presented in the table and plots suggest the following comments:

- The evolution of the performance with the segmental iterations shows the importance of the representation for pattern recognition problems. A small modification in the transformation meaningfully affects the recognition results.
- Even though the SGDFE transformations are computed using a very simple classifier, the recognition systems are improved in all the three tasks. The SGDFE transformations, optimized for a one Gaussian per class classifier improve significantly the HMM-based recognition system. This suggest that a representation space adapted for the discrimination makes easier the recognition process independently of the recognition system or task.
- The discriminative improvement of the representation space leads to a significant improvement with respect to the baseline representation. This improvement of the recognition systems is observed in the reduction of the error-rate, the number of active nodes and the recognition time.

Table 1. Summary of SGDFE recognition experiments

		128 centroids		256 centroids		512 centroids	
		baseline	SGDFE	baseline	SGDFE	baseline	SGDFE
Task 1	WER (%)	4.59	2.21	3.05	1.75	2.15	1.66
PrTh=60	ANN	1417.7	1073.1	1099.6	844.6	903.6	728.7
	RT/D	1.41	1.19	1.30	1.10	1.13	1.00
Task 1	WER (%)	4.44	2.09	2.81	1.75	2.03	1.63
PrTh=100	ANN	6106.2	4864.8	4974.9	3921.0	4180.9	3405.3
	RT/D	3.86	3.19	3.42	2.79	2.95	2.48
Task 2	WER (%)	8.47	6.45	8.24	6.16	6.68	5.86
PrTh=60	ANN	619.4	460.7	506.7	383.0	434.4	345.9
	RT/D	0.37	0.30	0.32	0.26	0.28	0.24
Task 2	WER (%)	8.47	6.45	8.24	6.05	6.51	5.77
PrTh=100	ANN	1759.2	1424.2	1529.1	1248.7	1379.2	1145.2
	RT/D	0.85	0.72	0.78	0.65	0.70	0.59
Task 3	PhER (%)	50.97	47.55	48.29	43.45	45.52	42.03

5. Conclusions

The DFE method provides a formalism for improving the representation for pattern recognition problems with a discriminative criterion. We have proposed a variant of this method which allows the application of the DFE formalism to improve the representation for Continuous Speech Recognition systems. In the proposed method, at every segmental iteration, the training data base is segmented by the current recognizer, and the transformation is updated by applying the SGDFE algorithm to the labelled sequences of vectors. The utility of the proposed method is demonstrated with some experiments including different recognition tasks. The application of the SGDFE transformation improves the representation space (from a discriminative point of view) which leads to an increment of the recognition accuracy and a reduction of the recognition time.

References

[1] B. H. Juang, L. R. Rabiner, and J. G. Wilpon, "On the use of bandpass liftering in speech recognition," *IEEE Trans. on ASSP*, vol. 35, pp. 947–954, July 1987.

[2] A. Biem, S. Katagiri, and B. H. Juang, "Pattern recognition using Discriminative Feature Extraction," *IEEE Transactions on Signal Processing*, vol. 45, pp. 500–504, Feb. 1997.

[3] A. Biem and S. Katagiri, "Cepstrum-based filter-bank design using Discriminative Feature Extraction training at various levels," in *Proc. of ICASSP'97*, vol. 2, pp. 1503–1506, 1997.

[4] A. de la Torre, A. M. Peinado, A. J. Rubio, V. E. Sánchez, and J. E. Díaz , "An application of Minimum Classification Error to feature space transformations for Speech Recognition," *Speech Communication*, vol. 20, pp. 273–290, Dec. 1996.

[5] Y. Tohkura, "A weighted cepstral distance measure for speech recognition," *IEEE Trans. on ASSP*, vol. 35, no. 10, pp. 1414–1422, Oct. 1987.

Dealing with Loss of Synchronism in Multi-Band Continuous Speech Recognition Systems

Christophe Cerisara

CRIN-CNRS & INRIA Lorraine
BP 239, F-54506 Vandoeuvre-les-Nancy Cedex, France
email: `cerisara@loria.fr`

Summary. In multi-band systems, the signal is decomposed into several frequency bands, which are processed separately. Then, the recombination part must compute a unique sentence from all these different solutions. The task is quite easy in isolated word recognition, each word ending at the same time, but it becomes more difficult in continuous speech recognition, where each band has a different segmentation. The problem here is to decide when the recombination should be done. Two major solutions have been tested: the first one introduces synchronism between the bands, and recombination is done when all the bands are synchronous. The second one leaves the sub-recognizers totally independent and tries to extract from their solutions a phonetic structure which will allow us to process the recombination part. We will briefly present an example of the first solution, then we will focus on the algorithm we have developed for the second one.

Key words: automatic speech recognition; continuous speech recognition; multi-band; recombination; acoustic processing.

1. Introduction

Since the issue of J. B. Allen's paper [1], who reactualized the work Fletcher did in the 50s about the human auditory system, several ASR systems using the Multi-Band paradigm have been created. This is due to three major reasons:

1. According to Fletcher, this principle is close to the one used in human speech recognition.
2. Robustness to noise and reverberation could be achieved using this principle.
3. Until now, most ASR systems only used dynamic algorithms over the temporal dimension to determine the most likely pronounced sentence. Recent advances (e.g. [7]) have shown that it would be better to use the feature dimension in the stochastic process as well (using 2D Markov Fields for example), but it is very difficult to find efficient algorithms such as Viterbi's for classical HMMs. The multi-band model is less general than 2D Markov Fields, but it allows us to take into account the feature dimension while keeping the classical one-dimensional HMMs.

In this paper, I will not present the multi-band paradigm; please refer to [3] or [2] for such a general presentation. I will rather focus on what is the key-issue of all multi-band systems: the recombination algorithm.

2. Forcing Synchronism Between the Bands

There are mainly two ways to recombine bands: The first one is to let each sub-recognizer be totally independent from the others. This solution fits best with the multi-band paradigm. Moreover, it allows each sub-recognizer to find its optimal path. The problem here is that each band proposes a sequence of phonemes with different boundaries, and even a different number of phonemes. Then, it is not easy to decide when the bands should be recombined. This problem will be dealt with in the next section. For the time being, let us look at the second possibility to recombine the bands, that is to reintroduce synchronism between them. In this case, recombination can simply be achieved at these points of synchronism.

2.1 First Approach

The first idea we have when we try to implement a multi-band system is to maximize the probability of the sequence of frames (X) given a sequence of models (Ω), over Ω.

$$P(X|\hat{\Omega}) = \underset{\Omega}{\mathrm{argmax}}\, P(X|\Omega) \tag{1}$$

We can define $P(X|\Omega)$ by

$$P(X|\Omega) = \prod_i P(X|\Omega, \text{band } i) \tag{2}$$

As we compute these probabilities with HMMs, we can define Ω like the sequence of states associated with each frame, i.e.:

$$\Omega = (s_i)_{1 \leq i \leq N} \text{ and } X = (x_i)_{1 \leq i \leq N} \tag{3}$$

N is the number of frames of the pronounced sentence.
Then, after some simplifications, we obtain:

$$
\begin{aligned}
P(X|\Omega) = & \left[\prod_i P(x_1|s_1, \text{band } i) \right] \times \\
& \prod_{j=2}^{N} \left[\prod_i P(x_j|s_j, \text{band } i) \right] \left[\prod_i P(s_{j-1} \to s_j | \text{band } i) \right]
\end{aligned}
\tag{4}
$$

$P(s_{j-1} \to s_j | \text{band } i)$ is the transition probability between states for band i.
This is the form of a classical HMM which is defined by:

1. The emission probabilities are equal to the product of all the corresponding emission probabilities of the sub-recognizers.
2. The transition probabilities are equal to the product of all the corresponding transition probabilities of the sub-recognizers.

2.2 Experiments

All the experiments have been carried out on the BREF database for the training corpus, and on the development corpus of Aupelf-Uref [4] for the test database. The last one has been cut into two pieces, one for the training of the classifiers (which are introduced later in this paper), and one for the test part. When no classifier is used -as it is the case for the experiment described in this paragraph-, only the last part of the corpus is used for testing. The obtained precision is less than ±0.5% for all the given figures.

The sub-recognizers are based on second-order HMMs [5] with three states and a mixture of gaussian density estimators in each state. Four bands are used: band 1 (BF) between 0 and 901Hz, band 2 between 797 and 1661Hz, band 3 between 1493 and 2547Hz, and band 4 (HF) between 2298 and 4000Hz. Each band roughly encompasses one formant. 6 MFCC $+\Delta + \Delta\Delta$ coefficients are used for each band, and 12 MFCC $+\Delta + \Delta\Delta$ coefficients are used for the referring Full-Band system.

The experiment carried out with the model described in the last paragraph offers 62% of accuracy on clean speech, which is to compare with the 73% obtained with the Full-Band system. The loss of accuracy is due to the fact that sub-recognizers are less good (especially on clean speech), because of the limited range of frequencies they can use and also because the four bands are forced to synchronize frame by frame. We will try to avoid this drawback in the next section.

Other systems using synchronism between the bands have been developed by Bourlard [2] and Hermansky [6].

3. Modeling Loss of Synchronism

3.1 Theoretical Approach

In order to model the fact that all recognizers are independent, we can rewrite the equation Eq-2:

$$P(X|\Omega) = \prod_i P(X|\Omega_i, \text{band } i) \tag{5}$$

But if the recognizers are independent, it is not the case of the sequences Ω_i, which depend on the pronounced sentence Ω_0. Then, a relation between these sequences exists, and it is possible to write:

$$\mathcal{R}(\Omega_1, \ldots, \Omega_\text{B}, \Omega_0) = c \tag{6}$$

B is the number of bands, and c is a constant.

To simplify the problem, we will only consider the relation between phonemes, and we will compute the relation between sentences using a dynamic algorithm through all possible associations of phonemes.

Let $r(\omega_1, \ldots, \omega_B, \omega)$ be the relation between a set of phonemes, Γ a list of associated phonemes of the sequences $\Omega_1, \ldots, \Omega_B, \Omega$, and Ξ the space of all possible Γ. For example, if $\Omega_1 = (a, b)$ and $\Omega_2 = (c, d)$, then $\Xi = \{[(a, c), (b, d)], [(a, c), (b, \emptyset), (\emptyset, d)], \ldots\}$, and one possible Γ is $\Gamma = [(a, c), (b, d)]$.

Let γ be the function which returns the phoneme of one sub-recognizer in one association of Γ. For example, for the first group of phonemes of Γ, $\gamma(1) = a$ and $\gamma(2) = c$.

To define r, we must keep in mind that $\gamma(1), \ldots, \gamma(B)$ are built by the sub-recognizers in order to approach the pronounced ω_0. So, r describes our knowledge concerning ω_0, given $\gamma(1), \ldots, \gamma(B)$. We chose to define r as the probability that the sub-recognizers have found $\gamma(1), \ldots, \gamma(B)$, assuming that ω is the pronounced phoneme. This will allow us to approach ω_0 by maximizing $r(\gamma(1), \ldots, \gamma(B), \omega)$. So,

$$
\begin{aligned}
r(\gamma(1), \ldots, \gamma(B), \omega) &= P(\gamma(1), \ldots, \gamma(B)|\omega)P(\omega) \\
&= \prod_i [P(\gamma(i)|\omega, \text{band } i)] P(\omega)
\end{aligned}
\tag{7}
$$

And,

$$
\mathcal{R}(\Omega_1, \ldots, \Omega_B, \Omega) = \max_{\Gamma \in \Xi} \prod_{\gamma \in \Gamma} r(\gamma(1), \ldots, \gamma(B), \gamma(0))
\tag{8}
$$

$\gamma(0)$ refers to the sequence Ω that we have to estimate.

Finally, we can approach the pronounced sentence Ω_0 by computing:

$$
\hat{\Omega} = \underset{\Omega}{\operatorname{argmax}} \mathcal{R}(\Omega_1, \ldots, \Omega_B, \Omega)
\tag{9}
$$

3.2 Experimental Approach

To solve Eq-7, we must first compute $P(\omega_i|\omega_j, \text{band } k)$ for all (i, j, k). This is done using the confusion matrix of each sub-recognizer. Eq-8 and Eq-9 are solved together using a best-path algorithm in the graph of all possible associations of phonemes. Each vertex of this graph is composed of B markers to the first free phoneme of each band. (A free phoneme is a phoneme which is not included in an association). Each transition corresponds to the creation of an association, by incrementing some or all of these markers. For example, with 4 bands, each vertex has 16 output-transitions. The final vertex is the one with all markers pointing just after the last phonemes of Ω_i.

The probability of each path is:

$$
P(\Gamma) = \prod_{\gamma \in \Gamma} \max_\omega \left\{ \prod_i \left\langle \begin{array}{c} P(\text{band } i \text{ found } \gamma(i)|\omega) \\ P(\text{band } i \text{ forgot } \omega) \end{array} \right\rangle \cdot P(\omega) \right\}
\tag{10}
$$

We have treated the case of a band being empty in an association by computing the probability that a sub-recognizer forgot a phoneme.

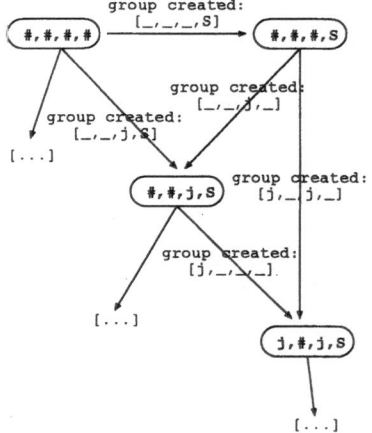

Fig. 1. Example of the graph used by the grouping algorithm

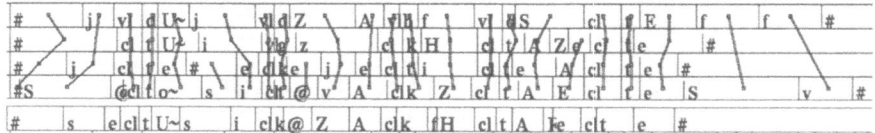

Fig. 2. Example of the solution proposed by the grouping algorithm

We can see on figure 1 the beginning of the graph for the phonetic sequence that is pronounced in the bottom line of figure 2. The top part of figure 2 is an example of the result proposed by the grouping algorithm.

Once the best path is found, we have managed to group together phonemes of the lists Ω_i which are likely to represent the same pronounced phoneme. This means that we are now able to recombine the bands within each of these groups. There are several ways to proceed. The easiest one is to realize a linear merging of the scores returned by the sub-recognizers for all phonemes in each group. But according to Bourlard, best results are achieved when using a non-linear recombiner, such as a Multi-Layer Perceptron. However, we do not have complete results yet, as we are currently working on this recombination part of the algorithm.

4. Conclusion

We have described here an original way to treat the problem of the lack of synchronism between the bands in sub-band based continuous speech recognition. On the one hand, we can force synchronism at some anchor points of the sentence, but this solution does not seem to be in complete harmony with the multi-band paradigm. On the other hand, we can let the sub-recognizers totally independent and try to find similarities between the bands, in order to recombine the recognizers. Such a method has been proposed in this paper. An advantage of this method is that the recombination part (which is usually done by a classifier) should be easier than in the first method, as more knowledge is used to group the data which will be proposed to the classifier. In the first case, these data are computed at regular time-intervals, with no regard for the data themselves. In the second case, they are grouped together according to their phonetic interpretation.

There are still a lot of work to be done in the field of multi-band systems, especially on the recombination part. Multi-band systems on noisy speech seem also worth testing, as it should improve classical HMM-based models. One of their advantage is also that they can easily be implemented in parallel systems, even though the task of each sub-recognizer is simplified (as less parameters are used). So, provided that the last recombination part would not consume too much time, it may be a new step toward real-time speech recognition.

References

[1] J. B. Allen. How do humans process and recognize speech ? *IEEE Trans. on Speech and Audio Processing*, 2(4), October 1994.

[2] H. Bourlard and S. Dupont. Subband-based speech recognition. In *Proc. ICASSP '97*, Munich, Germany, 1997.

[3] C. Cerisara, J.-P. Haton, J.-F. Mari, and D. Fohr. Multi-band continuous speech recognition. In *proc. EUROSPEECH '97*, Rhodes, Greece, 1997.

[4] J.-L. Gauvain, L.-F. Lamel, and M. Eskénazi. Bref, a large vocabulary spoken corpus for french. In *Proc. EUROSPEECH '91*, pages 505–508, Genova, Italy, 1991.

[5] J.-F. Mari, J.-P. Haton, and A. Kriouile. Automatic word recognition based on second-order hidden markov models. *IEEE Trans. on Speech and Audio Processing*, 5(1), January 1997.

[6] S. Tibrewala and H. Hermansky. Sub-band based recognition of noisy speech. In *Proc. ICASSP '97*, pages 1255–1258, Munich, Germany, Apr. 1997.

[7] D. Xu, C. Fancourt, and C. Wang. Multi-channel HMM. In *Proc. ICASSP '96*, pages 841–844, Atlanta, GA, May 1996.

K-Nearest Neighbours Estimator in a HMM-Based Recognition System

Fabrice Lefèvre, Claude Montacié and Marie-José Caraty

LIP6 - Université Pierre et Marie Curie - CNRS
4, place Jussieu - 75252 Paris Cedex 5 - France
e-mail : `Lefevre.Fabrice@lip6.fr`

Summary. For many years, the K-Nearest Neighbours method (K-NN) has been known as one of the best probability density function (pdf) estimator [2]. The development of fast K-NN algorithms allows to reconsider its use in applications with large sample sets. In this outlook, the K-NN decision principle has been assessed on a frame by frame phonetic identification on the TIMIT database. Thereafter, a method to integrate the K-NN pdf estimator in a HMM-based system is proposed and tested on an acoustic-phonetic decoding task.

Key words: nonparametric estimator; probability density function; HMM training; acoustic-phonetic decoding; SNALC.

1. Introduction

In continuous HMM, the state output distributions are usually represented by Gaussian mixture densities. But most of the time, the analysis vectors do not have a Gaussian distribution [5]. Theoretically, Gaussian mixtures could estimate any pdf. Practically, the number of Gaussian functions in the mixtures is derived from heuristics and its optimality is not warranted. This problem occur with any parametric pdf representing the state output distribution of a HMM. To address this difficulty, we propose to develop a HMM-based system using the nonparametric K-NN pdf estimator.

Basically, the K-NN decision rule assigns to an unclassified sample point the class of the majority of its K nearest neighbours of a set of previously classified points. This method is nonparametric since no assumption is made upon the joint distribution of the sample points. A simple modification of the K-NN rule gives a consistent pdf estimator [4] whose asymptotic performances are well known. The 1-NN error bound is at the most twice the optimal Bayes error, and the K-NN error decreases as K increases. The difficulty raised by the high computational demand of the K-NN estimator has been drastically reduced by the development of fast K-NN algorithms which could avoid up to 99.8% of the systematic computation [6].

2. K-NN Assessment

Two preliminary experiments aim at the assessment of the K-NN estimator. These experiments are carried out on TIMIT database ($1,124,823$ training-frames and $57,919$ test-frames). Computed per centi-second, a frame is represented by 12

MFCC and by the energy coefficient. The TIMIT reference labeling [7] provides the classification of the training-frames.

In the first experiment, for each test-frame, we compute the probability of its expected phonetic class according to a 50-NN estimator and an 8 Gaussian mixture estimator (8-Gaussian). Figure 1 gives the histograms of the expected class probability for all the test-frames.

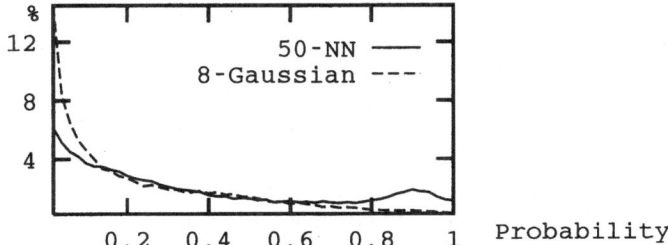

Fig. 1. Histograms of expected class probability for 50-NN and 8-Gaussian estimators

The average expected class probability is 0.36 for the 50-NN and 0.25 for the 8-Gaussian. For the 50-NN, it means that a test-frame has an average of 18 nearest neighbours of the expected class. Thus, three distinctive parts are distinguished. The first one ($k < 18$) represents the vectors far from their proper training set. These vectors have to be studied to analyze precisely the reasons for such spatial distortions. The second part ($k > 25$, i.e. *prob.* > 0.5) represents the vectors which identification is simple with, for instance, the majority rule. The last part ($18 \leq k \leq 25$) represents the vector requiring a more sophisticated identification rule.

In the second experiment, the phonetic identification frame rate has been computed using the 50-NN maximum maximorum decision principle. The global identification rate is about 50%. The explanation of this result is threefold : 1. the low occurrence of phonemes such as [ch, dh, jh, ng, uh, y], 2. the segmentation error inherent to any labeling, 3. the difficulty in identifying complex phonemes such as diphthongs or plosives using a single frame. These statements account for the introduction of the K-NN information in a global decision principle such as HMM.

3. K-NN estimator in HMM

Training and decoding techniques for HMM do not rely on the used pdf. In practice, some adaptations are required from a Gaussian pdf-based system to a K-NN pdf-based system. We have adapted the Forward Backward (FB) and Viterbi algorithms in *HMM ToolKit 1.4* [8] to handle the K-NN estimator.

3.1 Adaptation Principle

The state output probability calculation in HMM is adapted to the K-NN estimator. For this, we compute for each training vector its state occupation probabilities.

The state occupation probability of state s for the vector v of the vector sequence V is computed as :

$$P_{Occ}^{S}(v) = \frac{P(V, X(v) = s|M)}{P(V|M)} \tag{1}$$

where X is a state sequence of V and M the model of V according to a previous or reference alignment. Both numerator and denominator probabilities are derived from the FB or the Viterbi algorithm. The Viterbi algorithm finds the maximum likelihood state sequence, so each observation vector is assigned to a single state. That is, P_{Occ} is 1 for one state and 0 for the others. Whereas, in the FB, the full likelihood is computed on all possible state sequence and each observation vector is assigned to every state in proportion of the model being in that state when the vector was observed.

Thereafter, the state output probability of any vector is calculated as the normalized summation of the state occupation probabilities of its K-NN :

$$P_{Out}^{S}(v) = \frac{\sum_{k=1}^{K} P_{Occ}^{S}(k^{th} - NN(v))}{K} \tag{2}$$

where $k^{th} - NN(v)$ is the kth nearest neighbour of observation vector v.

We have introduced this state output probability calculation in the FB and Viterbi algorithms. Usually, the Viterbi algorithm is used to perform a first estimation before the re-estimation with FB. In our case, this first estimation with Viterbi does not fit since nearly all the vectors are finally assigned to one single state. The direct assignment of vectors to individual states in Viterbi creates an irreversible accumulation phenomenon. An original method is proposed to estimate the model parameters without the initial Viterbi stage.

3.2 HMM Estimation Improvement

A HMM training assumes initial estimates of the HMM parameters. Commonly, a rough guess of the initial pdf values is obtained by a uniform segmentation of the observations sequences, associating each successive segment with successive states. Then a first estimation is made with the Viterbi algorithm, which we saw is not adapted in our case. To address this, the first estimation of the models is obtained from a new method using the information provided by the K-NN.

The parameters are still initialized by a uniform segmentation. Then, a time to states projection is performed. That is, each vector is associated to every states proportionally to the number of its K-NN in this state considering the uniform segmentation. Then, the vector sequences are divided into time-proportional intervals.

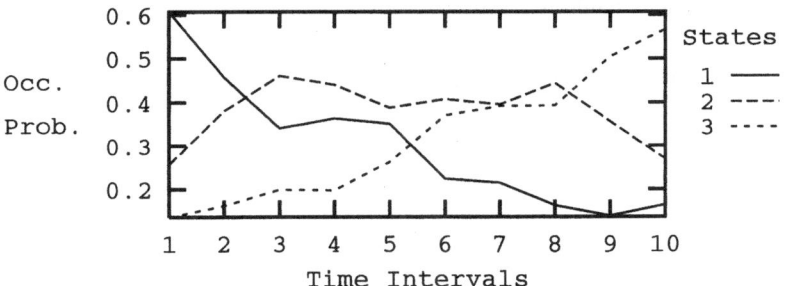

Fig. 2. Time/states projection for vowel /uw/

In this way, the time/states projections are averaged amongst all vectors of each interval. These average time/states projections are used to initialize state occupation probabilities.

Figure 2 illustrates the time/states projection in a three-states Bakis model for the vowel /uw/, with 50-NN. For instance, the state occupation probabilities of the observations of the fifth interval (i.e. included within 40 and 50% of their sequence duration) are 0.26 for the first state, 0.46 for the second one and 0.27 for the third one.

The trainings performed with the time/states projection do not show any significant gain in the accuracy of the model estimation (i.e. do not increase the average model likelihood). Nonetheless, it should be noted that a certain amount of observation sequences could have a null likelihood because of the incoherence between the paths imposed by two successive frames. This effect is rather eliminated using the time/states projection initialization. Thus, the FB convergence should be considered better as it involves more training sequences. The impact on the accuracy rate (+0.2%) is not significant. Others attempts, such as topology inference [1], will be investigated to address the problem of the first estimates of the parameters .

4. Evaluations

The experiments aim at comparing the Gaussian and the K-NN estimators in HMM-based system. The chosen task is the reference acoustic-phonetic decoding [3] on TIMIT database core-test. The basis HMM is a three-states Bakis. Both systems use a phonetic back-off bigram learned on the trainset. Two kinds of criterion are used : the recognition rates and the Segmental Normalized Acoustic Likelihood Coefficient (SNALC).

4.1 Recognition rates

Table 1 presents the recognition results for the 8-Gaussian and 50-NN systems. These results are low but are related to the simplicity of the involved models. The gain in identification with the 50-NN estimator is nearly 6%. This difference is

Table 1. Recognition results on TIMIT core-test

	Identification	Accuracy	Error Details		
			Deletion	Substitution	Insertion
50-NN	58.17	51.84	14.91	26.91	6.33
8-Gaussian	52.35	50.06	24.09	23.56	2.29

lowered to 1.8% in accuracy due to the high level of insertion in the 50-NN system (6.3%). This difference between accuracy rates is barely statistically meaningful since the confidence interval for these conditions is 1.1%. To compare the systems thoroughly, we introduce a new criterion : the SNALC.

4.2 SNALC Evaluation

The SNALC is an attempt to obtain a better evaluation of the pdf influence in a decision process such as HMM. This coefficient is computed for each frame during a Viterbi decoding as :

$$SNALC = 1 - \frac{L_{ref}/T_{ref}}{\sum_i L_i/T_i} \tag{3}$$

where L_i is the ith model likelihood for a segment of T_i frames. The reference phoneme corresponds to the TIMIT alignment.

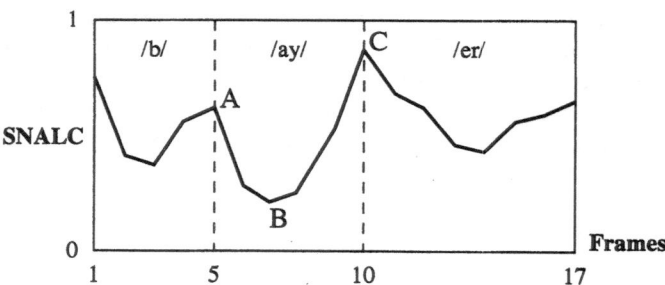

Fig. 3. Typical behaviour of SNALC

Figure 3 is an illustration of the typical behaviour of SNALC, here on an utterance of the world "buyer". From the beginning of the phoneme /ay/ (A), the SNALC decreases to B : the /ay/ model likelihood becomes preponderant. Then, the likelihood of /er/ increases and the SNALC raises up to the end of /ay/ (C). The SNALC values are relevant to confusions between phonemes or co-articulation phenomena. Thus, the average SNALC can be used as an assessment measure.

Experiments were performed on the TIMIT core test in the above-mentioned conditions. The average SNALC for the 50-NN (0.62) is better than the 8-Gaussian one (0.74). If the average SNALC provides a synthetic value for assessment, the

SNALC affords more detailed analysis of the decision process behaviour. Thus, it highlights a resonance effect in the computation of the state output probabilities with the Gaussian estimator : values overstepping 1. Actually, the Gaussian functions are pdf's. They should be integrated around the considered vector to become probabilities. We noted this on SNALC curves for phonetic classes such as /silence/, /f/ or even /s/. This point reveals an uncontrolled effect (the resonance) in the Gaussian HMM-based decision process.

5. Perspectives

For the first time, the K-NN estimator has been introduced in a HMM-based system. At this moment, its performances are comparable to the Gaussian ones. However, it is remarkable that most of the continuous HMM techniques are strongly adapted to Gaussian pdf's. Suitable techniques are to be found for the K-NN estimator.

Improvements of the K-NN HMM-based system could arise from the ability to retrace and to analyze the training vectors causing the identification errors. Another interesting propriety of the K-NN estimator is to provide a general framework in which we could combine information from different representation spaces, and thus use locally adapted spaces and their associated metrics.

Finally, as they behave differently, K-NN and Gaussian should favourably be used simultaneously in a composite pdf estimator.

References

[1] R. de Mori, M. Galler, and F. Brugnara. Search and learning strategies for improving hmm. *Computer Speech and Language*, 9:107–121, 1995.

[2] J. Goût. L'apprentissage en reconnaissance de la parole. Technical report, Université PARIS 6, 1993.

[3] K.-F. Lee and H.-W. Hon. Context-dependent phonetic hmm for speaker-independent continuous speech recognition. *IEEE Trans. ASSP*, 38(4):599–609, 1990.

[4] D. Lotfsgaarden and C. Quesenberry. A nonparametric estimate of a multivariate density function. *Annals Math. Stat.*, 36:1049–1051, 1965.

[5] C. Montacié, M.-J. Caraty, and C. Barras. Mixture splitting technic and temporal control in a hmm-based recognition system. In *Proc. ICSLP*, 1996.

[6] C. Montacié, M.-J. Caraty, and F. Lefèvre. K-NN versus gaussian in a HMM-based recognition system. In *Proc. Eurospeech*, 1997.

[7] S. Seneff and V. Zu. *Transcription and Alignment of the TIMIT Database*. NIST, 1988. CD-ROM TIMIT.

[8] S. Young. *HTK Version 1.4 : Reference Manual and User Manual*. CUED-Speech Group, 1992.

Robust Speech Recognition

Sadaoki Furui

Tokyo Institute of Technology
Department of Computer Science
2-12-1, Ookayama, Meguro-ku, Tokyo, 152 Japan
furui@cs.titech.ac.jp

Abstract. This paper overviews the main technologies that have recently been developed for making speech recognition systems more robust against acoustic variations. These technologies are reviewed from the viewpoint of a stochastic pattern matching paradigm for speech recognition. Improved robustness enables better speech recognition over a wide range of unexpected and adverse conditions by reducing mismatches between training and testing speech utterances.

Keywords. Mismatches, robustness, speech variation, noise, distortion, speaker, adaptation, Bayesian learning

1 Mismatches between Training and Testing

Speech recognition process is usually formulated as a *maximum a posteriori* (MAP) decoder [9]. That is, it estimates w' by maximizing *a posteriori* probability $p(w \mid x)$ given an acoustic parameter sequence x. Based on Bayes rule, the speech decoding problem is reformulated as

$$w' = \operatorname*{argmax}_{w \in \Gamma} p(w \mid x) = \operatorname*{argmax}_{w \in \Gamma} p(x \mid w)p(w). \qquad (1)$$

Γ is the set of all possible word sequences, $p(x \mid w)$ is the conditional probability that x is produced by speakers, given a particular sequence of words w, and $p(w)$ is the *a priori* probability that word sentence w is internally constructed by speakers for utterance. A *language model* is used to calculate $p(w)$, and an *acoustic model* is used to calculate $p(x \mid w)$. The stochastic pattern matching approach often assumes particular parametric forms for $p(x \mid w)$ and $p(w)$. Typically, hidden Markov models (HMMs) and stochastic language models, respectively, are used.

1.1 Speech Variation

Mismatches between the training data and the test utterances always occur in speech recognition. These mismatches are the main factor degrading the performance of speech recognition systems in practical situations. Even if a speech recognition system performs remarkably well in laboratory evaluations and during demonstrations to prospective clients, it often performs much worse in the "real world".

Two factors cause mismatches. First, speech signals are subject to acoustic and linguistic variations. The main causes of acoustic variations resulting from the speech production processes are shown in Fig. 1 [14]. They include speaker individuality, additive noise, room reverberation, and microphone and transmission characteristics. Although the physical phenomena of speech variations can be classified as either noise addition or distortion, the distinction between these two categories is not clear. (Additive noise also increases the difficulty of determining the speech period, but this problem is not addressed in this paper.) Nonlinear time expansion and contraction also occur, but these problems can be coped with by using the dynamic programming (DP) technique. When people speak in a noisy environment, not only does the loudness (energy) of their speech increase, but the pitch and frequency components also change. These speech variations are called the Lombard effect [18]. The linguistic variation includes environment-dependent speaking styles.

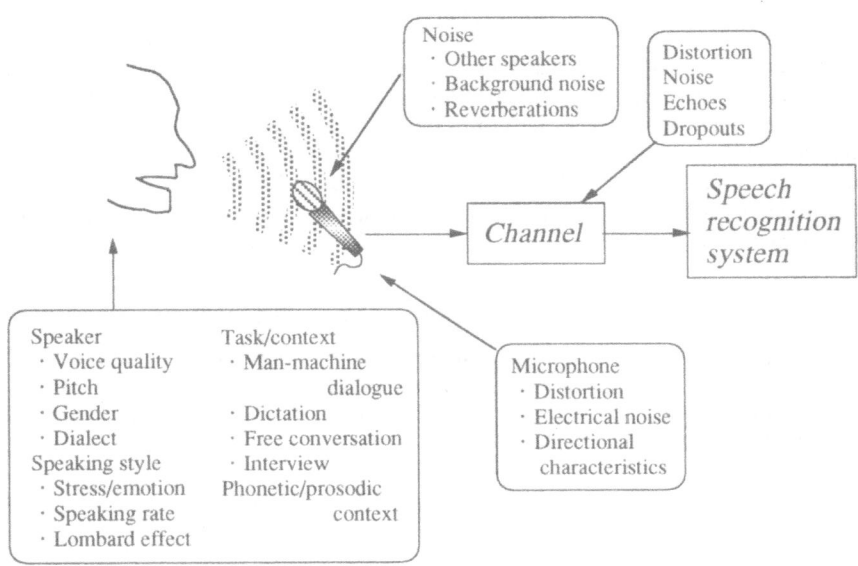

Fig. 1. Main causes of acoustic variation in speech.

The second factor causing mismatches is an incorrect modeling assumption and/or an insufficiency of training data to accurately estimate the model parameters. It is not practical to collect a huge set of speech and text data spoken and written by a large population over all possible combinations of signal conditions.

Recognition performance under adverse conditions is often affected by the degree of speech quality degradation rather than by the degradation itself. Problems are created, for example, by a mismatch in noise levels associated with variations in the distance between the speaker and the microphone. If training can be done under the same noise conditions as those under which the speech is to be recognized, better performance is attained than when the training is done under noise-free conditions.

Conversely, if training is done under noisy conditions and the speech to be recognized is clean, the performance is worse than when noisy speech is recognized.

1.2 Inter-Speaker Variation

The performance of speaker-independent speech-recognition systems has been greatly improved by training them with a large speech database spoken by many speakers and by incorporating stochastic models of speech variations. The acoustic models can be refined by partitioning the training data according to speaker groups, then estimating separate models for each group (for example, gender-dependent models).

It is still impossible, however, to accurately recognize the utterances of every speaker. A small percentage of people occasionally cause systems to produce exceptionally low recognition rates because of large mismatches between the models and the input speech. This is an example of the "sheep and goats" phenomenon. Inter-speaker variation is one of the most serious sources of variation in speech challenging today's speech recognition technologies [10] [11][32].

Differences among speakers arise from differences in the anatomy of vocal tracts and vocal cords, in dialects, in speaking idiosyncrasies, etc. Mutual effects have been observed between phonetic and speaker-specific characteristics; specifically, the differences in speaker-specific characteristics depend on the phonemes [7]. Experiments have shown that humans can adapt to a new speaker's voice after hearing just a few syllables [19]. Recent research has therefore explored the possibility of enabling recognition systems to adapt to individual speakers by automatically reducing the mismatches [10][13].

2 Reducing Mismatches to Improve Speech Recognition

Robust speech recognition refers to the problem of designing an automatic speech recognizer that works well over a wide range of unexpected and adverse conditions. Suppose the models Λ are used in speech recognition to decode acoustic parameters x by using the MAP decoder:

$$w' = \operatorname*{argmax}_{w \in \Gamma} p(x|w,\Lambda)p(w). \tag{2}$$

A mismatch between the training and testing environments causes a corresponding mismatch in the likelihood of x given Λ, as evaluated by Eq. 2, thus causing errors in decoded sequence w'. Decreasing this mismatch should reduce the recognition error rate [27].

The mismatch can be viewed in the signal space, the feature space, or the model space. Figure 2 shows the main methods for reducing mismatches [16][17] [12][13], along with the basic sequence of speech recognition processes. Since there is no significant theoretical difference in viewing the mismatch in these three spaces, we will consider the problems in the model space.

2.1 Principles of Adaptive Speech Recognition

Various adaptive speech recognition methods have been investigated, in which mismatch is decreased by automatically adapting models Λ. Adaptation methods are

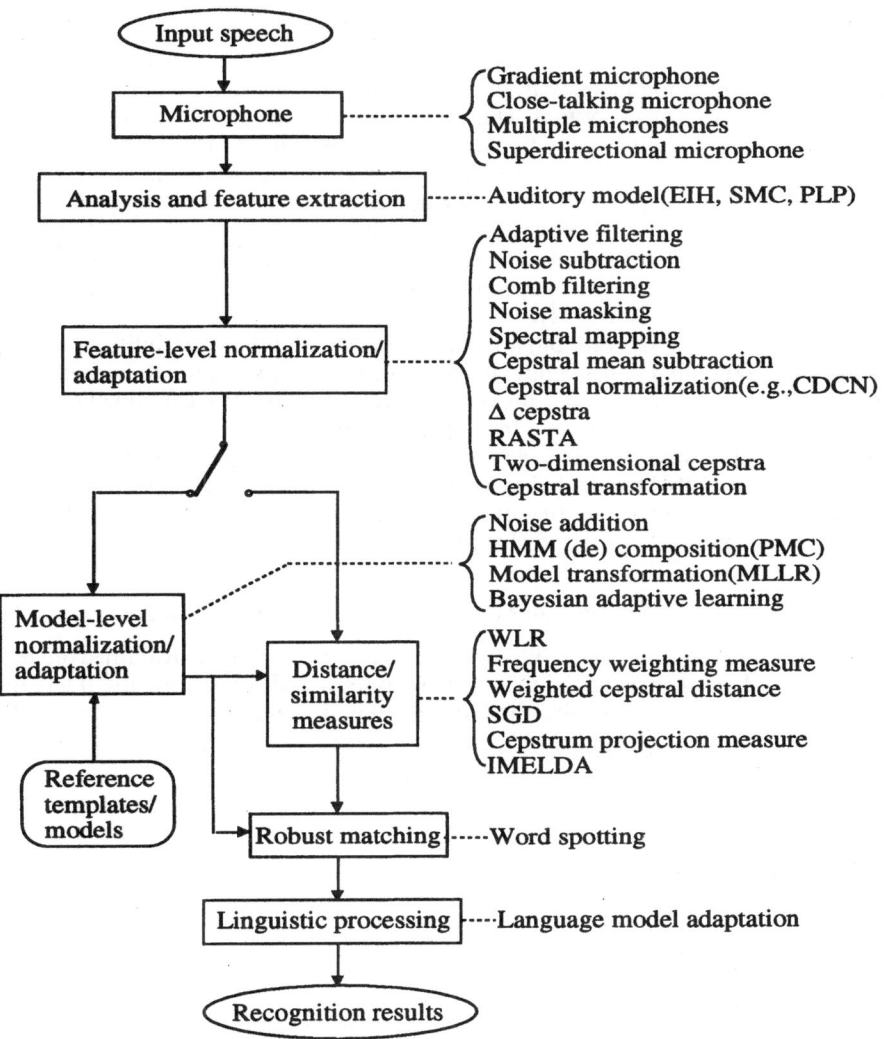

Fig. 2. Main methods for reducing mismatches to cope with speech variation in speech recognition.

generally classified either as supervised (text-dependent), in which training words or sentences are known, or as unsupervised (text-independent), in which arbitrary utterances are used. They are further classified into off-line and on-line modes. The off-line mode requires a new speaker to contribute a certain amount of adaptation speech; the on-line mode takes the (unknown) speech from the speaker as the adaptation data. Supervised on-line adaptation requires user feedback.

The most useful adaptation method is on-line instantaneous (unsupervised) adaptation. In this approach, adaptation is performed at runtime on the input speech in an unsupervised manner. Therefore, the recognition system does not require training speech to estimate the mismatch; it works as if it were a universal (speaker and environment-independent) system. This method is especially useful when the speakers or environmental conditions vary frequently. The most important issue in this method is how to perform phoneme-dependent adaptation without knowing the correct model sequence for the input speech. This is especially difficult for speakers whose utterances are error prone when using universal models, that is, for speakers who definitely need adaptation. Another useful adaptation method is on-line incremental adaptation, in which the recognition system continuously adapts to new adaptation data without using previous training data [24].

Because inter-speaker variability often interacts with other variabilities, such as allophonic contextual dependency, intra-speaker speech variation, environmental noise, and channel distortion, it is important to create methods that can simultaneously cope with these other variabilities. Inter-speaker variability is generally more difficult to cope with than noise and channel variability, since the former is non-linear whereas the latter can usually be modeled as a linear transformation in the time, spectral, or cepstral domain. Therefore, the algorithms proposed for speaker adaptation can generally be applied to noise and channel adaptation.

2.2 Three Principal Adaptation Methods for Reducing Mismatches

Let's consider a transformation with parameter η that maps Λ into transformed models. One approach to decreasing the mismatch between x and Λ is to find the η and the word sequence w that maximize the joint likelihood of x and w in Eq. 2, given the models Λ. We thus need to find η' and w' such that

$$(\eta', w') = \underset{\eta,\ w}{\mathrm{argmax}}\ p(x, w\ |\eta,\ \Lambda), \qquad (3)$$

which is equivalent to

$$(\eta', w) = \underset{\eta,\ w}{\mathrm{argmax}}\ p(x\ |\ w,\ \eta,\ \Lambda)p(w). \qquad (4)$$

Trying every possible combination of hypothesis w and η is very costly; three methods have been investigated to reduce this cost: *Method I* is an iterative method that sequentially estimates w and η, *Method II* is a joint maximization method that uses the N-best hypotheses, and *Method III* is a supervised adaptation method that uses given texts.

In Method I, a typical conventional unsupervised adaptation method, the *pseudo*-joint maximization over η and w is performed iteratively by keeping η fixed and maximizing over w, then keeping w fixed and maximizing over η. In unsupervised speaker adaptation, a speaker-independent model is used as the initial model Λ and hypothesis w maximizing the likelihood is chosen as an initial hypothesis. Since recognition hypothesis w governs the algorithm, a poor hypothesis can result in suboptimal performance, especially if the degree of freedom in the adaptation is large.

In Method II, proposed by Matsui et al. [22], the N-best hypotheses are used in parallel and the joint likelihood is iteratively maximized. This method will be described later.

In Method III, which is widely used for supervised adaptation (training), the texts for the set of word sequences w's used for adaptation are given. Maximization is thus performed only for η, with the expectation that the transformed models Λ will also decrease the mismatches between unseen input speech and the models. This method has difficulty generalizing the adaptation of the models to input speech not seen in training. More research is thus needed on using these adaptation data more efficiently in creating stochastic models that are capable of dealing with different acoustic variations. A better understanding of speech signals and how they are affected by the speaker characteristics and speaking environments will help researchers develop theories and techniques for coping with this generalization problem.

In general, it is not easy to estimate η directly. However, for some model transformations, we can use the expectation-maximization (EM) algorithm for iteratively improving the current estimate and obtaining new estimates such that the likelihood increases at each iteration.

2.3 Important Practical Issues

Important practical issues in using adaptation techniques include the specification of *a priori* parameters (information), the availability of supervision information, and the amount of adaptation data needed to achieve effective learning. Since it is unlikely that all the phoneme units will be observed enough times in a small adaptation set, especially in large-vocabulary continuous-speech recognition systems, only a small number of parameters can be effectively adapted. It is therefore desirable to introduce some parameter correlation or tying so that all model parameters can be adjusted at the same time in a consistent manner, even if some units are not included in the adaptation data.

The Bayesian learning framework offers a way to incorporate newly acquired application-specific data into existing models (*a priori* information) and to combine them in an optimal manner. It is therefore an efficient technique for handling the sparse training data problem typically found in model parameter adaptation. This framework has been used to derive MAP estimates of the parameters of speech models, including HMM parameters [15][20].

In the spectral mapping-based speaker adaptation method, speaker-adaptive parameters are estimated from speaker-independent parameters based on mapping rules. The mapping rules are estimated from the relationship between speaker-independent and speaker-dependent parameters in a supervised or unsupervised way [8][11][29].

If a correlation structure between parameters can be established, and the correlation parameters can be estimated when training the general models, the parameters of unseen units can be adapted accordingly [6][4]. To improve adaptation efficiency and effectiveness along this line, several techniques have been proposed, including probabilistic spectral mapping [28], hierarchical spectral clustering and smoothing [8][25], cepstral normalization [1], and spectrum bias and shift transformation [27].

In addition to correlation, a second type of constraint can be given to the model parameters so that all the parameters are adjusted simultaneously according to a predetermined set of transformations, e.g., a transformation based on multiple regression analysis [6]. Various methods have recently been proposed in which a linear transformation (affine transformation) between the reference and adaptive speaker-feature vectors is defined and then translated into a bias vector and a scaling matrix, which can be estimated using an EM algorithm. The transform parameters can be estimated from supervised adaptation data that form pairs with the training data [31][2][21][26][30] or estimated from unsupervised adaptation data using existing acoustic models [3][5].

In the speaker cluster selection method, it is assumed that speakers can be divided into clusters, within which the speakers are similar. From many sets of phoneme model clusters representing speaker variability, the most suitable set for the new speaker is automatically selected. This method is useful for choosing initial models, to which more sophisticated speaker adaptation techniques are applied.

2.4 N-Best-Based Unsupervised Adaptation

The N-best-based unsupervised adaptation method [22] uses the N most likely word sequences in parallel and iteratively maximizes the joint likelihood. The N-best hypotheses are created for each input speech by applying speaker-independent models; speaker adaptation based on constrained Bayesian learning is then applied to each hypothesis. Finally, the hypothesis with the highest likelihood is selected as the most likely sequence. Figure 3 shows the overall structure of such a recognition system. Without giving reasonable constraints based on models of inter-speaker variability, an input utterance can be adapted to *any* hypothesis with resulting high likelihood. To reduce this problem, constraints should be placed on the transformation so that it maintains a reasonable geometrical shape.

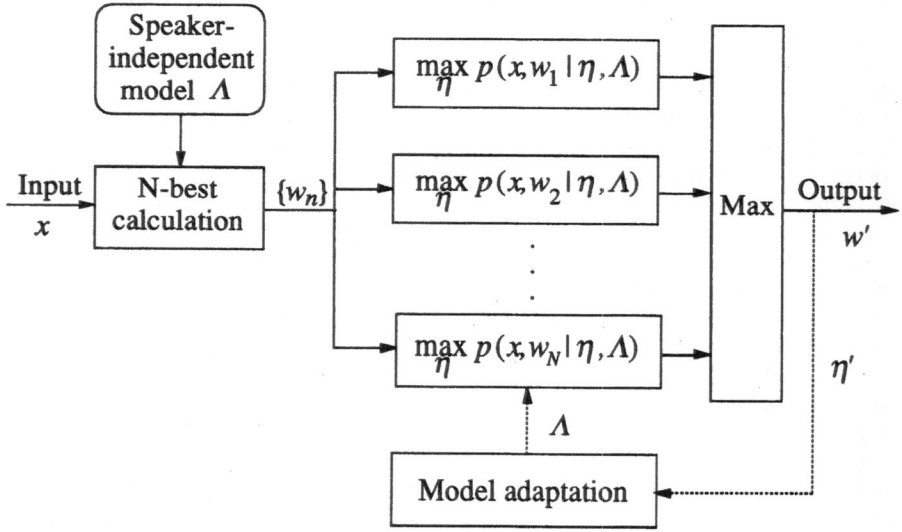

Fig. 3. Overall structure of recognition system including N-best-based hierarchical model parameter adaptation.

To achieve such a constraint, we apply constrained MAP estimation, including hierarchical spectral clustering and smoothing; we begin with a small number of bias parameters and increase the number with each iteration. This averages out the hypothesis errors for a smaller number of parameters. As the iterations continue, the hypotheses improve, allowing for an increased number of parameters.

Experimental results using Japanese connected digits recorded over a telephone network showed that the likelihood of a correct hypothesis in a low rank using speaker-independent models is increased by the adaptation, so that recognition accuracy is greatly improved even for speakers having very low accuracy. Other constraints, including linear transformation (linear regression), have also been investigated [23].

This method can also be interpreted as a multiple-pass search algorithm, where speaker-independent and speaker-adaptive models, respectively, are used in the first and second passes.

3 Conclusion

This paper has briefly reviewed the main approaches to improving the performance of speech recognition systems under actual conditions, in which speech signals are subject to many different kinds of variation. The approaches were reviewed from the viewpoint of reducing mismatches between the input speech and the acoustic models in a stochastic pattern matching framework.

Because many variations can simultaneously occur under actual conditions, it is crucial to create a universal framework that can cope with them. From the human-interface point of view, it is desirable to develop an unsupervised instantaneous adaptation technique that can simultaneously reduce multiple mismatches introduced by several variation sources. A better understanding of speech signals and how they interact with speaker characteristics and speaking environments is thus needed.

References

[1] Acero, A. and Stern, R. M., "Environmental robustness in automatic speech recognition," Proc. IEEE Int. Conf. Acoust., Speech, Signal Processing, Albuquerque, S15b.11, pp. 849-852 (1990).

[2] Bellegarda, J. R., De Sousa, P. V., Nadas, A. J., Nahamoo, D., Picheny, M. A. and Bahl, L. R., "The metamorphic algorithm, a speaker mapping approach to data augmentation," IEEE Trans. Speech and Audio Processing, Vol. 2, No. 3, pp. 413-420 (1994).

[3] Cox, S. J. and Bridle, J. S., "Unsupervised speaker adaptation by probabilistic fitting," Proc. IEEE Int. Conf. Acoust., Speech, Signal Processing, Glasgow, Scottland, S6.11, pp. 294-297 (1989).

[4] Cox, S. J., "Predictive speaker adaptation in speech recognition," Computer Speech and Language, Vol. 9, pp. 1-17 (1995).

[5] Digalakis, V. and Neumeyer, L., "Speaker adaptation using combined transformation and Bayesian methods," Proc. IEEE Int. Conf. Acoust., Speech, Signal Processing, Detroit, pp. I-680-683 (1995).

[6] Furui, S., "A training procedure for isolated word recognition systems," IEEE Trans. Acoust., Speech Signal Processing, Vol. 28, No. 2, pp. 129-136 (1980).

[7] Furui, S., "Research on individuality features in speech waves and automatic speaker recognition techniques," Speech Communication, Vol. 5, No. 2, pp. 183-197 (1986).

[8] Furui, S., "Unsupervised speaker adaptation method based on hierarchical spectral clustering," Proc. IEEE Int. Conf. Acoust., Speech, Signal Processing, Glasgow, S6.9, pp. 286-289 (1989).

[9] Furui, S., *Digital Speech Processing, Synthesis and Recognition*, Marcel Dekker, New York (1989).

[10] Furui, S., "Speaker-dependent-feature extraction, recognition and processing techniques," Speech Communication, Vol. 10, Nos. 5-6, pp. 505-520 (1991).

[11] Furui, S., "Speaker-independent and speaker-adaptive recognition techniques," in *Advances in Speech Signal Processing*, edited by S. Furui and M. M. Sondhi, pp. 597-622 (1992).

[12] Furui, S., "Toward robust speech recognition under adverse conditions," Proc. ESCA Workshop on Speech Processing in Adverse Conditions, Cannes-Mandelieu, France, pp. 31-42 (1992).

[13] Furui, S., "Flexible speech recognition," Proc. Eurospeech, Madrid, pp. 1595-1603 (1995).

[14] Furui, S., "Recent advances in robust speech recognition," Proc. ESCA-NATO Workshop on Robust Speech Recognition for Unknown Communication Channels, Pont-a-Mousson, pp. 11-20 (1997).

[15] Gauvain, J.-L. and Lee, C.-H., "Bayesian learning for hidden Markov models with Gaussian mixture state observation densities," Speech Communication, Vol. 11, Nos. 2-3, pp. 205-214 (1992).

[16] Juang, B. H., "Recent developments in speech recognition under adverse conditions," Proc. Int. Conf. Spoken Language Processing, Kobe, 25.1, pp. 1113-1116 (1990).

[17] Juang, B.-H., "Speech recognition in adverse environments," Computer Speech and Language, Vol. 5, pp. 275-294 (1991).

[18] Junqua, J. C. and Anglade, Y., "Acoustic and perceptual studies of Lombard speech: Application to isolated-words automatic speech recognition," Proc. IEEE Int. Conf. Acoust., Speech, Signal Processing, Albuquerque, S15b.9, pp. 841-844 (1990).

[19] Kato, K. and Furui, S., "Listener adaptability for individual voice in speech perception," Trans. Committee of Hearing Research, H85-5 (1985).

[20] Lee, C.-H. and Gauvain, J.-L., "Bayesian adaptive learning and MAP estimation of HMM," in *Advanced Topics in Automatic Speech and Speaker Recognition*, edited by C.-H. Lee, K. K. Paliwal and F. K. Soong, Kluwer Academic Publishers, pp. 83-107 (1995).

[21] Leggetter, C. J. and Woodland, P. C., "Maximum likelihood linear regression for speaker adaptation of continuous density hidden Markov models," Computer Speech and Language, Vol. 9, pp. 171-185 (1995).

[22] Matsui, T. and Furui, S., "N-best-based instantaneous speaker adaptation method for speech recognition," Proc. Int. Conf. Spoken Language Processing, Philadelphia, pp. 973-976 (1996).

[23] Matsui, T., Matsuoka, T. and Furui, S., "Smoothed N-best-based speaker adaptation for speech recognition," Proc. IEEE Int. Conf. Acoust., Speech, Signal Processing, Munich, pp. 1015-1018 (1997).

[24] Matsuoka, T. and Lee, C.-H., "A study of on-line Bayesian adaptation for HMM-based speech recognition," Proc. Eurospeech, Berlin, pp. 815-818 (1993).

[25] Ohkura, K., Sugiyama, M. and Sagayama, S., "Speaker adaptation based on transfer vector field smoothing with continuous mixture density HMMs," Proc. Int. Conf. Spoken Language Processing, Banff, We.fPM.1.1, pp. 369-372 (1992).

[26] Sankar, A. and Lee, C.-H., "Robust speech recognition based on stochastic matching," Proc. IEEE Int. Conf. Acoust., Speech, Signal Processing, Detroit, pp. I-121-124 (1995).

[27] Sankar, A. and Lee, C.-H., "A maximum-likelihood approach to stochastic matching for robust speech recognition," IEEE Trans. Speech and Audio Processing, Vol. 4, No. 3, pp. 190-202 (1996).

[28] Schwartz, R., Chow, Y.-L. and Kubala, F., "Rapid speaker adaptation using a probabilistic spectral mapping," Proc. IEEE Int. Conf. Acoust., Speech, Signal Processing, Dallas, 15.3, pp. 633-636 (1987).

[29] Shikano, K., Lee, K.-F. and Reddy, R., "Speaker adaptation through vector quantization," Proc. IEEE Int. Conf. Acoust., Speech, Signal Processing, Tokyo, 49.5, pp. 2643-2646 (1986).

[30] Zavaliagkos, G., Schwartz, R. and Makhoul, J., "Batch, incremental and instantaneous adaptation techniques for speech recognition," Proc. Int. Conf. Acoust., Speech, Signal Processing, Detroit, pp. I-676-679 (1995).

[31] Zhao, Y., "An acoustic-phonetic-based speaker adaptation technique for improving speaker-independent continuous speech recognition," IEEE Trans. Speech and Audio Processing, Vol. 2, No. 3, pp. 380-394 (1994).

[32] Zhao, Y., "Robust speaker characterization," Proc. IEEE Automatic Speech Recognition Workshop, Snowbird, pp. 101-102 (1995).

Channel Adaptation

Keith M. Ponting

Speech Research Unit, DERA Malvern
St. Andrew's Road, Great Malvern, Worcs. WR14 3PS, UK
email: ponting@signal.dera.gov.uk

Summary. Any mismatch between training and test conditions can cause difficulty for current automatic speech recognition systems. In recent years many approaches have been proposed for resolving this mismatch problem. These approaches can be divided broadly into three classes: model adaptation, channel adaptation and robust features. This paper presents a review and discussion of methods for channel adaptation and their relationship to methods in the other classes.

Key words: environment mismatch; speech classes; general speech model; noise; distortion.

1. Introduction

Most current large vocabulary automatic speech recognition systems use hidden Markov models (HMMs) and rely on large databases of acoustic training material to train those models. Typically databases are collected under clean conditions and the resulting laboratory recognition systems are evaluated using similarly clean data. However the conditions under which recognition systems are used in real applications are often far from "clean" and as a result recognition performance often falls short of that obtained in the laboratory.

1.1 Matched condition training

mismatched training and test environments One solution is to insist that the training conditions should match the conditions of use. For each different condition, models must be trained using either artificially distorted clean database material, or a similar large database collected under the target conditions. Such systems also require advance knowledge of the conditions of use in order that such tailoring can be performed, and may not perform as well if actual conditions differ.

1.2 Robust features

A second approach is to use features which are robust to general or particular types of signal degradation, such as the *Integrated MEl scale representation with Linear Discriminant Analysis* (IMELDA) [15] or the *RelAtive SpecTrAl* (RASTA) [13] representation.

1.3 Model adaptation

Much recent work has focussed on combining models trained on clean data with a model of the expected channel conditions either within the recognition algorithm (Decomposition [20, 33]) or prior to processing a particular utterance (Parallel Model Combination – PMC [34]).

Given some data recorded under the expected conditions, Maximum Likelihood Linear Regression (MLLR) [17], and other general adaptation techniques may also be used in a supervised or unsupervised manner to modify speech models trained on clean data.

1.4 Channel adaptation

Channel adaptation techniques, in contrast, aim to estimate a correction to be applied to speech frames or utterances prior to training or testing.

The likelihood calculations used in most speech recognition systems depend on input feature vectors $\mathbf{O}$ only as a function of the difference $(\mathbf{O} - \mu_j)$ for each model mean μ_j. Therefore there is a strong correspondence between "channel adaptation" techniques and those "model adaptation" techniques which alter only the model means. MLLR in particular includes an additive offset term, which is very similar to the class-specific correction vectors discussed in section 3.3.

This correspondence is only exact when the channel adaptation consists of adding a bias $\mathbf{b}$ to each observed feature vector. This bias may depend on the observed $\mathbf{O}$, which is one of the reasons for considering channel adaptation techniques. In rapidly changing conditions it may be prohibitively expensive to change all the model means (particularly in a large system), whereas the computation required to apply a bias to the input is trivial.

However a single additive correction is not sufficient to cope with all conditions, therefore much effort has been put in to obtaining more general mappings to transform speech obtained under the test conditions (usually "noisy") so that it better matches that used for training (usually "clean"). Section 3. reviews a number of the proposed methods.

1.5 Speech enhancement

There is a separate long tradition of work on enhancing noisy signals for human-human communication purposes (see [7] for a review, more recent examples are [2, 35]). The methods are usually evaluated subjectively as well as by output SNR and distortion measures. Such systems generally improve the quality (and thereby reduce listener fatigue) at the expense of some loss in intelligibility or occasionally vice versa [7].

2. Models of distortion

The most commonly used model of distortion is that the original speech signal may be distorted by both additive and convolutional noise. This can be represented in the log spectral domain as:

$$\mathbf{O}_t = \mathbf{H}_t + \log\left\{\exp\left(\mathbf{S}_t\right) + \exp\left(\mathbf{N}_t\right)\right\}, \tag{1}$$

where $\mathbf{S} = \{\mathbf{S}_1 \ldots \mathbf{S}_T\}$ is the entire input signal, $\mathbf{H}$ represents the effect of the convolutional noise and $\mathbf{N}$ that of the additive noise. Any sequence of such distortions may be reduced to this form by suitable redefinition of $\mathbf{H}_t$ and $\mathbf{N}_t$. Re-arranging (1) gives:

$$\mathbf{O}_t = \mathbf{S}_t + \mathbf{H}_t + \log\left(1 + \exp\left(\mathbf{N}_t - \mathbf{S}_t\right)\right) = \mathbf{S}_t + f\left(\mathbf{S}_t, \mathbf{N}_t, \mathbf{H}_t\right) \tag{2}$$

say, where f is sometimes known as the *environment function* [31] or *gain function* [35]. The various applications of this model rely on the assumption that $\mathbf{S}, \mathbf{N}$ and $\mathbf{H}$ are independent.

In particular multi-path (reverberant) environments and the (Lombard [16]) effect on the speaker of the noise environment lie outside the scope of this model, although Lombard induced "variations of the spectral tilt"[12] could be covered by the $\mathbf{H}$ term, which would then become dependent on $\mathbf{N}$.

2.1 Minimum mean square error

Minimum mean square error (MMSE) is a common theme in restoration of a noise-corrupted signal. The MMSE estimate is given by the conditional expectation of the clean representation given the noisy representation (over some time window or even the whole utterance) and the parameters and form of assumed distributions of the speech and the noise.

For purposes of speech recognition it is most appropriate to work in the feature domain used by the speech recogniser. However additional complexities are introduced by the log-spectral–to–feature transformation, so many MMSE methods minimise error in the log-spectral, power-spectral or other domains[25]. Computation of the MMSE estimate is still non-trivial because of the non-linearities involved in (1) and closed form solutions are only available in specific cases.

2.2 Additive noise estimation

The additive noise models used in channel adaptation range from a simple estimate of the current noise level to single or even multiple state HMMs. Assuming the additive noise $\mathbf{N}$ is relatively constant, the simplest mechanism for estimating that noise is to use some form of heuristic speech detector and calculate the noise statistics based on periods when no speech is present.

The histogram method described in [32] for "applications with significant positive signal-to-noise ratios" is an example of one such speech detector which does not require long pauses for measuring noise statistics but gathers them from a mixed speech and noise signal. Similarly [14] describes a recursive speech-noise detector combined with a noise estimation and subtraction algorithm.

3. Methods for channel adaptation

If the distortions $\mathbf{H}$ and $\mathbf{N}$ are independent of the original speech spectrum $\mathbf{S}$, then in principle the simplest compensation method is to obtain estimates $\hat{\mathbf{H}}$ and $\hat{\mathbf{N}}$ and modify the distorted signal to recover an estimate of the original:

$$\hat{\mathbf{S}}_t = \log \left\{ \exp \left(\mathbf{O}_t - \hat{\mathbf{H}}_t \right) - \exp \left(\hat{\mathbf{N}}_t \right) \right\} \tag{3}$$

The many proposed techniques can be grouped according to whether the actual or deduced speech content of an utterance is used in computing the corrections, applying the corrections or both.

Most of the techniques proposed determine the corrections based on a probabilistic assignment of the input frame $\mathbf{O}_t$ to one of a set of classes. Let p_{kt} be the posterior probability that a particular input speech frame $\mathbf{O}_t$ belongs to class (Gaussian, VQ cluster, HMM state or mixture component) k, an event denoted by $c_t = k$. Some techniques use only the most likely class at each time t, denoted by κ_t.

HMMs (with parameters denoted by λ) may be used as the basis of the speech classes as in [4, 29]. In this case, p_{kt} may be obtained from either the forward probabilities $\alpha_t(k)$ or the mixture component occupancies $\gamma_t(k)$ from the forward-backward algorithm:

$$\alpha_t(k) = P(O_1 \ldots O_t, c_t = k \,|\, \lambda) \tag{4}$$
$$\gamma_t(k) = P(O_1 \ldots O_T, c_t = k \,|\, \lambda) \tag{5}$$

The classes may also be based on a simpler model of typical speech patterns [27]. For such models Bayes' rule is usually used to obtain posterior probabilities of the class given the observation(s):

$$w_{kt} = P(c_t = k \,|\, \mathbf{O}_t) = P(\mathbf{O}_t |\, c_t = k) \,/ \left(\sum_k P(\mathbf{O}_t |\, c_t = k) \right) \tag{6}$$

3.1 Global transformations

The first group of methods attempts to compute a single $\hat{\mathbf{H}}$ and $\hat{\mathbf{N}}$ to apply to all speech frames, independent of the value of $\mathbf{O}_t$.

If $\hat{\mathbf{H}}_t$ is zero in (3), compensation reduces to subtraction of the noise estimate in the linear spectral domain. However the argument of the logarithm may become negative, hence the spectral subtraction (SS) algorithm (e.g. [3]) uses a thresholded difference. [18] discusses the relationship between SS and the optimal (least mean squared error among linear estimators) Wiener filter.

If $\hat{\mathbf{H}}_t$ is constant over time and $\hat{\mathbf{N}}_t$ is zero (3) reduces to the subtraction of a single correction vector from the input frame before further processing:

$$\hat{\mathbf{S}}_t = \mathbf{O}_t - \hat{\mathbf{H}} \tag{7}$$

To compensate for convolutional noise, cepstral mean subtraction (CMS [10]) relies on a speech detector to determine which frames in an input signal are speech frames. From the speech frames a mean vector for the whole of an utterance is computed. That mean is then subtracted (during both training and recognition) from the cepstral representation before further processing.

One key aspect of the RASTA processing [13] is the use of band pass filters (BPF) on each frequency channel in the logarithmic domain. If the convolutional noise is reasonably consistent, such filters will remove it from the signal altogether. In this context CMS can be regarded as a kind of non-causal high-pass filter (HPF) removing only the constant component. [5] reports that, in compensating for varying telephone handsets over the public telephone network, CMS outperformed RASTA.

CMS classifies frames into speech and non-speech but makes no attempt to distinguish among the speech frames. The remaining methods use a provisional assignment[1] of speech frames to classes in the process of computing the corrections. The work reported in [36] for $N = 0$ uses a provisional assignment to (Gaussian – $N(\mu_k, C_k)$ – based) classes giving:

$$\hat{\mathbf{H}} = \left(\sum_{t=1}^{T} \sum_{k} w_{kt} C_k^{-1} \right)^{-1} \sum_{t=1}^{T} \sum_{k} w_{kt} C_k^{-1} (\mathbf{O}_t - \mu_k) \tag{8}$$

for summation over all classes and a simpler form for summation using only the most likely class at each time t. The equivalent algorithm (signal bias removal – SBR) for a discrete density HMM is developed in [27] and extended to make the correction dependent on the input frame (hierarchical signal bias removal – HSBR). [26] reports that SBR outperformed CMS for a task with varying telephone handsets.

The stochastic matching approach [29, 30] results in ML solutions for more general transformations for reconstructing the clean speech from the noisy speech. For a function of the form (7) this reduces to:

$$\hat{\mathbf{H}} = \left(\sum_{t=1}^{T} \sum_{k} \frac{\gamma_t(k)}{\sigma_k^2} \right)^{-1} \sum_{t=1}^{T} \sum_{k} \frac{\gamma_t(k)}{\sigma_k^2} (\mathbf{O}_t - \mu_k) \tag{9}$$

[24] describes a similar algorithm embedded in a speech recognition system and using deterministic assignment of frames to classes based on the dynamic programming alignment.

3.2 Class-specific corrections

The non-linear effect of joint additive and convolutional noises leads to methods which apply a different compensation according to some preliminary, usually probabilistic, classification of the input speech:

$$\hat{\mathbf{S}}_t = \mathbf{O}_t - \sum_k p_t(k) \hat{\mathbf{H}}_k \tag{10}$$

[1] All such methods may be iterated, using the estimated $\hat{\mathbf{S}}$ to obtain a new assignment and re-estimate the correction terms.

where $\hat{\mathbf{H}}_k$ is a correction appropriate to the kth class and $p_t(k)$ is the posterior probability of that class at time t given the input (possibly more of the input than just $\mathbf{O}_t$) and the appropriate models.

The original MMSE approach [25] used a single general speech distribution as the prior pdf. That work demonstrated better performance from MMSE than from spectral subtraction for a task with white noise added to the original signal.

Subsequent work showed considerable further improvement by using classes and sums of class-specific MMSE estimates weighted by the class posterior probabilities. Minimum mean log spectral distance (MMLSD [8]) uses clusters in acoustic space as the classes. Energy-conditioned(EC)-MMSE [9] uses total frame energy to define the clusters.

Using an HMM for the clean speech pdf, [6] derives the MMSE estimate of a fairly general function $g(\mathbf{S}_t)$ of the clean speech given the input up to time t as a sum of class-specific corrections weighted by the normalised forward probability $\bar{\alpha}_{kt} = \alpha_t(k)/\sum_j \alpha_t(j)$

$$\widehat{g(\mathbf{S}_t)} = O_t - \sum_k \bar{\alpha}_{kt} \left(O_t - E\left[g(\mathbf{S}_t)\middle| O_t, c_t = k\right]\right) \tag{11}$$

Model-based (MB)-MMSE [7] extends this to multiple state models of the noise and conditioning on the whole of the input.

HMM inversion [19], although differently motivated, has the same general form. Here the classes and $P\left(c_t = k | \mathbf{O}_t\right)$ are derived from HMM speech models with Gaussian mixture densities and the forward-backward algorithm. The reconstructed speech is then given by:

$$\hat{\mathbf{S}}_t = \sum_k \gamma_t(k)\mu_k = \mathbf{O}_t - \sum_k \gamma_t(k)\left(\mathbf{O}_t - \mu_k\right) \tag{12}$$

3.3 Empirical methods based on stereo data

A series of empirical methods have been developed at CMU aimed at obtaining corrections additive in the logarithmic feature domain. Essentially class specific additive correction vectors are used, with increasingly complicated methods of defining the class, while the corrections are computed based on real or simulated stereo training data.

SNR dependent cepstral normalisation (SDCN [1]) uses deterministic assignment to classes based on total frame energy. Subsequent CMU extensions to this algorithm partition the input into distinct classes using:

- a VQ codebook in addition to SNR (Fixed Codeword Dependent–FCDCN);
- an uncompensated recognition pass (Phone Dependent–PDCN).

RATZ (multivaRiate gAussian based cepsTral normaliZation) [21] uses a mixture Gaussian distribution $\sum_k c_k N\left(\mu_k, \Sigma_k\right)$ to represent the statistics of clean speech. Additive corrections to the means and variances of this distribution are computed based on stereo recordings using the following equations:

$$\hat{\mathbf{r}}_k = \sum_{t=1}^{T} w_{kt} \left(\mathbf{O}_t - \mu_k \right) / \sum_{t=1}^{T} w_{kt} \qquad (13)$$

$$\hat{\mathbf{R}}_k = \sum_{t=1}^{T} w_{kt} \left(\mathbf{O}_t - \mu_k \right) \left(\mathbf{O}_t - \mu_k \right)^T / \sum_{t=1}^{T} w_{kt} \qquad (14)$$

where the values w_{kt} are obtained from the clean stereo material using (6). Blind RATZ (BRATZ) iterates those equations in a form of the EM algorithm, alternately estimating $\hat{\mathbf{r}}_k$ and $\hat{\mathbf{R}}_k$ and updating the mean μ_k and covariance matrix Σ_k. At recognition time the adapted clean speech distribution is used to compute w_{kt} and calculate:

$$\hat{\mathbf{S}}_t = \mathbf{O}_t - \sum_k w_{kt} \mathbf{r}_k \qquad (15)$$

Note that whereas [4] computes a single correction based on the weighted average difference of observations and Gaussian means, RATZ and BRATZ compute and use the frame-mean differences separately for each class k, according to the posterior weights w_{kt}. Comparison with (9) shows some similarities – if one bias is computed for each class k the result is:

$$\hat{\mathbf{H}}_{\mathbf{k}} = \sum_{t=1}^{T} \gamma_t(k) \left(\mathbf{O}_t - \mu_k \right) / \sum_{t=1}^{T} \gamma_t(k) \qquad (16)$$

which should be compared with (13).

An alternative mapping approach is probabilistic optimum filtering (POF) [23]. The speech is re-constructed as a combination of filters applied to a time window of speech frames, one filter per Gaussian, weighted by w_{kt}.

3.4 Model-based compensation

Codeword dependent cepstral normalisation (CDCN) [1] also calculates and uses class-specific offsets weighted by posterior probability of the class. Instead of training the offsets directly they are calculated, as $r_k = f(\mu_k, \hat{\mathbf{N}}, \hat{\mathbf{H}})$ in (2). $\hat{\mathbf{H}}$ and $\hat{\mathbf{N}}$ are maximum likelihood estimates obtained iteratively under an approximation that there is a single noise component primarily affected by $\mathbf{N}$ and the remaining components are primarily affected by $\mathbf{H}$. [31] reports that when compensating for microphone mismatch in quiet conditions CDCN outperformed CMS which in turn outperformed RASTA.

The vector Taylor series (VTS) [22] methods can be viewed as extensions of CDCN in which the single noise component approximation is removed and the additive noise follows a Gaussian distribution, whose parameters (mean μ_N and covariance matrix Σ_N) may be re-estimated. A Taylor series is used to expand the environment function about the class means μ_k and the current noise estimates in the iteration. For VTS-0 the MMSE correction which results has the form (15) with $r_k = f(\mu_k, \hat{\mu}_{\mathbf{N}}, \hat{\mathbf{H}})$.

The VTS method attempts to approximate the environment function directly, whereas the vector polynomial approximations (VPS) [28] approach uses various polynomials to approximate the distributions of the Gaussians representing the noisy speech. [28] reports that in added white noise VPS outperformed CDCN and RATZ and was at least as good as VTS. The final correction is based on the difference between the estimates $\bar{\mu}_k$ of the class means corrupted by the noise and the original means:

$$\hat{\mathbf{S}}_t = \mathbf{O}_t - \sum_k P\left(c_t = k \,|\, \mathbf{O}_t\right) \left(\bar{\mu}_k - \mu_k\right) \tag{17}$$

4. Conclusion

Many techniques have been proposed for preprocessing speech signals to improve performance. Most of these methods reduce to a sum of class specific corrections, weighted by posterior probabilities of each class given the data. The best are able to provide substantial reductions in error rates compared to uncompensated recognition.

The prime advantage and drawback of these preprocessing approaches lies in the separation of the channel adaptation and the main recognition pass. This allows compensation to be attempted with lower complexity and computational load than an integrated system but as a result the methods are also sub-optimal. The fully optimal approach involves simultaneous recognition and compensation, leading to techniques such as decomposition [33] and the popular PMC [11] approximation.

Generally model adaptation outperforms the preprocessing approaches discussed in this paper (e.g. [30]). The bulk of the benefit comes from adapting the model means, with a small additional gain from adapting model variances as well[34]. The more complicated mapping approaches bear some similarity to model adaptation, but:

- the (recognition) model variances are not changed;
- the adaptation applied depends only on the position of the feature vector (or sequence) in feature space, not of the recognition model;
- only a relatively small number of corrections need be computed.

Most of the proposed techniques make little explicit use of the structures in the speech signal. Those which are based on HMM state classes implicitly make no more and no less use of the speech properties than the HMM itself.

Given matched recordings in the clean and distorted conditions, a mapping may be constructed defining how clean speech is affected by particular distortions. Many of the techniques discussed here can be regarded as mappings of this form.

In the limit, speaker adaptation may also be achieved by regarding an individual speaker as an unknown "distortion" and successfully applying similar techniques. [4, 36] show that a simple additive correction can indeed give some degree of *speaker* adaptation.

References

[1] A. Acero and R. M. Stern. Environmental robustness in automatic speech recognition. In *Proc. IEEE ICASSP*, volume II, pages 849–852, 1990.

[2] Y. Cohen, A. Erell, and Y. Bistritz. Enhancement of connected words in an extremely noisy environment. *IEEE Trans Speech and Audio Processing*, 5(2):141–148, March 1997.

[3] D. V. Compernolle. Improved noise immunity in large vocabulary speech recognition with the aid of spectral subtraction. In *Proc. IEEE ICASSP*, pages 1143–1146, Dallas, 1987.

[4] S. J. Cox and J. S. Bridle. Unsupervised speaker adaptation by probabilistic spectrum fitting. In *Proc. IEEE ICASSP*, pages 294–297, Glasgow, 1989.

[5] J. de Veth and L. Boves. Comparison of channel normalisation techniques for automatic speech recognition over the phone. In *Proc. ICSLP*, Philadelphia, 1996.

[6] Y. Ephraim. A minimum mean square error approach for speech enhancement. In *Proc. IEEE ICASSP*, pages 829–832, Albuquerque, 1990.

[7] Y. Ephraim. Statistical-model-based speech enhancement systems. *Proc. IEEE*, 80:1562–1555, 1992.

[8] A. Erell and M. Weintraub. Estimation using log-spectral-distance criterion for robust speech recognition. In *Proc. IEEE ICASSP*, pages 853–856, Albuquerque, 1990.

[9] A. Erell and M. Weintraub. Energy conditioned spectral estimation for recognition of noisy speech. *IEEE Trans Speech and Audio Processing*, 1(1), 1993.

[10] S. Furui. Cepstral analysis technique for automatic speaker verification. *IEEE Trans Acoustics, Speech and Signal Processing*, 29(2):254–272, April 1981.

[11] M. J. F. Gales. "Nice" model-based compensation schemes for robust speech recognition. In *Robust Speech Recognition for Unknown Communication Channels*, pages 55–64, Pont-a-Mousson, France, April 1997.

[12] B. A. Hanson and T. H. Applebaum. Robust speaker-independent word recognition using static, dynamic and acceleration features: Experiments with lombard and noisy speech. In *Proc. IEEE ICASSP*, volume II, pages 857–860, Albuquerque, 1990.

[13] H. Hermansky and N. Morgan. Rasta processing of speech. *IEEE Trans Speech and Audio Processing*, 2(4):578–589, October 1994.

[14] H. G. Hirsch and C. Ehrlicher. Noise estimation techniques for robust speech recognition. In *Proc. IEEE ICASSP*, volume I, pages 153–156, Detroit, 1995.

[15] M. J. Hunt and S. M. Richardson. An investigation of PLP and IMELDA acoustic representations and of their potential combination. In *Proc. IEEE ICASSP*, volume 2, pages 881–884. IEEE, 1991.

[16] J.-C. Junqua. The Lombard reflex and its role on human listeners and automatic speech recognisers. *J. Acoust. Soc. Am*, 93(1):510–524, 1993.

[17] C. J. Legetter and P. C. Woodland. Speaker adaption of HMMs using linear regression. Technical Report CUED/F-INFENG/TR181, Cambridge University Engineering Department, 1994.

[18] B. P. Milner and S. V. Vaseghi. Comparison of some noise compensation methods for speech recognition in adverse environments. *Proc. IEE*, 141:280–288, 1994.

[19] S. Moon and J.-N. Hwang. Noisy speech recognition using robust inversion of hidden Markov models. In *Proc. IEEE ICASSP*, pages 145–148, Detroit, 1995.

[20] R. K. Moore. Signal decomposition using Markov modelling techniques. RSRE Memo 3931, DERA Malvern, St. Andrew's Road, Great Malvern, Worcs. WR14 3PS, UK, 1986.

[21] P. J. Moreno, B. Raj, E. Gouvêa, and R. M. Stern. Multivariate gaussian based cepstral normalisation for robust speech recognition. In *Proc. IEEE ICASSP*, pages 137–140, Detroit, 1995.

[22] P. J. Moreno, B. Raj, and R. M. Stern. A vector Taylor series approach for environment-independent speech recognition. In *Proc. IEEE ICASSP*, volume II, pages 733–736, Atlanta, 1996.

[23] L. Neumeyer and M. Weintraub. Probabilistic optimum filtering for robust speech recognition. In *Proc. IEEE ICASSP*, pages 417–420, Adelaide, 1994.

[24] K. M. Ponting. Automatic speech recognition for time-varying channels. In *Robust Speech Recognition for Unknown Communication Channels*, pages 175–178, Pont-a-Mousson, France, April 1997.

[25] J. E. Porter and S. F. Boll. Optimal estimators for spectral restoration of noisy speech. In *Proc. IEEE ICASSP*, page 18A2.1, San Diego, 1984.

[26] M. G. Rahim and B.-H. Juang. Signal bias removal for robust telephone based speech recognition in adverse environments. In *Proc. IEEE ICASSP*, volume I, pages 445–448, Adelaide, 1994. IEEE.

[27] M. G. Rahim, B.-H. Juang, W. Chou, and E. Buhrke. Signal conditioning techniques for robust speech recognition. *IEEE Signal Processing Letters*, 3(4):107–109, April 1996.

[28] B. Raj, E. B. Gouvêa, P. J. Moreno, and R. M. Stern. Cepstral compensation by polynomial approximation for environment-independent speech recognition. In *Proc. ICSLP*, Philadelphia, 1996.

[29] A. Sankar and C.-H. Lee. Robust speech recognition based on stochastic matching. In *Proc. IEEE ICASSP*, pages 121–124, Detroit, 1995.

[30] A. Sankar and C.-H. Lee. A maximum likelihood approach to stochastic matching for robust speech recognition. *IEEE Trans Speech and Audio Processing*, 4(3):190–202, May 1996.

[31] R. M. Stern, B. Raj, and P. J. Moreno. Compensation for environmental degradation in automatic speech recognition. In *Robust Speech Recognition for Unknown Communication Channels*, pages 33–41, Pont-a-Mousson, France, April 1997.

[32] D. van Compernolle. Noise adaptation in a hidden Markov model speech recognition system. *Computer Speech and Language*, 3(2):151–167, April 1989.

[33] A. P. Varga and R. K. Moore. Hidden Markov model decomposition of speech and noise. In *Proc. IEEE ICASSP*, pages 845–848, Albuquerque, 1990. IEEE.

[34] P. C. Woodland, M. J. F. Gales, and D. Pye. Improving environmental robustness in large vocabulary speech recognition. In *Proc. IEEE ICASSP*, volume 1, pages 65–68, Atlanta, 1996.

[35] F. Xie and D. V. Compernolle. Speech enhancement by spectral magnitude estimation – a unifying approach. *Speech Communication*, 19(2):89–104, 1996.

[36] Y. Zhao. An acoustic phonetic based speaker adaptation technique for improving continuous speaker independent speech recognition. *IEEE Trans Speech and Audio Processing*, 2(3):380–394, July 1994.

Speaker Characterization, Speaker Adaptation and Voice Conversion

Sadaoki Furui

Tokyo Institute of Technology
Department of Computer Science
2-12-1, Ookayama, Meguro-ku, Tokyo, 152 Japan
furui@cs.titech.ac.jp

Abstract. This paper discusses recent advances in and perspectives of research on speaker-dependent-feature extraction from speech waves, automatic speaker identification and verification, speaker adaptation in speech recognition, and voice conversion. Both supervised and unsupervised speaker adaptation algorithms for speech recognition have recently been actively investigated, and remarkable progress has been achieved in this field. Improving synthesized speech quality by adding natural characteristics of voice individuality, and converting synthesized voice individuality from one speaker to another, are as yet little exploited research fields to be studied in the near future.

Keywords. Speaker-dependent features, speaker adaptation, speaker normalization, spectral mapping, voice conversion

1 Introduction

Speaker-dependent information plays an important role in our daily life in addition to linguistic (phonetic) information [13]. For example, when we talk with our friends over the telephone, we can easily recognize the other party based on his/her voice without hearing the name. The human capability of distinguishing individual voices also plays a central role in the "cocktail party effect", supplementing the auditory binaural effect. Using these capabilities, we can pick up utterances of a specified speaker from the surrounding noise including other speakers' voices.

This paper presents recent progress in and the future direction of research on extraction of speaker-dependent features from speech waves, speaker adaptation techniques using these features, and voice conversion [11][13].

2 Speaker-Characterization

Speaker-dependent information exists both in the spectral envelope (vocal tract characteristics) and in the supra-segmental features (voice source characteristics) of speech. This individual information can be further classified into temporal and dynamic features. It arises both from hereditary individual differences in articulatory organs, such as the length of the vocal tract, the size of the nasal cavity, and vocal cord characteristics, and from acquired differences in manner of speaking, such as dialect and accent. Hereditary differences (anatomical differences) appear as variations in formant frequencies, bandwidth, mean fundamental frequency, inclination of overall spectrum pattern. Acquired differences (speaker-dependent speaking habits) appear as pitch, speech rate and loudness. They are represented by differences in time functions of fundamental frequency, formant frequencies and word duration. These variations are combined together in the speech waves, making it difficult to separately extract these two types of individual information. Therefore, feature parameters containing both types of information are usually used in individual feature extraction.

The average characteristics of variations in vowel formant frequencies due to gender and age can be modeled by the shift in the logarithmic frequency scale maintaining the relative location of the vowels. However, individual characteristics are extremely complicated. And it has also been observed that mutual effects exist between phonetic and individual information, specifically the differences in individual information depending on the phonemes produced [10]. This means that text-independent individual information is hard to extract.

3 Speaker Recognition

Speaker recognition can be principally divided into speaker verification and speaker identification. Speaker verification is the process of accepting or rejecting the identity claim of a speaker. Speaker identification is the process of determining from which of the registered speakers a given utterance comes. Speaker verification is applicable to various kinds of services which involve the use of voice as the key to confirming the identity claim of a speaker. These services include banking transactions using a telephone network, database access services, security control for confidential information areas, and so on. Speaker identification can be used in criminal investigations to determine which of the suspected persons produced the voice recorded at the scene of the crime. Speaker verification has generally larger application areas than speaker identification.

The difficulty of speaker identification increases with the size of the candidate population. On the other hand, the difficulty of speaker verification does not depend on the population because it involves only a binary decision of acceptance or rejection [11].

Speaker recognition methods can also be divided into text-dependent and text-independent methods. The former require the speaker to issue a predetermined utterance, whereas the latter do not rely on a specific text being spoken. Although text-dependent speaker verification techniques have almost reached the practical level, text-independent techniques are still in the fundamental research stage with various techniques being investigated.

One of the most difficult problems in speaker recognition is that intersession variability (variability over time) of speech waves and spectra for a given speaker significantly affects recognition accuracy. The spectral equalization (normalization) technique has been confirmed to be effective in reducing long-term spectral variation [7]. It is also essential to use physical features which are stable and not easily mimicked or affected by transmission characteristics.

Recent advances in speaker recognition technology are described in a separate chapter in this book (S. Furui: "Speaker Recognition").

4 Speaker-Adaptation Techniques for Speech Recognition

Recently, speaker-independent speech recognition methods using HMM techniques have been actively researched, and the recognition accuracy has been improved. However, one of the disadvantages of the speaker-independent approach is that it neglects various useful characteristics of the speaker. If these characteristics can be properly used, the recognition process is expected to be accelerated due to the narrowing of the search space. Another disadvantage is that when the distributions of feature parameters are very broad or multi-modal, such as in the cases of the combination of male and female voices and of various dialects, it is difficult to separate phonemes using speaker-independent methods. To cope with these problems, it is essential to introduce speaker-adaptation techniques [14].

4.1 Classification of Speaker-Adaptation/Normalization Methods

Speaker adaptation or normalization ("speaker adaptation" indicates both adaptation and normalization hereafter) is the method of automatically adapting reference templates or models to each new speaker or normalizing (reducing) interspeaker variation in each input speech based on the transformation rules obtained using a few training words or short sentences. In large vocabulary recognition systems, training by the utterances of all vocabulary words is too troublesome of the users and consequently unrealistic. Therefore, training by a few words or short sentences is a practical and realistic solution.

Speaker-adaptation methods are generally classified into supervised (text-dependent) methods in which training words or sentences are known, and unsupervised methods in which arbitrary utterances can be used. Both methods can also be classified into off-line methods in which training words or sentences must be uttered before the recognition, and on-line methods in which utterances for recognition are, at the same time, used for training.

Ideally for users, the system should work as if it were a speaker-independent system which requests no additional training utterance of each speaker. Also, the system should actually adapt to the speaker's voice automatically using utterances for recognition. Such a system can be realized by the unsupervised, on-line adaptation mechanism. Humans have been found to have a mechanism of unsupervised, on-line speaker adaptation [20].

Since individual information (individuality) varies depending on the phoneme classes, it is effective to introduce the mechanism of explicitly or implicitly recognizing the phonemes included in the training utterances into the unsupervised adaptation process. By recognizing the phonemes and utilizing the phoneme-dependent individual information, adaptation performances can be improved. In such processes, it is crucially important that phoneme decision errors do not cause inappropriate adaptation.

4.2 Speaker Cluster Selection Methods

One of the basic speaker-adaptation methods is the speaker cluster selection method. In this method, it is assumed that speakers can be divided into clusters, within which the speakers are similar. From many sets of phoneme template clusters representing speaker variability, the most suitable set for the new speaker is automatically selected.

In an HMM-based supervised method, a group of speakers is divided into speaker clusters, and a codebook and HMMs are generated for each cluster [22]. Cluster-specific HMM parameters are derived by a training process using speakers in the cluster, or by a training process using all speakers and then converting the parameters to cluster-specific parameters through probabilistic mapping. Cluster-specific parameters for the cluster selected to be the most suitable for the speaker are used in the recognition stage. It has been observed that when the size of training data is restricted, the possible number of clusters is limited, and therefore the adaptation performance is limited.

A neural network-based approach has also been investigated [17]. In this approach, a multi-network Time-Delay Neural Network (TDNN) -based connectionist architecture performs multi-speaker phone discrimination (/b, d, g/). The overall network called "Meta-Pi network" gates the phonemic decisions of modules trained on individual speakers to form its overall classification decision. In the Meta-Pi network, K speaker-dependent modules are linked by a multiplicative connection. The Meta-Pi paradigm implements a dynamically adaptive Bayesian MAP classifier, and it learns without supervision. By dynamically adapting to the input speech and focusing on a combination of speaker-specific modules, the network outperforms a single TDNN trained on the speech of all K speakers.

4.3 Interpolated Re-Estimation Algorithm

The deleted interpolation technique [19] has been studied for generating speaker-adaptive parameters by interpolating speaker-independent, fixed existing parameters and speaker-dependent parameters created from a small number of training sentences. The speaker-independent parameters are well-trained but less appropriate for the individual speaker. On the other hand, the speaker-dependent parameters are appropriate for the individual speaker but poorly trained because of limited training data. The deleted interpolation technique can be used for combining the two parameter sets into estimates that are more suitable than speaker-independent parameters, yet more robust than speaker-dependent ones.

Since the deleted interpolation technique needs a large amount of computation, the Bayesian learning framework which also offers a way to incorporate newly acquired application-specific data into existing models (*a priori* information) and to combine them in an optimal manner is now widely used. It is an efficient technique for handling the sparse training data problem typically found in model parameter adaptation. This framework has been used to derive MAP estimates of the parameters of speech models, including HMM parameters [16] [21].

4.4 Spectral Mapping Algorithm

Instead of interpolating speaker-independent and speaker-dependent parameters, speaker-adaptive parameters can be estimated from speaker-independent parameters based on mapping rules. The mapping rules are estimated from the relationship between speaker-independent and speaker-dependent parameters. Within the framework of VQ-based speech recognition, both supervised and unsupervised methods of adapting the speaker-independent (initial/reference) codebook to a new speaker or normalizing (adjusting) the input speech to the codebook have been proposed. Each word is represented in the word dictionary as single or multiple sequences of codebook entries. In the latter case, individual variations on how a word is uttered are modeled by multiple code sequences. The code sequences are not changed during the adaptation and are universally used for all speakers. Within the framework of HMM-based speech recognition, supervised and unsupervised techniques of mapping Gaussian mixture parameters (means and variances) have recently been actively investigated.

Since it is unlikely that all the phoneme units will be observed enough times in a small adaptation set, especially in large-vocabulary continuous-speech recognition systems, only a small number of parameters can be effectively adapted. It is therefore desirable to introduce some parameter correlation or tying so that all model parameters can be adjusted at the same time in a consistent manner, even if some units are not included in the adaptation data.

If a correlation structure between parameters can be established, and the correlation parameters can be estimated when training the general models, the parameters of unseen units can be adapted accordingly [8][4]. To improve adaptation efficiency and effectiveness along this line, several techniques have been proposed, including probabilistic spectral mapping [30], hierarchical spectral clustering and smoothing [11][26], cepstral normalization [2], and spectrum bias and shift transformation [29].

In addition to correlation, a second type of constraint can be given to the model parameters so that all the parameters are adjusted simultaneously according to a predetermined set of transformations, e.g., a transformation based on multiple regression analysis [8]. Various methods have recently been proposed in which a linear transformation (affine transformation) between the reference and adaptive speaker-feature vectors is defined and then translated into a bias vector and a scaling matrix, which can be estimated using an EM algorithm. The transform parameters can be estimated from supervised adaptation data that form pairs with the training data [32][3][23][28][31] or estimated from unsupervised adaptation data using existing acoustic models [5][6]. These methods are now widely called the MLLR (maximum-likelihood linear regression) methods.

4.4.1 Supervised Adaptation

In the case of supervised adaptation, the mapping rules are obtained through DTW or a forward-backward algorithm. The utterances by a reference speaker are initially used to create an initial codebook and these utterances are converted into sequences of codebook entries. In the training stage, training utterances by a new speaker are converted into code sequences and time-aligned with the same words or sentences uttered by the reference speaker, using the DTW technique. The spectral mapping function between codebook elements of the reference speaker and the new speaker is calculated using the alignment results for all training words or sentences. The mapping function is weighted by the histogram to find the best correspondence rule. In the recognition stage, input speech is vector-quantized and mapped to the reference speaker's spectrum at every frame using the mapping rules. The similarity between mapped input speech and each word of the reference speaker is then calculated and used for the recognition decision.

In HMM-based recognition, HMM speech models derived from one "prototype" speaker or multiple speakers are transformed so that they model the speech of the new speaker [30]. This transformation is accomplished by probabilistic spectral mapping from one speaker's spectral space to that of another. The transformation matrix, which represents the conditional probability of a quantized spectrum of the new speaker, given the quantized spectrum of the reference speaker, is computed by applying a modified forward-backward algorithm to training utterances.

An approach to speaker adaptation based on the use of a stochastic model, called the "speaker Markov model", has also been investigated for a large-vocabulary HMM-based speech recognition system [27]. The speaker Markov model indicates which prototype of the new speaker is likely to occur and what acoustic parameters are generated by the reference speaker if, at the same time, a certain prototype is generated by the new speaker.

4.4.2 Supervised Adaptation Using Neural Networks

An adaptation method using neural networks has also been studied [18][25]. In this method, nonlinear continuous mapping between the reference spectral space and spectral space of a new speaker is represented by a multi-layer neural network. The network training is executed using the back-propagation method so that the network generates reference template spectra at the output layer when corresponding training utterance spectra are given to the input layer. Corresponding data sets of reference templates and training utterance spectra are obtained using the DTW technique in the same way as the supervised adaptation method described in the previous subsection. VQ-distortion of training utterance spectra can be avoided and non-linear mapping can be realized by this neural network-based method. The problem, however, is that this method requires a large number of training words.

4.4.3 Unsupervised Adaptation

An unsupervised codebook adaptation method based on hierarchical spectral clustering has been proposed [12]. First, an initial codebook and a VQ-indexed word dictionary are prepared. The initial codebook is produced by clustering the voices of multiple speakers, and commonly serves as the initial condition for every new speaker.

In the adaptation process, a set of spectra from the training utterances of a new speaker and the reference codebook elements are clustered hierarchically in an increasing number of clusters. Using deviation vectors between centroids of the training spectra clusters and the corresponding codebook clusters, either codebook elements or input frame spectra are shifted so that the corresponding centroids coincide. Continuity between adjacent clusters is maintained by determining the shifting vectors as the weighted-sum of the deviation vectors of adjacent clusters. The size of the codebook is maintained throughout the adaptation process. Adaptation is thus performed hierarchically from global to local individuality.

Recently an N-best-based unsupervised adaptation method has been proposed [24], which uses the N most likely word sequences in parallel and iteratively maximizes the likelihood. The N-best hypotheses are created for each input speech by applying speaker-independent models; speaker adaptation based on constrained Bayesian learning is then applied to each hypothesis. Finally, the hypothesis with the highest likelihood is selected as the most likely sequence. Without giving reasonable constraints based on models of inter-speaker variability, an input utterance can be adapted to *any* hypothesis with resulting high likelihood. To reduce this problem, constrained MAP estimation, including hierarchical spectral clustering and smoothing is placed on the transformation so that it maintains a reasonable geometrical shape. This averages out the hypothesis errors for a smaller number of parameters. As the iterations continue, the hypotheses improve, allowing for an increased number of parameters.

5 Individuality Problems in Speech Synthesis and Coding

Controlling or adding the individuality in synthesized speech is one of the important problems. Naturalness of synthesized voice is highly related to its individuality. This means that an ultimate goal of speech synthesis is the production of voices

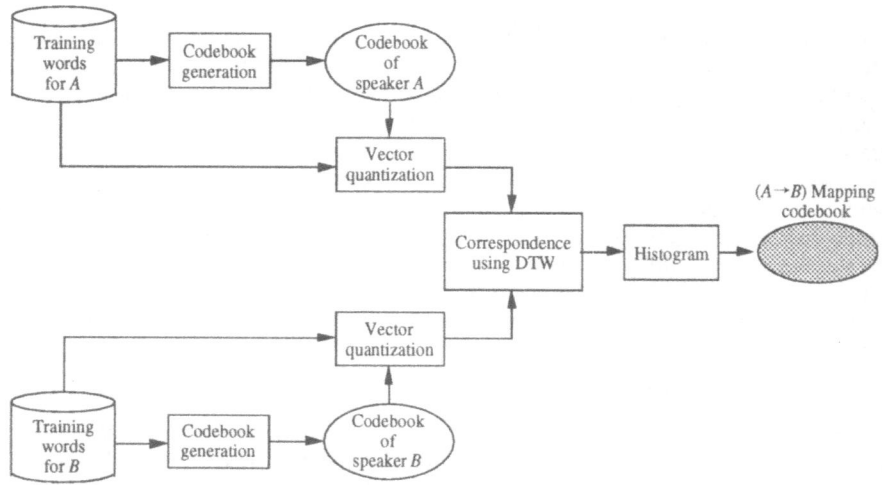

Fig. 1. Method for generating a mapping codebook for voice conversion.

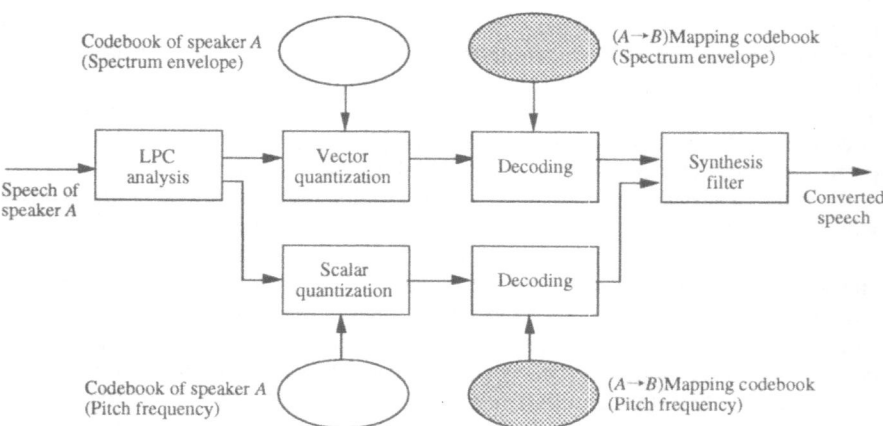

Fig. 2. Block diagram of voice conversion from speaker A to speaker B

having the individuality of real human beings. Thus far no system has been constructed that can precisely control the synthesized voice quality or imitate a desired speaker's voice.

Along this line, a technique of converting voice quality from one speaker to another through vector quantization and spectrum mapping has been investigated [1]. The basic idea of this technique is to make codebook mapping rules which represent the correspondence between different speakers' codebooks. A set of mapping rules for each codebook element is called a mapping codebook. Mapping codebooks for three parameter sets, that is, spectral parameters, energy values and pitch frequencies, are separately generated using training utterances. The block diagram in Figure 1 illustrates the method of generating a mapping codebook for spectral parameters, which is similar to the technique used in the supervised speaker adaptation described in Section 4.4.1. First, a separate training word set is pronounced by each of two speakers, A and B, and vector-quantized frame by frame. The correspondence between vectors of the same words from the two speakers is determined using DTW. The vector correspondences between two speakers are accumulated throughout the training word set to create a histogram. Using the histogram as weighting function, each element of the mapping codebook is defined as a weighted linear combination of the codebook elements.

Figure 2 shows a block diagram of a voice conversion from speaker A to speaker B. Synthesis is carried out by decoding (converting) the quantized parameters of the speaker A using the mapping codebooks between speakers A and B. To evaluate the performance of this technique, hearing tests have been carried out under two kinds of voice conversion conditions. One is a conversion between male and female, the other is a conversion between male speakers. In the male-to-female conversion experiment, all converted utterances were judged as female, and in the male-to-male conversion, 65% of the converted voices were identified as the target speaker's voices.

In speech coding, the dependency of the coded speech quality on the individuality of the original speech increases with the more sophisticated, high-compression-rate methods. Robustness to the variation of voice individuality is one of the essential issues in developing and evaluating advanced speech coding methods. For this purpose, it is necessary to standardize speech databases which cover a wide variety of voice individuality.

6 Conclusion

The algorithms for extracting individual information from speech waves, clarifying its mechanism, and controlling and normalizing the individuality are important future research topics for realizing advanced speech information processing systems, including speaker recognition, speech recognition, synthesis and coding systems. Among them, speaker recognition and speaker adaptation or normalization in speech recognition are particularly noteworthy research and development topics. Since these topics are two sides of the same problem: how best to separate the speaker's information and the phonetic information in speech waves, algorithms for them should be investigated using a common approach. For this purpose, it will be necessary to foster basic research on elucidating the mechanism of individuality in speech spectra and on inventing a sophisticated yet tractable model of speech variation based on a sufficiently large database.

References

[1] M. Abe, S. Nakamura, K. Shikano and H. Kuwabara , "Voice conversion through vector quantization", Proc. IEEE Internat. Conf. Acoust. Speech Signal Process., New York, S14.1 (1988).

[2] A. Acero and R. M. Stern, "Environmental robustness in automatic speech recognition," Proc. IEEE Int. Conf. Acoust., Speech, Signal Processing, Albuquerque, S15b.11, pp. 849-852 (1988).

[3] J. R. Bellegarda, P. V. De Sousa, A. J. Nadas, D. Nahamoo, M. A. Picheny and L. R. Bahl, "The metamorphic algorithm, a speaker mapping approach to data augmentation," IEEE Trans. Speech and Audio Processing, Vol. 2, No. 3, pp. 413-420 (1994).

[4] S. J. Cox, "Predictive speaker adaptation in speech recognition," Computer Speech and Language, Vol. 9, pp. 1-17 (1995).

[5] S. J. Cox and J. S. Bridle, "Unsupervised speaker adaptation by probabilistic fitting," Proc. IEEE Int. Conf. Acoust., Speech, Signal Processing, Glasgow, Scottland, S6.11, pp. 294-297 (1989).

[6] V. Digalakis and L. Neumeyer, "Speaker adaptation using combined transformation and Bayesian methods," Proc. IEEE Int. Conf. Acoust., Speech, Signal Processing, Detroit, pp. I-680-683 (1995).

[7] S. Furui, "An analysis of long-term variation of feature parameters of speech and its application to talker recognition", Trans. IECE, 57-A, Vol.12, pp. 880-887 (1974).

[8] S. Furui, "A training procedure for isolated word recognition systems", IEEE Trans. Acoust. Speech Signal Process., Vol. ASSP-28, No. 2, pp. 129-136 (1980).

[9] S. Furui, "Cepstral analysis technique for automatic speaker verification", IEEE Trans. Acoust. Speech Signal Process., Vol.29, No.2, pp. 254-272 (1981).

[10] S. Furui, "Research on individuality features in speech waves and automatic speaker recognition techniques", Speech Communication, Vol.5, No.2, pp. 183-197 (1986).

[11] S. Furui, Digital Speech Processing, Synthesis, and Recognition, Marcel Dekker, New York (1989).

[12] S. Furui, "Unsupervised speaker adaptation method based on hierarchical spectral clustering", Proc. IEEE Internat. Conf. Acoust. Speech Signal Process., Glasgow, Scotland, S6.9, pp. 286-289 (1989).

[13] S. Furui, "Speaker-dependent-feature extraction, recognition and processing techniques", Speech Communication, Vol. 10, Nos. 5-6, pp. 505-520 (1991).

[14] S. Furui, "Speaker-independent and speaker-adaptive recognition techniques," in Advances in Speech Signal Processing, edited by S. Furui and M. M. Sondhi, pp. 597-622 (1992).

[15] S. Furui, F. Itakura and S. Saito, "Talker recognition by longtime averaged speech spectrum", Trans. IECE, 55-A, Vol.10, pp. 549-556 (1972).

[16] J.-L. Gauvain and C.-H. Lee, "Bayesian learning for hidden Markov models with Gaussian mixture state observation densities," Speech Communication, Vol. 11, Nos. 2-3, pp. 205-214 (1992).

[17] J.B. Hampshire II and A.H. Waibel, "The Meta-Pi network: Connectionist rapid adaptation for high-performance multi-speaker phoneme recognition", Proc. IEEE Internat. Conf. Acoust. Speech Signal Process., Albuquerque, S3.9 (1990).

[18] K. Iso, M. Asogawa, K. Yosida and T. Watanabe, "Speaker adaptation using neural network", Proc. Spring Meeting of Acoust. Soc. Jap., 1-6-16 (in Japanese) (1989).

[19] F. Jelinek and R. L. Mercer , "Interpolated estimation of Markov source parameters from sparse data", in Pattern Recognition in Practice, edited by E.S. Gelsema and L. N. Kanal, North-Holland, Amsterdam, the Netherlands, pp. 381-397 (1980).

[20] K. Kato and S. Furui, "Listener adaptability for individual voice in speech perception", Trans. Commitee Hearing Res., Acoust. Soc. Japan, H-85-5 (in Japanese) (1985).

[21] C.-H. Lee and J.-L. Gauvain, "Bayesian adaptive learning and MAP estimation of HMM," in *Advanced Topics in Automatic Speech and Speaker Recognition*, edited by C.-H. Lee, K. K. Paliwal and F. K. Soong, Kluwer Academic Publishers, pp. 83-107 (1995).

[22] K. -F. Lee, *Large-vocabulary speaker-independent continuous speech recognition: The SPHINX system*, Ph. D. disseration, Compter Science Department, Carnegie Mellon Universitiy (1988).

[23] C. J. Leggetter and P. C. Woodland, "Maximum likelihood linear regression for speaker adaptation of continuous density hidden Markov models," Computer Speech and Language, Vol. 9, pp. 171-185 (1995).

[24] T. Matsui and S. Furui, "N-best-based instantaneous speaker adaptation method for speech recognition," Proc. Int. Conf. Spoken Language Processing, Philadelphia, pp. 973-976 (1996).

[25] C. Montacie, K. Choukri and G. Chollet, "Speech recognition using temporal decomposition and multi-layer feed-forward automata", Proc. IEEE Internat. Conf. Acoust. Speech Signal Process., Glasgow, Scotland, S8.6 (1989).

[26] K. Ohkura, M. Sugiyama and S. Sagayama, "Speaker adaptation based on transfer vector field smoothing with continuous mixture density HMMs," Proc. Int. Conf. Spoken Language Processing, Banff, We.fPM.1.1, pp. 369-372 (1992).

[27] G. Rigoll, "Speaker adaptation for large vocabulary speech recognition systems using 'speaker Markov models'", Proc. IEEE Internat. Conf. Acoust. Speech Signal Process., Glasgow, Scotland, S1.2 (1989).

[28] A. Sankar and C.-H. Lee, "Robust speech recognition based on stochastic matching," Proc. IEEE Int. Conf. Acoust., Speech, Signal Processing, Detroit, pp. I-121-124 (1995).

[29] A. Sankar and C.-H. Lee, "A maximum-likelihood approach to stochastic matching for robust speech recognition," IEEE Trans. Speech and Audio Processing, Vol. 4, No. 3, pp. 190-202 (1996).

[30] R. Schwartz, Y. -L. Chow and F. Kubala, "Rapid speaker adaptation using a probabilistic spectral mapping", Proc. IEEE Internat. Conf. Acoust. Speech Signal Process., Dallas, 15.3, pp. 633-636 (1987).

[31] G. Zavaliagkos, R. Schwartz and J. Makhoul, "Batch, incremental and instantaneous adaptation techniques for speech recognition," Proc. Int. Conf. Acoust., Speech, Signal Processing, Detroit, pp. I-676-679 (1995).

[32] Y. Zhao, "An acoustic-phonetic-based speaker adaptation technique for improving speaker-independent continuous speech recognition," IEEE Trans. Speech and Audio Processing, Vol. 2, No. 3, pp. 380-394 (1994).

Speaker Recognition

Sadaoki Furui

Tokyo Institute of Technology,
Department of Computer Science,
2-12-1, O-okayama, Meguro-ku, Tokyo, 152 Japan
furui@cs.titech.ac.jp

Abstract. This paper introduces recent advances in speaker recognition technology. The first part discusses general topics and issues. The second part is devoted to a discussion of more specific topics of recent interest that have led to interesting new approaches and techniques. They include VQ- and ergodic-HMM-based text-independent recognition methods, a text-prompted recognition method, parameter/ distance normalization and model adaptation techniques, and methods of updating models and *a priori* thresholds in speaker verification. The paper concludes with 16 open questions about speaker recognition and a short discussion assessing the current status and future possibilities.

Keywords. Speaker verification, speaker identification, text-independent, text-prompted

1 Principles of Speaker Recognition

Speaker recognition is the process of automatically recognizing who is speaking by using speaker-specific information included in speech waves [6][7][8][9][35]. Speaker recognition can be classified into speaker identification and speaker verification. Speaker identification is the process of determining from which of the registered speakers a given utterance comes. Speaker verification is the process of accepting or rejecting the identity claim of a speaker. Most of the applications in which voice is used to confirm the identity claim of a speaker are classified as speaker verification.

In the speaker identification task, a speech utterance from an unknown speaker is analyzed and compared with speech models of known speakers. The unknown speaker is identified as the speaker whose model best matches the input utterance. In speaker verification, an identity claim is made by an unknown speaker, and an utterance of this unknown speaker is compared with the model for the speaker whose identity is claimed. If the match is good enough, that is, above a threshold, the identity claim is accepted. A high threshold makes it difficult for impostors to be accepted by the system, but at the risk of falsely rejecting valid users. Conversely, a low threshold enables valid users to be accepted consistently, but at the risk of accepting impostors. To set the threshold at the desired level of customer rejection and impostor acceptance, it is necessary to know the distribution of customer and impostor scores.

There is also the case called "open set" identification, in which a reference model for the unknown speaker may not exist. In this case, an additional decision alternative, "the unknown does not match any of the models", is required. In either verification or identification, an additional threshold test can be applied to determine whether the match is close enough to accept the decision or ask for a new trial.

Speaker recognition methods can also be divided into text-dependent and text-independent methods. The former require the speaker to provide utterances of key words or sentences that are the same text for both training and recognition, whereas the latter do not rely on a specific text being spoken. The text-dependent methods are usually based on template-matching techniques in which the time axes of an input speech sample and each reference template or reference model of the registered speakers are aligned, and the similarity between them is accumulated from the beginning to the end of the utterance. Since this method can directly exploit voice individuality associated with each phoneme or syllable, it generally achieves higher recognition performance than the text-independent method.

There are several applications, such as forensic [16] and surveillance applications, in which predetermined key words cannot be used. Moreover, human beings can recognize speakers irrespective of the content of the utterance. Therefore, text-independent methods have recently attracted more attention. Another advantage of text-independent recognition is that it can be done sequentially, until a desired significance level is reached, without the annoyance of the speaker having to repeat the key words again and again.

Both text-dependent and independent methods have a serious weakness. That is, these systems can easily be defeated, because someone who plays back the recorded voice of a registered speaker uttering key words or sentences into the microphone can be accepted as the registered speaker. Another problem is that people often do not like text-dependent systems because they do not like to utter their identification number, such as their social security number, within the hearing of other people. To cope with these problems, some methods use a small set of words, such as digits, as key words, and each user is prompted to utter a given sequence of key words that is randomly chosen every time the system is used [14][34]. Yet even this method is not reliable enough, since it can be defeated with advanced electronic recording equipment that can reproduce key words in a requested order. Therefore, a text-prompted speaker recognition method has recently been proposed (See Section 3).

2 Text-Independent Speaker Recognition Methods

In text-independent speaker recognition, the words or sentences used in recognition trials generally cannot be predicted. Since it is impossible to model or match speech events at the word or sentence level, the following three kinds of methods shown in Fig. 1 have been actively investigated [9].

2.1 Long-Term-Statistics-Based Methods

As text-independent features, long-term sample statistics of various spectral features, such as the mean and variance of spectral features over a series of utterances, have been used (Fig. 1(a)). However, long-term spectral averages are extreme condensations of the spectral characteristics of a speaker's utterances and, as such, lack the discriminating power included in the sequences of short-term spectral features used as models in text-dependent methods. In one of the trials using the long-term averaged spectrum [3], the effect of session-to-session variability was reduced by introducing a weighted cepstral distance measure.

Studies on using statistical dynamic features have also been reported. Montacie et al. [27] applied a multivariate auto-regression (MAR) model to the time series of cepstral vectors to characterize speakers, and reported good speaker recognition results. Griffin et al. [12] studied distance measures for the MAR-based method, and reported that when ten sentences were used for training and one sentence was used for testing, identification and verification rates were almost the same as those obtained by an HMM-based method. It was also reported that the optimum order of the MAR model was 2 or 3, and that distance normalization using *a posteriori* probability was essential to obtain good results in speaker verification.

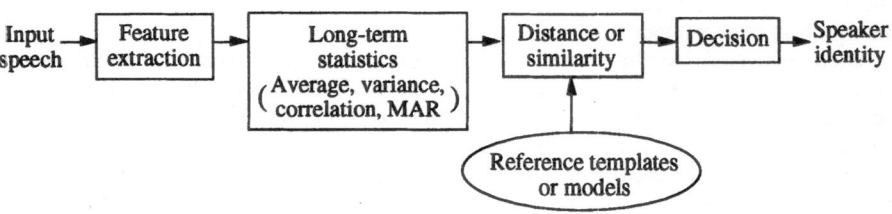

(a) Long-term-statistics-based method

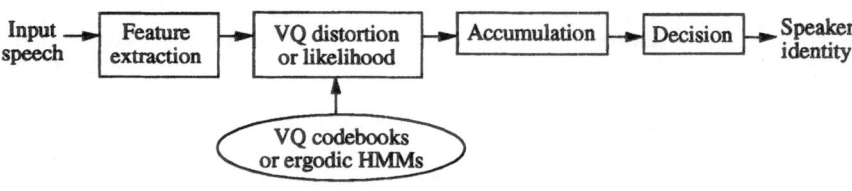

(b) VQ / HMM-based method

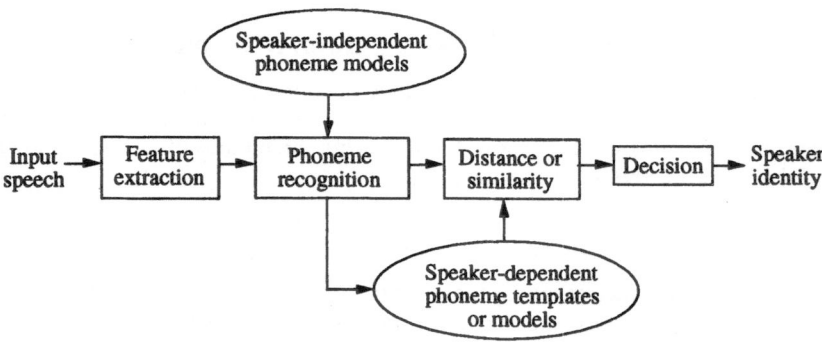

(c) Speech-recognition-based method

Fig. 1. Basic structures of text-independent speaker recognition methods.

2.2 VQ-Based Methods

A set of short-term training feature vectors of a speaker can be used directly to represent the essential characteristics of that speaker. However, such a direct representation is impractical when the number of training vectors is large, since the memory and amount of computation required become prohibitively large. Therefore, attempts have been made to find efficient ways of compressing the training data using vector quantization (VQ) techniques.

In this method, VQ codebooks, consisting of a small number of representative feature vectors, are used as an efficient means of characterizing speaker-specific features [17][19][20][33]. In the recognition stage, an input utterance is vector-quantized by using the codebook of each reference speaker; the VQ distortion accumulated over the entire input utterance is used for making the recognition determination (Fig. 1(b)).

Matsui et al. [19][20] tried a method using a VQ-codebook for long feature vectors consisting of instantaneous and transitional features calculated for both cepstral coefficients and fundamental frequency. Since the fundamental frequency cannot be extracted from unvoiced speech, there are two separate codebooks for voiced and unvoiced speech for each speaker. A new distance measure was introduced to take into account the intra- and inter-speaker variability and to deal with the outlier problem in the distribution of feature vectors. The outlier vectors correspond to intersession spectral variation and to the difference between phonetic content of the training texts and the test utterances. It was confirmed that, although the fundamental frequency achieved only a low recognition rate by itself, the recognition accuracy was greatly improved by combining the fundamental frequency with spectral envelope features.

In contrast with the memoryless VQ-based method, non-memoryless source coding algorithms have also been studied using a segment (matrix) quantization technique [15]. The advantage of a segment quantization codebook over a VQ codebook representation is its characterization of the sequential nature of speech events. Higgins and Wohlford [13] proposed a segment modeling procedure for constructing a set of representative time normalized segments, which they called "filler templates". The procedure, a combination of K-means clustering and dynamic programming time alignment, provided a means for handling temporal variation.

2.3 Ergodic-HMM-Based Methods

The basic structure is the same as the VQ-based method, but in this method an ergodic HMM is used instead of a VQ codebook. Over a long timescale, the temporal variation in speech signal parameters is represented by stochastic Markovian transitions between states. Poritz [29] proposed using a five-state ergodic HMM (i.e., all possible transitions between states are allowed) to classify speech segments into one of the broad phonetic categories corresponding to the HMM states. A linear predictive HMM was adopted to characterize the output probability function. He characterized the automatically obtained categories as strong voicing, silence, nasal/liquid, stop burst/post silence, and frication.

Gauvain et al. [11] investigated a statistical modeling approach, where each speaker was viewed as a source of phonemes, modeled by a fully connected Markov chain. Maximum *a posteriori* (MAP) estimation was used to generate speaker-specific models from a set of speaker-independent seed models. The lexical and syntactic structures of the language were approximated by local phonotactic constraints. The unknown speech is recognized by all of the speakers' models in parallel, and the hypothesized identity is that associated with the model set having the highest likelihood. Since phonemes and speakers are simultaneously recognized by using speaker-specific Markov chains, this method can be considered as an extension of the ergodic-HMM-based method.

Matsui et al. [21] compared the VQ-based method with the discrete/continuous ergodic HMM-based method, particularly from the viewpoint of robustness against utterance variations. They found that the continuous ergodic HMM method is far superior to the discrete ergodic HMM method and that the continuous ergodic HMM method is as robust as the VQ-based method when enough training data is available. However, when little data is available, the VQ-based method is more robust than the continuous HMM method. They investigated speaker identification rates using the continuous HMM as a function of the number of states and mixtures. It was shown that the speaker recognition rates are strongly correlated with the total number of mixtures, irrespective of the number of states. This means that the information on transitions between different states is ineffective for text-independent speaker recognition.

The case of a single-state continuous ergodic HMM corresponds to the technique based on the maximum likelihood estimation of a Gaussian-mixture model representation investigated by Rose et al. [31]. Furthermore, the VQ-based method can be regarded as a special (degenerate) case of a single-state HMM with a distortion measure being used as the observation probability.

2.4 Speech-Recognition-Based Methods

The VQ- and HMM-based methods can be regarded as methods that use phoneme-class-dependent speaker characteristics in short-term spectral features through implicit phoneme-class recognition. In other words, phoneme-classes and speakers are simultaneously recognized in these methods. On the other hand, in the speech-recognition-based methods (Fig. 1(c)), phonemes or phoneme-classes are explicitly recognized, and then each phoneme (-class) segment in the input speech is compared with speaker models or templates corresponding to that phoneme (-class).

Savic et al. [37] used a five-state ergodic linear predictive HMM for broad phonetic categorization. In their method, after frames that belong to particular phonetic categories have been identified, feature selection is performed. In the training phase, reference templates are generated and verification thresholds are computed for each phonetic category. In the verification phase, after phonetic categorization, a comparison with the reference template for each particular category provides a verification score for that category. The final verification score is a weighted linear combination of the scores for each category. The weights are chosen to reflect the effectiveness of particular categories of phonemes in discriminating between speakers and are adjusted to maximize the verification performance. Experimental results showed that verification accuracy can be considerably improved by this category-dependent weighted linear combination method.

Rosenberg et al. have been testing a speaker verification system using 4-digit phrases under field conditions of a banking application [34]. In this system, input speech is segmented into individual digits using a speaker-independent HMM. The frames within the word boundaries for a digit are compared with the corresponding speaker-specific HMM digit model and the Viterbi likelihood score is computed. This is done for each of the digits making up the input utterance. The verification score is defined to be the average normalized log-likelihood score over all the digits in the utterance.

Newman et al. [28] used a large vocabulary speech recognition system for speaker verification. A set of speaker-independent phoneme models were adapted to each speaker. The speaker verification consisted of two stages. First, speaker-independent speech recognition was run on each of the test utterances to obtain phoneme segmentation. In the second stage, the segments were scored against the adapted models for a particular target speaker. The scores were normalized by those with speaker-independent models. The system was evaluated using the 1995 NIST-administered speaker verification database, which consists of data taken from the Switchboard corpus. The results showed that this method could not out-perform Gaussian mixture models.

3 Text-Prompted Speaker Recognition

In this method, key sentences are completely changed every time [22][24]. The system accepts the input utterance only when it determines that the registered speaker uttered the prompted sentence. Because the vocabulary is unlimited, prospective impostors cannot know in advance the sentence they will be prompted to say. This method can not only accurately recognize speakers, but can also reject an utterance whose text differs from the prompted text, even if it is uttered by a registered speaker. Thus, a recorded and played back voice can be correctly rejected.

It uses speaker-specific phoneme models as basic acoustic units. One of the major issues in this method is how to properly create these speaker-specific phoneme models when using training utterances of a limited size. The phoneme models are represented by Gaussian-mixture continuous HMMs or tied-mixture HMMs, and they are made by adapting speaker-independent phoneme models to each speaker's voice.

In the recognition stage, the system concatenates the phoneme models of each registered speaker to create a sentence HMM, according to the prompted text. Then the likelihood of input speech against the sentence model is calculated and used for the speaker verification determination.

4 Normalization and Adaptation Techniques

How can we normalize the intra-speaker variation of likelihood (similarity) values in speaker verification? The most significant factor affecting automatic speaker recognition performance is variation in signal characteristics from trial to trial (intersession variability or variability over time). Variations arise from the speaker him/herself, from differences in recording and transmission conditions, and from noise. Speakers cannot repeat an utterance precisely the same way from trial to trial. It is well known that samples of the same utterance recorded in one session are much more highly correlated than tokens recorded in separate sessions. There are also long term trends in voices [3][4].

It is important for speaker recognition systems to accommodate these variations. Adaptation of the reference model as well as the verification threshold for each speaker is indispensable to maintain a high recognition accuracy for a long period. In order to compensate for the variations, two types of normalization techniques have been tried

— one in the parameter domain, and the other in the distance/similarity domain. The latter technique uses the likelihood ratio or *a posteriori* probability. To adapt HMMs for noisy conditions, the HMM composition (PMC: parallel model combination) method has been successfully tried.

4.1 Parameter-Domain Normalization

As one typical normalization technique in the parameter domain, spectral equalization, the so-called "blind equalization" method, has been confirmed to be effective in reducing linear channel effects and long-term spectral variation [1][5]. This method is especially effective for text-dependent speaker recognition applications using sufficiently long utterances. In this method, cepstral coefficients are averaged over the duration of an entire utterance, and the averaged values are subtracted from the cepstral coefficients of each frame. This method can compensate fairly well for additive variation in the log spectral domain. However, it unavoidably removes some text-dependent and speaker-specific features, so it is inappropriate for short utterances in speaker recognition applications. It was shown that time derivatives of cepstral coefficients (delta-cepstral coefficients) are resistant to linear channel mismatch between training and testing [5][38].

4.2 Likelihood Normalization

Higgins et al. [14] proposed a normalization method for distance (similarity or likelihood) values that uses a likelihood ratio. The likelihood ratio is the ratio of the conditional probability of the observed measurements of the utterance given the claimed identity is correct to the conditional probability of the observed measurements given the speaker is an impostor.

This normalization method is, however, unrealistic because conditional probabilities must be calculated for all the reference speakers, which costs a lot. Therefore, a set of speakers, "cohort speakers", that are representative of the population near the claimed speaker has been chosen for calculating the normalization term. Another way of choosing the cohort speaker set is to use speakers who are typical of the general population. Reynolds [30] reported that a randomly selected, gender-balanced background speaker population outperformed a population near the claimed speaker.

Matsui et al. [22][23] proposed a normalization method based on *a posteriori* probability. The difference between the normalization method based on the likelihood ratio and that based on *a posteriori* probability is whether or not the claimed speaker is included in the impostor speaker set for normalization; the cohort speaker set in the likelihood-ratio-based method does not include the claimed speaker, whereas the normalization term for the *a posteriori*-probability-based method is calculated by using a set of speakers including the claimed speaker. Experimental results indicate that both normalization methods almost equally improve speaker separability and reduce the need for speaker-dependent or text-dependent thresholding, compared with scoring using only the model of the claimed speaker [23][36].

Carey et al. [2] proposed a method in which the normalization term is approximated by the likelihood for a world model representing the population in general. This method has an advantage that the computational cost for calculating the normalization term is much smaller than the original method since it does not need to sum the likelihood values for cohort speakers. Matsui et al. [23] recently proposed a new method based on tied-mixture HMMs in which the world model is made as a pooled mixture model representing the parameter distribution for all the registered speakers.

Since these normalization methods neglect the absolute deviation between the claimed speaker's model and the input speech, they cannot differentiate highly dissimilar speakers. Higgins et al. [14] reported that a multilayer network decision algorithm makes effective use of the relative and absolute scores obtained from the matching algorithm.

4.3 HMM Adaptation for Noisy Conditions

Rose et al. [32] applied the HMM composition (PMC) method [10][18] to speaker identification under noisy conditions. The HMM composition is a technique to combine a clean speech HMM and a background noise HMM to create a noise-added speech HMM. In order to cope with the problem of the variation of the signal-to-noise ratio (SNR), Matsui et al. [26] proposed a method in which several noise-added HMMs with various SNRs were created and the HMM that had the highest likelihood value for the input speech was selected. A speaker decision was made using the likelihood value corresponding to the selected model. Experimental application of this method to text-independent speaker identification and verification in various kinds of noisy environments demonstrated considerable improvement in speaker recognition.

4.4 Updating Models and *A Priori* Threshold for Speaker Verification

How do we deal with long-term variability in people's voices? How should we update the speaker models to cope with the gradual changes in people's voices? Since we cannot ask every user to utter many utterances at many different sessions in real situations, it is necessary to build each speaker model based on a small amount of data collected at a few sessions, and then the model must be updated using speech data collected when the system is used. How do we adequately retrain the models?

How should we set the *a priori* decision threshold for speaker verification? In most laboratory speaker recognition experiments, the threshold is set *a posteriori* so that the equal error rate (EER) is achieved. Since the threshold cannot be set *a posteriori* in real situations, we have to have reasonable ways to set the threshold before verification. It must be set according to the importance of the two errors, which depends on the application.

These two problems are highly related each other. Furui [5] proposed methods for updating reference templates and the threshold in DTW-based speaker verification. An optimum threshold was estimated based on the distribution of overall distances between each speaker's reference template and a set of utterances of other speakers (interspeaker distances). The interspeaker distance distribution was approximated by a normal distribution, and the threshold was calculated by the linear combination of its mean value and standard deviation. The intraspeaker distance distribution was not taken into account in the calculation, mainly because it is difficult to obtain stable estimates of the intraspeaker distance distribution for small numbers of training utterances. The reference template for each speaker was updated by averaging new utterances and the present template after time registration. Matsui et al. [25] extended these methods and applied them to text-independent and text-prompted speaker verification using HMMs.

5 Open Questions and Concluding Remarks

Although many recent advances and successes have been achieved as described in the previous sections, there are still many problems for which good solutions remain to be found. Sixteen major problems are listed below [9].

- How can human beings correctly recognize speakers?
- Is it useful to study the mechanism of speaker recognition by human beings?
- Is it useful to study the physiological mechanism of speech production to get new ideas for speaker recognition?
- What feature parameters are appropriate for speaker recognition?
- How can we model the relatively macro-transitional features covering the interval between 200 to 300 ms?
- How can we fully exploit the clearly evident encoding of identity in prosody and other suprasegmental features of speech?
- Is the "sheep and goats" problem (a small percentage of speakers account for the majority of errors) universal?
- Can we ever reliably cluster speakers on the basis of similarity/dissimilarity?
- How do we acquire realistically sized (viable) databases that still adequately model inter- and intra-speaker variability?
- How do we deal with long-term variability in people's voices?
- Can we model or develop strategies for dealing with factors that significantly alter a person's voice (short-term transitory)?
- How can we extract text-independent speaker-specific features?
- How can we deal with deliberately disguised voices?
- How does speaker recognition compare to other means of personal identification, both now and in the future?
- What are the conditions that speaker recognition systems must satisfy in order for them to be utilized in the field?
- What about combining speech and speaker recognition?

Most of these problems arise from variability, including speaker-generated variability and variability in channel and recording conditions. It is very important to investigate feature parameters that are stable over a long period, insensitive to variations in speaking manner, including speaking rate and level, and robust against variations in voice quality such as those due to voice disguise or colds. It is also important to develop a method to cope with the problems of distortion due to telephone sets and channels and background and channel noises.

References

[1] B. S. Atal, "Effectiveness of Linear Prediction Characteristics of the Speech Wave for Automatic Speaker Identification and Verification," J. Acoust. Soc. Am., Vol. 55, No. 6, pp. 1304-1312 (1974).

[2] M. J. Carey and E. S. Parris, "Speaker Verification Using Connected Words," Proc. Institute of Acoustics, Vol. 14, Part 6, pp. 95-100 (1992).

[3] S. Furui, F. Itakura and S. Saito, "Talker Recognition by Longtime Averaged Speech Spectrum," Trans. IECE, 55-A, Vol. 1, No. 10, pp. 549-556 (1972).

[4] S. Furui, "An Analysis of Long-Term Variation of Feature Parameters of Speech and its Application to Talker Recognition," Trans. IECE, 57-A, Vol. 12, pp. 880-887 (1974).

[5] S. Furui, "Cepstral Analysis Technique for Automatic Speaker Verification," IEEE Trans. Acoust. Speech, Signal Processing, Vol. 29, No. 2, pp. 254-272 (1981).

[6] S. Furui, *Digital Speech Processing, Synthesis, and Recognition*, Marcel Dekker, New York (1989).

[7] S. Furui, "Speaker-Dependent-Feature Extraction, Recognition and Processing Techniques," Speech Communication, Vol. 10, No. 5-6, pp. 505-520 (1991).

[8] S. Furui, "An Overview of Speaker Recognition Technology," ESCA Workshop on Automatic Speaker Recognition, Identification and Verification, pp. 1-9 (1994).

[9] S. Furui, "Recent Advances in Speaker Recognition", First Int. Conf. Audio- and Video-based Biometric Person Authentication, Crans-Montana, Switzerland, pp. 237-252 (1997)

[10] M. J. F. Gales and S. J. Young, "HMM Recognition in Noise Using Parallel Model Combination," Proc. Eurospeech, Berlin, pp. II-837-840 (1993).

[11] J. L. Gauvain, L. F. Lamel and B. Prouts, "Experiments with Speaker Verification over the Telephone," Proc. Eurospeech, Madrid, pp. 651-654 (1995).

[12] C. Griffin, T. Matsui and S. Furui, "Distance Measures for Text-Independent Speaker Recognition Based on MAR Model," Proc. IEEE Int. Conf. Acoust. Speech, Signal Processing, Adelaide, 23.6, pp. I-309-312 (1994).

[13] A. L. Higgins and R. E. Wohlford, "A New Method of Text-Independent Speaker Recognition," Proc. IEEE Int. Conf. Acoust., Speech, Signal Processing, 17.3, pp. 869-872 (1986).

[14] A. Higgins, L. Bahler and J. Porter, "Speaker Verification Using Randomized Phrase Prompting," Digital Signal Processing, Vol. 1, pp. 89-106 (1991).

[15] B. -H. Juang and F. K. Soong, "Speaker Recognition Based on Source Coding Approaches," Proc. IEEE Int. Conf. Acoust., Speech, Signal Processing, S5.4, pp. 613-616 (1990).

[16] H. J. Kunzel, "Current Approaches to Forensic Speaker Recognition," ESCA Workshop on Automatic Speaker Recognition, Identification and Verification, pp.135-141 (1994).

[17] K. -P. Li and E. H. Wrench Jr., "An Approach to Text-Independent Speaker Recognition with Short Utterances," Proc. IEEE Int. Conf. Acoust., Speech, Signal Processing, 12.9, pp. 555-558 (1983).

[18] F. Martin, K. Shikano and Y. Minami, "Recognition of Noisy Speech by Composition of Hidden Markov Models," Proc. Eurospeech, Berlin, pp. II-1031-1034 (1993).

[19] T. Matsui and S. Furui, "Text-Independent Speaker Recognition Using Vocal Tract and Pitch Information," Proc. Int. Conf. Spoken Language Processing, Kobe, 5.3, pp. 137-140 (1990).

[20] T. Matsui and S. Furui, "A Text-Independent Speaker Recognition Method Robust Against Utterance Variations," Proc. IEEE Int. Conf. Acoust. Speech Signal Processing, S6.3, pp. 377-380 (1991).

[21] T. Matsui and S. Furui, "Comparison of Text-Independent Speaker Recognition Methods Using VQ-Distortion and Discrete/Continuous HMMs," Proc. IEEE Int. Conf. Acoust. Speech, Signal Processing, San Francisco, pp. II-157-160 (1992).

[22] T. Matsui and S. Furui, "Concatenated Phoneme Models for Text-Variable Speaker Recognition," Proc. IEEE Int. Conf. Acoust. Speech, Signal Processing, Minneapolis, pp. II-391-394 (1993).

[23] T. Matsui and S. Furui, "Similarity Normalization Method for Speaker Verification Based on a Posteriori Probability," ESCA Workshop on Automatic Speaker Recognition, Identification and Verification, pp. 59-62 (1994).

[24] T. Matsui and S. Furui, "Speaker Adaptation of Tied-Mixture-Based Phoneme Models for Text-Prompted Speaker Recognition," Proc. IEEE Int. Conf. Acoust. Speech, Signal Processing, Adelaide, 13.1 (1994).

[25] T. Matsui and S. Furui, "Robust Methods of Updating Model and A Priori Threshold in Speaker Verification," Proc. IEEE Int. Conf. Acoust. Speech, Signal Processing, Atlanta, pp. I-97-100 (1996).

[26] T. Matsui and S. Furui, "Speaker Recognition Using HMM Composition in Noisy Environments," Computer Speech and Language, Vol. 10, pp. 107-116 (1996)

[27] C. Montacie et al., "Cinematic Techniques for Speech Processing: Temporal Decomposition and Multivariate Linear Prediction," Proc. IEEE Int. Conf. Acoust. Speech, Signal Processing, San Francisco, pp. I-153-156 (1992).

[28] M. Newman, L. Gillick, Y. Ito, D. McAllaster and B. Peskin, "Speaker Verification through Large Vocabulary Continuous Speech Recognition," Proc. Int. Conf. Spoken Language Processing, Philadelphia, pp. 2419-2422 (1996).

[29] A. B. Poritz, "Linear Predictive Hidden Markov Models and the Speech Signal," Proc. IEEE Int. Conf. Acoust., Speech, Signal Processing, S11.5, pp. 1291-1294 (1982).

[30] D. Reynolds, "Speaker Identification and Verification Using Gaussian Mixture Speaker Models," ESCA Workshop on Automatic Speaker Recognition, Identification and Verification, pp.27-30 (1994).

[31] R. C. Rose and R. A. Reynolds, "Text Independent Speaker Identification Using Automatic Acoustic Segmentation," Proc. IEEE Int. Conf. Acoust., Speech, Signal Processing, S51.10, pp. 293-296 (1990).

[32] R. C. Rose, E. M. Hofstetter, and D. A. Reynolds, "Integrated Models of Signal and Background with Application to Speaker Identification in Noise," IEEE Trans. Speech and Audio Processing, Vol. 2, No. 2, pp. 245-257 (1994).

[33] A. E. Rosenberg and F. K. Soong, "Evaluation of a Vector Quantization Talker Recognition System in Text Independent and Text Dependent Modes," Computer Speech and Language, 22, pp. 143-157 (1987).

[34] A. E. Rosenberg, C. -H. Lee, and S. Gokcen, "Connected Word Talker Verification Using Whole Word Hidden Markov Models," Proc. IEEE Int. Conf. Acoust. Speech, Signal Processing, Toronto, S6.4, pp. 381-384 (1991).

[35] A. E. Rosenberg and F. K. Soong, "Recent Research in Automatic Speaker Recognition," in Advances in Speech Signal Processing (eds. S. Furui and M. M. Sondhi), Marcel Dekker, New York, pp. 701-737 (1991).

[36] A. E. Rosenberg, "The Use of Cohort Normalized Scores for Speaker Verification," Proc. Int. Conf. Spoken Language Processing, Banff, Th.sAM.4.2, pp. 599-602 (1992).

[37] M. Savic and S. K. Gupta, "Variable Parameter Speaker Verification System Based on Hidden Markov Modeling," Proc. IEEE Int. Conf. Acoust., Speech, Signal Processing, S5.7, pp. 281-284 (1990).

[38] F. K. Soong and A. E. Rosenberg, "On the Use of Instantaneous and Transitional Spectral Information in Speaker Recognition," IEEE Trans. Acoust. Speech, Signal Processing, Vol. ASSP-36, No. 6, pp. 871-879 (1988).

Application of Acoustic Discriminative Training in an Ergodic HMM for Speaker Identification

Leandro Rodríguez Liñares and Carmen García Mateo

Dpto. de Tecnoloxías das Comunicacións, Universidade de Vigo
36200-Vigo (Pontevedra), Spain
email: leandro@tsc.uvigo.es

Summary. We present a novel architecture for a Speaker Recognition system over the telephone. The proposed system introduces acoustic information into a HMM-based recognizer. This is achieved by using a phonetic classifier during the training phase. Three broad phonetic classes: voiced frames, unvoiced frames and transitions, are defined. We design speaker templates by the combination of four single state HMMs into a four state HMM after re-estimation of the transition probabilities. Experiments conducted with two databases are reported, and the results show that this architecture performs better than others without phonetic classification.

1. Introduction

Continuous HMM (Hidden Markov Model) based systems are presently the state of art for speaker recognition purposes. It is well known [3] that their performance relies on the total number of Gaussian mixtures of the model and not so much on how many states are used. Thus, one single-state multiple-Gaussian mixtures model called GMM is one of the preferred algorithms for this task [4]. After training it is supposed that the Gaussian components have learned the more relevant aspects regarding the distinctive phonetic features that characterize a person voice. The higher the number of Gaussian mixtures, the better. The important point is then to decide how many mixtures must be used to achieve a good compromise between representation accurateness and the amount of data required for training. Computational complexity must be kept as low as possible as well. As the vocal tract exhibits widely articulatory configurations during the production of distinct sounds, an average set of features does not represents a speaker's voice characteristics accurately. To include acoustic discrimination helps to improve performance. The point is to select the best sound classes and how to perform a robust automatic speech classification.

We investigate the role played by several phonetic characteristics regarding speaker recognition using HMM based systems. Namely, voiced part, unvoiced part and transitions are considered. Eventually, we propose an ergodic HMM model that combines these phonetic classes. We have called this architecture PTE-HMM (Phonetically Trained Ergodic Hidden Markov Model). This way, we force the learning capability of the model and reduce the required number of Gaussian when com-

---

This work has been partially supported by the Spanish CICYT under the project TIC96-0956-C04-02

pared with a complexity equivalent system that makes no use of a priori phonetic information.

The rest of the chapter is organized as follows: Section 2 presents the databases we used. In Section 3 we present the architecture of the model and in Section 4 the results obtained with this model. Finally, Section 5 presents some conclusions and guidelines for future work.

2. Experimental Conditions

The experiments were conducted using two databases. The first is called "TelVoice" and has been designed for Speaker Recognition purposes. Its goal is to have at least 50 Spanish speakers with 10 sessions each, recorded over a period of one month an a half. Thus, time interval between sessions may vary from one speaker to another, but it is never less than three days. The other is "PolyCost" [1], recorded by the COST250 action during January-March 1996 by 134 subjects from 14 countries. Both databases consist of telephone speech sampled at 8 kHz.

In order to asses the performance of the proposed system, we have set up an experiment making some choices about recording conditions and speech parameterization. Mel-cepstrum and Δ-mel-cepstrum coefficients were computed using a frame length of 20ms, and a frame period of 10 ms. Energy and the first derivative of energy were appended to the parameters of each frame. The performance was evaluated for a speaker identification application using 12 mel-cepstral coefficients using an order for the cepstral coefficients of 14. The number of Gaussian mixtures varies from 1 to 32. We use covariance-tied models across all the experiments.

The recording work of TelVoice is still in progress, so in this experiment we use a subset of the database that contains 5 sessions uttered by 20 Spanish speakers (10 males and 10 females). The material we use from each session consists of 4 repetitions of the Spanish Identity Card number, made up of 8 digits (approximately 5 seconds each). The speakers were addressed to pronounce it naturally (digit by digit, grouping digits, or as a whole, as they usually do). The utterances recorded in one session were used for training and the other four were used for testing. The session used for training was rotated.

In the experiments performed with PolyCost, we used the seven-digit client codes (CLI01-CLI04) from 5 sessions recorded by 110 speakers. The utterances recorded in the first session (four repetitions of the client code) were used for training and the sessions 2 to 5 were used for testing. In this case, there was no rotation of the sessions.

3. System Architecture

3.1 Acoustic Segmentation

The phonetic classifier identifies the type of speech frame. In this implementation, we use a phonetic classifier that we have previously developed for speech coding purposes. It considers three distinct sound classes:

- Voiced sounds which have quasi-periodic waveforms and fairly harmonic spectra.
- Unvoiced frames which have aperiodic waveforms and irregular spectra; their energy is usually lower than that of voiced sounds.
- Transitions defined as the two first voiced frames after an unvoiced segment and the two last voiced frames before an unvoiced segment. This type of frames is characterized by a non stationary waveform.

One important point is that we also use a Voice Activity Detector (VAD) to identify the noise segments. The phonetic classifier uses an algorithm close to the one in [5] with some modifications to improve its behavior in noisy environments and to work in an Multiband Excitation Speech Coder [2].

The labeling of the training utterances is performed in a completely automatic way by this phonetic classifier. We train three HMMs per speaker with the frames corresponding to each phonetic class. All the non-voice material of the training session is used to train a noise HMM that is the same for all speakers.

With the purpose of checking how the training of this HMMs works, we have done some tests with a parallel system. That is, in the testing phase we use for each speaker a grammar that allows all the transitions among the four HMMs (three for the voice and one for the noise). This means that the phonetic segmentation is embedded in the testing procedure.

3.2 The PTE-HMM Model

The PTE-HMM Model consists in an Ergodic Model like the one shown in Fig. 1. As a starting point in the construction of this model, we take the four previous HMMs trained with the different acoustic segments. We combine these models into an ergodic one (all the transitions between states are allowed) and retrain this model with the restriction that only the transition probabilities can be modified.

With the purpose of studying the influence of the training of the transition probabilities in the final performance of the PTE-HMM, we also tested the same architecture with a totally free retraining. That means that, after building the Ergodic Model, we allow all the parameters of the model (transition probabilities, means and variances) to vary.

4. Experimental Results

Fig. 2 presents the result of the embedded segmentation in the parallel system for a utterance during the testing phase with 1 Gaussian mixture per model. In this

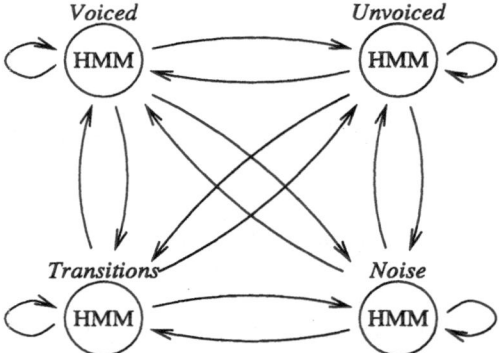

Fig. 1. The ergodic model

graphic, the height of the dashed line means the output of the phonetic segmentation: the maximum height corresponds to the voiced segments, the minimum to the unvoiced ones and the intermediate to the transitions; the chunks without dashed line are classified as noise. One important conclusion than can be derived from this figure is that the segmentation is quite correct with a number as low as one Gaussian mixture per model.

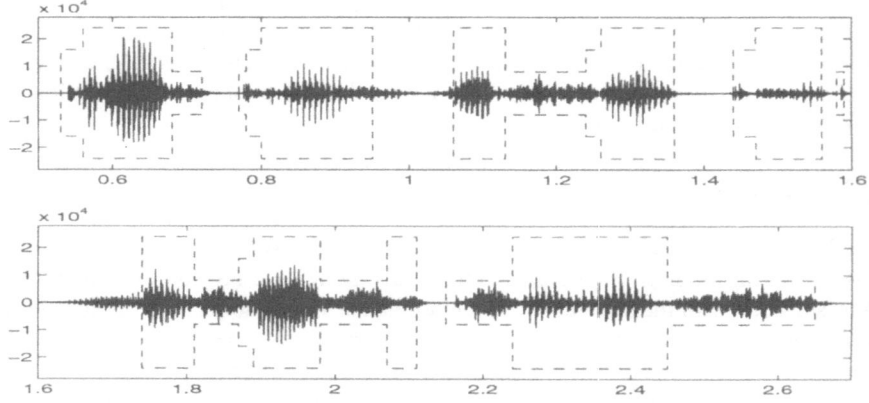

Fig. 2. Embedded segmentation in parallel system (1 mixture per model)

In figs. 3 (TelVoice) and 4 (PolyCost) the speaker identification rate for the PTE-HMM and for a system without Acoustic Segmentation can be compared. There is a dramatic improvement in the identification rate, as can be seen. In these figures, the top line corresponds to the PTE-HMM. In fig. 3, the dashed line corresponds to the system without Acoustic Segmentation.

With the purpose of studying the influence of the training of the transition probabilities in the final performance of the PTE-HMM, we also tested the same architecture with a totally free retraining. That means that, after building the Ergodic Model,

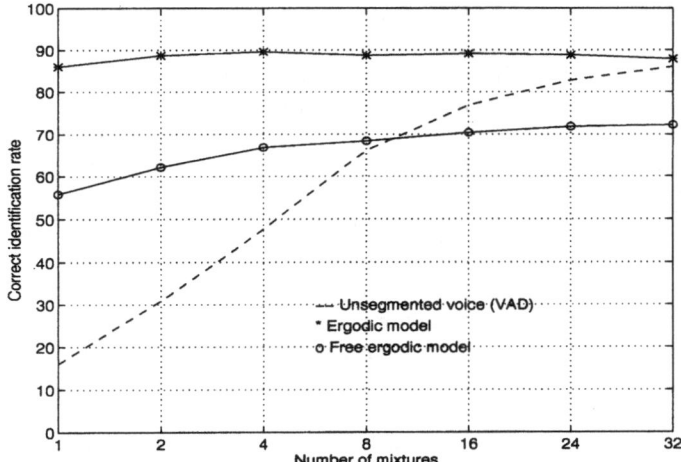

Fig. 3. Speaker identification rate - ergodic models - TelVoice

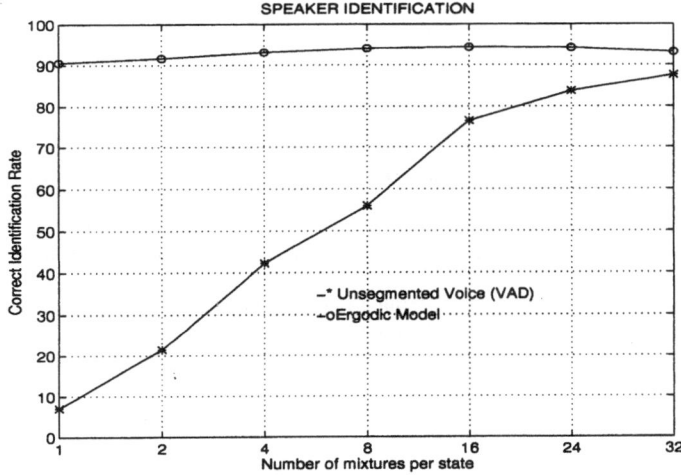

Fig. 4. Speaker identification rate - ergodic model - PolyCost

we allow all parameters of the model (transition probabilities, means and variances) to vary. The recognition rate obtained experiments a decrease between 15 and 30%, as it can be seen in fig. 3.

5. Conclusions

In [3] it is stated that for HMM based systems, identification scores are highly correlated with the total number of mixtures independently of the number of states. Our results show that the performance of the PTE-HMM is very high even with a number of mixtures per state as low as one, an it maintains approximately constant independently of the number of mixtures. The explanation for this behavior can be derived from fig. 4: with one mixture per state the representation of the general characteristics of the phonetic classes is accurate enough. When we increase the number of mixtures, we are improving the representation of the boundaries between these phonetic classes and the overall effect is that the recognition rate increases. When we build an Ergodic Model and retrain the transition probabilities, we are introducing this information in the model in a different and much more efficient way.

We have, as well, tested the performance of the proposed system with two different databases, and we found that the behavior is similar. It is important to notice that, when the size of the database grows, while the performance of the system without Acoustic Segmentation decreases, the correct identification rate of the Ergodic Model maintains approximately constant.

The high recognition rate of this architecture, along with its low computational load, encourages us to think that it can be a suitable choice for a real-world application as a speaker verification system over the telephone line.

References

[1] Complete information on PolyCost at *http://circhp.epfl.ch/polycost/*.
[2] C. García Mateo and D. Docampo Amoedo. Modeling Techniques for Speech Coding: a Selected Survey. In A. Figueiras Vidal, editor, *Digital Signal Processing in Telecommunications*. Springer Verlag, 1996.
[3] T. Matsui and S. Furui. Comparison of Text-Independent Speaker Recognition Methods Using VQ-Distortion and Discrete/Continuous HMM's. *IEEE Trans. Speech and Audio Processing*, 2:456–459, 1994.
[4] D. A. Reynolds. Speaker Identification and Verification using Gaussian Mixture Speaker Models. *Speech Communication*, 17:91–108, August 1995.
[5] R. Tucker. Voice Activity Detection Using a Periodicity Measure. *Proceedings of the IEEE*, 139, August 1992.

Comparison of Several Compensation Techniques for Robust Speaker Verification

Laura Docío-Fernández and Carmen García-Mateo

E.T.S.I. de Telecomunicación, Dpto. Tecnologías de las Comunicaciones
Campus Universitario de Vigo, 36200 - VIGO (Pontevedra), SPAIN
e-mail: `ldocio@tsc.uvigo.es`, `carmen@tsc.uvigo.es`

Summary. It is well known that the performance of speaker recognition systems degrade rapidly as the mismatch between the training and test conditions increases. Thus, for example, in real-world telephone-based speaker recognition systems, both, additive and convolutional noise influence the error rate considerably. In this paper, different techniques which make a speaker verification system more robust against noise are described and compared. Some of these techniques have already been successfully applied in Robust Speech Recognition, and our preliminary results show that they are also very encouraging for Robust Speaker Verification.

1. Introduction

Speaker identity is important for many applications such as access control, automatic money transfer, telephone shopping, etc. In this field, it is known that the recognition systems can perform well if the training and testing conditions are comparable, but problems arise when there is a mismatch between the environments for training and testing, which is generally true in most applications. The goal of robust speaker recognition is to remove or to compensate the effect of this mismatch between training and testing conditions, so that the recognition performance would be as close as possible to that of the matched conditions.

In actual telephone-based speaker verification applications, both, additive and convolutional noise are present causing in general a degradation in performance. The distortion, or mismatch, due to the channel and ambient noise is assumed to be linear either, in the cepstral domain or in the spectral domain, respectively. This can be expressed as follows:

$$Y = H(X + N) \qquad \text{linear spectral domain,}$$
$$log Y = log H + log(X + N) \quad \text{log spectral domain,}$$
$$c_y(k) = c_h(k) + c_{xn}(k) \qquad \text{cepstral domain}$$

where, X is the spectrum of the original clean speech, H is the channel response, Y is the spectrum of the received signal, $c_y(k)$, $c_h(k)$, and $c_{xn}(k)$ denotes the kth cepstral coefficient of Y, H, and $(X + N)$, respectively. Thus, the global mismatch

---

This work has been partially supported by the Spanish CICYT under the project TIC96-0956-C04-02

between training and test can be expressed as a linear transform in the cepstral domain:

$$\hat{c} = A * c + b \tag{1}$$

where $\hat{c}$ is the cepstral vector of the distorted signal, c is the cepstral vector of the original signal, and the matrix A and the vector b model the transformation.

In this paper, we will describe and compare three different techniques to make the speaker verification system more robust against the possible mismatch between training and test.

In speaker verification systems, the goal is to determine from a given utterance whether a person is who he or she claims to be. Although it requires only a binary decision it is a difficult task to come up with the most suitable score to perform the test. For HMM recognition based systems, the log-likelihood of the given utterance is the easiest-to-use score. Thus, what it is called, **unnormalized log-likelihood score**, $\Lambda_u(X; C)$, takes the form

$$\Lambda_u(X; C) = logp(X|\lambda_C)$$

where $p(X|\lambda_C)$ is the likelihood of the spoken utterance X conditioned to come from the claimed speaker C and λ_C is the claimed speaker's HMM.

Since the system must decide whether the claimed identity is true or false the problem can be formulated in a hypothesis testing frame-work as: Given a test utterance X from a speaker with a claimed identity, we test the *null Hypothesis H_0* against the *alternative Hypothesis H_1*, where: H_0: X is from the claimed speaker, and H_1: X is *not* from the claimed speaker. This lead us to use the ratio between both hypothesis as the verification score. Thus, in the log domain, the above ratio , that we call the **normalized log-likelihood score**, $\Lambda_n(X; C)$, becomes

$$\Lambda_n(X; C) = logp(X|\lambda_C) - logp(X|\lambda_{\bar{C}})$$

where $p(X|\lambda_C)$ is the likelihood of the test utterance is from the claimed speaker and $p(X|\lambda_{\bar{C}})$ is the likelihood of the test utterance is not from the claimed speaker. The likelihood ratio $\Lambda_i(X; C)$ $(i = u, n)$, is compared to a threshold η and the claimed speaker is accepted if $\Lambda_i(X; C) > \eta$ and rejected if $\Lambda_i(X; C) < \eta$. The choice of a *anti-speaker* or *background* model, $\lambda_{\bar{C}}$, can significantly affect the verification performance. Several methods have been proposed for the construction of this model, i.e., to approximate the denominator of the likelihood ratio. In this paper, the background model is obtained as described in [4].

The paper is organized as follows. In section 2 we give a description of our baseline speaker verification system. Section 3 gives a description of the three proposed compensation methods. Experimental results and comparative performance among the different compensation techniques are given in section 4. In this section also experiments comparing the unnormalized and normalized scores are reported. Finally, some conclusions are given in section 5.

2. The HMM recognition system

In our baseline recognition system speakers are represented by an ergodic four-state continuous density and multiple-Gaussian HMM trained by using an acoustic discriminative procedure [1]. Briefly, this training procedure consists of two stages: first, training frames are a priori segmented into four categories: voiced, unvoiced, transitions, and non-speech frames by means of a phonetic classifier. Then, all frames assigned to a particular category contribute to the estimation of the mixture Gaussian parameters of a particular state. Variances are tied across the mixtures due to the scarcity of training data. Second, transition probabilities are estimated keeping up fixed means and variances of the Gaussian mixtures.

In all experiments a database from 20 Spanish speakers (10 male and 10 female) recorded on the telephone network have been used. Each speaker provided five sessions collected over a period of about one month. The speakers were encouraged to use different handsets, telephone lines, locations, etc., so that a range of conditions are sampled over the different sessions. Each session consists of four repetitions of the speaker Spanish Identity Card. One session is used for training and the remainder four sessions for test.

The speech input is sampled at 8 KHz and preemphasized using a first-order filter with a coefficient of 0.97. Then, 15th-order LPC analysis is performed each 10 ms time interval with a 20 ms Hamming window. For each frame we extract a 26 element feature vector, which consists of 12 linear prediction cepstral coefficients, 12 delta-cepstrum coefficients, the normalized log energy, and the delta normalized log energy.

3. Mismatch Compensation Techniques

We have analyzed three different techniques that cope the mismatch between training and test conditions: Cepstral Mean Subtraction (**CMS**), Signal Bias Removal (**SBR**) by ML estimation as described in [3] that we will call Stochastic Method 1 (**SM1**), and the ML Stochastic Matching procedure described in [5] that we will call Stochastic Method 2 (**SM2**).

3.1 CMS

Long-term CMS can be considered as a classical technique for decrease the channel influence both in speech and speaker recognition. This approach is very simple. It subtracts from each frame of the observed utterance the average cepstrum over the entire utterance

$$\hat{c} = c - \mathcal{E}(c)$$

where the expectation, $\mathcal{E}(c)$, is approximated by time average.

3.2 SM1

This method of SBR was originally applied to robust speech recognition by Rahim and Juang [3]. It has been integrated as a part of a discrete-density HMM recognition system, that use a vector quantization (VQ) codebook based model to estimate the bias process b. As we have said in section 2, our recognition system is a continuous-density HMM architecture, then in order to apply this method we need to compute and to store the SBR VQ separately from the model.

Assuming a single additive bias term, b, its estimation is accomplished in the training phase by an iterative procedure. First we obtain an estimation of the bias process, and then it is subtracted from the signal. Second, a new VQ is created with the processed data. During the recognition phase only step 1 is carried out [2].The key issue of SM1 is the VQ required to compute the bias estimation.

3.3 SM2

This technique is also a maximum likelihood (ML) approach to decrease the acoustic mismatch between a test utterance and a given set of HMM's. This mismatch can be reduced in two domains: in the observation or feature space mapping the observed utterance Y into an utterance X which matches better with the models given, and in the model space mapping the original models to a transformed models that matched better with the observed utterance Y. An important aspect is that the algorithm work only on the given test data and the given set of speaker models, and no additional training data is required for the estimation of the mismatch prior to actual testing. The bias can be modeled as either a fixed bias or a state-dependent bias. For a fuller system description the reader is referred to [5] and [2].

4. Experiments and Results

A series of speaker verification experiments were conducted. The goal of the experiments was:

1. Examine the efficacy of the above compensation techniques in improving the performance of the baseline verification system in the presence of mismatch due to different transducers and channels.
2. Compare the performance of the system using the normalized log-likelihood scoring method with that using unnormalized log-likelihood method.
3. Compare the performance of different *anti-speaker* or *background* model selection methods.

Two kinds of verification trials are performed for each speaker, I, in the database. First, the set of verification utterances spoken by I are compared with I's model to provide a set of the true or "customer" scores. Second, verification utterances spoken by speakers other than I are compared with I's model to provide a set of false speaker or "impostor" scores. Background members were excluded from the impostor population of each speaker. Thus, for a verification experiment on the given database, each speaker is used as a claimant with the remaining speakers (excluding the claimant's background speakers) acting as impostors, and rotating through all speakers. Results are reported as the equal-error rate (EER).

First, we have carried out experiments over the baseline system in order to analyze and compare the effect of using either, the normalized score function as proposed by [4] or the unnormalized score function. To obtain the normalized scores we have used a cohort set of size 7 and a background size of 4. Note that using large background speaker sets decreases the number of impostor tests. Figure 1 shows verification performance as a function of number of mixtures per state in the HMM for both unnormalized and normalized scoring functions. We can see in this figure that normalized log-likelihood scoring method got better performance than unnormalized one. The equal error rate is significantly reduced by almost 93% using the normalized method for a HMM architecture with 4 mixtures per state. We also observe that the EER using the normalized scores is less insensitive to the number of mixtures per state than that of using the unnormalized scores.

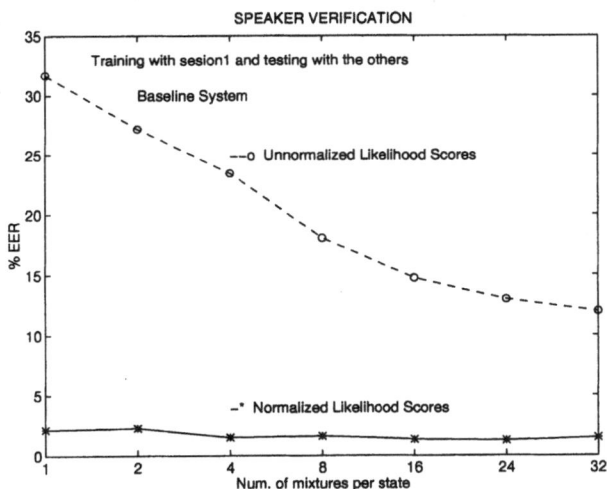

Fig. 1. Verification performance using normalized and unnormalized likelihood scores for different HMM architectures

Figures 2 and 3 shows histograms of customer (true) and impostor (false) verification scores for the two approaches: unnormalized and normalized scoring methods, respectively. We can see the big overlap between the true and false scores when the unnormalized scoring method is used. This results in a large equal-error rate.

As we said, the performance of the normalized methods depends on the selection of the background speakers. Thus, for a dedicated impostor situation (for example, impostors with the same gender) the selection of background speakers as *close cohorts* (msc) performs well, but for a casual impostor scenario (for example, impostors are the opposite sex) the selection of background speakers as *close cohorts and far cohorts* (msc + msf) gets better performance. This effect is shown in Figure 4. We can see that for Cross-Sex EER (i.e., opposite sex impostors) the

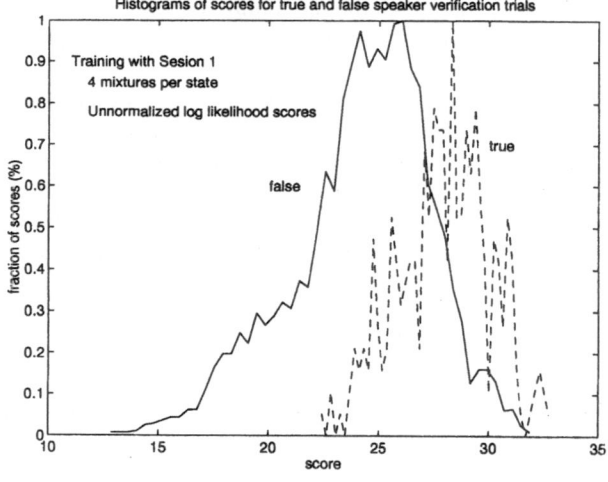

Fig. 2. Histograms of scores for true and false speaker verification trials for unnormalized likelihood scores

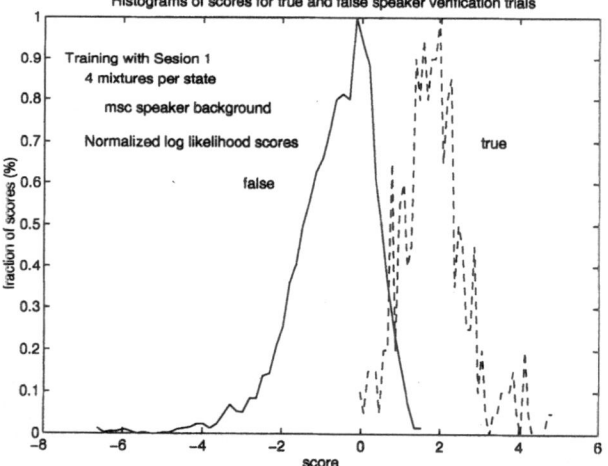

Fig. 3. Histograms of scores for true and false speaker verification trials for normalized likelihood scores

performance by the "msc + msf background speaker set" is better than the "msc background speaker set". Nevertheless, for Same-Sex EER (i.e., same gender impostors) the "msc background speaker set" is the best.

In [2] we have shown the improvement in performance achieved in a *Speaker Identification* task when the compensation techniques are applied. Here we will show the obtained performance of those same techniques for a *Speaker Verification* task. Figures 5 shows the performance comparison for the different compensation methods when normalized log-likelihood score is used. As a reference we also plot the verification results obtained by the baseline system. This figure clearly shows that the performance of the baseline speaker verification system improves when these techniques are applied. Furthermore, CMS method is the best one providing

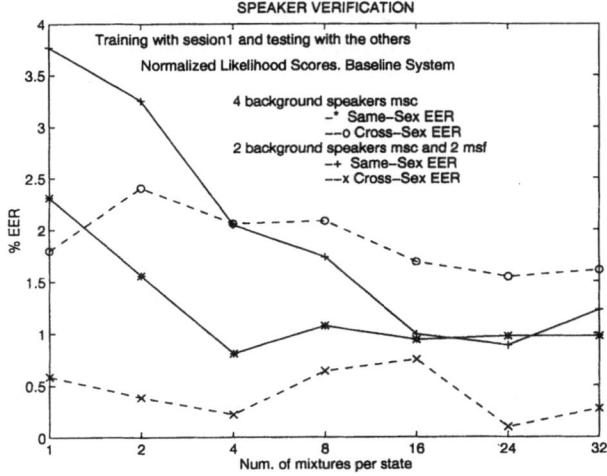

Fig. 4. Verification equal-error rates (%) for two different choices of the background speaker set and different EER measures

a drop in the error rate by almost 67% for a HMM architecture of 4 mixtures per state.

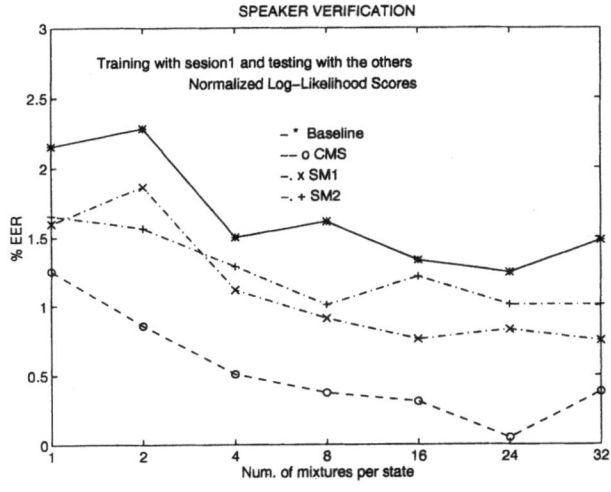

Fig. 5. Verification performance with normalized log-likelihood scores – comparison of different compensation techniques

To apply the SM1 technique we have built a VQ of 8 centroids for each speaker using his training data. In testing, the VQ of the claimed speaker is used to estimate and remove the bias. We can see the improvement after applying SM1 method. We have applied the SM2 technique in the feature space, and we have considered the approach of a single bias vector for the whole utterance. The bias vector was initialized to zero. It can be observed that this method also outperforms the baseline system. The results show then that SM techniques significantly reduce the error rate compared with the baseline system, but neither SM1 nor SM2 improve the CMS performance when normalized scores are used. Taking into account these results, we can conclude that the improvement achieved by the normalized scoring method overcome the improvement achieved by the SM techniques. In [2] the results over a *speaker identification task* can be seen.

5. Discussion and Conclusion

In this paper, we have presented a series of experiments with two stochastic compensation techniques to increase performance in a robust speaker verification task by dealing with the mismatch between training and testing conditions. They have proved to be effective, but do not outperform CMS. We have also explored how much the performance of the system improves using normalized log-likelihood scores rather than unnormalized scores. The results show that the normalized scoring method got better performance. Other remarkable point is the selection and design of the *anti-speaker* model to use in the normalized method. The kind of impostors (dedicated or casual) should be taken into account.

References

[1] C. García-Mateo and L. Rodriguez-Liñares. Speaker recognition based on a weighted acoustic discrimination. In *EUSIPCO 96*, volume III, pages 1047–1050, Trieste, September 1996.
[2] C. G.-M. L. Docío-Fernández. Application of several channel and noise compensation techniques for robust speaker recognition. In *EUROSPEECH 97*, September 1997.
[3] M. Rahim and B.-H. Juang. Signal bias removal by maximum likelihood estimation for robust telephone speech recognition. *IEEE Trans. on ASP*, 1(4):19–30, January 1996.
[4] D. Reynolds. Speaker identification and verification using gaussian mixture speaker models. In *Proc. ESCA Workshop on Automatic Speaker Recognition*, pages 27–30, 1994.
[5] A. Sankar and C.-H. Lee. Robust speech recognition based on stochastic matching. In *ICASSP 95*, volume I, pages 125–124, 1995.

Segmental Acoustic Modeling for Speech Recognition

Mari Ostendorf

Electrical and Computer Engineering Department, Boston University
8 St. Mary's St., Boston, MA 02215 USA
email: mo@bu.edu

Summary. In recent years, several alternative acoustic models have been proposed that attempt to represent trends or correlation of observations over time. These models, which can be broadly classified as segment models, are surveyed in this chapter and presented in a general probabilistic framework that includes the hidden Markov model (HMM) as a special case. The overview gives options for modeling assumptions in terms of correlation structure and parameter tying and outlines the extensions to HMM recognition and training algorithms needed to handle segment models.

Key words: speech recognition; acoustic modeling; trajectory modeling; time correlation model; segment model.

1. Introduction

To date, the most successful speech recognition systems have been based on the hidden Markov model (HMM), and the use of HMMs for acoustic modeling dominates the continuous speech recognition (CSR) field. Although HMMs will continue to play a role in most recognition systems for a long time to come, many alternative models have been proposed in recent years to address some of the shortcomings of HMMs. In particular, a variety of models that could be broadly classified as segment models have been described for representing a variable-length observation sequence in CSR applications. These higher-order models tend to require more computation than HMMs, but with increases in computational power and the use of multi-pass search techniques, they are viable and of interest for current systems. This chapter reviews the general framework of segment modeling and surveys a variety of specific examples.

Broadly speaking, there are three HMM limitations that different models have tried to address: weak duration modeling, the assumption of conditional independence of observations given the state sequence, and the restrictions on feature extraction imposed by frame-based observations. The limitation that an HMM state duration model implicitly defines a geometric distribution has been addressed by introducing models with explicit state duration distributions [1, 2]. Relaxation of the assumption of conditional independence of observations, widely recognized to be useful in practice but physically implausible, has been the subject of several studies. A simple mechanism for capturing time dependence is to augment the observation space with feature derivatives. In addition, several variations of HMMs have been proposed to explicitly model correlation, including conditionally Gaussian HMMs [3, 4, 5] and "segmental" HMMs [6, 7]. Finally, the goal of using segmental rather than frame-based features, probably the initial motivating factor for development of segment models, led to a variety of studies dominated by the work of Zue and colleagues, e.g. [8, 9].

In this chapter, we bring many of the different proposed models under a common framework that extends stochastic modeling beyond HMMs. In Section 2., we introduce the segment model (SM) as a generalization of an HMM, and then describe specific distribution assumptions for modeling the dynamics of frame-based feature vectors. General recognition and training algorithms for segment models are described in Section 3. Alternatives for modeling segmental features are described in Section 4. Finally, Section 5. concludes with a discussion of several questions in segment modeling that are unresolved by current studies.

2. Segmental and Hidden Markov Models

Taking a statistical approach to the problem of acoustic modeling, our goal is to find component distributions with which we can represent the probability $p(y_1^T | a_1^N)$, where $a_1^N = \{a_1, \ldots, a_N\}$ is an N-length sequence of phone labels and $y_1^T = \{y_1, \ldots, y_T\}$ is a T-length sequence of D-dimensional feature vectors. In an HMM, the observation distribution model operates at the frame level: $b_s(y) = p(y|s)$ for $s \in S$, where S is the set of discrete HMM states. In segment modeling, the observation distribution is at the segment level, representing an l-length segment $y_1^l = [y_1, \ldots, y_l]$ with $b_{a,l}(y_1^l) = p(y_1^l | a, l)$ for $a \in \mathcal{A}$ and $l \in \mathcal{L}$, where $\mathcal{A}$ is the set of segment labels and $\mathcal{L}$ is the set of allowable segment lengths. Figure 1 illustrates this difference between the HMM and the SM from a generative perspective.

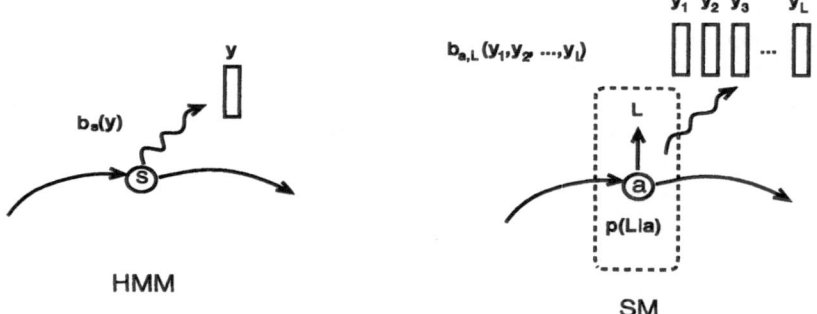

Fig. 1. An HMM and an SM illustrated as generative processes: one frame (y) is generated by an HMM state, and a variable-length sequence of frames ($y_1, \ldots, y_l$) is generated by a SM "state" associated with random length l.

In coding, quantizing sequences of observations achieves lower distortion for a given bit rate than quantizing individual samples. If a segmental coding strategy is more efficient than a frame-based coding strategy, it seems reasonable to expect that a segmental recognition strategy will be more effective than a frame-based (e.g.

HMM) recognition strategy, since minimum distortion is related to maximum likelihood in a Gaussian model. In fact, an early segment model [10] was motivated by work in segment coding [11], and there are strong similarities between other segment coding algorithms [12] and models developed for recognition.

In acoustic modeling, a "segment" might correspond to a phone-sized unit, but the units might also be sub-phone units [13, 14, 15], diphones [16], syllables [17], or acoustically derived units [18]. We therefore use the term "segment" in a more general sense than the typical linguistic association of "segment" with phonetic units. The unit size does not affect the probabilistic formalism, though it does have an impact on the computational costs of the model because of the greater length variability that must be accounted for in longer units.

With an HMM, there are several options for modeling the distribution $p(y|s)$, including discrete distributions, full or diagonal covariance Gaussian densities, Gaussian mixtures, and Laplacian distributions. Similarly, there are many possible distribution assumptions with SMs, in fact many more options because of the large number of degrees of freedom in the model. In order to highlight similarities across the different distribution assumptions, we begin with a description of a general form of the segment model, then move to specific distribution assumptions that have been proposed for capturing dynamics of frame-based features.

2.1 General Modeling Framework

A general segment model represents a random-length sequence of observations $y_1^l = [y_1, \ldots, y_l]$, generated by unit a, according to the density

$$p(y_1, \ldots, y_l, l|a) = p(y_1, \ldots, y_l|l, a)p(l|a) = b_{a,l}(y_1^l)p(l|a). \tag{1}$$

A segment model for label $a \in \mathcal{A}$ is characterized by a *duration distribution* $p(l|a)$, and a *family of output densities* $\{b_{a,l}(y_1^l); l \in \mathcal{L}\}$ that describes observation sequences of different lengths. In addition, a Markov assumption for sequences of a_i is made either implicitly or explicitly, e.g. by embedding phone segments in a word pronunciation network.

The segment duration distribution $p(l|a)$ can be either parametric or nonparametric. Parametric models explored include the Poisson distribution [1], the Gamma distribution [2], and extensions conditioning on context. The non-parametric model simply uses smoothed relative frequencies. For phone-sized units, any reasonable assumption works well empirically, since the contribution of the duration model is small relative to the higher dimensional segment observation probability.

The family of output densities $\{b_{a,l}(y_1^l); l \in \mathcal{L}\}$ represents l-length trajectories in D-dimensional vector space. It is impractical to have a separate collection of parameters $\theta_{a,l}$ for every l when $|\mathcal{L}|$ is large, so $b_{a,l}(\cdot)$ is defined by thinking of y_1^l as a random *process* rather than treating the different lengths as unrelated random vectors. In specifying a random process distribution, two key issues that must be addressed are whether the features are correlated in time and whether the distribution parameters are time varying.

The simplest process distribution assumption is that successive observations are independent and identically distributed. In this case, the probability of the segment given label a and length l is the product of the probability of each observation y_t,

$$b_{a,l}(y_1^l) = \prod_{t=1}^{l} p(y_t|a), \qquad (2)$$

and the segment model reduces to a one-state HMM with an explicit duration model $p(l|a)$ (as opposed to the implicit geometric HMM duration model). This model is known as a hidden "semi-Markov" model [1] and a continuously-variable duration HMM [2], as well as a segment model [19]. By introducing an explicit state duration distribution, these models have the added complexity of hypothesizing segmentations in recognition and training. If one can accept this cost, then it is natural to move beyond this simple case to less restrictive distribution assumptions.

To make the model slightly more complex, we can use time-varying parameters and a deterministic time warping function, but retain a conditional independence assumption, as in [20, 21]. In this case, the probability of a segment given label a and duration l becomes

$$b_{a,l}(y_1^l) = \prod_{t=1}^{l} p(y_t|a, r_t), \qquad (3)$$

where the observations are assumed conditionally independent given a "region" sequence r_1^l specified by the time warping function. The regions partition the segment in time and are analogous to HMM states except that their progression can be length-dependent rather than random.

When time-varying parameters are used, the time warping function, or distribution mapping, $\{T_l(i); i = 1, \ldots, l; \ l \in \mathcal{L}\}$ is a key component needed to specify the distribution family. T_l can be deterministic or dynamic. Two variations of the *deterministic mapping* are possible, either a) to a fixed number of distributions using a table look-up, $T_l(\cdot) \in \{1, \ldots, R\}$, or b) to a continuum of models determined by sampling a segment trajectory, $T_l(\cdot) \in \Re$, as shown in Figure 2. Trade-offs are discussed in Section 2.2. A *dynamic mapping*, as used in [16, 22], is implemented using dynamic programming to find the maximum likelihood mapping of frames in a particular segment observation to distributions or regions. If the distribution family is given by Equation 3, then a segment model with an unconstrained dynamic mapping is equivalent to an HMM network except for the explicit duration distribution. For the complex distribution assumptions that will be described next, deterministic mappings have the advantage of reduced computation relative to dynamic programming, and for phone-sized units and smaller, they work quite well in practice.

The use of time-varying parameters gives the segment model an additional mechanism for representing temporal variability that is not available in an HMM. In an HMM, temporal variability is modeled by the state transition probabilities, which is essentially a collection of geometric length distributions. In the segment model, the length distributions are more general, and there is the option for finer-level control of temporal structure through the distribution mapping. Thus, the SM

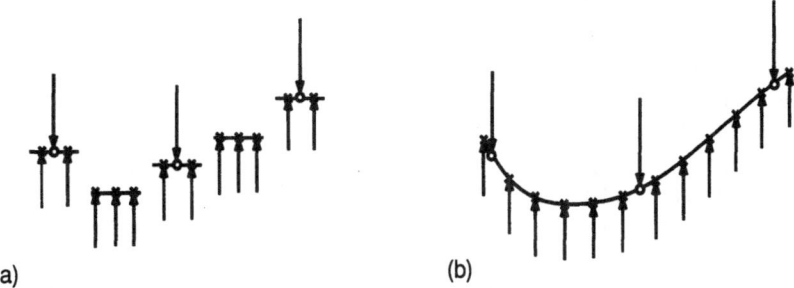

Fig. 2. Distribution mapping to (a) a fixed number of model regions (five) vs. (b) a continuum of distributions via trajectory sampling, illustrated in both cases for a linear-time warping with three-frame (o, down-arrows) and twelve-frame (x, up-arrows) observations. The line shows a one-dimensional distribution mean as a function of time.

has the advantage of accounting for temporal variability in both the state sequence and the observation distribution, compared to the HMM which only weakly models temporal variability through the state sequence.

2.2 Models of Feature Dynamics

Since an HMM is a special case of a segment model, segment models are capable of achieving at least the same level of performance as an HMM, and experiments have shown that performance is similar for equivalent distribution assumptions and numbers of free parameters [23]. However, the segment model allows for more general families of distributions than does an HMM, particularly distributions that implicitly or explicitly model feature dynamics. Below, we outline several different alternatives, including constrained mean, Gauss-Markov and more general linear models, as well as segmental mixture models.

Constrained Mean Trajectory Model. When a deterministic distribution mapping is used, Equation 3 defines a constrained mean trajectory model. The particular type of distribution mapping, either an indexing function to a fixed set of regions or a trajectory sampling function, specify whether the mean trajectory is parametric or non-parametric. If parametric, the mean is specified by a constant, linear or higher order polynomial trajectory in D-dimensional space, and distributions for specific regions are determined by points along the trajectory. Non-parametric trajectory models, on the other hand, have distribution parameters that are separately estimated for each model region. Both cases have extensions of HMM techniques for distribution clustering [24, 25] and MAP adaptation [26, 27].

The first frame-based stochastic segment models were non-parametric, i.e. [10] for fixed-length observations and [20] for variable-length observations, and used a deterministic time warping. Non-parametric trajectory models using dynamic mappings include [16, 22]. Although the non-parametric approach has been relatively successful, it is also the least explicit model of feature dynamics. Parametric segment models were introduced separately by Gish and Ng [21] (as a segment model) and Deng *et al.* [15] (as a non-stationary HMM). The two approaches differ in the representation of the complete trajectory, i.e. relative time sampling of a fixed trajectory [21] vs. absolute time increments in a length-dependent trajectory [15]. Using absolute time has the advantage of efficient recognition and segmentation algorithms, since the Markov assumption holds at the frame level, but it is only reasonable for sub-phonetic units. Both sets of researchers originally proposed a common covariance for all samples in the segment, but time-varying covariance frameworks have since been developed [28, 29].

The parametric and non-parametric approaches each have advantages. The parametric approach is motivated by the smooth trajectories in many speech units, but the non-parametric approach has computational advantages since score caching can be used. Parametric models tend to have fewer parameters, but non-parametric models allow distribution mapping estimation and more flexible parameter tying. Further research is needed to fully assess the relative benefits.

Conditionally Gaussian Model. After conditional independence of observations, the next simplest distribution assumption is the Markov property. For Gaussian distributions, this corresponds to a Gauss-Markov assumption within and optionally across segment regions or HMM states, e.g. for segments:

$$b_{a,l}(y_1^l) = \prod_{t=1}^{l} p(y_t | y_{t-1}, a, r_t, r_{t-1}). \tag{4}$$

Researchers have long observed that the HMM assumption of conditional independence is not valid and have investigated alternative assumptions. Early work with Markov assumptions, referred to here as conditionally Gaussian HMMs, was due to Wellekens [3], who described extensions to the Viterbi and Baum-Welch algorithms for this case, and Brown [4] who also explored these models experimentally. Conditionally Gaussian models were rediscovered by Kenny *et al.* [5], who (like Brown) found a benefit to the conditionally Gaussian model for simple cepstral features but not for features augmented with derivatives. The Gauss-Markov assumption for segment models was explored by Digalakis *et al.* [23] with similar conclusions. Deng extended his parametric trajectory model [15] to the Gauss-Markov case [30], but did not assess its performance in recognition experiments. More recently, the HMM results of Takahashi *et al.* [31] indicate that improved performance can be obtained by using the Gauss-Markov assumption in a mixture framework. Segment modeling research, on the other hand, took a different approach to solving this problem by adding an observation noise term, as described next.

Dynamical System Model. A stochastic, linear dynamical system (DS) is described by the equations

$$x_{t+1} = F_t x_t + w_t \tag{5}$$
$$y_t = H_t x_t + v_t \tag{6}$$

where x_t is an unobserved state vector, y_t is the observed feature vector, and w_t and v_t are uncorrelated Gaussian vector processes with mean and autocovariance functions: μ_W, Q_W, μ_V, Q_V. The initial state x_0 is Gaussian with mean μ_0 and covariance Σ_0. The dynamical system model is widely used for estimation and control problems with non-stationary signals, but was introduced as a speech recognition model by Digalakis *et al.* [13, 32]. A segment model with time-varying parameters has a finite number of regions with a parameter set $\Theta = \{H, \mu_V, Q_V, F, \mu_W, Q_W, \mu_0, \Sigma_0\}$ defined for each region, assuming that parameters are locally time-invariant within a region. The probability of a segment is computed using the innovation sequence $\{e_t\}$

$$b_{a,l}(y_1^l) = \prod_{t=1}^{l} p(e_t|a, r_t, r_{t-1}), \tag{7}$$

where $p(e_t|r_t)$ is Gaussian with zero mean and covariance Σ_{e_t}, and e_t and Σ_{e_t} are found using Kalman filter recursions. Within a region, the DS model can be viewed as a continuous-state HMM, since the hidden trajectory vectors x_t are continuous valued. Thus, the DS segment model combines a continuous unobserved state x_t with a discrete state (a, l).

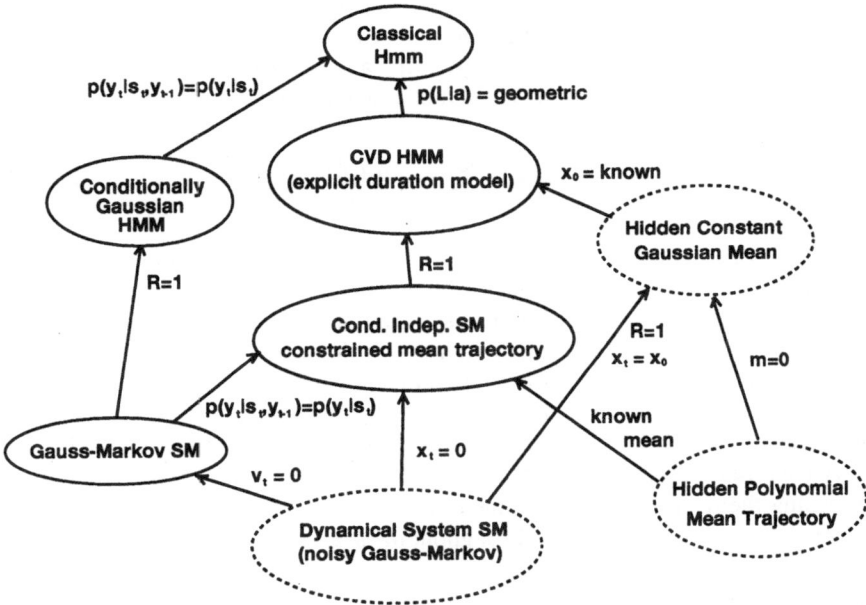

Fig. 3. Family tree of stochastic models for a variable-length frame-based observation sequence. The arrows indicate model simplifications, with s_t representing both HMM state and SM region.

Viewing the hidden trajectory as a filtered series of targets (μ_W), the DS model is similar to that proposed by Bakis [33]. It can be thought of as a scaled, noisy observation of a Gauss-Markov process. Thus, the DS includes the Gauss-Markov process as a special case ($H = I, v_t = 0, y_t = x_t$). Many of the other modeling assumptions described here can also be viewed as special cases of the DS model, as shown in Figure 3. For example, if the unobserved state x_t is taken to be zero, then the terms v_t provide the distributions for the regions ($y_t = v_t$), and the DS model corresponds to the constrained-mean trajectory assumption. In context-independent phone classification experiments [32], one form of the DS model compared favorably to other variations that included the non-parametric constrained mean and Gauss-Markov assumptions. Presumably, the benefit of the DS model would be greater for context-dependent models because of the reduced variance of the initial state.

Segment-level Mixtures. Since mixture distributions have been used so successfully in hidden Markov models, a natural extension of any of the models described so far is to segmental mixtures. If the advantage of frame-level mixture distributions stems from systematic variation in speech, then segmental mixtures may be able to represent the systematic component by keeping the mixture mode constant across the segment. The direct analogy to HMM Gaussian mixtures uses a discrete mode to specify which mixture component generated the observed segment. Alternatively, one can envision a continuous mixture mode by defining a prior on a parametric trajectory. Both cases are described below.

The *discrete-mode* segmental mixture model generalizes the SM to have a finite collection of segment-level distributions, which are combined with mixture weights that correspond to the probability of observing a particular trajectory in a segment. Specifically, the probability of y_1^l given unit a and length l is:

$$b_{a,l}(y_1^l) = \sum_{j=1}^{m} \lambda_{aj}\, p_j(y_1^l|a, l). \tag{8}$$

For each of the m mixture components, $p_j(y_1^l|a, l)$ gives the probability of the complete segment and λ_{aj} is the mixture weight. Mixtures can also be used at the sub-segment level [13, 14]. Initial studies used constrained mean trajectory models as the mixture components [34, 21], though in principle any of the above distribution assumptions could be used. Experiments with non-parametric trajectory models [34] give good performance for context-independent phone modeling, but parameter tying is probably needed for good performance with context-dependent models.

A *continuous-mode* segmental mixture model puts a prior on some parameter describing the mean in the constrained-mean trajectory model. The simplest such model assumes a constant mean throughout the segment $y_t \sim N(\mu, \Sigma)$, where the mean is modeled by a Gaussian prior $\mu \sim N(\mu_0, \Sigma_0)$.[1] This model was proposed by Russell [6] and Gales and Young [7] as a "segmental HMM", and by Digalakis [13] as a "target state SM". The constant Gaussian mean model is again a special case of the dynamical system model, where the state is constant for all t, $x_t = x_0 = \mu$, i.e. $F = I$, $\mu_V = 0$, $w_k = 0$ and $Q_V = \Sigma$. In addition, it can be viewed as a sophisticated version of variable frame rate analysis, as shown in [6], where the segment mean is used rather than the first observed value. The model can also be

---

[1] Note that the use of a prior on the mean here differs from the use of a prior in adaptation in that a different mean is found for each instance of a unit, rather than a common mean shared by all instances of the unit as in adaptation.

generalized to have a mean that is both time varying and random, with a joint prior on the offset and slope [35, 36]. In addition, a mixture distribution can be used to represent the intrasegmental variance, where the mixture components share the same mean but have different variances [36]. Initial experiments with context-independent models and constant mean trajectories did not outperform HMMs, but good results have been achieved with more complex models.

3. Recognition and Training

All of the different distribution assumptions described above have in common the need to hypothesize and evaluate different segmentations in both recognition and training, the general algorithms for which are described below.

3.1 Recognition Algorithms

The standard recognition solution for HMMs involves finding the most likely state sequence via Viterbi decoding (dynamic programming), and then mapping the state sequence to the appropriate word sequence. Assuming that the state network used in decoding has been defined by a word/pronunciation network, then the mapping gives a unique word sequence. For segment models, the solution is analogous, except that the "state" includes both the segment label and duration. In other words, using the notation q for a state in general, the segment state is $q = (a, l) \in \mathcal{A} \times \mathcal{L} = \mathcal{Q}$, whereas the HMM state is $q = s \in \mathcal{S} = \mathcal{Q}$. Segment-based recognition then involves finding

$$(\hat{N}, \hat{a}_1^{\hat{N}}) = \underset{N, a_1^N}{\operatorname{argmax}} \{ \max_{l_1^N} p(y_1^T | l_1^N, a_1^N) p(l_1^N | a_1^N) p(a_1^N) \}, \qquad (9)$$

using a dynamic programming algorithm, and then mapping the segment label sequence $\hat{a}_1^N$ to the appropriate word sequence $\hat{w}_1^K$ as in HMM decoding. The key difference between the SM and HMM search algorithms is the explicit evaluation of different segmentations, which adds an extra dimension to the dynamic programming search. The increase in the state space corresponds to an increase in the search cost: where an HMM computes one vector likelihood for each state at each time, the SM may compute $|\mathcal{L}|$ vector *sequence* likelihoods for each state at each time. Combined with the added complexity of trajectory modeling, this can greatly increase the computational cost of recognition.

There are two main mechanisms for reducing the cost of the search. First, segment evaluation costs can be reduced by caching sub-segment scores when segments of different lengths share partial distribution mappings, and/or by using within-segment pruning to eliminate hypothesized segments based on partial segment likelihoods [37]. Second, the search space can be reduced by restricting the set of segmentations considered and/or using the SM only in later passes of a multi-pass search framework. The segmentation space can be restricted to a fixed set based on either hierarchical clustering of frames according to their acoustic similarity (resulting in a dendrogram representation of a set of possible segmentations [8]) or restricting segment boundaries to fall within some window of the hypothesized boundaries in a previous stage of a multipass search. The segmentation search space can also be changed dynamically, restricted to some neighborhood of the current segmentation [37]. All methods of reducing the search space introduce errors, and it is an empirical question as to which approach is better. However, HMM hypothesis rescoring is generally more effective for word recognition than a single-pass search with a reduced segmentation space, both in terms of accuracy and speed.

3.2 Parameter Estimation Algorithms.

The hidden state component that is common to all the acoustic models presented in Section 2.1 complicates parameter estimation, requiring some form of iterative algorithm with steps that: 1) characterize the hidden state given the current parameter estimates, and 2) estimate new parameters based on that characterization. Possible options for these two steps are described next, followed by a discussion of algorithms for estimating the distribution mapping.

Hidden State Characterization. The two iterative schemes used in speech recognition are generalized below to include the different SM variations presented here. One estimates the conditional probability distribution of the hidden state and is an instance of the expectation-maximization (EM) algorithm [38]. The other finds the most likely hidden state sequence at each iteration and is sometimes called "Viterbi training"; the HMM segmental k-means algorithm [39] is an example. For some models, the state has both a discrete and a continuous component, in which case the state is characterized using a combination of these two algorithms.

The EM algorithm for training HMMs is the standard Baum-Welch training algorithm, and extensions have been derived for many of the new variations of HMMs that have been proposed, including continuously variable duration HMMs [2], segmental HMMs [7], and conditionally Gaussian HMMs [3, 4, 5]. A general solution for discrete-state segmentation models, referred to as the *generalized forward-backward algorithm*, is derived in [40] by defining an unobserved state sequence that is not necessarily time-synchronous with the observation sequence, assuming that the state sequence is Markov and that observations are conditionally Markov given the state sequence. For a segment model, the discrete state takes values in the product space $\mathcal{A} \times \mathcal{L}$, so computation and storage costs are proportional to $T|\mathcal{A}||\mathcal{L}|$. In most CSR applications, the storage cost is prohibitive, but can be reduced by "pruning", or only keeping track of the most likely state sequences.

A less costly means of characterizing the hidden state (or, an extreme version of pruning) is to jointly estimate the *most likely state sequence* (MLSS) and the model parameters. In the case of segment models, we must find the most likely segment label sequence and segmentation jointly. The solution is a dynamic programming algorithm that differs from the Viterbi recognition search in that the possible label sequence is constrained according to the known word sequence, effectively reducing the set of active segment labels. This MLSS re-estimation procedure has been applied to both HMMs [39] and segment models [10]. Starting from an HMM segmentation, segment model training requires only a few iterations of MLSS training.

For the dynamical system and continuous mode segmental mixture distribution assumptions, there is a continuous component to the hidden state as well as a discrete component. If the discrete state space is large, it is impractical to implement a full EM training solution. As a compromise, the MLSS algorithm is first used to "estimate" the discrete state, and then the E-step of an EM algorithm is used for the continuous state, which involves computing the conditional expectations of the sufficient statistics for the hidden state. For the dynamical system and continuous mode segmental mixture models, the statistics are the expected first and second order moments given the observation set, discrete state, and current model parameters.

Parameter Estimation. Once a solution to the hidden-state problem is provided, the second step of the iterative algorithm involves estimation of the state transition probabilities and observation distribution parameters. Maximum likelihood parameter update formulae can be found in specific references where the models are introduced and are summarized in [40] for the case where the MLSS state characterization step is used. Extension to the full EM solution is simply a weighted version of the MLSS solution.

Although the standard maximum likelihood estimates are straightforward, obtaining robust context-dependent models can be difficult, since some of the interesting SM distribution assumptions are not amenable to simple smoothing techniques, such as variance clipping and Bayesian smoothing. An alternative solution is parameter tying, i.e. assuming that some model parameters are shared across models and/or regions. Parameters can be tied based on heuristic rules using knowledge of the application, as in [17], or can be determined automatically through distribution clustering. Divisive distribution clustering has been used extensively in hidden Markov modeling, e.g. [41, 42, 43], where a greedy search successively adds models through binary splits of the data that at each step maximize the likelihood of the training data represented by the clustered models (or, equivalently, to minimize entropy). Each possible split is evaluated in terms of the likelihood ratio of one vs. two distributions for representing the data at that node of the tree. In principle, the algorithm can be extended to any segment model that assumes conditional independence across but not necessarily within regions, but so far the node evaluation functions have only been determined for the constrained mean cases [24, 25].

Distribution Mapping Estimation. For frame-based segment models that have time-varying parameters, the mapping $T_l(i)$ must be either specified according to heuristics or estimated automatically. Heuristics that have been used successfully include: linear time warping for phones (e.g. [20, 10]), linear sampling of the cepstral vector trajectory for phones [10], and functions of consonant-vowel structure for syllables [17]. However, there is evidence that intra-phone timing, though systematic, is non-linear [44], so better performance may be obtained by deriving the mapping automatically. One approach uses ML distribution clustering as described above, but in the temporal domain. Starting with one region per segment, data is successively partitioned to add regions in the segment model where they most increase the overall training likelihood. Contextual and temporal clustering can also be combined, resulting in an overall algorithm very similar to maximum likelihood successive state splitting [45]. Another approach assumes a known number of regions for each model, and finds T_l separately for each length $l \in \mathcal{L}$ to maximize the likelihood of the training data of that length, using dynamic programming [46]. A disadvantage of the second algorithm is that it has no generalization mechanism for unobserved lengths l.

4. Segmental Features

One of the initial motivations for considering segmental models is the potential for incorporating segmental features. A segmental feature is any transformation $f(y_1^l, l)$ of the variable-length segment observation sequence y_1^l to a fixed-length vector v. Examples include cepstral vector regression coefficients [47], a vector of average formant values and energies in different frequency bands, and a sampled version of the observation sequence [10, 48, 49]. Unfortunately, the fixed-length feature mapping changes the dimensionality of the probability space of the whole sequence so that it is proportional to the number of hypothesized segments, and different segmentations are not comparable. The dimensionality problem can be addressed heuristically with a length-dependent weighting factor [10], but this use of segmental features is still theoretically problematic because it involves conditioning on different events for different segmentations. The foundation of statistical detection theory is lost, since the theory holds for comparing $p(a|z)$ to $p(a'|z)$, not for the comparison of $p(a|z)$ to $p(a'|w)$.

One approach for dealing with this problem is the use of posterior distributions in combination with a segmentation score, in which case equation 9 becomes

$$(\hat{N}, \hat{a}_1^{\hat{N}}) = \operatorname*{argmax}_{N, a_1^N} \{ \max_{l_1^N} p(a_1^N | y_1^T, l_1^N) p(l_1^N | y_1^T) \} \tag{10}$$

where $p(l_1^N | y_1^T)$ is the segmentation score. A version based on neural networks was first described in [50]. Other work shows the segmentation score to be critical to the segmental feature approach [51]. Most work assumes that segments are independent, i.e. $p(a_1^N | y_1^T, l_1^N) = \prod_{i=1}^N p(a_i | v_i)$, where v_i is the transformation of the i-th segment. This assumption is not valid for context-dependent models, and we conjecture that better performance can be achieved by operating on a wider window of features y_t (see [40] for further discussion).

A second approach for using segmental features solves the problem of inconsistent conditioning events by including features v_i for all possible segmentations in evaluating a particular hypothesis, but using a "not-a-segment" distribution for those observations not in the current hypothesis [52]. With some simplifying assumptions, they show that the likelihood of a complete hypothesis reduces to the product of likelihood ratios for the features that correspond to that hypothesis: $\prod_{i=1}^N p(v_i | a_i) / p_0(v_i)$, where $p_0(v_i)$ is the "not-a-segment" model.

With all of the approaches that use segmental features, the general solutions to the recognition search and hidden segmentation training problems are essentially the same. Given a hypothesized segmentation or segmentation likelihood, the details of the parameter estimation step in training will change depending on the distribution assumption, as with the frame-based modeling approaches. In addition, if the posterior distribution approach is used, the parameter estimation step must include re-estimation of segmentation likelihood parameters.

5. Summary

In summary, segment models can be thought of as a higher dimensional version of a hidden Markov model, where Markov states generate random sequences rather than a single random vector observation. The basic segment model includes an explicit segment-level duration distribution and a family of length-dependent joint distributions specified via a random process model. Since segment models are a generalization of HMMs, the standard HMM training and recognition algorithms can be easily extended to handle segment models but with a higher computational cost due to the expanded state space.

The advantage of segment models is that there are many alternatives for representing a family of distributions, allowing for explicit trajectory and/or correlation modeling. Several distribution assumptions proposed in the literature have been described here. Looking at the group of options as a whole, the key modeling assumptions include: 1) whether the trajectory is hidden (as for the dynamical system and segmental mixture models) vs. observed (as for the constrained-mean trajectory and Gauss-Markov models); 2) whether correlation is modeled explicitly through Gauss-Markov assumptions or a mixture mode, vs. implicitly through the distribution mapping constraints; and 3) whether the parameters are time-varying, and if so whether that variation is represented parametrically or non-parametrically. In addition, there are options for representing intra-segmental timing, as encoded in the distribution mapping. Some of these alternatives have been explored in isolated experiments, but more work is needed to assess the relative benefits of the different modeling assumptions as a whole. Of course, the answer to questions about structure will depend on the particular feature vectors used and the units represented, which raises further questions about what problems segmental models are best suited for.

In addition to better understanding the empirical behavior of the different segmental models, further algorithmic and theoretical developments are needed on several fronts. For example, frame-level divisive distribution clustering and adaptation has proved to be very useful for both HMMs and simple SMs, but new techniques are needed for more complex distribution assumptions. In addition, the distribution assumptions described here are all based on linear models, and there are many possible non-linear extensions. Finally, the use of segmental features is still in its early stages and much can be done to advance these models.

In conclusion, we note that much of the theoretical framework of the segment model can also be applied to other time series modeling problems, including language modeling, where a "segment" corresponds to a phrase and the "observations" are words. SMs can also applied to synthesis problems, as in [17], where trajectory modeling and structural constraints make the DS model more useful than HMMs.

Thus, further development of segment models will have implications beyond acoustic modeling for speech recognition.

Acknowledgements. The author wishes to thank V. Digalakis, O. Kimball, A. Kannan and J. R. Rohlicek for many valuable discussions on segment modeling and this survey. This work was supported by the United States DoD, grant ONR-N00014-92-J-1778.

References

[1] M. Russell and R. Moore, "Explicit modeling of state occupancy in hidden Markov models for automatic speech recognition," *Proc. Int'l. Conf. on Acoust., Speech and Signal Processing,* 1985, pp. 2376-2379.

[2] S. Levinson, "Continuously variable duration hidden Markov models for automatic speech recognition," *Computer Speech and Language,* vol. 1, pp. 29-45, 1986.

[3] C. J. Wellekens, "Explicit time correlation in hidden Markov models for speech recognition," *Proc. Int'l. Conf. on Acoust., Speech and Signal Processing,* 1987, pp. 384-386.

[4] P. F. Brown, "The acoustic modeling problem in automatic speech recognition," Ph.D. Thesis, Computer Science Department, CMU, May 1987.

[5] P. Kenny, M. Lennig and P. Mermelstein, "A linear predictive HMM for vector-valued observations with applications to speech recognition," *IEEE Trans. on Acoust., Speech, and Signal Processing,* vol. ASSP-38, no. 2, pp. 220-225, 1990.

[6] M. Russell, "A segmental HMM for speech pattern matching," *Proc. Int'l. Conf. on Acoust., Speech and Signal Processing,* vol. II, 1993, pp. 499-502.

[7] M. Gales and S. Young, "The theory of segmental hidden Markov models," Cambridge University Engineering Department, Technical Report, CUED/F-INFENG/TR.133, 1993.

[8] V. Zue, J. Glass, M. Philips and S. Seneff, "Acoustic segmentation and phonetic classification in the SUMMIT system," *Proc. Int'l. Conf. on Acoust., Speech and Signal Processing,* 1989, pp. 389-392.

[9] H. Meng and V. Zue, "Signal representation comparison for phonetic classification," in *Proc. Int'l. Conf. on Acoust., Speech and Signal Processing,* May 1991, pp. 285-288.

[10] M. Ostendorf and S. Roukos, "A stochastic segment model for phoneme-based continuous speech recognition," *IEEE Trans. on Acoust., Speech, and Signal Processing,* vol. 37, no. 12, pp. 1857-1869, 1989.

[11] S. Roucos, R. Schwartz and J. Makhoul, "Segment quantization for very low rate speech coding," *Proc. Int'l. Conf. on Acoust., Speech and Signal Processing,* pp. 1565-1568, 1982.

[12] Y. Shiraki and M. Honda, "LPC speech coding based on variable-length segment quantization," *IEEE Trans. on Acoust., Speech, and Signal Processing,* vol. 36, no.9, pp. 1437-1444, 1988.

[13] V. Digalakis, "Segment-based stochastic models of spectral dynamics for continuous speech recognition," Ph.D. Thesis, E.C.S. Department, Boston University, January 1992.

[14] A. Kannan and M. Ostendorf, "A comparison of trajectory and mixture modeling in segment-based word recognition," *Proc. Int'l. Conf. on Acoust., Speech and Signal Processing,* vol. II, April 1993, pp. 327-330.

[15] L. Deng, M. Aksmanovic, D. Sun and J. Wu, "Speech recognition using hidden Markov models with polynomial regression functions as nonstationary states," *IEEE Trans. on Speech and Audio Proc.,* vol. 2, no. 4, pp. 507-520, 1994.

[16] O. Ghitza and M. M. Sondhi, "Hidden Markov models with templates as non-stationary states: an application to speech recognition," *Computer Speech and Language,* vol. 2, pp. 101-119, 1993.

[17] K. Ross and M. Ostendorf, "A dynamical system model for generating F_0 for synthesis," *Proc. of the ESCA/IEEE Workshop on Speech Synthesis*, 1994, pp. 131-134.

[18] M. Bacchiani, M. Ostendorf, Y. Sagisaka and K. Paliwal, "Design of a speech recognition system based on acoustically derived segmental units," *Proc. Int'l. Conf. on Acoust., Speech and Signal Processing*, vol. I, pp. 443-446, 1996.

[19] H. Gish, K. Ng and J. R. Rohlicek, "Secondary processing using speech segments for an HMM word spotting system," *Proc. Int'l. Conf. on Spoken Language Proc.*, vol. I, 1992, pp. 17-20.

[20] S. Roucos, M. Ostendorf, H. Gish, and A. Derr, "Stochastic segment modeling using the estimate-maximize algorithm," *Proc. Int'l. Conf. on Acoust., Speech and Signal Processing*, 1988, pp. 127-130.

[21] H. Gish and K. Ng, "A segmental speech model with applications to word spotting," *Proc. Int'l. Conf. on Acoust., Speech and Signal Processing*, 1993, vol. II, pp. 447-450.

[22] J. He and H. Leich, "A unified way in incorporating segmental feature and segmental model into HMM," *Proc. Int'l. Conf. on Acoust., Speech and Signal Processing*, 1995, pp. 532-535.

[23] V. Digalakis, M. Ostendorf and J. R. Rohlicek, "Improvements in the stochastic segment model for phoneme recognition," *Proc. DARPA Workshop on Speech and Natural Language*, 1989, pp. 332-338.

[24] A. Kannan, M. Ostendorf and J. R. Rohlicek, "Maximum likelihood clustering of Gaussians for speech recognition," *IEEE Trans. on Speech and Audio Proc.*, vol. 2, no. 3, pp. 453-455, 1994.

[25] A. Kannan and M. Ostendorf, "A comparison of constrained trajectory models for large vocabulary speech recognition," *IEEE Trans. on Speech and Audio Proc.*, vol. 6, no. 3, pp. 303-306, May 1998.

[26] A. Kannan and M. Ostendorf, "Adaptation of polynomial trajectory segment models for large vocabulary speech recognition," *Proc. Int'l. Conf. on Acoust., Speech and Signal Processing*, vol. II, pp. 1411-1414, 1997.

[27] C. Rathinavelu and L. Deng, "Speaker adaptation experiments using nonstationary-state hidden Markov models: a MAP approach," *Proc. Int'l. Conf. on Acoust., Speech and Signal Processing*, vol. II, pp. 1415-1418, 1997.

[28] H. Gish and K. Ng, "Parametric trajectory models for speech recognition," *Proc. Int'l. Conf. on Spoken Language Proc.*, vol. 1, pp. 466-469, 1996.

[29] T. Fukada, Y. Sagisaka and K. Paliwal, "Model parameter estimation for mixture density polynomial segment models," *Proc. Int'l. Conf. on Acoust., Speech and Signal Processing*, vol. II, pp. 1403-1406, 1997.

[30] L. Deng, "A stochastic model of speech incorporating hierarchical nonstationarity," *IEEE Trans. on Speech and Audio Proc.*, vol. 1, no. 4, pp. 471-474, 1993.

[31] S. Takahashi, T. Matsuoka, Y. Minami and K. Shikano, "Phoneme HMMs constrained by frame correlations," *Proc. Int'l. Conf. on Acoust., Speech and Signal Processing*, vol. II, 1993, pp. 219-222.

[32] V. Digalakis, J. R. Rohlicek and M. Ostendorf, "A dynamical system approach to continuous speech recognition," *IEEE Trans. on Speech and Audio Proc.*, vol. 1, no. 4, pp. 431-442, 1993.

[33] R. Bakis, "An articulatory-like speech production model with controlled use of prior knowledge," notes from *Frontiers in Speech Processing: Robust Speech Recognition*, CD-ROM, 1993.

[34] O. Kimball, "Segment modeling alternatives for continuous speech recognition," Ph.D. Thesis, E.C.S. Department, Boston University, September 1994.

[35] W. Holmes and M. Russell, "Speech recognition using a linear dynamic segmental HMM," *Proc. European Conf. on Speech Commun. and Technology*, pp. 1611-1614, 1995.

[36] W. Holmes and M. Russell, "Linear dynamic segmental HMMs: Variability representation and training procedure," *Proc. Int'l. Conf. on Acoust., Speech and Signal Processing*, vol. II, pp. 1399-1402, 1997.

[37] V. Digalakis, M. Ostendorf and J. R. Rohlicek, "Fast search algorithms for phone classification and recognition using segment-based models," *IEEE Trans. on Signal Proc.*, vol. 40, no. 12, pp. 2885-2896, 1992.

[38] A. P. Dempster, N. M. Laird and D. B. Rubin, "Maximum likelihood estimation from incomplete data," *Journal of the Royal Statistical Society (B)*, vol. 39, no. 1, pp. 1-38, 1977.

[39] L. R. Rabiner, J. G. Wilpon and B.-H. Juang, "A segmental *k*-means training procedure for connected word recognition," *AT&T Technical Journal*, vol. 65, no. 3, pp. 21-40, 1986.

[40] M. Ostendorf, V. Digalakis and O. Kimball, "From HMMs to segment models: A unified view of stochastic modeling for speech recognition," *IEEE Trans. on Speech and Audio Proc.*, vol. 4, no. 5, pp. 360-378, 1996.

[41] L. R. Bahl, P. V. de Souza, P. S. Gopalakrishnan, D. Nahamoo and M. A. Picheny, "Decision trees for phonological rules in continuous speech," *Proc. Int'l. Conf. on Acoust., Speech and Signal Processing*, 1991, pp. 185-188.

[42] M.-Y. Hwang and X. Huang, "Shared-distribution hidden Markov models for speech recognition," *IEEE Trans. on Speech and Audio Proc.*, vol. 1, no. 4, pp. 414-420, 1993.

[43] S. J. Young, J. J. Odell and P. C. Woodland, "Tree-based state tying for high accuracy acoustic modeling," *Proc. ARPA Workshop on Human Language Technology*, 1994, pp. 307-312.

[44] J. van Santen, "Segmental duration and speech timing," *Computing Prosody*, ed. Y. Sagisaka, N. Campbell, and N. Higuchi, Springer, pp. 225-250, 1997.

[45] M. Ostendorf and H. Singer, "HMM topology design using maximum likelihood successive state splitting," *Computer Speech and Language*, vol. 11, no. 1, pp. 17-42, 1997.

[46] M. Afify, Y. Gong, J.-P. Haton, "Non-linear time alignment in stochastic trajectory models for speech recognition," *Proc. Int'l. Conf. on Spoken Language Proc.*, 1994, pp. 291-293.

[47] S. Krishnan and P. V. S. Rao, "Segmental phoneme recognition using piecewise linear regression," *Proc. Int'l. Conf. on Acoust., Speech and Signal Processing*, vol. I, 1994, pp. 49-52.

[48] W. Goldenthal and J. Glass, "Statistical trajectory models for phonetic recognition," *Proc. Int'l. Conf. on Spoken Language Proc.*, 1994, pp. 1871-1874.

[49] Y. Gong and J.-P. Haton, "Stochastic trajectory modeling for speech recognition," *Proc. Int'l. Conf. on Acoust., Speech and Signal Processing*, vol. I, 1994, pp. 57-60.

[50] H. C. Leung, I. L. Hetherington and V. Zue, "Speech recognition using stochastic segmental neural networks," *Proc. Int'l. Conf. on Acoust., Speech and Signal Processing*, vol. I, 1992, pp. 613-616.

[51] J. Verhasselt, I. Illina, J.-P. Martens, Y. Gong and J.-P. Haton, "The importance of segmentation probability in segment-based speech recognizers," *Proc. Int'l. Conf. on Acoust., Speech and Signal Processing*, vol. II, pp. 1407-1410, 1997.

[52] J. Glass, J. Chang and M. McCandless, "A probabilistic framework for feature-based speech recognition," *Proc. Int'l. Conf. on Spoken Language Proc.*, vol. 4, pp. 2277-2280, 1996.

Trajectory Representations and Acoustic Descriptions for a Segment-Modelling Approach to Automatic Speech Recognition

Wendy J. Holmes

Speech Research Unit, DERA Malvern
St. Andrew's Road, Great Malvern, Worcs. WR14 3PS, UK
email: holmes@signal.dera.gov.uk

Summary. This paper discusses some of the possibilities for modelling speech segment trajectories in a domain which is more directly correlated with the mechanisms of speech production than the typical mel-cepstrum representation. Initial developments are described towards using linear dynamic segmental HMMs [12] to model underlying (unobserved) trajectories of features which closely reflect the nature of articulation. So far, this work has involved calculating segment probabilities using an approach which is different from that used in earlier studies (e.g. [4]), and is more consistent with the idea of treating the trajectory as unobserved. In parallel, experiments have demonstrated that formant features can be useful for HMM-based automatic speech recognition [3].

1. Introduction

In recent years there has been a growing interest in approaches to automatic speech recognition which overcome speech-modelling limitations of the conventional frame-based HMM approach. These alternative approaches model observation sequences or "segments" (see [8] for a comprehensive review of segment models). The temporal dynamic characteristics of speech are typically represented by incorporating the concept of a trajectory to describe how acoustic features change over time. At the Speech Research Unit we have developed a *segmental HMM* [12] which incorporates an underlying parametric trajectory, together with a distinction between variability across different instantiations of a sub-phonemic speech segment and variability within one example. Thus, extra-segmental variability in the trajectory parameters is treated separately from intra-segmental variation in the observations given any one underlying trajectory. Recognition experiments [4] have demonstrated that a linear dynamic segmental HMM can give performance advantages over conventional HMMs for the same acoustic representation.

Our work on segmental HMMs, in common with most other work on trajectory-based models [8], has so far applied the trajectory representation to mel-cepstrum features of the type successfully used for conventional-HMM systems. However, the underlying motivation for the trajectory-modelling concept is to take better account of the nature of speech production. The process of speaking produces continuously changing articulatory configurations, with movements which are quite constrained and mostly smooth. Inspection of a spectrogram reveals immediately that speech dynamics are most apparent across, rather than within, channels (as changing formant frequencies for example). Although the changes are somehow incorporated in any sensible spectral representation of the acoustic signal, the characteristics of observed trajectories will depend critically on details of the spectral representation.

Time-domain behaviour more similar to that of articulatory movements should be provided by trajectories represented in a domain which is more directly correlated with the mechanisms of speech production. This approach represents a progression away from simply representing the surface manifestation of the speech generation process and towards modelling the underlying mechanisms, as advocated in [1] and in [11] for example. A model based on an accurate representation of the speech production process has the potential to significantly improve the capabilities of automatic speech recognition systems by reducing the reliance on explaining variations in the acoustic signal in terms of random behaviour [7]. This paper considers some of the benefits and issues for modelling trajectories in a domain more closely related to speech production, discusses how a segmental HMM could provide the basis for such a model, and describes initial experiments in modelling formant features.

2. Modelling Trajectories in Speech

There are several potential advantages to be gained from modelling trajectories in a domain which is closely related to speech production, as an alternative to spectral representations such as the mel-cepstrum. The cepstrum transformation is useful for reducing the correlation between various spectrum features, and the lowest-order features offer quite a good representation of gross articulatory changes, but the higher-order cepstrum features are related to articulation in a very complicated way. Typical sequences of high-order cepstrum coefficients therefore show frame-to-frame fluctuations which do not have a simple relation with phonetic properties of the speech [5]. In attempting to represent such features with a simple stochastic model, a large part of the variation is therefore treated as random rather than as a proper consequence of articulatory gestures.

Appropriateness of the trajectory assumption is obviously important, but there are also opportunities for new ways of using the models themselves. For example, at an underlying level there may be considerable benefits from representing components of the acoustic signal as asynchronous parallel streams using techniques such as HMM decomposition [6]. This approach can be used to systematically model the effect of noise on the resulting signal, and also to allow asynchrony between different components of the speech itself. The technique has recently been successfully applied to mel-cepstrum features [14], but could be more beneficial when applied at a level at which it could more directly represent the different rates of movement of different articulators. Effects of speaking rate could also be represented in a meaningful way, as could speaker effects. Basic vocal tract configurations are similar for all speakers, while differing in aspects of detail. By representing the similarities and differences in a way closely related to articulation, it might eventually be possible to automatically recognise speech from any speaker in a way which does not simply represent all speech from all speakers in a single acoustic model. Furthermore, a trajectory model that is closely linked with production mechanisms should be applicable to synthesis as well as to recognition.

Given the many potential opportunities arising from adopting a representation of speech which is closely related with the mechanisms, an appropriate representation needs to be found. Obvious possibilities are some form of articulatory description (e.g. [9]), or a formant-based description (e.g. [13]). Such a representation could in principle replace the typical time-frequency description of the surface acoustics currently used in conventional systems. However, it is likely to be more successful and more robust if used at an unobserved intermediate level between the surface acoustic representation and the underlying deeper message structure, thus avoiding the need for early decision making in the explicit extraction of higher-level features. This approach suggests the use of a two-stage probabilistic model, with an unobserved trajectory which is responsible for generating observed acoustic sequences.

3. Representing an Unobserved Trajectory with Segmental HMMs

The distinction between extra- and intra-segmental variability in the linear dynamic segmental HMM [12, 4] provides a two- stage model of the type alluded to above. The concept of the trajectory is thus an underlying characteristic of the observations, although in our work so far the probability calculations have involved first identifying the most likely location of the trajectory. In recent work, success has been achieved with an approach whereby segment probabilities are calculated without requiring explicit identification of the trajectory, and this development is described below.

3.1 Calculating segment probabilities

So as to simplify notation, it is assumed in the following discussion that the observations to be modelled are one-dimensional. In a linear Gaussian segmental HMM (GSHMM), the distributions of trajectory parameters for a given state are described by Gaussian distributions $N_{(\mu,\gamma)}$ and $N_{(\nu,\eta)}$ for the slope m and mid-point c respectively. The intra-segmental distribution is assumed to be Gaussian with variance τ. Ignoring any duration probability, the joint probability of a segment of observations $y = y_0, ..., y_T$ and a trajectory $f_{(m,c)}$ is specified as

$$P(y, m, c) = N_{(\mu,\gamma)}(m) . N_{(\nu,\eta)}(c) . \prod_{t=0}^{T} N_{(f_{(m,c)},\tau)}(y_t)$$

In our work so far, we have defined the probability of a segment given a linear GSHMM state as being the above quantity for the single *optimal trajectory*, which is defined by a maximum *a posteriori* estimate of the slope and mid-point. An alternative is to compute a "trajectory-independent" probability which allows for all possible values of the trajectory parameters, thus:

$$P(y) = \int_{-\infty}^{\infty} \int_{-\infty}^{\infty} P(y, m, c) \, dc \, dm.$$

This approach represents an extension of the approach used by Gales and Young [2] in their work on static segmental HMMs. By integrating over all possible values of trajectory parameters the trajectory is in effect modelled as a distribution, and this distribution depends on the segment length. This approach therefore automatically takes into account the fact that, as the length of the data sample increases, the most likely trajectory becomes a more reliable estimate of the true trajectory and so accounts for a greater proportion of the total probability. In the case of the optimal-trajectory approach, the consequence of this effect was accommodated by introducing a more complicated intra-segment distribution model with two mixture components [4].

3.2 Recognition experiment

The trajectory-independent approach was tested using the same speaker-independent connected-digit recognition task with three-state phone models and a maximum segment duration of 100ms (ten frames) as was used in earlier studies [4]. The performance of (single-Gaussian) linear GSHMMs with trajectory-independent probability calculations was compared with that of conventional HMMs and with simpler segmental models. These models involved first introducing a ten-frame maximum segment duration constraint to the HMM, and then also incorporating a linear trajectory representation (but without distinguishing between extra- and intra-segment variation).

Table 1. Digit recognition results for linear GSHMMs (trajectory-independent probabilities), compared with simpler segmental models and with HMMs

Model Set	%Subs.	%Del.	%Ins.	%Err.
Standard HMM	5.6	1.2	1.0	7.8
Constrain maximum segment duration	4.9	0.5	0.6	6.0
Linear trajectory representation	3.8	0.4	0.5	4.7
Linear GSHMM	2.0	0.8	0.1	2.9

The results demonstrated some advantage from simply introducing the duration constraint, with further improvement from the linear trajectory but best performance when also making the distinction between the different types of variability (see Table 1). The word error rate for the linear GSHMMs was less than half that which was achieved when using conventional HMM output probability calculations. The trajectory-independent linear GSHMM approach has given somewhat better performance than the best achieved with the optimal-trajectory approach (an error rate of 3.3% [4]). This better performance was obtained using a simpler model with less parameters, which is also theoretically more attractive as a representation of underlying (unobserved) trends. So far both the trajectory and the observations have been represented in the cepstrum domain, but they could both be described using other features (such as formants). Alternatively, with the addition of some form of mapping component such as the one suggested in [10], the observations could be represented in terms of simple surface-level acoustic features while the trajectory is described by features at some deeper level.

4. HMM Recognition with Formant Features

Before attempting to represent trajectories of features more related to production mechanisms, it is important to demonstrate that features of this type can provide the necessary information for an automatic system to discriminate between different speech sounds. It is therefore useful to evaluate HMM recognition performance using formant features, which are closely related with the underlying articulation but also have a more direct relation with the surface acoustics than do articulatory features. However, formants do not provide the necessary information for making certain distinctions (such as identifying silence), and are frequently badly defined in some speech sounds. It is therefore difficult to identify formants independently from the recognition process, and systems using formant features have generally not shown a performance advantage when used as an alternative to more conventional cepstral features (e.g. [5, 13, 15]). Recently, a method of formant analysis has been developed [3] which includes techniques to largely overcome the difficulties normally associated with extracting and using formant information. To deal with cases where the formants are not well defined by the spectrum shape, a confidence measure is provided, represented as the variance of a notional Gaussian distribution of the true formant frequency about the estimated value. In those cases where there is uncertainty about how the formants should be allocated to spectral peaks, alternative sets of formant frequencies are offered to the recognition process.

Again using the connected-digit recognition task, results using formant features for describing fine spectral detail with a standard HMM were compared with those obtained using a more typical mel-cepstrum representation. In order to assess the usefulness of the formants directly, the same total number of features was used for both feature sets and exactly the same low-order cepstrum features were used for describing general spectral shape. The output of the FFT was therefore used both to estimate formant frequencies with associated confidence measures and to compute a mel-cepstrum. Experiments were carried out to compare a representation comprising the first eight cepstrum coefficients and an overall energy feature with a feature set in which cepstrum coefficients 6, 7 and 8 were replaced by three formant features.

Table 2. HMM digit recognition results using cepstrum features compared with a feature set using formant features in place of the higher-order cepstrum features

Model Set	%Subs.	%Del.	%Ins.	%Err.
8 cepstrum features+energy	2.6	0.6	0.6	3.7
5 cepstrum features+energy+3 formants	4.6	0.7	12.0	17.3
add confidence measure for formants	2.4	0.5	0.4	3.3
include second choice formants	2.1	0.4	0.4	2.9

The results given in Table 2 show that, provided the degree of reliability in the formant estimation is taken into account, recognition performance is better when using formant features than when using only mel-cepstrum features. Also including alternative formant sets gave a further small improvement in performance. These results have hence shown that formants can provide useful information for discrimination with HMMs.

5. Modelling trajectories of cepstrum and formant features

In a trajectory-based approach, it is important that the trajectory model is applied to features for which the assumptions of that model are appropriate. Some initial studies were carried out on the temporal characteristics of the mel-cepstrum and formant features which had been used for the HMM recognition experiments described in the previous section. The data were labelled at the segment level by using the three-state-per-phone standard HMMs to perform a Viterbi alignment with the known transcription. For each identified segment, a trajectory vector was estimated as the best-fitting straight line parameters.

For an utterance of the digit sequence "five zero", Figure 1 shows trajectories for the energy feature, first eight mel-cepstrum coefficients and three formant frequencies. It is evident that a linear trajectory can follow general trends over time for all the features. However, there are regions for which the cepstrum coefficients change quite erratically from frame to frame but there is relatively little articulatory change (in the /z/ of "zero" for example). The changes observed here in the cepstral coefficients were more erratic than for the filterbank-based analysis used in earlier experiments [4], including those described in Section 3 above. Although lack of temporal smoothness does not appear to be a problem for the standard HMM, this characteristic may be less appropriate for the trajectory assumptions of the segmental HMM. This potential difficulty is avoided with formant trajectories, which are by nature smooth. Experiments are currently in progress to compare performance with formant features and different types of cepstral features, in order to assess the influence of the features' temporal characteristics on GSHMM performance.

6. Conclusions

Recent developments in dynamic segmental HMMs have been described, and the usefulness of formant features as a basis for automatic speech recognition has been demonstrated. The next stage is to determine how features such as formants can be best used for representing trajectories in the segmental HMM. This approach could then be further developed to explore the range of potential advantages from adopting an underlying trajectory model that is more closely linked with production mechanisms.

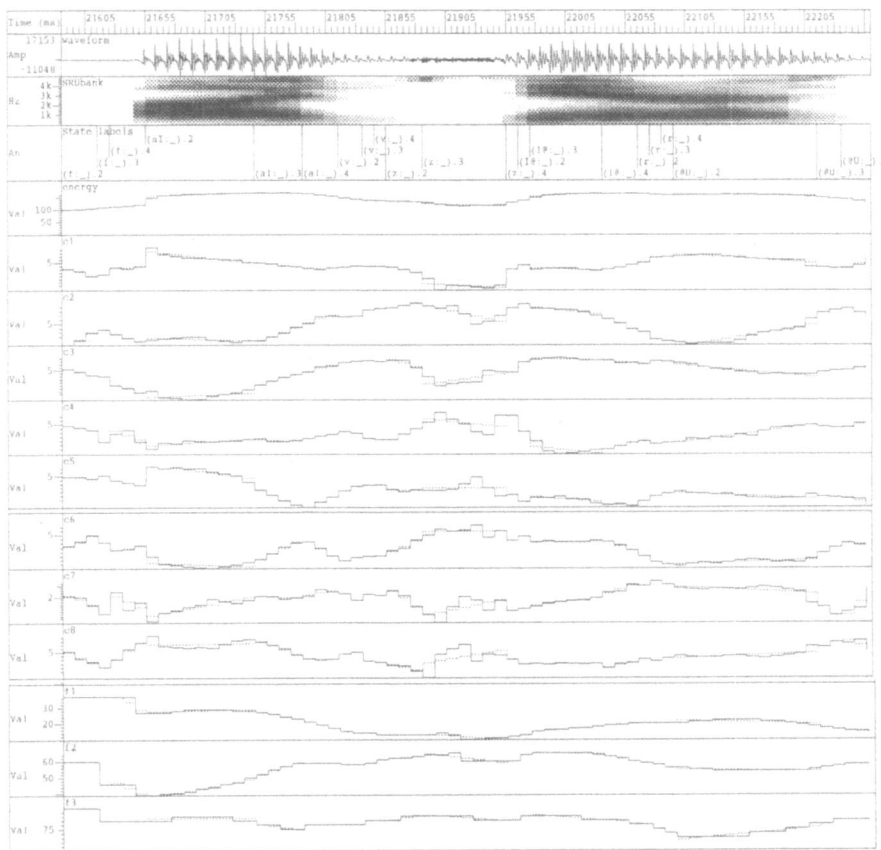

Fig. 1. Frame-by-frame values (solid lines) superimposed on calculated trajectories (dotted lines) for 12 different acoustic features — starting from the top, feature values are shown for energy, the first eight mel-cepstrum coefficients and the first three formant frequencies

References

[1] L. Deng, G. Ramsay, and H. Sameti. From modeling surface phenomena to modeling mechanisms: Towards a faithful model of the speech process aiming at speech recognition. In *Proc. IEEE Automatic Speech Recognition Workshop*, pages 183–184, Snowbird, 1995.

[2] M. J. Gales and S. J. Young. Segmental hidden Markov models. In *EUROSPEECH*, pages 1611–1614, Berlin, 1993.

[3] J. N. Holmes, W. J. Holmes, and P. N. Garner. Using formant frequencies in speech recognition. In *EUROSPEECH*, Rhodes, 1997.

[4] W. J. Holmes and M. J. Russell. Linear dynamic segmental HMMs: Variability representation and training procedure. In *ICASSP*, pages 1399–1402, Munich, 1997.

[5] A. Hu and E. Barnard. Smoothness analysis for trajectory features. In *ICASSP*, pages 979–982, Munich, 1997.

[6] R. K. Moore. Signal decomposition using Markov modelling techniques. RSRE Memo 3931, RSRE, Malvern, UK, 1986.

[7] R. K. Moore. Twenty things we still don't know about speech. In *Proceedings CRIM/FORWISS Workshop on Progress and Prospects of Speech Research Technology*, 1994.

[8] M. Ostendorf, V. V. Digalakis, and O. A. Kimball. From HMM's to segment models: A unified view of stochastic modeling for speech recognition. *IEEE Trans Speech and Audio Processing*, 4(5):360–378, 1996.

[9] G. Ramsay and L. Deng. Maximum-likelihood estimation for articulatory speech recognition using a stochastic target model. In *EUROSPEECH*, pages 1401–1404, Madrid, 1995.

[10] H. B. Richards, J. S. Bridle, M. J. Hunt, and J. S. Mason. Vocal tract shape trajectory estimation using MLP analysis-by-synthesis. In *ICASSP*, pages 1287–1290, Munich, 1997.

[11] M. J. Russell. Advances in speech recognition. In *Proceedings: Institute of Acoustics, Vol. 18:Part 9*, pages 267–274, 1996.

[12] M. J. Russell and W. J. Holmes. Linear trajectory segmental HMM's. *IEEE Signal Processing Letters*, 4(3):72–74, 1997.

[13] P. Schmid and E. Barnard. Explicit, N-best formant features for vowel classification. In *ICASSP*, pages 991–994, Munich, 1997.

[14] M. J. Tomlinson, M. J. Russell, R. K. Moore, A. P. Buckland, and M. A. Fawley. Modelling asynchrony in speech using elementary single-signal decomposition. In *ICASSP*, pages 1247–1250, Munich, 1997.

[15] L. Welling and H. Ney. A model for efficient formant estimation. In *ICASSP*, pages 797–800, Atlanta, 1996.

Suprasegmental Modelling

E. Nöth[1], A. Batliner[1], A. Kießling[1,2], R. Kompe[1,3] and H. Niemann[1]

[1] Lehrstuhl für Mustererkennung (Informatik 5)
   Universität Erlangen–Nürnberg, Martensstr. 3, 91058 Erlangen, Germany
   email: noeth@informatik.uni-erlangen.de

[2] now with Ericsson Eurolab, Nürnberg

[3] now with Sony Stuttgart Technology Center, Fellbach

Summary. We show how prosody can be used in speech understanding systems. This is demonstrated with the VERBMOBIL speech–to–speech translation system, the world wide first complete system, which successfully uses prosodic information in the linguistic analysis. Prosody is used by computing probabilities for clause boundaries, accentuation, and different types of sentence mood for each of the word hypotheses computed by the word recognizer. These probabilities guide the search of the linguistic analysis. Disambiguation is already achieved during the analysis and not by a prosodic verification of different linguistic hypotheses. So far, the most useful prosodic information is provided by clause boundaries. These are detected with a recognition rate of 94%. For the parsing of word hypotheses graphs, the use of clause boundary probabilities yields a speed–up of 92% and a 96% reduction of alternative readings.

Key words: prosody; speech understanding; syntax; parsing; suprasegmental.

1. Introduction

In human decoding of speech, suprasegmental information plays a major role. The term *Suprasegmentals* was introduced by [22] as a cover term for speech phenomena which are attributed to speech segments larger than phonemes. Examples for such segments are syllables, words, phrases, and whole turns of a speaker. To these segments we attribute perceived properties like *pitch, loudness, speaking rate, voice quality, duration, pause, rhythm,* and so on. Even though there generally is no unique feature in the speech signal corresponding to these perceived properties, we can find features which highly correlate with them; examples are the acoustic feature *fundamental frequency* (F_0), which correlates to *pitch*, and the *short time signal energy* correlating to *loudness*. Other and probably more commonly used names for these suprasegmental phenomena are *prosody* and *intonation*, where the latter is mostly used in connection with pitch related suprasegmental phenomena. In the following we will use the term *prosody*.

   The listener extracts information out of these perceived phenomena and this means that we can attribute certain functions to them. The prosodic functions which are generally considered to be the most important ones in human–human communication are phrase boundaries, accents and sentence mood. Already Lea [21] has proposed the use of this prosodic information in automatic speech understanding (ASU) systems. Illustrations for their use are given in the examples below (Section 4.), cf. also [21, 35, 25, 17]. For several reasons, the extraction of prosodic features, their classification into prosodic classes, and the use of these classes in ASU is not an easy task. Thus, even though the number of research projects on prosody in the context of automatic speech recognition/understanding has increased steadily over the past ten years, it took 17 years from [21] to the presentation of the VERBMOBIL system [36], which is the world wide first complete speech understanding system, where prosody is really used. Moreover with VERBMOBIL it can be demonstrated that prosody leads to drastic performance improvements. We see the following main reasons for this gap between the amount of research on prosody and its use in complete systems:

The major role of prosody in human–human–communication is segmentation and disambiguation. In systems for restricted tasks, the user utterances might be so short that these segmentation capabilities of prosodic information would not lead to a system improvement. For example, the average length of a user utterance length in a field test with a travel information system was 3.5 words [12].

In the speech–to–speech translation task of VERBMOBIL, the communication form is human–(computer)–human whereas it is human–computer in almost all other ASU applications. Thus, in VERBMOBIL spontaneous, "real–life" utterances have to be processed. A corpus analysis of VERBMOBIL data, which were collected in human–human dialogs, showed that about 70 % of the utterances contain more than a single sentence [34]; on the average an utterance comprises about 20 words. Furthermore, spontaneous speech phenomena like elliptic constructions and interruptions or restarts are frequent and increase the amount of ambiguities a lot. Our results show that the most important contribution of prosody lies in the understanding rather than in the recognition phase. This shows up clearly in a system like VERBMOBIL which is one of the few systems where the end–to–end performance (including a deep linguistic analysis) is the optimization criterion. The current version of the VERBMOBIL research prototype translates more than 70% approximatively correct [36].

In this paper, we want to show how prosodic information can be computed and used in a speech understanding system. Since the authors developed the prosody module of the VERBMOBIL system and since the use of prosody is implemented on all levels of linguistic processing in this speech–to–speech translation system, most examples will be taken from there.

After a short description of the VERBMOBIL architecture (Section 2.) we will describe how prosodic information is computed in our system (Section 3.). This is divided into the steps feature extraction (Subsection 3.1), description of classes to be recognized (Subsections 3.2 and 3.3), classification into these classes (Subsection 3.4), and improvement of the classification results with stochastic language models (Subsection 3.5). Finally in Subsection 3.6 we show, how these prosodic classes are calculated in a word hypotheses graph (WHG) rather than in the spoken word sequence. Following this we will show how we use prosodic information at different linguistic levels (Section 4.). We will concentrate on the use of prosodic information in syntactic analysis (Subsection 4.1) since for this topic, we can present results of extensive experiments. With respect to the other linguistic levels, we will show *how* prosodic information is used in VERBMOBIL (Subsection 4.2). However, we currently cannot present systematic experimental results, which show the performance improvement caused by prosodic information, as is the case on the syntax level. The paper ends with an outlook to future work and a concluding summary.

2. The Verbmobil System

VERBMOBIL is a speech–to–speech translation project [37, 6] in the domain of appointment scheduling dialogs, i.e., two persons try to fix a meeting date, time, and place. Currently the emphasis lies on the translation of German utterances into English. In October 1996 a research prototype was successfully presented to the public; an overview of the architecture of this VERBMOBIL prototype is shown in Figure 1. After the recording of the spontaneous utterance, a WHG is computed by a standard *Hidden Markov Model* word recognizer and enriched with prosodic information (cf. Section 3.). The WHG is parsed by one of two alternative syntactic modules, i.e., the best scored syntactically correct word chain together with its different possible parse trees (readings) is passed onto the semantic analysis. Also governed by the dialog module, the utterance is translated on the semantic level (transfer module) and an English utterance is generated and synthesized. Parallel to the *deep* analysis performed by these modules, the dialog module conducts a *shallow* processing, i.e., the important dialog acts are detected in the utterance and are roughly translated. A more detailed account of the architecture can be found in [10, 37].

Figure 1 shows the interaction of the prosody module with the other modules in the VERBMOBIL architecture. The solid lines point out interfaces and the dashed lines mark additional flow of information. For the time being, the following modules use prosodic information: syntactic analysis, semantic construction, dialog processing, transfer, and speech synthesis. In the following section, we will describe the computation of prosodic information.

3. Computation of Prosodic Information

Basically, there are two approaches to the extraction of features which represent the prosodic information contained in the speech signal:

1. The prosody module only uses the speech signal as input. This means that the module has to segment the signal into the appropriate suprasegmental units (syllables, words, ...) and calculate features for them.
2. The prosody module takes the output of the word recognition module in addition to the speech signal as input. In this case the time–alignment of the recognizer and the information about the underlying phoneme classes (like *long vowel*) can be used by the prosody module.

The first approach has the advantage that the computation of the prosodic information can be done immediately and in parallel to the word recognition and that the

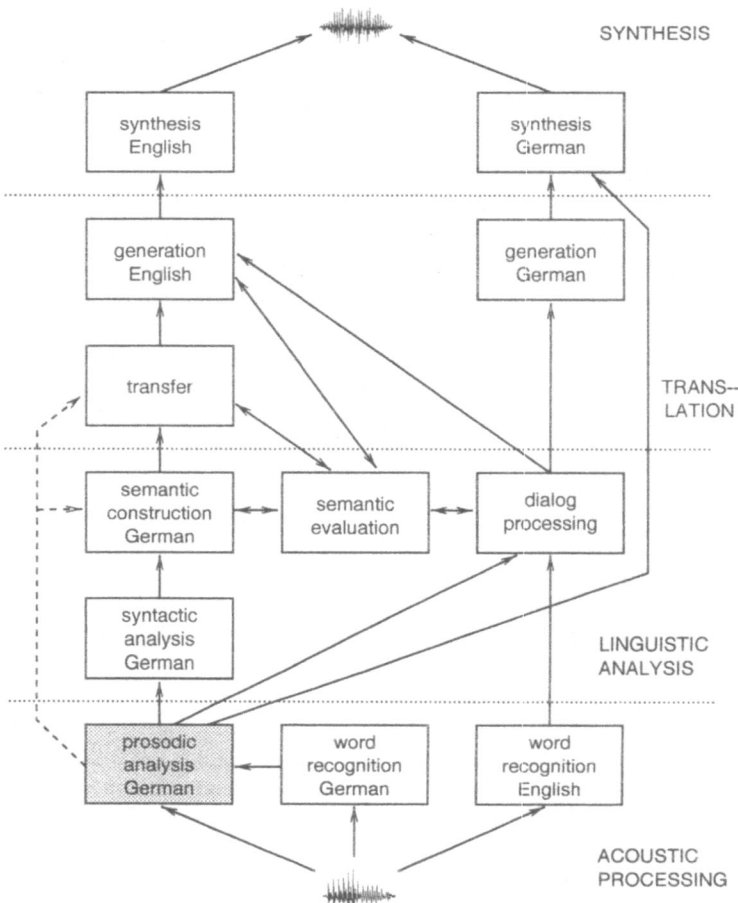

Fig. 1. The VERBMOBIL architecture at a glance

module can be optimized independently. In the second approach, the prosody module has to wait for the output of the word recognition module but no synchronization of the segmentation results of these two modules is necessary at a later stage and the prosody module is more informed.

We decided for the second approach: input to the module is the WHG and the speech signal. Output is a prosodically scored WHG [20], i.e., to each of the word hypotheses, probabilities for prosodic accent, for prósodic clause boundaries, and for sentence mood are attached. We will now describe the individual steps towards the calculation of these probabilities for the word hypotheses.

3.1 Extraction of Prosodic Features

It is still an open question, which prosodic features are the most relevant for the different classification problems and how the different features are interrelated. We try therefore to be as exhaustive as possible, and leave it to the statistic classifier to find out the relevant features and the optimal weighting of them. As many relevant prosodic features as possible are therefore extracted over a prosodic unit (here: the word final syllable) and composed into a large feature vector which represents the prosodic properties of this and of several surrounding units in a specific context.

We investigated different contexts of up to ± 6 syllables (± 3 words, resp.) to the left and to the right of the actual word final syllable. For every classification problem investigated, many different subsets of these features were analyzed. The best results so far were achieved by using 276 features computed for each word considering a context of ± 2 syllables (± 2 words, resp.).

In more detail the features used here are:

- duration (absolute and normalized as in [38]) for each syllable/syllable nucleus/word
- for each syllable and word in this context
  - minimum and maximum of fundamental frequency (F_0) and their positions on the time axis relative to the position of the actual syllable as well as the F_0-mean
  - maximum energy (also normalized) + positions and mean energy (also normalized)
- F_0-offset + position for actual and preceding word
- F_0-onset + position for actual and succeeding word
- for each syllable: flags indicating whether the syllable carries the lexical word accent or whether it is in a word final position
- length of the pause preceding/succeeding actual word
- linear regression coefficients of F_0–contour and energy contour over 11 different windows to the left and to the right of the actual syllable
- for an implicit normalization of the other features, measures for the speaking rate are computed over the whole utterance based on the absolute and the normalized syllable durations (as in [38]).

3.2 Prosodic Classes

As much as it is an open question, what features to use, it is open what prosodic classes to look for, i.e. which reference labels should be used to train the classifiers that perform the transformation from the acoustic features to the functional prosodic classes: how many levels of accentuation should be distinguished? Should we try to detect events which *can* be marked prosodically (i.e. *all questions*) or only those, which really *are* marked prosodically? Who decides on the classes — a panel of naive listeners or phonetic experts?

In VERBMOBIL, we started with the different types of perceptual–prosodic reference labels provided by the University of Braunschweig, cf. [28]:

Prosodically marked phrasal accents

Four different types of syllable based phrasal accent labels (*primary accent, secondary accent, emphatic or contrastive accent,* and *no accent*). For the experiments described below, these labels were mapped onto word–based labels denoting if a word is accented (A) or not ($\neg$A).

Prosodically marked boundaries

Four different types of boundary labels (*full intonational boundary* with strong intonational marking, *intermediate phrase boundary* with weak marking, *normal word boundary,* and *"agrammatical" boundary* like, e.g., hesitation, repair). For the experiments described below, these labels were mapped onto word–based labels denoting if after a word a full intonational boundary (B) or one of the other three classes ($\neg$B) occurs.

Prosodically marked sentence mood

We distinguish between the prosodically marked sentence moods *statement, question,* and *continuation rise.*

Disadvantage of perceptual classes in automatic speech understanding

There are some drawbacks in these reference labels if one wants to use prosodic information in the later linguistic analysis which are best explained with respect to the use of prosodic boundary information in parsing:

- Prosodic labeling by hand is very time consuming, the labeled data-base up to now is therefore rather small.
- A perceptual labeling of prosodic boundaries is not an easy task and possibly not very robust.
- Prosodic boundaries do not only mirror syntactic boundaries but are influenced by other factors as rhythmic constraints and speaker specific style. In the worst case, clashes between prosody and syntax might be lethal for a syntactic analysis if the parser goes on the wrong track and never returns.

Earlier experiments on a large corpus with read speech showed that syntactic–prosodic labels can be successfully used for the training of prosodic classifiers (cf. [20]). This and the work with pure syntactic boundaries together with our colleagues from IBM (Heidelberg) [14, 1] encouraged us to develop a new labeling scheme which is described in the following section.

3.3 New Boundary Labels: The Syntactic–prosodic M–labels

Our new labels should fulfill the following requirements:

- It should allow for fast labeling. Therefore, the labeling scheme should be rather rough, because the more precise it is the more complicated and the more time consuming the labeling will be. A "small" amount of labeling errors can be tolerated, since it will be used to train statistical models, which should be robust to cope for these errors.

- Prosodic tendencies and regularities should be taken into account. In this context, it is suboptimal to label a syntactic boundary that is most of the time not marked prosodically with the same label as an often prosodically marked boundary. Since large quantities of data should be labeled within a short time, only expectations about prosodic regularities based on the textual representation of a turn (transliteration) can be considered.
- The specific characteristics of spontaneous speech have to be incorporated in the scheme.
- It should be independent of particular syntactic theories but at the same time, it should be compatible with syntactic theory in general.

According to these requirements, we defined nine different syntactic–prosodic boundary classes. For the experiments described below we only used the distinction between the main classes *clause boundary* (M3) and *no clause boundary* (M0) that are for the time being robust enough and most relevant for the linguistic analysis in VERBMOBIL. Nevertheless, the distinction of the nine classes was considered to be useful, because their automatic discrimination might become important in the future. Furthermore, these boundary classes might be marked prosodically in a different way; for a detailed discussion of the M labels see [3, 4]. 7286 VERBMOBIL turns (17 hours of speech, 149514 word tokens counting word fragments but not non–verbals) were labeled by one person in about four months.

3.4 Classification of Prosodic Events

Given a feature set and a training database of hand labeled classes to be recognized, pattern recognition offers a large variety of classifiers for supervised learning. Here we will only report results obtained with MLPs which turned out to be superior compared to Gaussian distribution classifiers and polynomial classifiers in similar investigations [16, 5]. Different MLP topologies were analyzed for the various classification problems. As training procedure the Quickpropagation algorithm [13] with the sigmoid activation function was used. Experiments were performed with different feature sets. In any case the MLP had as many input nodes as the dimension of the specific feature vector and one output node for each of the classes to be recognized. During training the desired output for each of the feature vectors is set to one for the node corresponding to the reference label; the other one is set to zero. With this method in theory the MLP estimates a posteriori probabilities for the classes under consideration. In order to balance for the a priori probabilities of the different classes, during training the MLP was presented with an equal number of feature vectors from each class.

Table 1 shows the confusion matrix for the two class problem B vs. ¬B.

3.5 Improving the Classification Results with Stochastic Language Models

Let w_i be a word out of a vocabulary where i denotes the position in the utterance; v_i denotes a symbol out of a predefined set V of prosodic symbols. These can be for

Table 1. Confusion matrix for the classification of prosodic boundaries (¬B|B)

reference	#	B	¬B
B	165	84.8	15.2
¬B	1284	11.2	88.8

example {B, ¬B}, {¬A, A}, or a combination of both {¬B¬A, ¬BA, B¬A, BA} depending on the specific classification task. For example, $v_i = $ B means that the ith word in an utterance is succeeded by a full intonational boundary.

Ideally one would like to model the following a priori probability

$$P(w_1 v_1 w_2 v_2 \ldots w_m v_m)$$

which is the probability for strings, where words and prosodic labels alternate (m is the number of words in the utterance).

In [18] we used a language model similar to this one to score chains containing words and prosodic labels. In the following, we are interested in the recognition of prosodic classes given a (partial) word chain (which in the case of WHGs is obtained from the best path through the word hypothesis to be classified). When determining the appropriate label to substitute v_i the labels at positions v_{i-k} and v_{i+k} are not known ($k = 1, 2, \ldots$). Thus, we used the following probabilities:

$$P(w_1 \ldots w_i v_i w_{i+1} \ldots w_m) = P_l P_v P_r \tag{1}$$

where P_l, P_v, and P_r are defined as follows:

$$
\begin{aligned}
P_l &= P(w_1)P(w_2|w_1) \cdot \ldots \cdot P(w_i|w_1 \ldots w_{i-1}) &\quad (2)\\
P_v &= P(v_i|w_1 \ldots w_i) &\quad (3)\\
P_r &= P(w_{i+1}|w_1 \ldots w_i v_i) \\
&\quad \cdot \ldots \cdot P(w_m|w_1 \ldots w_i v_i w_{i+1} \ldots w_{m-1}) &\quad (4)
\end{aligned}
$$

Terms like $w_1 \ldots w_i$ in $P(v_i|w_1 \ldots w_i)$ are called *history*. As usual in stochastic language modelling, the history has to be restricted to a certain length [24]. The stochastic language model approach we used is the so called *polygram* [31], where the histories have variable length depending on the available training data. A maximum history length can be defined.

For each word boundary in the training corpus, a sufficient number of context words (according to the maximum history length) and the corresponding prosodic reference label are extracted from the text corpora and used to estimate the probabilities of the equations above by counting the frequencies (maximum likelihood estimation) as is usually done when training stochastic language models. In fact, the above probabilities are not used, rather the words are put into 150 categories.

We used the thus trained polygrams for the classification of prosodic labels. Given a word chain $w_1 \ldots w_i \ldots w_m$, the appropriate prosodic class v_i^* is determined by maximizing the probability of equation 1:

$$v_i^* = \operatorname*{argmax}_{v_i \in V} P(w_1 \ldots w_i v_i w_{i+1} \ldots w_m)$$

Note, that the probability P_l is independent of v_i (equation 2). Thus this maximization (and v_i^*) is independent from P_l. Note also, that v_i^* does not only depend on the left context (probability P_v, equation 3) but also on the words succeeding the word w_i (probability P_r, equation 4). In practice, the context is restricted to the maximum history length h_l used during training of the polygram:

$$v_i^* = \operatorname*{argmax}_{v_i \in V} P(w_{i-h_l} \ldots w_i v_i w_{i+1} \ldots w_{i+h_l}) \tag{5}$$

Table 2 shows the recognition rates for the two class problem M3 vs. M0.

3.6 Prosodic scoring of WHGs

A WHG is a directed acyclic graph [26]. Each edge corresponds to a word hypothesis which has attached to it its acoustic probability, its first and last time frame, and a time alignment of the underlying phoneme sequence. The graph has a single start node (corresponding to time frame 1) and a single end node (the last time frame in the signal). Each path through the graph from the start to the end node forms a sentence hypothesis. Each edge in the graph lies on at least one such path. In the following the term *neighbors* of a word hypothesis in a graph refers to all its predecessor and successor edges.

With *prosodic scoring of WHGs* we mean in fact the annotation of the word hypotheses in the graph with the probabilities for the different prosodic classes. These probabilities are used by the other modules during linguistic analysis, e.g. by the parser in the syntax module. Note, that also in the case of phrase boundaries we do not compute the probability for a prosodic boundary located at a certain node in the graph, but for each of the word hypotheses in the graph the probability for a boundary being after this word is computed. This is important, since the acoustic–prosodic features also include the duration of syllable nuclei; these are most robustly obtained from the time alignment of the phoneme sequence underlying a word hypothesis computed with the word recognizer, and these durations have to be normalized with respect to the intrinsic phoneme duration.

The following steps have to be conducted for each word hypothesis w_i:

1. Determine recursively appropriate neighbors of the word hypothesis until a word chain $w_{i-k} \ldots w_{i+l}$ is built which contains enough syllables to compute the acoustic–prosodic feature vector and where $k \geq h_l, l \geq h_l$.
2. For each $v_i \in V$ and for each syllable s in the word w_i compute the probabilities

$$P_{v_i} = \frac{Q_{v_i}}{\sum_{v_i \in V} Q_{v_i}} \qquad \text{where}$$

$$Q_{v_i} = P(v_i|\mathbf{c}_{is})P^\xi(w_{i-h_l} \ldots w_i v_i w_{i+1} \ldots w_{i+h_l})$$

Note, that in the case of boundaries only the word final syllable is considered.

c_{is} denotes the acoustic–prosodic feature vector, ξ is a weight for the combination of the acoustic–prosodic model probability $P(v_i|c_{is})$ which is computed by the MLP trained with B boundaries and the prosodic–syntactic language model probability which is computed by the polygram trained with M boundaries. The value of ξ is determined empirically on a validation set.

In the current implementation we just select that hypothesis as the "appropriate" neighbor of w_i, which is most probable according to the acoustic model. Note, that this is suboptimal, because the context words in a path through the WHG may differ from the spoken words. An exact solution would be a weighted sum of all probabilities P_{v_i} computed on the basis of all the possible contexts. However, this does not seem to be feasible under real–time constraints. As a trade–off the neighbors could be determined on the basis of the best of the paths through the graph which contain the hypothesis w_i. The best path could be determined efficiently with dynamic programming using acoustic and language model scores.

The evaluation of the prosodic scores only makes sense for the WHGs which contain the spoken word chain:

1. Score the WHG prosodically with the probabilities P_{v_i}. Note, that this is based on the best paths through the hypotheses which may be different from the spoken word chain.
2. For each word contained in the (best) path corresponding to the spoken word chain determine the prosodic class with the largest probability P_{v_i} (i.e. the recognized class).
3. Compare the recognized classes with the reference labels and determine the recognition error.

In Table 2 the recognition rates for different experiments on 160 WHGs are presented. Each WHG contained all the spoken words, the density of the graphs was about 13 words per spoken word, for details see [17]. LM_h denotes the polygram–classification as described in Section 3.5, where h specifies the maximum context allowed during training of the polygram. The column 'word chain' refers to experiments conducted on the time alignment of the spoken word chain, i.e. with optimal context. Keep in mind that the MLP is trained on perceptual Band ¬Bclasses and evaluated on prosodic syntactic M3 and M0 classes.

In the next section we will see, how the prosodic information is used during linguistic analysis.

4. The Use of Prosodic Information

4.1 Prosody and Syntax — Interaction with the TUG–Grammar

In this subsection, we describe the interaction of prosody with the syntax–module developed by Siemens (Munich). The adaptations of the syntax–module were done by Siemens and are described in [19]. For the interaction with the VERBMO-BIL syntax–module developed by IBM (Heidelberg) cf. [1, 2]. In the module de-

Table 2. Recognition rates ($\mathcal{RR}$) for the classification·of syntactic–prosodic boundaries (M3 | M0) on 160 WHG, which contain the spoken words. The averages of the class-dependent recognition rates ($\mathcal{RR}_{\overline{C}}$) are given in parenthesis.

	word chain		WHG	
	$\mathcal{RR}$	$\mathcal{RR}_{\overline{C}}$	$\mathcal{RR}$	$\mathcal{RR}_{\overline{C}}$
MLP	89.3	(82.5)	77.5	(78.0)
LM$_2$	91.0	(77.6)	90.6	(76.5)
LM$_3$	93.5	(84.8)	91.9	(81.3)
MLP + LM$_3$	94.0	(90.0)	92.2	(86.6)

scribed here, we use a Trace and Unification Grammar (TUG) [7] and a modification of the parsing algorithm of Tomita [33]. The basis of a TUG is a context free grammar augmented with PATR-II-style feature equations. The Tomita parser uses a graph-structured stack as central data structure [32]. After processing word w_i the top nodes of this stack keep track of all partial derivations for $w_1 \ldots w_i$. The parsing–scheme uses an A^*–search and is able to combine different knowledge sources in order to find the optimal word sequence in a WHG with respect to these knowledge sources. It is presented in [30].

When searching the WHG, partial sentence hypotheses are organized as a tree. A graph-structured stack of the Tomita parser is associated with each node. In the search an agenda of score–ranked orders to extend a partial sentence hypothesis (hypo$_i$ = hypo($w_1, \ldots, w_i$)) by a word w_{i+1} or a symbol for a clause boundary, which we will call PSCB (prosodic–syntactic clause boundary), respectively, is processed. The best entry is taken; if the associated graph–structured stack of the parser can be extended by w_{i+1} or by PSCB, respectively, new orders are inserted in the agenda for combining the extended hypothesis hypo$_{i+1}$ with the words, which then follow in the graph, and, furthermore, the hypothesis hypo$_{i+1}$ is extended by the PSCB symbol. Otherwise, no entries will be inserted. Thus, the parser makes hard decisions and rejects hypotheses which are ungrammatical.

The acoustic, prosodic and trigram knowledge sources deliver scores which are combined to give the score for an entry of the agenda. In the case the hypothesis $hypo_i$ is extended by a word w_{i+1}, the score of the resulting hypothesis is computed by

$$
\begin{aligned}
score(hypo_{i+1}) \quad = \quad & score(hypo_i) \\
& + acoustic\_score(w_{i+1}) \\
& + \alpha \cdot trigram\_score(w_{i-1}, w_i, w_{i+1}) \\
& + \beta \cdot prosodic\_score(w_{i+1}, B) \\
& + 'score\ of\ optimal\ continuation' \,,
\end{aligned}
$$

where B can be PSCB or ¬PSCB. $prosodic\_score(w, \text{PSCB})$ is a 'good' score if the prosodic classifier detected M3 after word w with a high probability, a 'bad' score otherwise. $prosodic\_score(w, \neg\text{PSCB})$ is 'good' if the prosodic classifier shows high evidence for M0 after word w, 'bad' otherwise.

Table 3. Grammar 1 for multiple phrase utterances

(rule1)	input	→	phrase	input .
(rule2)	phrase	→	s	PSCB .
(rule3)	phrase	→	s_ell	PSCB .
(rule4)	phrase	→	np	PSCB .
(rule5)	phrase	→	excl	PSCB .
(rule6)	phrase	→	excl	.

The weights α and β are determined heuristically. Prior to parsing, a Viterbi–like backward pass approximates the scores of optimal continuations of partial sentence hypotheses (A^*–search). After a certain time has elapsed, the search is abandoned. With these scoring functions, hard decisions about the positions of clause boundaries are only made by the grammar but not by the prosody module. If the grammar rules are ambiguous given a specific hypothesis $hypo_i$, the prosodic score guides the search by ranking the agenda.

In order to make use of the prosodic information, the grammar had to be slightly modified. The best results were achieved by a grammar that neatly designed the occurrence of PSCBs between the multiple phrases of the utterance. A context–free grammar for spontaneous speech has to allow for a variety of possible input phrases following each other in a single utterance, cf. (rule1) in Table 3. Among those count normal sentences (rule2), sentences with topic ellipsis (rule3), elliptical phrases like PPs or NPs (rule4), or pre-sentential ('excl') particles (rule5 and rule6). Those phrases were classified as to whether they require an *obligatory* or *optional* PSCB behind them. The grammar fragment in Table 3 says that the phrases s, s_ell and np require an obligatory PSCB behind them, whereas excl(amative) may also attach immediately to the succeeding phrase (rule 6). The segmentation of utterances according to a grammar like in Table 3 is of relevance to the text understanding components that follow the syntactic analysis, cf. the following two examples which differ w.r.t. the attachment of the 'exclamative' (pre-sentential) particle *ja*. In the first example it is followed immediately by a sentence (rule6), whereas in the second it is separated by a PSCB from the following sentence (rule5). Semantic analysis or dialog can make use of these different rules. The exclamative particle in example (1) might be identified as introduction, in example (2) it might be interpreted as affirmation.

(1) [ja,also,bei,mir,geht,prinzipiell,jeder,Montag,und,
jeder, Donnerstag,PSCB]
Well, as far as I'm concerned, in principle every Monday or Thursday is possible.

(2) [ja,PSCB,das,pa"st,mir,Dienstag,PSCB,ist,der,
f"unfzehnte,PSCB]
Yes. This Tuesday, that suits me. That is the fifteenth.

Table 4. Grammar 2 for multiple phrase utterances

(rule 7)	`input`	$\rightarrow$	`phrase` ,	`PSCB` ,	`input`	.
(rule 8)	`phrase`	$\rightarrow$	`s` .			
(rule 8)	`phrase`	$\rightarrow$	`s_ell` .			
(rule 9)	`phrase`	$\rightarrow$	`np` .			
(rule 10)	`phrase`	$\rightarrow$	`excl` .			

Table 5. Parsing statistics for 594 WHGs

	with PSCBs	without PSCBs
# successful analyses	359	368
⊘# syntactic readings	5.6	137.7
⊘ parse time (secs)	3.1	38.6

The occurrence of the second PSCB in example (2) does not mirror the intention of the speaker: Here the PSCB divides the subject *Dienstag* from its matrix clause *ist der fünfzehnte*. A hesitation in the input that did not get detected as false alarm might be responsible for this. However (2) is a syntactically correct segmentation since a grammar for spoken language has to allow for topic ellipsis and the phrase *ist der fünfzehnte* constitutes a correct sentence according to (rule 3). The grammar therefore retrieves the interpretation for this lattice as indicated by the English translation.[1]

In experiments using a preliminary version of the sub–grammars for the individual types of phrases, we compared the grammar explained above with a grammar that *obligatorily* required a PSCB behind every input phrase, see Table 4.

With the grammar shown in Table 3, 149 WHGs could successfully be analyzed; with the one given in Table 4, only 79 WHGs were analyzed. This indicates that often the prosody module computes a high score for ¬PSCB after exclamative particles so that parsing fails if a PSCB is obligatorily required as in the grammar of Table 4.

With an improved version of the grammar for the individual phrases, we repeated the experiments using the grammar of Table 3 and compared them with the parsing results using a grammar *without* PSCBs. For the latter, we took the category PSCB out of the grammar and allowed all input phrases to adjoin recursively to each other. The graphs were parsed without taking notice of the prosodic PSCB information contained in the lattice. In this case, the number of readings increases and the efficiency decreases drastically, cf. Table 5. The statistics show that on the average, the number of readings decreases by 96% when prosodic information is

---

[1] For this word chain, it would make no difference for the text understanding component, whether the PSCB is before or after *Dienstag*. Actually, the spoken word chain is: *Ja, das paßt. Nur Dienstag ist der fünfzehnte.* and the dialog goes like this: A: *What about Tuesday the sixteenth?* B: *Yes. That's ok. But Tuesday is the fifteenth.* A: *Sorry. Then let's say Wednesday the sixteenth.* B: *OK. Fine.* B thus only confirms *the sixteenth*, but not *Tuesday*.

used, and the parse time drops by 92%. If the lattice parser does not pay attention to the information on possible PSCBs, the grammar has to determine by itself where the phrase boundaries in the utterance might be. It may rely only on the coherence and completeness restrictions of the verbs that occur somewhere in the utterance. These restrictions are furthermore softened by topic ellipsis, etc. Any simple utterance like *Er kommt morgen* results therefore in a lot of possible segmentations, see Table 6.

Table 6. Syntactically possible segmentations

[er,kommt,morgen]	*He comes tomorrow.*
[er] , [kommt,morgen]	*He? Comes tomorrow!*
[er kommt] , [morgen]	*He comes. Tomorrow!*
[er] , [kommt] , [morgen]	*He? Comes! Tomorrow.*

The reason why 9 WHGs (i.e. 2%) could not be analyzed with the use of prosody is that the search space is explored differently and that a preset time limit has been reached before the analysis succeeded. However, this small number of non-analyzable WHGs is negligible considering the fact that without prosody, the average real–time factor is 6.1 for the parsing. With prosodic information the real–time factor drops to 0.5; the real–time factor for the computation of prosodic information is 1.0 (with WHGs of about 10 hypotheses per spoken word).

Empty categories are an even more serious problem. They are used by the grammar in order to deal with verb movement and topicalisation in German. The binding of these empty categories has to be checked inside a single input phrase, i.e., the main sentence. No movement across phrase boundaries is allowed. Now, whenever a PSCB signals the occurrence of a boundary, the parser checks whether all binding conditions are satisfied and accepts or rejects the path that was found so far. This mechanism works efficiently in the case prosodic information is used. For the grammar without PSCBs, no signal where to check the binding restrictions is available. Therefore, the uncertainty about segmentation of multiple phrase utterances led to indefinite parsing time for some of the lattices in the corpus. Those lattices were analyzed correctly with PSCBs.

In the following section, we will indicate how prosody is used by other linguistic modules. The use of prosodic information can be demonstrated but so far no systematic tests have been conducted.

4.2 Prosody and the Other Linguistic Modules

Semantic construction:

The VERBMOBIL semantic module receives a parse tree, the underlying word chain and the prosodic scores for accentuation from the syntax module. Based on these, underspecified *Discourse Representation Structures* (DRS) [15, 9] are created. These yield assertions, representing the direct meaning of a sentence, and presuppositions. If several DRSs are plausible due to ambiguities, accent information is used to rule out the wrong DRS. Context information might also be used to disambiguate the interpretation, however, prosodic information can be utilized at much lower cost [8]. This use of prosody can be illustrated by the following examples from the VERBMOBIL corpus where the meaning of both sentences is the same. However, the position of the primary accent changes the scope and thereby the presupposition of the utterances, which results in a different translation of the particle *noch* (*still, another*).

(3) *"Dann müssen wir noch einen __Termin__ ausmachen."*
    "Then we still have to fix a __date__."

(4) *"Dann müssen wir __noch__ einen Termin ausmachen."*
    "Then we have to fix __another__ date."

Dialog processing:
One of the tasks of the dialog module [27] is to keep track of the state of the dialog in terms of dialog acts. Dialog act recognition is done by statistical classifiers. Dialog acts are, e.g., *greeting, confirmation of a date, suggestion of a place*. In VERBMOBIL, a turn of a user can consist of more than one dialog act. Currently, the processing is done in two steps: First, the best path in the WHG (extracted by a Viterbi search using acoustic and trigram scores) is segmented into dialog act units. Second, these units are classified into dialog acts. For the segmentation into dialog acts, we use the same prosodic clause boundary information as used by the syntax modules. Due to less amount of training data, the use of a different classifier trained directly on dialog act boundaries did not improve the recognition rate. Further details can be found in [17, 23].

Transfer:
The transfer module of the VERBMOBIL system translates DRSs representing the semantic information underlying the utterance into DRSs corresponding to English sentences [11]. This task might involve pragmatic analysis and disambiguation which is partly done by the semantic evaluation module. The transfer module uses accent and sentence mood information for a few tasks. The sentence mood information is used to distinguish between questions and non–questions if grammatical indicators are missing; e.g., questions and declaratives with topic elision can have an identical word order. The accent information disambiguates mainly the interpretation of particles. In the following examples, the same word chain has different meanings depending on whether the accent is on *schon* or on *finde*. For further use of prosodic information in the VERBMOBIL transfer module cf. [29].

(5) *"Finde ich __schon__."*   *"I really believe that."*
(6) *"__Finde__ ich schon."*   *"I'll find it certainly."*

Speech synthesis:
For a better user acceptance, the synthesized output of a translation system should be adapted to the voice of the original speaker (especially in a multi–party scenario). With respect to prosody this means that parameters like the pitch level and the speaking rate should be adapted. So far, the speech synthesis of the VERBMOBIL

system is only switched to a male or a female voice according to the F0 contour of the original user utterance.

5. Concluding Remarks

Prosodic information is known to play a major role in human speech understanding; a growing number of research projects within the last ten years dealt with this topic. The German speech–to–speech translation system VERBMOBIL is, however, the first complete ASU system where prosody is used. Currently, this use is mainly confined to the prosodic scoring of WHGs. We have shown that by that, a substantial speed up of parse time and a substantial reduction of syntactic readings could be achieved. Other applications are, e.g., the prosodic marking of accents (center of information for dialog act classification), and the prosodic marking of emotions, e.g., neutral state vs. arousal and anger that might trigger the reaction of the system. These are, amongst others, topics that are currently being addressed within the second phase of the VERBMOBIL project lasting from 1997 to 2000.

Although it might be possible that segmentation is really the most important contribution of prosody to speech understanding, we are still at the very beginning of an integration of prosody into automatic speech understanding systems. Further improvements are therefore very likely.

Acknowledgements. This work was funded by the German Federal Ministry of Education, Science, Research and Technology (*BMBF*) in the framework of the VERBMOBIL Project under Grants 01 IV 102 H/0 and 01 IV 102 F/4. The responsibility for the contents of this study lies with the authors. We wish to thank all VERBMOBIL partners who integrated the prosodic information into their analysis modules.

References

[1] A. Batliner, A. Feldhaus, S. Geissler, A. Kießling, T. Kiss, R. Kompe, and E. Nöth. Integrating Syntactic and Prosodic Information for the Efficient Detection of Empty Categories. In *Proc. of the Int. Conf. on Computational Linguistics*, volume 1, pages 71–76, Kopenhagen, 1996.

[2] A. Batliner, A. Feldhaus, S. Geißler, T. Kiss, R. Kompe, and E. Nöth. Prosody, Empty Categories and Parsing — A Success Story. In *Int. Conf. on Spoken Language Processing*, volume 2, pages 1169–1172, Philadelphia, 1996.

[3] A. Batliner, R. Kompe, A. Kießling, M. Mast, and E. Nöth. All about Ms and Is, not to forget As, and a comparison with Bs and Ss and Ds. Towards a syntactic–prosodic labeling system for large spontaneous speech data bases. Verbmobil Memo 102, 1996.

[4] A. Batliner, R. Kompe, A. Kießling, H. Niemann, and E. Nöth. Syntactic–prosodic Labelling of Large Spontaneous Speech Data–bases. In *Int. Conf. on Spoken Language Processing*, volume 3, pages 1720–1723, Philadelphia, 1996.

[5] A. Batliner, R. Kompe, A. Kießling, E. Nöth, H. Niemann, and U. Kilian. The Prosodic Marking of Phrase Boundaries: Expectations and Results. In A. Rubio Ayuso and J. López Soler, editors, *Speech Recognition and Coding. New Advances and Trends*, volume 147 of *NATO ASI Series F*, pages 325–328. Springer, Berlin, 1995.

[6] H. Block. The Language Components in Verbmobil. In *Proc. Int. Conf. on Acoustics, Speech and Signal Processing*, volume 79–82, page 1, München, 1997.

[7] H. Block and S. Schachtl. Trace & Unification Grammar. In *Proc. of the Int. Conf. on Computational Linguistics*, volume 1, pages 87–93, Nantes, 1992.

[8] J. Bos. Personal communication, July 1996.

[9] J. Bos, B. Gambäck, C. Lieske, Y. Mori, M. Pinkal, and K. Worm. Compositional Semantics in Verbmobil. In *Proc. of the Int. Conf. on Computational Linguistics*, volume 1, pages 131–136, Kopenhagen, 1996.

[10] T. Bub and J. Schwinn. Verbmobil: The Evolution of a Complex Large Speech-to-Speech Translation System. In *Int. Conf. on Spoken Language Processing*, volume 4, pages 1026–1029, Philadelphia, 1996.

[11] K. Eberle. Disambiguation by Information Structure in DRT. In *Proc. of the Int. Conf. on Computational Linguistics*, volume 1, pages 334–339, Kopenhagen, 1996.

[12] W. Eckert, E. Nöth, H. Niemann, and E. Schukat-Talamazzini. Real Users Behave Weird — Experiences made collecting large Human–Machine–Dialog Corpora. In P. Dalsgaard, L. Larsen, L. Boves, and I. Thomsen, editors, *Proc. of the ESCA Tutorial and Research Workshop on Spoken Dialogue Systems*, pages 193–196, Vigsø, Denmark, 1995. ESCA.

[13] S. Fahlman. An Empirical Study of Learning Speed in Back–Propagation Networks. Technical Report CMU-CS-88–62, Carnegie Mellon University, Pittsburgh, 1988.

[14] A. Feldhaus and T. Kiss. Kategoriale Etikettierung der Karlsruher Dialoge. Verbmobil Memo 94, 1995.

[15] H. Kamp and U. Reyle. *From Discourse to Logic and DRT; An Intorduction to Modeltheoretic Semantics of Natural Language*. Kluwer Academic Publishers, Dordrecht, 1993.

[16] A. Kießling, R. Kompe, H. Niemann, E. Nöth, and A. Batliner. Detection of Phrase Boundaries and Accents. In H. Niemann, R. de Mori, and G. Hanrieder, editors, *Progress and Prospects of Speech Research and Technology: Proc. of the CRIM / FORWISS Workshop*, PAI 1, pages 266–269, Sankt Augustin, 1994. Infix.

[17] R. Kompe. *Prosody in Speech Understanding Systems*. Lecture Notes for Artificial Intelligence. Springer–Verlag, Berlin, 1997.

[18] R. Kompe, A. Batliner, A. Kießling, U. Kilian, H. Niemann, E. Nöth, and P. Regel-Brietzmann. Automatic Classification of Prosodically Marked Phrase Boundaries in German. In *Proc. Int. Conf. on Acoustics, Speech and Signal Processing*, volume 2, pages 173–176, Adelaide, 1994.

[19] R. Kompe, A. Kießling, H. Niemann, E. Nöth, A. Batliner, S. Schachtl, T. Ruland, and H. Block. Improving Parsing of Spontaneous Speech with the Help of Prosodic Boundaries. In *Proc. Int. Conf. on Acoustics, Speech and Signal Processing*, volume 2, pages 811–814, München, 1997.

[20] R. Kompe, A. Kießling, H. Niemann, E. Nöth, E. Schukat-Talamazzini, A. Zottmann, and A. Batliner. Prosodic Scoring of Word Hypotheses Graphs. In *Proc. European Conf. on Speech Communication and Technology*, volume 2, pages 1333–1336, Madrid, 1995.

[21] W. Lea. Prosodic Aids to Speech Recognition. In W. Lea, editor, *Trends in Speech Recognition*, pages 166–205. Prentice–Hall Inc., Englewood Cliffs, New Jersey, 1980.

[22] I. Lehiste. *Suprasegmentals*. MIT Press, Cambridge, MA, 1970.

[23] M. Mast, R. Kompe, S. Harbeck, A. Kießling, H. Niemann, E. Nöth, and V. Warnke. Dialog Act Classification with the Help of Prosody. In *Int. Conf. on Spoken Language Processing*, volume 3, pages 1728–1731, Philadelphia, 1996.

[24] H. Ney, U. Essen, and R. Kneser. On Structuring Probabilistic Dependences on Stochastic Language Modelling. *Computer Speech & Language*, 8(1):1–38, 1994.

[25] E. Nöth. *Prosodische Information in der automatischen Spracherkennung — Berechnung und Anwendung*. Niemeyer, Tübingen, 1991.

[26] M. Oerder and H. Ney. Word Graphs: An Efficient Interface between Continuous Speech Recognition and Language Understanding. In *Proc. Int. Conf. on Acoustics, Speech and Signal Processing*, volume 2, pages 119–122, Minneapolis, MN, 1993.

[27] N. Reithinger, E. Maier, and J. Alexandersson. Treatment of Incomplete Dialogues in a Speech–to–speech Translation System. In P. Dalsgaard, L. Larsen, L. Boves, and I. Thomsen, editors, *Proc. of the ESCA Tutorial and Research Workshop on Spoken Dialogue Systems*, pages 33–36. ESCA, Vigsø, Denmark, 1995.

[28] M. Reyelt and A. Batliner. Ein Inventar prosodischer Etiketten für Verbmobil. Verbmobil Memo 33, 1994.

[29] B. Ripplinger and J. Alexandersson. Disambiguation and Translation of German Particles in Verbmobil, Verbmobil Memo 70, 1996.

[30] L. Schmid. Parsing Word Graphs Using a Linguistic Grammar and a Statistical Language Model. In *Proc. Int. Conf. on Acoustics, Speech and Signal Processing*, volume 2, pages 41–44, Adelaide, 1994.

[31] E. Schukat-Talamazzini, T. Kuhn, and H. Niemann. Speech Recognition for Spoken Dialogue Systems. In H. Niemann, R. de Mori, and G. Hanrieder, editors, *Progress and Prospects of Speech Research and Technology: Proc. of the CRIM / FORWISS Workshop*, PAI 1, pages 110–120, Sankt Augustin, 1994. Infix.

[32] N. Sikkel. *Parsing Schemata*. CIP-GEGEVENS KONINKLIJKE BIBLIOTHEEK, 1993.

[33] M. Tomita. *Efficient Parsing for Natural Language: A Fast Algorithm for Practical Systems*. Kluwer Academic Publishers, Dordrecht, 1986.

[34] H. Tropf. Spontansprachliche syntaktische Phänomene: Analyse eines Korpus aus der Domäne "Terminabsprache". Technical report, Siemens AG, ZFE ST SN 54, München, 1994.

[35] J. Vaissière. The Use of Prosodic Parameters in Automatic Speech Recognition. In H. Niemann, M. Lang, and G. Sagerer, editors, *Recent Advances in Speech Understanding and Dialog Systems*, volume 46 of *NATO ASI Series F*, pages 71–99. Springer–Verlag, Berlin, 1988.

[36] W. Wahlster. Presseerklärung zum Verbmobil-Forschungsprototypen am 25.10.1996 in München, 1996. http://www.dfki.uni-sb.de/verbmobil.

[37] W. Wahlster, T. Bub, and A. Waibel. Verbmobil: The Combination of Deep and Shallow Processing for Spontaneous Speech Translation. In *Proc. Int. Conf. on Acoustics, Speech and Signal Processing*, volume 1, pages 71–74, München, 1997.

[38] C. Wightman. *Automatic Detection of Prosodic Constituents*. PhD thesis, Boston University Graduate School, 1992.

Computational Models for Speech Production

Li Deng

Department of Electrical and Computer Engineering
University of Waterloo, Waterloo, Ontario, Canada N2L 3G1
email: deng@crg6.uwaterloo.ca

Summary. Major speech production models from speech science literature and a number of popular statistical "generative" models of speech used in speech technology are surveyed. Strengths and weaknesses of these two styles of speech models are analyzed, pointing to the need to integrate the respective strengths while eliminating the respective weaknesses. As an example, a statistical task-dynamic model of speech production is described, motivated by the original deterministic version of the model and targeted for integrated-multilingual speech recognition applications. Methods for model parameter learning (training) and for likelihood computation (recognition) are described based on statistical optimization principles integrated in neural network and dynamic system theories.

1. Introduction

In the past thirty years or so, the same physical entity of human speech has been studied and modeled using drastically different approaches undertaken by largely distinct communities of speech scientists and speech engineers. Models for how speech is generated in human speech production system developed by speech scientists typically have rich model structures [17, 24]. The structures have embodied detailed multi-level architectures which transform the high-level symbolic phonological construct to acoustic streams via intermediate stages of phonetic task specification, motor command generation, and articulation. However, these models are often underspecified due to 1) deterministic nature which does not accommodate random variabilities of speech and only weakly accommodates systematic variabilities; 2) lack of comprehensiveness in covering all classes of speech sounds (with some exceptions); and 3) lack of strong computational formalisms allowing for automatic model learning from data and for optimal choice of decision variables necessary for high-accuracy speech classifications.

On the other hand, models for how speech patterns are characterized by statistical generative mechanisms, which have been developed in speech technology, notably by speech recognition researchers, typically contain weak and poor structures. These models often simplistically assume direct, albeit statistical, correlates between phonological constructs and the surface acoustic properties. This causes recognizers built from these models to perform poorly for unconstrained tasks, and to break down easily when porting from one task domain to another, from one speaking mode to another, and from one language to another. Empirical system tuning and ever-more increasing data appear to be the only options for making the systems behave reasonably if no fundamental changes are made to the speech models underlying the recognizers. However, a distinct strength associated with these models is that they are equipped with powerful statistical formalisms, based solidly on mathematical optimization principles , and are suitable for implementation with flexible, integrated architecture. The precise mathematical framework, despite poor and simplistic model structure, gives rise to ease of automatic parameter learning (training) and to optimal decision rules for speech-class discrimination.

Logically, there are strong reasons to expect that a combination of the strengths of the above two styles of models, free from the respective weaknesses, will ultimately lead to superior speech recognition. In this paper, some versions of a statistical dynamic speech production model[1], with theoretical motivation, mathematical formulation, and procedures for parameter learning are described, aiming at achievement of such superiority.

2. Speech production models in science/technology literatures

In this section I will briefly survey major speech production models in science and technology literatures, with emphasis on drawing parallels and contrasts between these two styles of models developed by largely separate research communities. The purpose of scientific speech production models is to provide adequate representations to account for the conversion of a linguistic (mainly phonological) message to the actions of the production system and to the consequent articulatory and acoustic outputs. Critical issues addressed by these models are the serial-order problem, the degrees-of-freedom problem, and the related context-sensitivity or coarticulation problem in both articulatory and acoustic domains. The models can be roughly classified into categories of global and component models. Within the category of the global production models, the major classes (modified from the classification of [17] where a large number of references are listed) are: 1) Feedback-feedforward models, which enable both predictive and adaptive controls to operate, ensuring achievement of articulatory movement goals; 2) Motor program and generalized motor program (schema) models, which use pre-assembled forms of speech movement (called goals or targets) to regulate the production system; 3) Integrated motor program and feedback models, which combine the characteristics and modeling assumptions of 1) and 2); 4) Dynamic system and gestural patterning models, which employ coordinative structures with a small number of degrees of freedom to create families of functionally equivalent speech movement patterns, and use dynamic vocal tract constriction variables (in the "task" space) to directly define speech movement tasks or goals; 5) Models based on equilibrium point hypothesis, which use shifts of virtual target trajectories, arising from interactions among central neural commands, muscle properties, and external loads, in the articulatory space ("body" rather than "task" space) to control and produce speech movement; and finally 6) Connectionist models, which establish nonlinear units interconnected into a large network to functionally account for a number of prominent speech behaviors including serial order and coarticulation.

---

[1] Based mainly on the task-dynamic model originally developed by speech scientists [28] and on our earlier work on overlapping articulatory features [8].

Within the category of the component or subsystem models of speech production are the models for respiratory subsystem, laryngeal subsystem, and supra-laryngeal (vocal tract) subsystem. In addition, composite models have also been developed to integrate multiple subsystem models operating in parallel or in series [17, 24].

All of the scientifically motivated speech production models briefly surveyed above have focused mainly on explanatory power for speech behaviors (including articulatory movements and its relations to speech acoustics), and paid relatively minor attention to computation issues. Further, in developing and evaluating these models, comprehensiveness in covering speech classes is often seriously limited (e.g. CV, CVC, VCV sequences only). In stark contrast, speech models developed by technologists usually cover all classes of speech sounds, and computation issues are given a high priority of consideration with no exception. Another key character of the technology-motivated speech models is the rigorous statistical frameworks for model formulation which permit automatic learning of model parameters from realistic acoustic data of speech. On the negative side, however, the structures of these models tend to be oversimplified, often deviating significantly from the true stages in the human speech generation mechanisms which the scientific speech production models are aiming to account for. To show this, let us view the HMM (which forms the theoretical basis of the modern speech recognition technology) as a primitive (very inaccurate) generative model of speech. To show such inaccuracy, we simply note that the unimodal Gaussian HMM generates its sample paths which are piecewise constant trajectories embedded in temporally uncorrelated Gaussian noise, and, since variances of the noise are estimated from time-independent (except with the HMM state-bound) speech data from all speakers, the model could freely allow generation of speech from different speakers over as short as every 10 msec.[2]

A simplified hierarchy of statistical generative or production models of speech developed in speech technology is briefly reviewed here. Under the root node of conventional HMM there are two main classes of its extended or generalized models: 1) nonstationary-state HMM[3] whose sample paths are piecewise, explicitly defined stochastic trajectories (e.g. [3, 14, 11]);   2) multi-region dynamic system model whose sample paths are piecewise, recursively defined stochastic trajectories (e.g. [10]). [4] The parametric form of Class-1 models typically uses polynomials to constrain the trajectories, with the standard HMM as a special case when the polynomial order is set to zero. This model can be further divided according to whether the polynomial trajectories or trends can be observed directly [5] or the trends are hidden due to assumed randomness in the polynomial coefficients. For the latter, further

---

[2] For the mixture Gaussian HMM, the sample paths are erratic and highly irregular due to lack of temporal constraints forcing fixed mixture component within each HMM state; this deviates significantly from speech data coming from heterogeneous sources such as from multiple speakers and multiple speech collection channels.

[3] also called trended HMM, segmental HMM, stochastic trajectory model, etc., with slight variations in technical detail according to whether the parameters defining the trend functions are random or not.

[4] Intimate relations and implementational differences between the explicitly defined and recursively defined time series models are discussed in [19].

[5] This happens when the model parameters (i.e. polynomial coefficients) are deterministic.

classification gives discrete and continuous mixtures of trends depending on the assumed discrete or continuous nature of the model parameter distribution[6, 16, 12]. For Class-2, recursively defined trajectory models, the earlier linear model aiming at dynamic modeling at the acoustic level [10, 23] has been generalized to nonlinear models taking into account detailed mechanisms of speech production. Subclasses of such nonlinear dynamic models are 1) articulatory dynamic model; and 2) task-dynamic model. They differ from each other by distinct objects of dynamic modeling, one at the level of biomechanic articulators (body space) and the other at the level of more abstract task variables (task space)[6] [1, 2, 9, 4, 5, 25]. Depending on the assumptions about whether the dynamic model parameters are deterministic or random, and whether these parameters are allowed to change within phonological state boundaries, further subclasses can be categorized (see details in the following sections).

3. Derivation of discrete-time version of statistical task-dynamic model

In this section, a discrete-time statistical task-dynamic model , which is implementable and trainable for use in speech recognition, is derived from the original deterministic, continuous-time task-dynamic model well established in speech science literature [28]. Starting with the original model but incorporating random noise $\mathbf{w}(t)$:

$$\frac{d^2\mathbf{z}(t)}{dt^2} + 2\mathbf{S}(t)\frac{d\mathbf{z}(t)}{dt} + \mathbf{S}^2(t)(\mathbf{z}(t) - \mathbf{Z}^0(t)) = \mathbf{w}(t),$$

where $\mathbf{S}^2$ is normalized, gesture-dependent stiffness parameter (which controls fast or slow movement of tract variable $\mathbf{z}(t)$), and Z^0 is gesture-dependent point-attractor parameter of the dynamical system (which controls the target and hence direction of the movement). Here, for generality, we assume that the model parameters are (slowly) time-varying.

Rewrite the above into a canonical form (where $\dot{\mathbf{z}}(t) = \frac{d\mathbf{z}(t)}{dt}$):

$$\frac{d}{dt}\begin{pmatrix} \mathbf{z}(t) \\ \dot{\mathbf{z}}(t) \end{pmatrix} = \begin{pmatrix} 0 & 1 \\ -\mathbf{S}^2(t) & -2\mathbf{S}(t) \end{pmatrix}\begin{pmatrix} \mathbf{z}(t) \\ \dot{\mathbf{z}}(t) \end{pmatrix} + \begin{pmatrix} 0 \\ \mathbf{S}^2(t)\mathbf{Z}^0(t) \end{pmatrix} + \begin{pmatrix} 0 \\ \mathbf{w}(t) \end{pmatrix}$$

This, in matrix form, is:

$$\frac{d}{dt}Z(t) = F(t)Z(t) - F(t)T(t) + W(t),$$

where composite state is defined by

---

[6] They also have distinct origins, one from scientific speech production models based on equilibrium point hypothesis, the other from the deterministic version of the task-dynamic model.

$$Z(t) \equiv \begin{pmatrix} \mathbf{z}(t) \\ \dot{\mathbf{z}}(t) \end{pmatrix},$$

system matrix by

$$F(t) \equiv \begin{pmatrix} 0 & 1 \\ -\mathbf{S}^2(t) & -2\mathbf{S}(t) \end{pmatrix},$$

and attractor vector for the system dynamics by

$$T(t) \equiv -F^{-1}(t) \begin{pmatrix} 0 \\ \mathbf{S}^2(t)\mathbf{Z}^0(t) \end{pmatrix}.$$

Explicit solution to the above task-dynamic equation is [21]:

$$Z(t) = \Phi(t, t_0)Z(t_0) + \int_{t_0}^{t} \Phi(t, \tau)[-F(\tau)T(\tau) + W(\tau)]d\tau,$$

where $\Phi(t, t_0)$ (state transition matrix) is the solution to the following matrix homogeneous differential equation: [7]

$$\dot{\Phi}(t, \tau) = F(t)\Phi(t, \tau); \quad init.\ cond. : \Phi(t, t) = I.$$

Setting $t_0 = t_k, t = t_{k+1}$, we have

$$Z(t_{k+1}) \approx \Phi(t_{k+1}, t_k)Z(t_k) \quad - \quad \left[\int_{t_k}^{t_{k+1}} \Phi(t_{k+1}, \tau)F(\tau)d\tau\right] T(t_k)$$
$$+ \quad \int_{t_k}^{t_{k+1}} \Phi(t_{k+1}, \tau)W(\tau)d\tau,$$

which leads to the discrete-time form of the task-dynamic state equation:

$$Z(k + 1) = \Phi(k)Z(k) + \Psi(k)T(k) + W_d(k), \tag{1}$$

where

$$\Phi(k) \approx \Phi(t_{k+1}, t_k) = exp(F_k \Delta t), \quad \Delta t \equiv t_{k+1} - t_k$$
$$\Psi(k) \approx - \left[\int_{t_k}^{t_{k+1}} \Phi(t_{k+1}, \tau)F(\tau)d\tau\right] \approx - \int_{t_k}^{t_{k+1}} exp[F_k(t_{k+1} - \tau)]F(\tau)d\tau$$
$$\approx -F_k\ exp(F_k\ t_{k+1}) \int_{t_k}^{t_{k+1}} exp[-F_k\tau]d\tau = I - exp(F_k\Delta t) = I - \Phi(k),$$

and $W_d(k)$ is discrete-time white Gaussian sequence which is statistically equivalent through its first and second moments to $\int_{t_k}^{t_{k+1}} \Phi(t_{k+1}, \tau)W(\tau)d\tau$.

For a speech recognizer which has only acoustic data sequences at its disposal, the dynamic associated with task variables $Z(k)$ described in Eqn.(1) is a hidden or unobservable process. Following the treatment of task-dynamic model which uses intermediate model-articulator variables $\mathbf{x}(k)$ to link the task variables $Z(k)$ to acoustic variables $O(k)$ via static nonlinear functions, $O = \mathcal{O}(\mathbf{x})$ and $Z = \mathcal{Z}(\mathbf{x})$,

---

[7] The solution can be written in matrix exponential form: $\Phi(t, \tau) = exp[(t - \tau)F_k]$ if $F(t) = F_k$ for $t_k \leq t \leq t_{k+1}$)

we treat the hidden task-variable dynamic as observed through noisy (i.i.d. noise $V(k)$) nonlinear relation $\mathbf{h}(\cdot)$ between task-variable $Z(k)$ and acoustic observation $O(k)$:

$$O(k) = \mathcal{O}(\mathbf{x}(k)) + V(k) = \mathcal{O}[\mathcal{Z}^{-1}(Z(k))] + V(k) = \mathbf{h}[Z(k)] + V(k). \quad (2)$$

The above global nonlinearity has been implemented numerically in the deterministic version of task-dynamic model [28, 20] by geometric relationships in an improved version of Mermelstein-type articulatory model ($Z = \mathcal{Z}(\mathbf{x})$) together with a configurable articulatory synthesizer ($O = \mathcal{O}(\mathbf{x})$) [27]. For intended use in statistical speech recognition, we need to parameterize this nonlinearity with trainable sets of parameters and with the numerical simulation only serving as parameter initialization. While many possibilities exist for the parameterization based on well established statistical and neural-network techniques, in this tutorial I will describe a straightforward method based on use of Multi-Layer Perceptron (MLP) neural network which functions as a universal multidimensional functional approximation device [15]. Let i denote the element index of the acoustic-variable vector O as output of MLP, and l index of the task-variable vector Z as input to MLP. Then, each vector component of the MLP output can be parameterized by [8]

$$h_i[Z] = \sum_j W_{ij} x_j = \sum_j W_{ij} g(\sum_l w_{jl} g(Z_l)), \quad i = 1, 2, \dots, I. \quad (3)$$

The observation equation (2) can then be written in the parameterized form:

$$O_i(k) = \sum_j W_{ij} g(\sum_l w_{jl} g(Z_l)) + V_i(k), \quad i = 1, 2, \dots, I. \quad (4)$$

4. Algorithms for learning task-dynamic model parameters and for likelihood computation

The deterministic version of the task-dynamic model [28], although well developed and tested for use as an effective research tool in accounting for and in understanding dynamic behaviors of the speech process, is unlikely to be directly useful for engineering applications in speech recognition. Its lack of statistical structure does not allow the model to effectively capture variabilities, either systematic or random, in the observed speech data. Within the modeling framework of [28], there also seem to be no principled ways to devise an optimal decision rule for speech classification by matching the model's output to speech data. In contrast, the statistical

---

[8] In this tutorial, the MLP is further simplified to contain only one hidden layer. The output (acoustic variables $O_1, O_2, \dots$) layer is linear with weights W_{ij}; input (task variables $Z_1, Z_2, \dots$) and hidden layers are sigmoid nonlinear: $g(v) = 1/[1 + exp(-v)]$, with weights w_{jl}. In actual implementation, since the hidden layer $\mathbf{x}$ is intended to represent model-articulator variables, and since the relationship between $\mathbf{x}$ and Z and that between $\mathbf{x}$ and O are known to be strongly nonlinear, two more hidden layers are placed between the three layers illustrated here.

version of the task-dynamic model derived in the previous section permits the use of computable likelihoods to construct the optimal decision rule with the optimality guaranteed by Bayesian decision theory .

In this section, algorithms for learning parameters of three versions of statistical task-dynamic model, with increasing complexity in the model construct consistent with differing assumptions invoked by various phonetic theories, are outlined. Some main steps of algorithm derivation, based on statistical optimization principles integrated in neural network and dynamic system theories, are included also.

4.1 Model with deterministic, time-invariant parameters

This is the simplest version of statistical task-dynamic model where deterministic, unconstrained, time-invariant parameters are assumed in the state equation Eqn.(1), rewritten as

$$Z(k+1) = \Phi Z(k) + (I - \Phi)T + W_d(k).$$

The nonlinearity in the observation equation is parameterized by a form of MLP according to Eqn.(4).

With use of chain rule twice and use of $\frac{d}{dv}g(v) = g(v)(1 - g(v))$, the Jacobian matrix (needed for parameter learning) of the MLP-parameterized nonlinear mapping (Eqn.(3)) can be computed in an analytical form:

$$\mathbf{H}_z(Z) \equiv \frac{d}{dZ}\mathbf{h}(Z) = [H_{il}(Z)] = \begin{pmatrix} \frac{\partial h_1}{\partial Z_1} & \frac{\partial h_1}{\partial Z_2} & \cdots & \frac{\partial h_1}{\partial Z_L} \\ \frac{\partial h_2}{\partial Z_1} & \frac{\partial h_2}{\partial Z_2} & \cdots & \frac{\partial h_2}{\partial Z_L} \\ \vdots & \vdots & \vdots & \vdots \\ \frac{\partial h_I}{\partial Z_1} & \frac{\partial h_I}{\partial Z_2} & \cdots & \frac{\partial h_I}{\partial Z_L} \end{pmatrix}$$

where

$$H_{il}(Z) = \sum_j W_{ij}g[\sum_m w_{jm}g(Z_m)][1 - g(\sum_m w_{jm}g(Z_m))]w_{jl}\, g(Z_l)[1 - g(Z_l)].$$

In developing a parameter-learning procedure, the joint log-likelihood for acoustic observation sequence $O = [O(1), O(2), \ldots, O(N)]$ and hidden task-variable sequence $Z = [Z(1), Z(2), \ldots, Z(N)]$ is first written out as

$$\log L(Z, O, \Theta) =$$
$$-\tfrac{1}{2}\sum_{k=1}^{N-1}\{\log Q + [Z(k+1) - \Phi Z(k) - (I-\Phi)T]^T Q^{-1}[Z(k+1) - \Phi Z(k) - (I-\Phi)T]\}$$
$$-\tfrac{1}{2}\sum_{k=1}^{N}\{\log R + [O(k) - \mathbf{h}(Z(k))]^T R^{-1}[O(k) - \mathbf{h}(Z(k))]\} + const.$$

Then a pseudo-EM algorithm is used for learning model parameters including those in task-dynamics and those in MLP nonlinear mapping: $\Theta = \{T_{\mathcal{F}}, \Phi_{\mathcal{F}}, W_{ij}, w_{jl}, i = 1, 2, \ldots, I; j = 1, 2, \ldots, J; l = 1, 2, \ldots, L\}$.

E-step of the EM algorithm involves computation of the following conditional expectation: [9]

$$Q(Z, O, \Theta) = E\{\log L(Z, O)|O, \Theta\} = -\tfrac{N-1}{2}\log Q - \tfrac{N}{2}\log R - \tfrac{1}{2}\sum_{k=1}^{N}$$
$$E_N\{[Z(k+1) - \Phi Z(k) - (I - \Phi)T]^T Q^{-1}[Z(k+1) - \Phi Z(k) - (I - \Phi)T]|O, \Theta\}$$
$$-\tfrac{1}{2}\sum_{k=1}^{N-1} E_N\{[O(k) - \mathbf{h}(Z(k))]^T R^{-1}[O(k) - \mathbf{h}(Z(k))]|O, \Theta\}.$$

This can be simplified by standard algebraic manipulations to

$$Q(Z, O, \Theta) = Q_1(Z, O, \Phi, T) + Q_2(Z, O, W_{ij}, w_{jl}) \tag{5}$$

$$= -\frac{N-1}{2}\log\{\frac{1}{N-1}\sum_{k=1}^{N-1} E_N\{[Z(k+1) - \Phi Z(k) - (I - \Phi)T]^2|O, \Theta\}\}$$

$$- \frac{N}{2}\log\{\frac{1}{N}\sum_{k=1}^{N} E_N\{[O(k) - \mathbf{h}(Z(k))]^2|O, \Theta\}\} + const.$$

Note that the task-dynamic parameters (Φ, T) contained in Q_1 only and the MLP weight parameters (W_{ij}, w_{jl}) of the observation equation contained in Q_2 only can be optimized independently in the subsequent M-step which is discussed now.

M-step of the EM algorithm aims at optimizing the Q function in Eqn.(5) with respect to model parameters $\Theta = \{T, \Phi, W_{ij}, w_{jl}\}$. For the model at hand, it seeks solutions for

$$\frac{\partial Q_1}{\partial \Phi} \propto \sum_{k=1}^{N-1} E_N[\frac{\partial}{\partial \Phi}\{[Z(k+1) - \Phi Z(k) - (I - \Phi)T]^2|O, \Theta\} = 0 \tag{6}$$

$$\frac{\partial Q_1}{\partial T} \propto \sum_{k=1}^{N-1} E_N[\frac{\partial}{\partial T}\{[Z(k+1) - \Phi Z(k) - (I - \Phi)T]^2|O, \Theta\} = 0 \tag{7}$$

$$\frac{\partial Q_2}{\partial W_{ij}} \propto \sum_{k=1}^{N} E_N[\frac{\partial}{\partial W_{ij}}\{[O(k) - \mathbf{h}(Z(k))]^2|O, \Theta\} = 0 \tag{8}$$

$$\frac{\partial Q_2}{\partial w_{jl}} \propto \sum_{k=1}^{N} E_N[\frac{\partial}{\partial w_{jl}}\{[O(k) - \mathbf{h}(Z(k))]^2|O, \Theta\} = 0. \tag{9}$$

Eqns.(6) and (7) are third-order nonlinear algebraic equations (in Φ and T):

$$N\Phi T^2 - 2\{\sum_k E_N[Z(k)|O]\}\Phi T - NT^2 + \{\sum_k E_N[Z^2(k)|O]\}\Phi +$$

$$\{\sum_k E_N[Z(k) + Z(k+1)|O]\}T - \{\sum_k E_N[Z(k+1)Z(k)|O]\} = 0,$$

$$N\Phi^2 T - 2N\Phi T - \{\sum_k E_N[Z(k)|O]\}\Phi^2 +$$

$$\{\sum_k E_N[Z(k) + Z(k+1)|O]\}\Phi + NT - E_N[Z(k+1)|O]\} = 0.$$

---

[9] Together with a set of related sufficient statistics needed to complete evaluation of the conditional expectation.

The coefficients in the above algebraic equations constitute the sufficient statistics, which can be obtained by the standard technique of Iterated Extended Kalman Filtering (IEKF) with fixed-interval smoothing [21, 10], and the equations are solved for (Φ, T) by numerical methods. Alternatively, optimization of Q_1 can be found using gradient decent with explicit expressions of gradients given by Eqns.(6) and (7).

Solutions to Eqns.(8) and (9) for finding (W_{ij}, w_{jl}) to maximize Q_2 in Eqn.(5) have to rely on approximation (due to the complexity in the $\mathbf{h}(.)$ function). The approximation involves first finding smoothed estimates of hidden variables $Z(k)$, $Z(k|N)$, via the IEKF fixed-interval smoother. Given such estimates, the conditional expectations can be approximated (pseudo-EM) to give

$$\frac{\partial Q_2}{\partial W_{ij}} \propto \sum_k [O(k) - \mathbf{h}(Z(k|N))]^T \frac{\partial \mathbf{h}(Z(k|N))}{\partial W_{ij}}$$

$$\frac{\partial Q_2}{\partial w_{jl}} \propto \sum_k [O(k) - \mathbf{h}(Z(k|N))]^T \frac{\partial \mathbf{h}(Z(k|N))}{\partial w_{jl}},$$

where the partial derivatives can be evaluated using the MLP structure to give

$$\frac{\partial \mathbf{h}(Z(k|N))}{\partial W_{ij}} = \begin{pmatrix} 0 \\ \vdots \\ g[\sum_m w_{jm}g(Z_m(k))] \\ \vdots \\ 0 \end{pmatrix} \leftarrow (i^{th} \; element \; non-zero)$$

$$\frac{\partial \mathbf{h}(Z(k|N))}{\partial w_{jl}} = g[\sum_m w_{jm}g(Z_m(k))][1 - g(\sum_m w_{jm}g(Z_m(k)))]g(Z_l(k))] \begin{pmatrix} W_{1j} \\ W_{2j} \\ \vdots \\ W_{Ij} \end{pmatrix}.$$

Given the explicit expressions for $\frac{\partial Q_2}{\partial W_{ij}}, \frac{\partial Q_2}{\partial w_{jl}}$ derived above, gradient-decent algorithm is effectively used to obtain optimized parameters (W_{ij}, w_{jl}) in the M-step.

4.2 Model with random, time-invariant parameters

Statistical motivation for developing a model with random parameters (contrasting the deterministic model parameters discussed in the previous section) is a Bayesian one [21]; that is, we, as modelers, desire to achieve robustness in model parameter estimation and have some prior knowledge to use about the model parameters. Phonetic motivation for allowing for random parameters in task-dynamic model is that different classes of speech sounds have well known, systematic variations in their production strategies (which can be directly quantified in terms of the task-dynamic model's parameter variations), and that speakers tend to use a great degree of (constrained) freedom in choosing their production strategies (plasticity of phonetic gestures advocated in H&H theory).

For example, since parameters $T = (T_1, ..., T_l, ..., T_L)$ (l is index to utterance token) in task-dynamic model represents attractor constriction properties (degree and location) of the vocal tract, it is possible to use well established speech production knowledge (e.g. articulation-acoustics relationships described in quantal theory of speech [29]) to construct the prior in the form of inverse Gaussian distribution (non-negatively valued):

$$f(T_l; \mu_l, \lambda_l) = \sqrt{\frac{\lambda_l}{2\pi}} \times T_l^{-\frac{3}{2}} \times exp[-\frac{\lambda_l(T_l - \mu_l)^2}{2\mu_l^2 T_l}], \quad T_l > 0$$

Note that various broad classes of speech sounds have systematically different hyperparameters μ's and λ's, which are largely predictable, in the above inverse Gaussian distribution . Hence, such prior information can be effectively used to initialize these parameters in subsequent automatic MAP training.

The EM algorithm similar to the earlier deterministic-parameter case applies here for parameter estimation, except now three additional terms are needed in the auxiliary function Q_1 of the E-step due to use of the additional prior distribution:

$$Q_1(Z, O, \Phi, T) \quad \propto \quad -\{\frac{1}{N-1} \sum_{k=1}^{N-1} E_N\{[Z(k+1) - \Phi Z(k) - (I - \Phi)T]^2 | O, \Theta\}\}$$

$$+ \sum_{l=1}^{L} \{\frac{1}{2} log\lambda_l - \frac{3}{2} logT_l - \lambda_l \frac{(T_l - \mu_l)^2}{2\mu_l^2 T_l}\}$$

M-step of the EM algorithm then gives MAP (empirical Bayes) estimates of both hyper and random model parameters by solving

$$\frac{\partial Q_1}{\partial T} = 0 \quad and \quad \frac{\partial Q_1}{\partial \Phi} = 0,$$

where the second equation above is identical to that in the earlier deterministic-parameter case (Eqn.(7)). Solutions for the first one require that hyper parameters be given, and can be obtained by jointly or iteratively solving $\frac{\partial Q_1}{\partial \mu} = 0$ and $\frac{\partial Q_1}{\partial \lambda} = 0$, which gives optimal estimates for the hyper-parameters. Alternatively, M-step can be accomplished by gradient-decent methods.

4.3 Model with random, smoothly time-varying parameters

This further extension of the statistical task-dynamic model is again motivated by H&H theory of speech gesture plasticity: speakers have significant freedom in articulation, only to be constrained by tradeoffs between the speech-economy principle and by the listener's demand for clarity or sufficient perceptual contrast. In addition to using random parameters, the speaker's freedom in articulation can be further quantitatively represented in the task-dynamic model by allowing for time-varying parameters (within phonological-unit boundaries). On the other hand, the speech-economy principle can be simultaneously quantified by smoothness prior constraints imposed on possible sample paths of these random, time-varying parameters.

For example, the task dynamic with time-varying state transition matrix can be described by

$$Z(k+1) = \Phi(k)Z(k) + [I - \Phi(k)]T(k) + W_d(k), \qquad (10)$$

where $\Phi(k)$ is a random parameter (matrix) and is constrained to change slowly over time. A stochastically perturbed difference-equation (order r) model is used to quantitatively provide smoothness prior constraints for random time variation of $\Phi(k)$:

$$\nabla^r \Phi(k) = v(k), \quad v(k) \sim \mathcal{N}(0, \sigma^2).$$

When $r = 1$, $\Phi(k)$ has locally constant trend (i.e. random walk); when $r = 2$ and 3, $\Phi(k)$ has locally linear and quadratic trends, respectively, etc.

For the special case of $T = 0$ in Eqn.(10), a constrained least-square solution to parameter estimation of $\Phi(k)$ has been proposed in [18] as an optimization problem for the following objective function:

$$\sum_k [Z(k+1) - \Phi(k)Z(k)]^2 + \lambda^2 \sum_k [\nabla^r \Phi(k)]^2,$$

where λ^2 is the tradeoff parameter which balances the infidelity of the model to the data $Z(k)$ and the infidelity of the model to the smoothness constraint. [10]

Difficulties in solving the above least-square problem have prompted statisticians to devise more elegant solutions. Results contained in [19, 13] have shown that the smoothness-constraint problem on polynomial parameter trajectories can be equivalently treated as optimal smoothing problem using state-space model formulation. To see this, the smoothness polynomial constraint $\nabla^r \Phi(k) = v(k)$ is rewritten as an equivalent state-space (Gauss-Markov) system:

$$r = 1: \quad \Phi(k+1) = \Phi(k) + v(k)$$
$$r = 2: \quad \Phi(k+1) = 2\Phi(k) - \Phi(k-1) + v(k)$$
$$r = 3: \quad \Phi(k+1) = 3\Phi(k) - 3\Phi(k-1) + \Phi(k-2) + v(k)$$

$$\cdots \qquad \cdots$$

This, via the state-augmentation technique, can be equivalently written as a time-invariant linear system:

$$\tilde{\Phi}(k+1) = G\tilde{\Phi}(k) + H\tilde{v}(k),$$

where

$$for \ r = 1: \qquad G = I \qquad (constant \ trajectory)$$

$$for \ r = 2: \quad G = \begin{pmatrix} 2I & -I \\ I & 0 \end{pmatrix} \qquad (linear \ trajectory)$$

$$for \ r = 3: \quad G = \begin{pmatrix} 3I & -3I & I \\ I & 0 & 0 \\ 0 & I & 0 \end{pmatrix} \qquad (quadratic \ trajectory)$$

$$\cdots \qquad \cdots$$

---

[10] Bayesian interpretation of the above least-square problem is also given in [18].

and

$$H = \begin{pmatrix} 1 \\ 0 \\ \vdots \\ 0 \end{pmatrix}, \quad \tilde{\Phi}(k) = \begin{pmatrix} \Phi(k) \\ \Phi(k-1) \\ \vdots \\ \Phi(k-r+1) \end{pmatrix}.$$

This state-space formulation for time variation of parameter $\Phi(k)$, together with the task dynamic state equation, allows all the EM estimation results for the model with time-invariant parameters (discussed earlier) apply to the current time-varying parameter case.

4.4 Discriminative learning of production models' parameters

Discriminative model learning, as opposed to the maximum likelihood one discussed so far, can be theoretically motivated by the argument expressed in H&H theory that the listener's perceptual contrast is the primary objective of human speech communication while employing speech economy in speech production. For possible speech recognition applications in the context of task-dynamic model discussed so far, such an objective can be quantitatively formulated as the problem of minimizing speech recognition errors subject to a tradeoff principle of "least effort" implemented by smoothness constraint on time-varying, random model parameters. Analogies can be made here between the above machine speech recognition strategy and human speech perception: the smoothness constraint on model parameter variations implemented in the recognizer is analogous to minimizing speaker's efforts of production, and the criterion of minimizing speech recognition errors is analogous to maximizing human perceptual contrasts across different phonetic or lexical classes.

The basis of computational formalisms for carrying out the above constrained optimization in the framework of task-dynamic model is already established by speech technologists. The major step involves computation of the gradient of a smoothed estimate of the empirical recognition error with respect to all parameters in the model. For efficiency of training, the gradient has to be expressed in an analytical form. The computation of the gradient is lengthy and laborious and is not included here, but the general spirit of such computation can be gleaned from the work published in [26] applied to a far less sophisticated speech model.

5. Other types of computational models of speech production

So far in this paper I have concentrated on a specific type of speech production model, i.e., task dynamic one. This is a functional model, with no direct representation of biomechanical properties of the vocal tract and with dynamic properties of the system residing only in the "controller". One main virtue of this model is its uniform definition of the goal of speech production across all consonant and vowel classes in terms of vocal tract constriction properties. This has greatly facilitated algorithmic developments which enable implementation of the model for speech recognition applications. A number of other, non-task-dynamic types of computational models of speech production have been developed, intending to incorporate dynamics either at the biomechanical articulator level or more directly at the acoustic observation level. Within the former class or articulatory-dynamic models, an articulatory stochastic target model was developed which aims at accounting for detailed movement behaviors of biomechanical articulators guided by the highly complex, multi-dimensional target distributions defined in the biomechanical articulator coordinate [9, 25]. Correlations among subsets of articulators in such target distributions are essential because it is the articulator coordinate, rather than the task-variable coordinate, in which the targets are defined. Some more empirical methods used to model articulatory dynamics for purpose of speech recognition include those in [1, 2], where the dynamic of a set of pseudo-articulators is realized by FIR filtering from sequentially placed, phoneme-specific target positions or by applying trajectory-smoothness constraints.

Within the class of acoustic-dynamic models, the model which attempts to condition the properties of the dynamic directly on specific feature-coded speech production mechanisms is described in [7]. In that model, the underlying articulatory-feature based phonological units are used to determine dynamic or static trajectories (order of polynomials) that describe the acoustic correlates of the phonological units, and substantial phonetic recognition performance improvements have been demonstrated. An earlier version of this model, using piecewise-static trajectories (conventional HMM) to approximate continuous trajectories in speech acoustics, is described in [8].

Along the line of acoustic-dynamic model of speech production, there exists a further possibility of choosing more appropriate parametric forms than polynomials to describe production-correlated acoustic variables. For example, the polynomial trajectories (as used in many earlier segmental models) do not entail the concept of formant target since they do not have the asymptotic property which allows the trajectory to slowly and smoothly relax to an asymptotic value such as the formant target. Exponential form of trajectories, however, has such an asymptotic property; e.g. $f(t) \propto f^0 \times (1 - \alpha t \times exp[-\gamma \times t])$. But some serious difficulties would arise if this exponential form of the trajectory model were to be used for formants directly. Because many consonants do not show acoustically measurable formants (due to full or partial pole-zero cancellation in vocal-tract acoustics caused by supra-glottal excitation sources), the trajectory model has to be generalized from that describing measurable formant trajectories (applicable only to vowels) to that describing hidden vocal-tract resonance dynamics (applicable to all types of speech sounds). Smoothness and continuity constraints can then be naturally applied to the hidden vocal-tract resonance trajectories through entire utterances, thus naturally producing speech undershoot phenomena characteristic of casual, fast speech (as observed in Switchboard data). Due to the hidden nature of vocal-tract resonances, especially for consonants, it will be appropriate to use MFCCs as speech observations and to empirically build noisy nonlinear mappings from vocal-tract resonances (poles of vocal-tract transfer function) to MFCCs.

6. Summary and discussions

In this tutorial, major classes of speech models developed by two largely separate, scientific and technological, communities are surveyed, compared, and analyzed. Similar comparisons and analyses from a more global perspective have been made earlier in [22]. A particular type of speech production model, task-dynamic one, is developed which integrates the strengths of the two previously separate styles of production models. This integration owes much to successful use of the smoothness-prior (Bayesian equivalent) technique, motivated by gesture-plasticity and movement-economy principles in human speech production, in establishing the statistical task-dynamic model. Both maximum likelihood (via EM) and minimum classification error (via gradient descent) criteria are used for model parameter learning, justified in terms of various versions of phonetic theories. In either case, optimization of the likelihood or empirical classification error rate for a small number of hyperparameters in the model permits robust modeling of true dynamic behaviors of human speech. Modeling such behaviors requires a complex structure, but the technique we adopted enables use of a most compact set of hyperparameters which encompass a large number of implicitly inferred parameters.

Acknowledgements. Over the past several years and on the subject matter of the three tutorial papers written by the author in this book, many discussions with or experimental contributions from the following individuals are gratefully acknowledged: K. Stevens, J. Perkell, V. Zue, R. McGowan, E. Saltzman, C. Browman, L. Goldstein, C. Lee, G. Chollet, G. Ramsay, D. Sun, H. Sameti, J. Wu, H. Sheikhzadeh, and I. Kheirallah.

References

[1] Bakis R. (1993), "An articulatory-like speech production model with controlled use of prior knowledge," notes from *Frontiers in Speech Processing*, CD-ROM.
[2] Blackburn C., and Young. S. (1995), "Towards improved speech recognition using a speech production model," *Proc. Eurospeech*, vol. 2, pp. 1623-1626.
[3] Deng L. (1992) "A generalized hidden Markov model with state-conditioned trend functions of time for the speech signal," *Signal Processing*, vol.27, pp. 65-78.
[4] Deng L. (1993) "Design of a feature-based speech recognizer aiming at integration of auditory processing, signal modeling, and phonological structure of speech." *JASA*, vol. 93(4) Pt.2, pp. 2318.
[5] Deng L. (1992-1993) "A Computational Model of the Phonology-Phonetics Interface for Automatic Speech Recognition," Summary Report of Research in Spoken Language Systems, Laboratory for Computer Science, MIT.
[6] Deng L. and Aksmanovic M. (1997) "Speaker-independent phonetic classification using hidden Markov models with mixtures of trend functions," *IEEE Trans. Speech Audio Processing*, vol. 5, pp. 319-324.
[7] Deng L. and Sameti H. (1996) "Transitional speech units and their representation by the regressive Markov states: Applications to speech recognition," *IEEE Trans. Speech Audio Proc.*, vol. 4(4), pp. 301–306.
[8] Deng L. and Sun D. (1994), "A statistical approach to automatic speech recognition using the atomic speech units constructed from overlapping articulatory features," *JASA*, vol. 95, pp. 2702-2719.

[9] Deng L., Ramsay L., and Sun D. (1997) "Production models as a structural basis for automatic speech recognition," *Speech Communication*, August issue.

[10] Digalakis V., Rohlicek J., and Ostendorf M., (1993) "ML estimation of a stochastic linear system with the EM algorithm and its application to speech recognition", *IEEE Trans. Speech Audio Processing*, pp. 431-442.

[11] Ghitza O., and Sondhi M. (1993) "Hidden Markov models with templates as nonstationary states: an application to speech recognition," *Computer Speech and Language*, vol. 7, pp. 101–119.

[12] Gales M. and Young S. (1993) "Segmental HMMs for speech recognition," *Proc. Eurospeech*, pp. 1579-1582.

[13] Gersch W. (1992) "Smoothness priors," in *New Directions in Time Series Analysis*, D. Brillinger et al. (eds.), Springer, New York, pp. 111-146.

[14] Gish H. and Ng K. (1993) "A segmental speech model with applications to word spotting," *Proc. ICASSP*, pp. 447-450.

[15] Haykin S. (1994) *Neural Networks — A Comprehensive Foundation*, Maxwell Macmillan, Toronto.

[16] Holmes W. and Russell M. (1995) "Speech recognition using a linear dynamic segmental HMM," *Proc. Eurospeech*, pp. 1611-1641.

[17] Kent R., Adams S. and Turner G. (1995) "Models of speech production," in *Principles of Experimental Phonetics*, Ed. N. Lass, Mosby: London, pp. 3-45.

[18] Kitagawa G. and W. Gersch W. (1996) *Smoothness Priors Analysis of Time Series*, Springer, New York.

[19] Kohn R. and Ansley C. (1988) "Equivalence between Bayesian smoothness priors and optimal smoothing for function estimation," in *Bayesian Analysis of Time Series and Dynamic Models*, J. Spall (ed.), Marcel Dekker, New York, pp. 393-430.

[20] McGowan R. (1994) "Recovering articulatory movement from formant frequency trajectories using task dynamics and a genetic algorithm: Preliminary model tests," *Speech Communication*, 14, pp. 19-48.

[21] Mendel J. (1995) *Lessons in Estimation Theory for Signal Processing, Communications, and Control*, Prentice Hall, New Jersey.

[22] Moore R. (1994) "Twenty things we still don't know about speech," *Proc. CRIM/FORWISS Workshop on Speech Research and Technology*, pp. 1-9.

[23] Ostendorf M. (1996) "From HMMs to segment models," in *Automatic Speech and Speaker Recognition – Advanced Topics*, C. Lee, F. Soong, and K. Paliwal (eds.), Kluwer Academic Publishers, pp. 185-210.

[24] Perrier P. et al. (eds.) *Proceedings of the First ESCA Tutorial & Research Workshop on Speech Production Modeling*, Autrans, France, May 24-27, 1996.

[25] Ramsay G. and Deng L. (1996) "Optimal filtering and smoothing for speech recognition using a stochastic target model," *Proc. ICSLP*, pp. 1113-1116.

[26] Rathinavalu C. and Deng L. (1997) "HMM-based speech recognition using state-dependent, discriminatively derived transforms on Mel-warped DFT features", *IEEE Trans. Speech Audio Processing*, pp. 243-256.

[27] Rubin P. et al (1996) "CASY and extensions to the task-dynamic model," *Proc. 4th European Speech Production Workshop*, Autrans, France, pp. 125-128.

[28] Saltzman E. and Munhall K. (1989) "A dynamical approach to gestural patterning in speech production," *Ecological Psychology*, 1, 333-382.

[29] Stevens K. (1989) "On the quantal nature of speech," *J. Phonetics*, vol.17, 1989, pp. 3–45.

Articulatory Features and Associated Production Models in Statistical Speech Recognition

Li Deng

Department of Electrical and Computer Engineering
University of Waterloo, Waterloo, Ontario, Canada N2L 3G1
email: deng@crg6.uwaterloo.ca

Summary. A statistical approach to speech recognition is outlined which draws close parallel with closed-loop human speech communication schematized as a joint process of encoding and decoding of linguistic messages. The encoder consists of the symbolically-valued overlapping articulatory feature model and of its interface to a nonlinear task-dynamic model of speech production. A general speech recognizer architecture based on optimal decoding strategy incorporating encoder-decoder interactions is described and discussed.

1. Introduction

The general concept of closed-loop speech chain underlying human speech communication has been known for many years [2]. However, engineering construction of automatic speech recognition machines, which have been known to perform orders of magnitude worse than human, so far has hardly been able to capitalize on any significant properties of the closed-loop human speech communication . This situation arises due to a number of important factors including 1) (justifiable) desires for short-term engineering success in limited tasks; 2) lack of interactions between scientific and technological research communities and hence lack of integration of the respective research accomplishments; 3) fragmentary and incomplete nature of our understanding of the closed-loop human speech communication process; and 4) lack of suitable computational formalisms which would allow the scientific understanding to be readily useful in computation-intensive speech technology applications.

The purpose of this tutorial paper is to describe the general nature of the closed-loop human speech chain as an encoding-decoding process (analogous to information-theoretic design of engineering communication systems), and to show how within this framework computational formalisms can be established enabling graceful integration of engineering modeling-decoding techniques with scientific models and theories intended to faithfully describe the human speech process.

2. Functional description of human speech communication as an encoding-decoding process

At the global and functional level, human speech communication can be viewed as an encoding-decoding process, where the decoding process or perception is an active process consisting of auditory reception followed by phonetic/linguistic interpretation. As an encoder implemented by the speech production system, the speaker uses knowledges of meanings of words (or phrases), of grammar in a language, and of the sound representations for the intended linguistic message. Such knowledges can be made analogous to the keys used in engineering communication systems. The phonetic plan, derived from the semantic, syntactic, and phonological processes, is then executed through the motor-articulatory system to produce speech waveforms.

As a decoder which aims to accomplish speech perception, the listener uses a key, or the internal "generative" model , which must be compatible with (may not be identical to) the key used by the speaker to interpret the speech signal received and transformed by the auditory system. This enables the listener to reconstruct, via (probabilistic) analysis-by-synthesis strategies, the linguistic message intended by the speaker. Such an encoding-decoding view of human speech communication, where the observable speech acoustics plays the role of carrier of deep, linguistically meaningful messages , is strikingly similar to the modulation-demodulation scheme in electronic digital communication and to the encryption-decryption scheme in secure electronic communication.

Since the nature of the key used in the phonetic-linguistic information decoding or speech perception/understanding lies in the strategies used in the production or encoding process, speech production and perception are intimately linked in the closed-loop speech chain. The implication of such a link for speech recognition technology is the need to develop functional and computational models of human speech production for use as an "internal model" in the decoding process by machines.

3. Overview of theories of speech perception

With respect to the above encoding-decoding review of human speech communication which advocates intimate links between speech production and perception, a number of popular theories and models of speech perception are reviewed here.

Motor theory of speech perception, addressing the issue of ubiquitous acoustic variability of speech, experienced two main stages of development, both emphasizing a specialized phonetic module mediating speech production and perception. The early version of the theory asserts existence of phonetic invariance at the levels of articulatory gesture or motor command [13]. Due to the failure of finding such invariance experimentally, this earlier version was modified to move the proposed phonetic invariance to higher, vaguely specified levels of speech production [14]. The abstract nature of the modified motor theory renders it practically useless for possible speech recognition applications.

Closely related to motor theory, the analysis-by-synthesis model [11] of speech perception adopted a more tangible, hypothesis-and-test approach to phonetic decoding by human. Elements of this model include the proposal of active internal synthesis of comparison signals, use of generative rules to convert lexical items into phonetic parameters (which describe the behavior of structures controlling the vocal-tract configuration and vocal-cords activities), and rules to convert these phonetic parameters into time-varying speech spectra.

Sharing partial views with motor theory, direct-realist theory of speech perception proposes that listener directly perceives the articulatory gestures of the speaker via the structure that the gestures pass on to the common acoustic medium between listener and speaker. The theory does not require specialized phonetic module [10].

Contrary to motor theory, acoustic-auditory theory of speech perception asserts existence of phonetic invariance not in any internal levels of speech production, but in the acoustic-auditory domain, which is the outcome of speech production and determines the object of speech perception. In this theory, speech production and perception are indirectly linked by virtue of common acoustic goals or targets [20, 12, 9].

A drastically different theory of speech perception from all the above ones proposes that it is the interactions of speaker and listener based on balances between speaker's efforts and listener's contrastive perceptual goals , not the phonetic invariance at any levels of the speech chain, which are essential properties of speech perception. This theory is called Hyper-Hypo or H&H theory [15, 16]. H&H theory proposes that the distal object of speech perception is the speaker's intention (shared with motor theory), but that such an intention has both articulatory-gesture production component and contrastive perceptual component, and is determined by short-term, dynamic interactions of the two components. An essential concept of the theory is plasticity of phonetic gestures — speakers adaptively tune phonetic gestures to the needs of speaking situations under motor and perceptual constraints, and phonetic gestures are not invariant but are adaptations to constraints on production mechanisms for least "efforts" (or speech economy , low-cost behavior, or "hypo" speech) and on perceptual mechanisms for achieving sufficient contrast ("hyper" speech). These mechanisms are language independent and not special to speech; "invariance" must be defined according to the global purpose of speech communication (e.g. lexical access and speech comprehension).

One supporting evidence of H&H theory is the phenomenon of compensatory articulation where speakers are capable of re-organizing articulation to reach fixed acoustic and perceptual goals under both artificial bite-block condition and natural loud, clear, fast or spontaneous speaking conditions. Another evidence comes from the formation of the phonetic system with "quantal" properties which can be shown as being driven by a demand for sufficient perceptual contrast. Speech communication system is established via constant interaction between speaker and listener: the listener force the speaker to make sufficient phonetic distinctions (negative control), and the speaker tries to use least "efforts" but is simultaneously constrained by the listener's demand.

4. A general framework of statistical speech recognition

The Bayesian framework is adopted as a general framework for intended incorporation of scientifically motivated speech models in statistical speech recognition. Let $O = O_1, O_2, ..., O_T$ be a sequence of observable acoustic data of speech, and let $W = w_1, w_2, ..., w_n$ be the sequence of words intended by the speaker who produces the acoustic record O. The goal of a speech recognizer is to "guess" the most likely word sequence $\hat{W}$ given the acoustic data $\mathbf{O}$. The problem can be formulated as a top-down search problem over the allowable word sequences:

$$\hat{W} = arg \max_W P(W|\mathbf{O}) = arg \max_W P(\mathbf{O}|W)P(W), \qquad (1)$$

Decomposition of the word-to-acoustics probability $P(\mathbf{O}|\mathbf{W})$ above is accomplished by using law of total probability:

$$P(\mathbf{O}|W) = \sum_{\mathcal{F}} P(\mathbf{O}|\mathcal{F})P(\mathcal{F}|W) \approx \max_{\mathcal{F}} P(\mathbf{O}|\mathcal{F})P(\mathcal{F}|W), \qquad (2)$$

where $\mathcal{F}$ is a discrete-valued *phonological* construct (or "pronunciation" model), which specifies, according to probability $P(\mathcal{F}|W)$, how words and word sequences W can be expressed in terms of a particular organization of a small set of fundamental phonological units; $P(\mathbf{O}|\mathcal{F})$ is the probability that a particular organization $\mathcal{F}$ of phonological units produces the acoustic data $\mathbf{O}$. This probability is determined by the phonetic interface model .

According to phonetic theories, the interface model ideally should consist of at least three hierarchical levels of mapping: from phonological symbols ($\mathcal{F}$) to motor commands ($\mathcal{M}$), from motor commands to articulation ($\mathcal{A}$), and from articulation to acoustics ($\mathbf{O}$). That is, one can further decompose the probability $P(\mathbf{O}|\mathcal{F})$ associated with the global interface model into:

$$P(\mathbf{O}|\mathcal{F}) = \sum_{\mathcal{M},\mathcal{A}} P(\mathbf{O}|\mathcal{A})P(\mathcal{A}|\mathcal{M})P(\mathcal{M}|\mathcal{F}) \approx \max_{\mathcal{M},\mathcal{A}} P(\mathbf{O}|\mathcal{A})P(\mathcal{A}|\mathcal{M})P(\mathcal{M}|\mathcal{F}).$$

$$(3)$$

For efficient engineering construction of speech recognizers, an approximation to the above layered, multi-level mapping is necessary and can be made by one-level or two-level mappings from the phonological level $\mathcal{F}$ to the acoustic level $\mathbf{O}$. Any approximation must faithfully retain the dynamic character of the speech production process. [1]

5. Brief analysis of weaknesses of current speech recognition technology

Despite some success in highly constrained recognition tasks, the current HMM-based, data-driven speech recognition technology is fundamentally limited in its ability to achieve human-like speech recognition. Such a limitation stems from its weak theoretical foundations from both phonological and phonetic perspectives. First, nearly all currently popular speech recognition strategies use more or less the same set of phone-like phonological speech units (e.g. triphones) arranged in

---

[1] Some work done in our research group included three types of approximation (differing by three distinct levels at which lies the object of dynamic modeling): 1) Acoustic-dynamic model based on nonstationary-state or trended HMM [8]; 2) Articulatory-dynamic or stochastic target model [7, 18]; and 3) Task-dynamic model [3, 4, 5].

strictly linear sequences, like "beads-on-a-string". This, however, is not how human language faculty organizes its phonological primitives. Second, the weak theoretical foundation of the current speech recognition technology from phonetic perspective is reflected in the weak structure of the HMM in use and in the simplistic strategy of surface data fitting to the observable acoustics (equipped with virtually no underlying data generation mechanisms). A consequence of this weakness is that the sample paths of the HMM as a nonstationary stochastic process deviate significantly from true speech data trajectories.

The above weaknesses associated with the current speech recognition technology lead to speech recognizers which inherently lack robustness, and cannot generalize from training data to mismatched test data. The problem is particularly serious when little supervised adaptation data are available to recognizers, as in most real-world speech recognition applications. Such recognizers inevitably break down when moving from read or clear speech style to casual, fast and spontaneous speaking mode, switching from "sheep" speakers to "goat" speakers, or porting from one language to another or from one task to another. When new tasks or new languages are involved, re-design and re-training of the recognizers are undesirably needed.

It appears that the ultimate success of human-like speech recognition will require not only extensions of existing recognizer architectures, but fundamental changes in the statistical models of speech underlying speech recognizers. Such new models must at a functional level faithfully characterize essential properties of human behaviors in closed-loop speech communication (production and perception) and be equipped with effective computational formalisms and model learning strategies.

6. Phonological model: Overlapping articulatory features and related HMMs

Motivations of using vocal-tract constriction based articulatory features as the phonological primitive can be succinctly summarized by a quote from modern phonology literature [1]: "Phonetic Interpretation of the Feature Hierarchy: ...the basic organizing principle of the feature hierarchy is the *vocal tract constriction....* The place features define constriction location and the articulator-free features define constriction degree. The notion "constriction" is central to many current theories of speech production, both acoustic and articulatory. It is therefore not surprising that phonological representations may be organized in terms of (vocal tract) constrictions as well."

In the work described briefly in [5], a compact set of universal phonological/articulatory features across world languages is designed. The resulting feature specification systems, one for each language, share intensively among the component features. Through appropriate combinations of component features, new sounds or segments in new, target languages from the sounds in the source language(s) can be reliably predicted. The phonological model uses hierarchically or-

ganized articulatory features as the primitive phonological units motivated by articulatory phonology and feature-geometry theory.

The phonological model further entails a statistical scheme to allow probabilistic, asynchronous but constrained temporal overlapping among components (symbols) in the sequentially placed feature bundles. A set of feature overlapping and
constraining rules are designed based on syllable structure and other prosodic factors . Examples of the rules for English are: 1) overlap among consonant clusters
in syllable onset and coda, and in consonant sequence across connected words; 2)
overlap between syllable onset and nucleus; 3) overlap between syllable nucleus and
coda; 4) overlap of the tongue-dorsum feature between two adjacent syllable nuclei;
5) except for Lips and Velum features, no overlap between onset and coda within
the same syllable.

The phonological model finally contains a crucial component which converts
the above probabilistic feature overlap pattern to a finite state automaton (FSA)
or HMM state topology. This FSA represents ensemble sequences of phonological
units composed of the overlapped features, serving as the phonetic plan which controls lower (phonetic) levels of speech production resulting in dynamic patterns of
speech acoustics.

7. Task-dynamic model of speech production

Before I discuss how the above overlapping-articulatory feature based phonological model can be interfaced to the phonetic variables (including ultimately speech
acoustics), the (deterministic) task-dynamic model of speech production developed
in speech science [19] is briefly reviewed. This is a most comprehensive speech production model, well developed and tested, based originally on a general model of
skilled movement control. In this model, the control signal is derived from abstract
gestural units defined in articulatory phonology; these gestural units are organized
into utterance-specific "gestural scores". Each gesture is correlated with two task
variables , vocal-tract constriction degree and constriction location. At any point in
time, only a small subset (fewer than 3 or 4 usually) of the gestures are co-occurring
(overlapping or "blending") during speech production. When blending occurs, competitive blending rules are used to determine the final values of the correlated task
variables.

The intrinsic dynamics for each task variable is modeled as critically damped
second order system, which is characterized by the gesture-dependent (normalized)
stiffness and by the gesture-dependent point-attractor of the dynamical system. The
relation between the task variables and the model-articulators is characterized by a
static and nonlinear function, which is constructed by vocal-tract geometry [2].

Using piecewise locally linear approximation of the nonlinear relation between
task variables and model-articulators, the linear dynamics for the former is converted into quasi-linear dynamics in the latter. Jacobian transformation matrix de-

---

[2] Many-to-one nature of the relation gives rise to compensatory articulation (or motor
equivalence) and to coordinative structure in this model.

rived from the nonlinear relation becomes a component of the linearized dynamics. Finally, given the time-varying model-articulator motions, a further static, nonlinear relation maps the model-articulators into speech acoustics using Haskins Lab's configurable articulatory synthesizer.

8. Interfacing overlapping features to task-dynamic model and a general architecture for speech recognition

Our computational approach to phonology-to-phonetic interface is based on a discrete-time task-dynamic model derived from the continuous-time model reviewed above, with a statistical structure imposed [4]. In this approach, each individual articulatory feature (as symbolic phonological unit) is made to associate with the task-dynamic model parameters including stiffness and the point or region of the attractor (continuous phonetic variables). The stiffness and attractor parameters corresponding to simultaneously overlapped or blended features are determined from those of individual component features according to either empirical rules or automatic learning. The nonlinear relation between the task variables and the model-articulators is approximated by trainable Multi-Layer Perceptron (MLP) neural nets, which serve as a generic device for data interpolation in a multi-dimensional space. Another trainable MLP is used to approximate the nonlinear articulator-to-acoustics mapping.

An architecture for speech recognition is presented in Fig. 1 based on the approach described above using the overlapping articulatory feature model interfaced to task-dynamic model of speech production. Language-universal components include the feature primitives in the phonological model and in most subcomponents in the task-dynamic (phonetic) model. The decoding strategy (recognition search) is similar to that in the current HMM-based technology, but the way the acoustic likelihood is computed is drastically different. The structured, probabilistic phonological and phonetic models used here have a high degree of parameterization, which allows parsimonious yet accurate characterization of the recognizer capable of language independent, speaker independent, speaking-style independent, and unlimited-vocabulary speech recognition.

One most crucial component of the task-dynamic model is the nonlinear relation between task variables and model-articulators and that between model-articulators and acoustics. In the speech recognition architecture of Fig.1, a trainable neural network is used for approximating these relations. The general topology of the network is shown in Fig.2, with the network unit connections strongly constrained by speech production mechanisms .

9. Discussions: Machine speech recognition

The above sections described a feature-based phonological model interfaced with a task-dynamic model of speech production. While many details need to be specified,

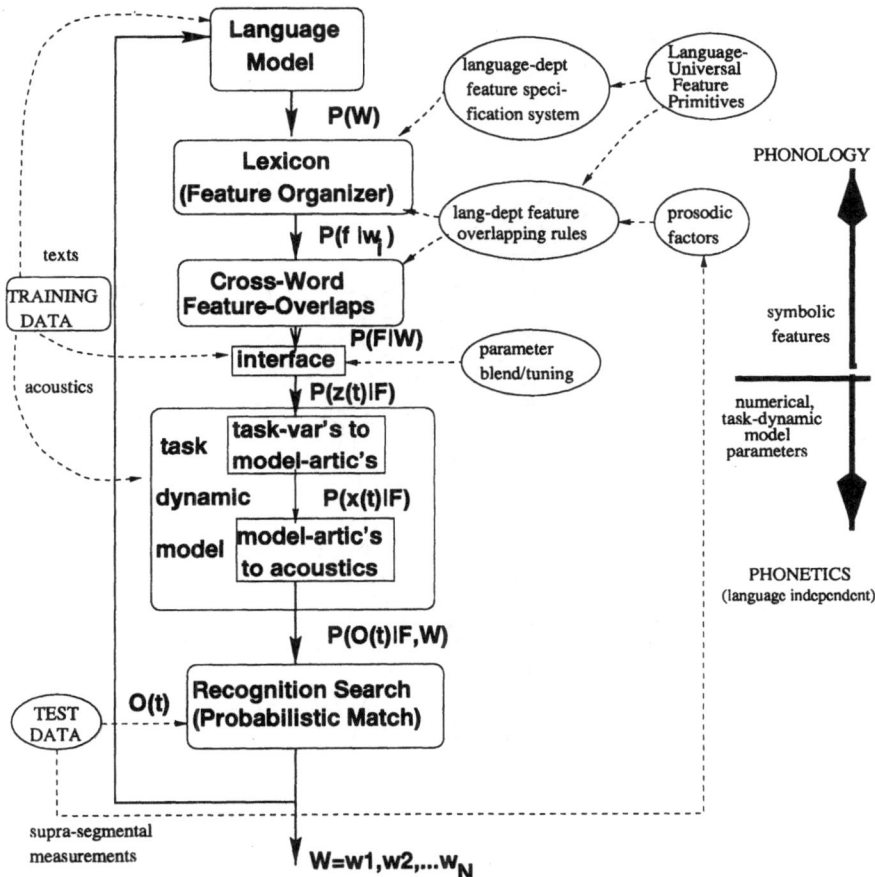

Fig. 1. An architecture of speech recognition using feature-based phonological model interfaced to statistical task-dynamic model of speech production

this model can be regarded as the a simple version of the "key" used by speaker to encode phonological messages and simultaneously as a functionally compatible "internal" model used by human listener in decoding speech (perception). In machine recognition of speech, this production-oriented model is used by the recognizer to accurately and succinctly characterize the dynamic pattern of the observed speech signal, thereby providing an accurate term in class probability $P(\mathbf{O}|W)$ of Eqn.(1). According to Bayesian decision which we adopt as an engineering strategy for the analogous human listener's cognitive interpretation of the auditorily received speech information, optimal recognition performance would be achieved given accurate estimate of $P(\mathbf{O}|W)$.

The proposed production-oriented approach to speech recognition is based on statistical characterization, via functional approximation, of the signals at various levels of the human speech "chain" — phonological, motor-task, articulatory, acous-

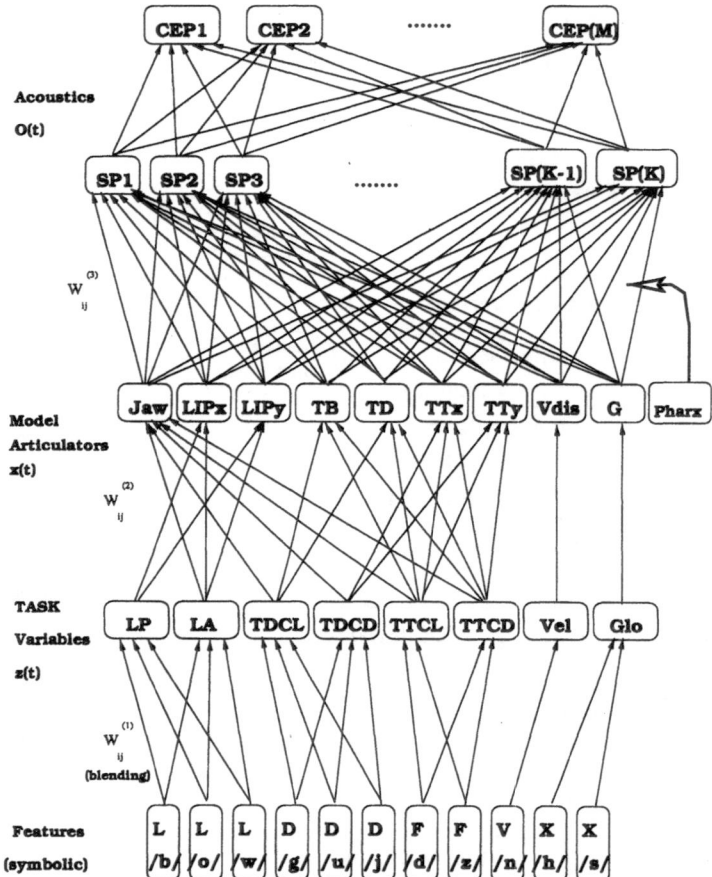

Fig. 2. Neural-net implementation of static nonlinear mappings in the task-dynamic model

tic, and auditory levels. In particular, statistical relations among the signals at these levels are functionally approximated. This, therefore, contrasts sharply with motor theory, direct-realist theory, and acoustic-auditory theory of speech perception (section 3) which all insist on existence of phonetic invariance either at particular level(s) of the speech production process, or at the output of such a process. It also contrasts sharply with the analysis-by-synthesis model of speech perception (section 3) in that the speech recognition decision is made in an integrated manner by evaluation and comparison of a posteriori probability $P(W|O)$ given possible candidate hypotheses[3], rather than performing step-by-step inversions from the auditorily received signal to the final perceptual object of linguistic messages.

Similar to the well established practice in modern speech recognition research, the functional form of the speech encoder described in this paper, which comprises

---

[3] This is the same philosophy permeating all modern speech recognition research.

the feature-based phonological model interfaced to a task-dynamic model, is fixed a priori, while the parameters of the speech encoder are automatically trained from observable speech data [4]. While the Bayesian-based optimal decoding strategy remains the same, different training methods have different implications in terms of the various phonetic theories of speech perception. Maximum-likelihood training implies phonetic invariance in the parameters of the speech production model characterizing systematic changes of the phonetic variables[5]. For Bayesian-style training (including but not limited to the MAP learning), the implication is that phonetic invariance exists in the probability distribution classes, consistent with the proposal in a recent theoretical framework that speech production goals are specified in terms of regions (distributions) rather than of points [17]. Finally, minimal classification error (discriminative) training will imply non-existence of any kind of underlying phonetic invariance; rather, perceptual contrasts are the primary objective in determining the speech encoder's parameters. If some type of constraints on speech economy are incorporated into the speech model as an encoder[6], then proper balance between the degree of the constraints and the discriminative objective would give a way of implementing the concept of H&H theory (section 3) advocating encoder-decoder or speaker-listener interactions and mutual constraints in the human speech communication process.

References

[1] Clements N. and Hume E. (1995) "The internal organization of speech sounds," in *The Handbook of Phonological Theory*, J. Goldsmith (ed.), Blackwell, Cambridge, 206-244.

[2] Denes P. and Pinson E. (1973) *The Speech Chain — The Physics and Biology of Spoken Languages*, New York, N.Y., Doubleday Press.

[3] Deng L. (1993) "Design of a feature-based speech recognizer aiming at integration of auditory processing, signal modeling, and phonological structure of speech." *JASA*, vol. 93(4) Pt.2, pp. 2318.

[4] Deng L. (1992-1993) "A Computational Model of the Phonology-Phonetics Interface for Automatic Speech Recognition," Summary Report of Research in Spoken Language, Laboratory for Computer Science, Massachusetts Institute of Technology.

[5] Deng L. (1997) "Integrated-multilingual speech recognition using universal phonological features in a functional speech production model," *Proc. ICASSP*, Munich, Germany, vol. 2, pp. 1007-1010.

[6] Deng L. (1998) "Computational models for speech production," In *Computational Models of Speech Pattern Processing*, (this volume) in *NATO ASI Series F*, pp. 199–213, Springer–Verlag, Berlin, 1998.

[7] Deng L., Ramsay L., and Sun D. (1997) "Production models as a structural basis for automatic speech recognition," *Speech Communication*, August issue.

[8] Deng L. and Sameti H. (1996) "Transitional speech units and their representation by the regressive Markov states: Applications to speech recognition," *IEEE Trans. Speech Audio Proc.*, vol. 4(4), pp. 301–306.

---

[4] For details, see [6].

[5] To be differentiated from the claim of motor theory that phonetic invariance is in the actual phonetic variables themselves.

[6] See one example in [6] based on the smoothness-prior Bayesian approach developed originally by statisticians.

[9] Diehl R. and Kluender K. (1989) "On the object of speech perception," *Ecological Psychology*, vol 1, pp. 1-45.

[10] Fowler C. (1986) "An event approach to the study of speech perception from a direct-realist perspective," *J. Phonetics*, vol. 14, pp. 3-28.

[11] Halle M. and Stevens K. (1962) "Speech recognition: A model and a program for research," *IRE Trans. Information Theory*, vol. 7, pp. 155-159.

[12] Klatt D. (1989) "Review of selected models of speech perception," in *Lexical Representation and Process*, W. Marslen-Wilson (ed.), pp. 169-226.

[13] Liberman A., Cooper F., Shankweiler D. and Studdert-Kennedy M. (1967) "Perception of the speech code," *Psychology Review*, vol. 74, pp. 431-461.

[14] Liberman A. and Mattingly I. (1985) "The motor theory of speech perception revised", *Cognition*, vol. 21, pp. 1-36.

[15] Lindblom B. (1990) "Explaining phonetic variation: A sketch of the H&H theory," in *NATO Workshop on Speech Production and Speech Modeling*, W. Hardcastle and A. Marchal (eds.), pp. 403-439.

[16] Lindblom B. (1996) "Role of articulation in speech perception: Clues from production," *JASA*, vol. 99(3), pp. 1683-1692.

[17] Perkell J.S., Matthies M.L., Svirsky M.A., and Jordan M.I. (1995) "Goal-based speech motor control: a theoretical framework and some preliminary data," *J. Phonetics*, vol.23, pp. 23-35.

[18] Ramsay G. and Deng L. (1995) "Maximum likelihood estimation for articulatory speech recognition using a stochastic target model," *Proc. Eurospeech*, vol. 2, pp. 1401-1404.

[19] Saltzman E. and Munhall K. (1989) "A dynamical approach to gestural patterning in speech production," *Ecological Psychology*, 1, pp. 333-382.

[20] Stevens K. and Blumstein S. (1981) "The search for invariant acoustic correlates of phonetic features," in *Perspectives on the Study of Speech*, P. Eimas and J. Miller (eds.), pp. 1-38.

Talker Normalization with Articulatory Analysis-by-Synthesis

Richard S. McGowan

Sensimetrics Corporation, 48 Grove Street
Somerville, MA 02144 USA
email: mcgowan@sens.com

Summary. Internal articulatory models are used in analysis-by-synthesis to recover the movement of the speech articulators from speech acoustics. The kind of articulatory information that is recovered depends on the application and the available data. In the laboratory some articulatory data may be available along with acoustic data, and in automatic speech recognition only acoustic data is available. While there is more data available in the former than in the latter case, the amount of information sought in recovery is different in the two cases. In the laboratory physically realistic articulatory trajectories are sought, while recovery in automatic speech recognition may simply require transforming the acoustic signal to an abstract articualtory representaion employed by statistical models for subsequent categorization. Both applications require that the internal articulatory models be normalized for each talker, either for realistic recovery or for robust statistical behavior. A method for constructing mappings between the human and the internal model, while simultaneously adjusting the internal model for acoustic matching is presented. The method is tested on x-ray microbeam data taken on human subjects.

Key words: Normalization, adaptation, articulatory recovery, analysis-by-synthesis

1. Introduction

The recovery of articulatory movement from speech acoustics, or the speech inverse problem, is a long-standing problem [1]. Many of the proposed methods of solution employ analysis-by-synthesis procedures, where an internal articulatory model or representation produces speech that can be compared with speech data, and whose control parameters can be iteratively adjusted in an optimization procedure to obtain acoustic outputs close to the speech data. Automatic learning techniques that result in codebooks or neural networks, and which have been trained using analysis-by-synthesis, are included in such methods. The recovered articulation may be represented in terms of vocal tract area functions [2], articulatory positions and trajectories [3], or articulatory control parameters [4] [5] [6]. The particular model or representation chosen for the internal model, and the degree to which acoustic matching is required, depends on the application (e.g. speech coding) and the amount of data available (e.g. acoustics and facial movement).

Recovery of articulation from speech acoustics may serve to supplement articulatory information that can be obtained by direct measurement. For instance, in the laboratory, lip movement of a talker may have been measured simultaneously with the speech acoustics, and tongue movement is to be recovered from these same data. Another possibility is that tongue position and the acoustic speech signal have been recorded from a talker in a previous experiment, and one would like to recover the articulatory movement in an experiment with the same talker where only speech acoustics is available. While this is a data-rich situation, the experimentalist hopes to recover as accurate a picture of the physical motion of the articulators as possible. If the parameters of the articulatory model, including dynamic parameters, are conceived of as forming a Cartesian space, then the experimentalist would be attempting to recover the detailed geometry of the space.

Statistical models can be based on articulatory representations for speech pattern recognition purposes [7], and may be specialized to automatic speech recognition (ASR) [8]. The only available data is the speech acoustics itself, but, on the other hand, it is not important to obtain a physically accurate picture of the articulatory movement to satisfy the requirements of this application. (This application would require that some kind of learning of the relation between the acoustic signal and the articulatory representation has occurred.) It may only be necessary to capture some of the relations between points in the Cartesian space of articulatory parameters, and not the detailed geometry, to obtain the correct categorization of the speech sounds.

Vocal tract normalization is required for both laboratory articulatory recovery and the use of an articulatory representation in ASR. For laboratory recovery it is necessary to account for differences in the model vocal tract's anatomy with human subject's vocal tract. This will ensure that the differences in anatomy are not attributed to the movement of the articulators or vice versa. The differences between talkers in the way they use their articulators to produce speech sounds, or idiolect, would be preserved and not "normalized out" in the recovery process. In the case of ASR both idiolectical and anatomical differences would, ideally, be normalized out in speaker independent training to keep distributions belonging to different linguistic units well separated. Therefore, like articulatory recovery itself, the way normalization is performed depends on the application and the amount of data available.

For both laboratory recovery and articulatory representation in ASR, articulatory movements in regions of the midsagittal plane are of most interest, because many of the linguistic properties and acoustic consequences of articulation depend on the midsagittal aspects of articulation. Thus, the job of a normalization procedure is to enable recovery of midsagittal articulatory movement via analysis-by-synthesis.

A normalization procedure has been implemented using a particular articulatory model, the Haskins articulatory synthesizer, ASY, shown in Figure 1 [9] [10]. For purposes of illustration, normalization will be discussed in relation to this implementation, and the ASY vocal tract will be known as the *standard vocal tract*. The ASY vocal tract is not advocated as an articulatory representation for ASR applications, but can be useful as a experimental tool when considering such applications. After the method of normalization has been discussed, illustrative experiments will be reported.

2. Normalization Procedure

The normalization procedure has two aspects. One is a mapping that transforms shapes in the region of the midsagittal plane of the human vocal tract (the domain

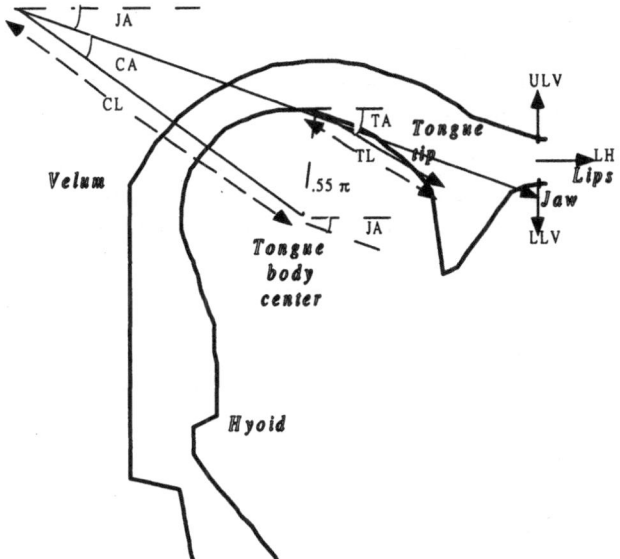

Fig. 1. The midsagittal plane of the ASY vocal tract

space) to corresponding shapes in the standard vocal tract (the range space). This mapping, known as the normalizing map (*NM*), codifies the equivalence of shapes in the midsagittal planes of the two vocal tracts. This mapping needs to be invertible, and it is presumed to map biologically homologous points to one another. The inverse of the NM will be used to obtain the motion of the human articulation in the midsagittal plane once the equivalent motion in the standard vocal tract has been obtained in the recovery procedure from the human speech data. The other aspect of the normalization procedure is the adjustment of parameters in the standard vocal tract not affected by the NM. These parameters, such as the average vocal tract length and average transverse pharyngeal dimension, can be considered to be outside the range space of the NM, and are called *PONMs*. The PONMs are adjusted to obtain similar acoustic outputs from the human and standard vocal tracts for articulatory movements that are equivalent under the NM. This allows the recovery procedure to use the standard vocal tract as part of an internal model with human speech data in order to recover the equivalent articulatory movement of the standard vocal tract. As noted above, the human articulatory movement is recovered simply by applying the inverse NM to each time frame of the recovered standard vocal tract articulatory movement. The PONMs characterize some of the individual aspects of the talker and serve to reshape the standard vocal tract to fit the human vocal tract. What this fitting actually means and how the PONMs characterize the talker will depend on exactly how the normalization is performed.

The particular steps of the normalization procedure are now discussed, in both the context of laboratory articulatory recovery and for automatic speech recognition.

In the case of laboratory recovery, it is assumed that articulatory point information, in terms of pellet data for the tongue, lips, and jaw are available, simultaneous with acoustics for several vowels. In the case of ASR there are only acoustic data and transcriptions available for several vowels.

1. Choose a representation (coordinate system) in each the domain space and the range space of the NM. This makes it possible to define the NM as a mathematical function from one coordinate system to another, when there are sufficient data. Thus, the mapping of coordinates is equivalent to mapping shapes. For example:
   a) In the case of laboratory recovery, the coordinate system for the human may be a Cartesian coordinate system of x-ray microbeam pellets, and the coordinate system for the standard vocal tract consists of the articulatory coordinates of ASY (see Figure 1).
   b) In the case of ASR, there is no mathematical coordinate system for the human vocal tract and articulatory positions are denoted only by phone category. The coordinate system for the standard vocal tract, again, consists of the articulatory coordinates of ASY.

2. Choose a set of pairs of midsagittal shapes, one member of a pair corresponding to the human vocal tract and the other to the standard vocal tract, that are to be equivalent under the NM. These shapes, or pairs of coordinates, will initialize the NM and are known as *initial corresponding vocal tracts*. For example:
   a) In the case of laboratory recovery, the initial corresponding vocal tracts are the vowels for which human articulatory and acoustic data is available. The midsagittal vocal tract coordinates of ASY are set for each vowel so that there is a close visual match between the outlines of the ASY vocal tract and the pellet positions. This is called *visual matching*.
   b) In the case of ASR, the initial corresponding vocal tracts are the vowels for which there is human acoustic data available. The midsagittal vocal tract coordinates of ASY are set for each vowel so that ASY produces an acoustic output which can be identified as belonging in the correct category by a human listener. This is called *auditory matching*.

3. Optimize for acoustic match over the set of initial corresponding vocal tracts, by adjusting the PONMs. These PONMs consist of parameters such as nominal vocal tract length, transverse pharyngeal dimensions, and the parameters used to convert midsagittal tongue-to-palate distances to cross-sectional areas. In both the laboratory recovery case and the ASR case, optimize for least squares matching in the first three formant frequencies over all initial corresponding vocal tract pairs simultaneously.
   a) In the case of laboratory recovery, the resulting PONMs should accurately reflect the anatomical features of the human talker, because the initial corresponding vocal tracts were initialized to provide a realistic midsagittal match.
   b) In the case of ASR, the resulting PONMs will not necessarily reflect the anatomical features of the human talker. The acoustic matching achieved

by adjusting the PONMs will compensate for any mismatch in the midsagittal fits between the human and standard vocal tracts, as well as other anatomical features.

4. Generalize the NM from the initial corresponding vocal tracts by using regression or interpolation. This step only needs to be performed for subsequent laboratory recovery. Steps 1-3 serve as a method of normalization or adaptation in both laboratory recovery and ASR.

3. Experiments

Experiments were conducted to test part of the procedure for constructing an NM and obtaining PONMs. The purpose of the two experiments was to test whether it is possible to obtain reasonable and effective PONMs. That is, the talker characterization aspects of the procedure were tested here. The first experiment involves using PONMs to segregate talkers according to gender. The second experiment was used to test whether the PONMs derived from a limited set of phones could be used to predict the acoustic properties of a novel phone. Also, in the second experiment, the differences between visual and auditory matching were considered.

In both experiments, vowel data from the Wisconsin x-ray microbeam facility [11] were used to provide both the acoustic and articulatory data for the initial corresponding vocal tracts. In some cases, initial corresponding vocal tracts for these vowels were set by auditory matching, and in others they were set by visual matching. Also, a genetic algorithm was used as the optimization algorithm in the analysis-by-synthesis procedure in both experiments [12]. For each optimization a set of PONMs was chosen to be varied to obtain the best acoustic fit over the set of vowels of interest. The goodness of fit between the acoustic output produced by the standard vocal tract with a proposed set of PONMs and the acoustics produced by the human was measured by a fitness function. This fitness function was defined as the inverse of the sum of the squared difference between the human data formant frequencies and standard vocal tract formant frequencies (normalized by the data formant frequency), where the sums are over the formant numbers (one to three) and over the set of vowels. Thus, the optimization was done over several phones (in this case vowels) simultaneously, with the articulator positions for the standard vocal tract set either according to auditory or visual matching.

For experiment 1, the vowels /aa/, /uw/, /iy/, /eh/, and /ae/ produced by four females and four males were used to set the standard vocal tract articulatory positions by auditory matching. (The articulatory positions for these vowel categories had been set by colleagues at Haskins Laboratories at an earlier date.) The PONMs that were varied in this experiment were vocal tract length and transverse pharyngeal dimensions, as given by a linear fit. (These transverse dimensions are combined with the tongue-to-pharynx distance in the midsagittal plane to calculate a cross-sectional area in the standard vocal tract.) The PONMs were derived for each talker by the procedure described in the previous paragraph.

The results, presented in Table 1, are represented as the ratio of derived PONM value to the default PONM value for the standard vocal tract. To obtain a single scalar for the pharyngeal transverse dimension, the sum of transverse dimensions over all pharyngeal sections was calculated.

Table 1. Results of Experiment 1 with vocal tract length and pharyngeal transverse dimensions as PONMs

Subject F=female, M=male)	Ratio of recovered to default length	Ratio of recovered to default sum of pharyngeal dimensions
F1	.92	.79
F2	.89	.80
F3	.92	.75
F4	.90	.78
M1	1.01	.86
M2	1.04	.99
M3	1.09	.98
M4	1.13	.83

The males and females can be segregated based either on length or on the sum of transverse pharyngeal dimension, as derived in this experiment. It should be noted that auditory matching (a necessity when there is no articulatory data available) should produce less accurate results than visual matching, where the standard vocal tract is set according to pellet data. This intuition was tested in the following experiment.

In the second experiment, the x-ray microbeam data from one female and one male were used. In some cases auditory matching was used, and in others visual matching was used to test whether a visual matching procedure produced better PONM values than did auditory matching. To test the goodness of the derived PONM values, three vowels, /aa/, /uw/, and /iy/ were used as initial corresponding vocal tracts (3-vowel optimization) and the subsequently derived PONMs were used to produce the vowel /ae/. The derived PONMs can be judged to be good if the fitness of /ae/ increases after optimization. A "best" possible improvement was provided by the condition when /ae/ was included with the other three vowels in the set of initial corresponding vocal tracts (4-vowel optimization). There was a larger set of PONMs allowed to vary for optimization in this experiment than in the first experiment. Not only were the vocal tract length and transverse pharyngeal dimensions allowed to vary, but also allowed to vary were the coefficients and powers for exponential functions that convert the midsagittal distance from the tongue to the palate to cross-sectional areas.

The results are shown in Tables 2 and 3 for the fitness of /ae/ produced by the standard vocal tract when there is no optimization, when there is 3-vowel optimization, and when there is 4-vowel optimization.

It can be noted that before optimization visual matching always provides a better fitness for /ae/ than does auditory matching. The same is true for the 4 vowel optimization. There seems to be an anomaly for the 3 vowel optimization for the female

Table 2. Fitness of /ae/ for the female subject

matching type	before optimization	after 3 vowel optimization	after 4 vowel optimization
auditory	7.3	50.5	14.5
visual	15.6	27.1	49.6

Table 3. Fitness of /ae/ for the male subject

matching type	before optimization	after 3 vowel optimization	after 4 vowel optimization
auditory	11.5	?	38.5
visual	24.3	55.1	101.0

subject (a data point is missing for the male subject). However, the high fitness for /ae/ in the 3 vowel optimization with auditory matching appears to have occurred by chance, because, when /ae/ is included in the set of initializing utterances for the 4 vowel optimization, the fitness is much smaller. While visual matching does appear to perform better than auditory matching, auditory matching does provide some advantage in the fitness of the vowel /ae/. In fact, the fitness of all four vowels using auditory matching increases in 4 vowel optimization from 2.6 to 6.7 for the male, and from 1.5 to 5.3 for the female.

4. Conclusion

A normalization procedure that also characterizes a talker's vocal tract appears to be possible. That characterization can go beyond simple vocal tract length to include pharyngeal dimensions and area functions. While visual matching appears to be best, auditory matching appears to improve the talker characterization as well. This approach to normalization lends itself to on-line adaptation where recognition and vocal tract characterization are performed simultaneously.

References

[1] Schroeter, J. and Sondhi, M.M: Techniques for estimating vocal-tract shapes from the speech signal, *IEEE Transactions on Speech and Audio Processing, 2*, 133-150 (1994)
[2] Wakita, H. Direct estimation of the vocal tract shape by inverse filtering of acoustic speech waveforms, *IEEE Transactions on Audio and Electroacoustics, 21*, 417-427, (1973)
[3] Parthasarthy, S. and Coker, C. H.: On automatic estimation of articulatory parameters in a text-to-speech system, *Computer Speech and Language, 6*, 37-75 (1992)
[4] Shirai, K. and Kobayashi, T.: Estimating articulatory motion from the speech wave, *Speech Communication, 5*, 159-170 (1986)
[5] McGowan, R.S.: Recovering articulatory movement from formant frequency trajectories using task-dynamics anda genetic algorithm, *Speech Communication, 14*, 19-48 (1994)

[6] McGowan, R.S. and Lee, M.: Task-dynamic and articulatory recovery of lip and velar approximations under model mismatch conditions: *Journal of the Acoustical Society of America, 99,* 595-608 (1996)

[7] Jung, T.-P., et al.: Deriving gestural scores from articulator-movement records using weighted temporal decomposition, *IEEE Transactions on Speech and Audio Processing, 4,* 2-18 (1996)

[8] Deng, L. and Sun, D. X.: A statistical approach to automatic speech recognition using atomic units constructed from overlapping articulatory features, *Journal of the Acoustical Society of America, 95,* 2702-2719 (1994)

[9] Mermelstein, P.: Articulatory model for the study of speech production, *Journal of the Acoustical Society of America, 53,* 1070-1082 (1973)

[10] Rubin, P., Baer, T. and Mermelstein, P.: An articulatory synthesizer for perceptual research, *Journal of the Acoustical Society of America, 70,* 321-328 (1981)

[11] Westbury, J.R.: *X-ray microbeam speech production database user's handbook.* Waisman Center on Mental Retardation and Human Development, University of Wisconsin.

[12] Goldberg, D.E.: *Genetic algorithms,* Reading, MA: Addison-Wesley 1989

The Psycholinguistics of Spoken Word Recognition

Cynthia M. Connine and Thomas Deelman

The State University of New York at Binghamton

Summary. The process of mapping acoustic-phonetic level input to a lexical representation is multi-faceted. Models of spoken word recognition provide a variety of processing architectures and make different assumption(s) regarding the unit(s) of representation used in the exchange of information from signal-to-word and the nature of information flow through the system. The current models provide a backdrop for a discussion of some of the advances and debates in the field. Some of the issues considered are: early versus delayed commitment to a lexical hypothesis, consequences of multiple activation, segmentation and lexical access, the processing and representation of phonological variants, and the role of attention in spoken word recognition.

Key words: psycholinguistics; speech processing; lexical competition; phonological variation; lexical access; attention; sentence context; representation; interactive activation.

1. Introduction

Auditory word recognition is a remarkably robust aspect of language processing. The acoustic signal may be distorted and/or disrupted in a wide variety of ways yet spoken word recognition appears to proceed with little difficulty. Among those potentially problematic aspects of spoken language are misarticulations in the form of speech errors, the presence of acoustic-phonetic variability due to coarticulation processes and other dynamic processes among successive phonemes, and masking of speech sounds due to environmental noise. Beyond the inherent variability in the acoustic signal, a second observation about spoken language is that the beginnings and endings of words or syllables are not delineated in ongoing speech by obvious discontinuities. One specific observation that has been emphasized in early models of spoken word recognition is that the speech signal is distributed over time. Unlike visual language in which information may be obtained in parallel, information in the speech signal is available to the listener in a time dependent manner. Simply stated, the beginnings of words reach the listeners' ear prior to the ends of words. We begin by providing an overview of three influential models of spoken word recognition. We then turn to a discussion of major issues that have emerged and the evidence offered in support of opposing views.

2. Overview: Models of spoken word recognition

A number of models have been developed to account for the ease and efficiency of spoken word recognition. A particularly influential model, the cohort model, proposed that the information that reaches the listener first (word-initial information) serves to activate a set of lexical hypotheses that are consistent with it [1]. The set of words that are activated early in spoken word recognition is determined by a strict mapping based on phoneme overlap of word-initial information. Lexical hypotheses are eliminated as soon as they deviate from the acoustic-phonetic input. Word recognition is claimed to occur when there is only one lexical hypothesis remaining in the cohort. In this version of the model, no cost to word recognition is incurred by the size of the cohort in that the set of words that are activated based on the input do not compete with each other during recognition. The cohort model has undergone several significant revisions [2]; [3]; [4]; [5]. The model now includes assumptions concerning activation and competition; candidates are activated to the degree that they match the acoustic-phonetic input and activation of any lexical hypothesis is evaluated with reference to other activated words. There are no direct effects of competition among simultaneously activated words, rather these effects are located at a decision phase of the recognition process. Reliance on word-initial information and elimination of lexical hypotheses from the cohort based on sequential mismatches have been retained.

The TRACE model of spoken word recognition assumes feature, phoneme and word levels [6]. The three levels are connected hierarchically via excitatory connections. Units within a level are inhibitory and the degree of lateral inhibition from any given unit is directly related to the degree of activation of the unit and the similarity between the activated and inhibited units. The model is an interactive activation model in that activated lexical representations can feed back to boost the activation of component phonemes. In this model, word recognition occurs when a pre-set level of activation is reached for a given word node. The model is able to account for a large number of phenomena in speech perception including co-articulatory effects, categorical perception and trading effects (among others). The interactive connections also recommend the model as it permits the influence of lexical structure on lower levels to emerge naturally from the model's architecture.

TRACE and the cohort model have been influential in the development of other models of spoken word recognition and aspects of both have been incorporated into the SHORTLIST model [7]. The SHORTLIST model is a connectionist model that consists of two stages. The initial stage involves activation of a limited set of lexical hypotheses that are consistent with the input. In contrast to TRACE, activation of lexical candidates is based strictly on the acoustic-phonetic input with no feedback connections between levels. A second stage is organized as an interactive activation network where the set of lexical hypotheses compete. More recently, the model has been revised to incorporate a distinction between strong and weak vowels. A word containing a strong, initial vowel that is incorporated into the short list is favored during the second stage. This modification was included to permit modeling of lexical segmentation and competitor effects (see discussion below).

A number of major issues have emerged as the focus of empirical and theoretical efforts. We begin by discussing the nature of the units proposed for mapping from the speech signal to a lexical representation. In this discussion, important questions concerning the way in which these units are extracted are finessed. Also addressed are alternative conceptions of lexical representation with a focus on processing phonological variants. When possible, we will refer to explicit assumptions of various models and the degree to which these assumptions are supported by empirical results.

3. Currency of mapping: units and the nature of lexical representations

A basic question in spoken word recognition has been the unit of processing that is extracted from the speech signal to activate a lexical representation. The proposed units vary in terms of their abstractness but the goal is to provide a representation that is of a comparable vocabulary as the lexicon. One early proposal by Dennis Klatt [8] was a spectral template that served to activate lexical representations through a lattice of possible templates. According to this view, there is no intermediate representation from the signal to lexical activation in the form of more abstract units. More abstract units of features and phonemes have been assumed to intervene between speech and lexical activation in models such as TRACE where separate nodes form hierarchical links to the lexicon (see [9] for arguments in support of a phoneme representation). More recently, Marslen-Wilson and Warren [10] have argued explicitly against a phoneme level of representation and for a feature-based mapping. Evidence and logical arguments in favor of units larger than a phoneme have been proposed for units that assume structure at a level that spans multiple phonemes or feature clusters (see [11] for discussion). One important consideration in evaluating any of these proposals is the functional utility of the representation. What motivated representation permits the emergence of important computational properties in word recognition? As we discuss below, a fundamental characteristic of the architecture of spoken word recognition is activation of a lexical representation based on similarity: lexical representations are activated to the degree that there is a match relative to the input. This architecture lends itself more naturally to a feature based representation where mapping from an abstracted input to a lexical representation is in a currency of features. This illustrates the general property that the representational system of the lexicon must interface with the representation constructed from the signal. (see [12]).

A related but separate question is the way in which predictable variations of a word are represented and processed. Predictable variations in language consist of instances where a language users' knowledge of the sound structure of their language may come into play. A class of variants of this type are phonological processes that result in systematic variations in the speech signal that are licensed for specific environments. Despite the ubiquitous nature of phonological variation, the topic has received relatively little empirical or theoretical attention. A recent proposal by Lahiri and Marslen-Wilson [13] that has been extended by Gaskell (e.g. [14]) has been guided by representational assumptions of some current approaches in linguistic theory. Some linguists have proposed that only those phonetic features that are marked are represented (cf. [15]). In the formulation of the representational theory for the lexicon, Lahiri and Marslen-Wilson suggest that underspecified representations form the target representation for spoken word recognition. In their work, they demonstrated that oral (unmarked) vowels presented in a gating paradigm generated lexical candidates with nasalized as well as oral vowels. The presence of candidates with nasalized vowels was taken as evidence that the oral vowel successfully activated lexical representations of both categories (nasal and oral) since nasality is unmarked in the lexical representation. In a related set of experiments, Gaskell and Marslen-Wilson [14] have argued that phonological variants are recognized via an underspecified representation plus phonological inference rules that assess the viability of a variant in context. Cross modal priming showed comparable activation for LEAN and its possible variant LEAM in isolation, hence supporting recognition of the variant and its base word via the same underspecified representation. However, no activation was found for the variant LEAM when it was presented in a context that did not permit its occurrence (LEAN GAMMON). Gaskell and Marslen-Wilson argued that these latter results support a phonological inferencing mechanism.

In our own work, we have investigated processing of phonological variants that are not context conditioned, in particular, the common variant of flapping in American English [15]. A flap can occur when a voiceless alveolar stop /t/ follows a strong vowel (e.g. the word PRETTY, /prIti/, may be realized as /prIDi/). The production of a flap is a very common variant in conversational speech while the medial /t/ variant is rare and typically produced only when a speaker articulates a word more carefully. Our research investigated how flapped and canonical forms are processed and represented and in particular whether the very frequent flapped form is lexically represented. In a semantic cross modal priming study we found that both forms (canonical and flap) showed comparable priming. However, an asymmetry emerged for minimal nonwords where the initial phoneme was altered by a single feature (e.g. /brIti/ and /brIDi/) - a reduced priming effect was found only for the flapped case. Canonical nonwords showed no significant priming. The word effects suggested that both forms were represented while the nonword effects suggested an asymmetry in representation that favored the flapped form. A second experiment was designed to investigate the functional relationship between these highly similar, yet distinct phonological representations using a form priming paradigm. Words sharing a high degree of acoustic/phonetic overlap typically compete (see Section 5.) and it is possible that this characterizes flapped and canonical forms. However, the unique relationship enjoyed by two forms that map onto a single meaning would favor a mutually facilitory relationship. If so, the asymmetry found in cross modal priming should be mirrored in form priming with activation of the more frequent flapped form exerting the larger effect on recognition of the canonical form. Further, if excitatory connections among lexical forms mirror the architecture of inhibitory connections, the strength of the connection may be defined by frequency. The results supported both of these predictions. An asymmetric facilitory relationship between variants was found: a flapped target showed a 25 ms priming effect when preceded by their canonical forms and a canonical target showed a 77 ms priming effect when preceded by their flapped forms. The processing assumptions of these experiments and from context conditioned variants (Gaskell and colleagues) appear to support different mechanisms. However, a resolution may be found in the nature of the variant (context condition vs word internal) or its frequency of occurrence. The two sets of studies may reflect the evolution of a surface form variant in its lexical representation. Clearly, additional research on phonological variants is necessary.

An issue related to units of representation is the way in which word units are extracted from continuous speech. Under most circumstances, word boundaries are not marked by obvious discontinuities (but see [17]) and pauses between words are consistently found only between words at major syntactic boundaries. The solution for the segmentation problem proposed by the cohort model assumes that the segmentation is simply a by-product of recognition. A word boundary is implicit at the offset of the speech stream that defines a word in that the onset of the following word is a default of recognizing the previous word. In the TRACE model, segmentation involves a competitive process among potential words and the actual word defined sequentially by the input. Lexical hypotheses wax and wane depending upon their match with the input and competition from other lexical hypotheses. The SHORT-LIST model has recently been revised to incorporate an explicit strategy called the Metrical Stress Strategy (MSS). Originally proposed by Cutler and Norris [18], the MSS assumes that listeners of English hypothesize word boundaries for syllables containing strong vowels. Evidence supporting the MSS comes from word spotting experiments where a real word must be detected when embedded in nonsense syllables. In the crucial conditions, a word was followed by a nonsense syllable with a strong vowel or a weak vowel. Cutler and Norris found that word spotting was faster in the context of a strong nonsense syllable. They argued that the strong-strong sequence triggered a segmentation providing an unambiguous representation for the word spotting response. The empirical evidence supporting the MSS is consistent with computational analyses of English words. Cutler and Carter [19] have shown that the majority of English words begin with strong initial syllables.

4. Temporal nature of speech: early vs delayed commitment

The time dependent nature of speech may impact lexical activation in that as more of the stimulus is heard (over time), the amount of information that contributes to lexical activation is greater. For some long words, a subset of the potentially relevant information may be sufficient to uniquely identify it. As previously described, this characteristic of spoken language has been formalized in the cohort model in the concept of a uniqueness point which is the point at which a word is recognized [1]. In support of this, Marslen-Wilson [20] found that fluent restorations of mispronounced words in a shadowing task often occurred within 250 ms. Further, fluent restorations were more likely when the mispronunciation occurred latter in a word. Marslen-Wilson (1984) found that nonword detection times were independent of the point at which the stimulus deviated from a word and of word length (see also [21], [2]).

Support for the uniqueness point as the 'magic' point of word recognition is mixed. Some researchers found that nonword detection times were longer for stimuli with a late nonword point compared to stimuli with an early nonword point [22]; see also [23]. Some recent research has demonstrated that the influence of early vs late uniqueness point can be accounted for by a sub-optimal lexical strategy [24]. Radeau et al found that effects of uniqueness point location were greater when early and late uniqueness point stimuli were presented in homogeneous blocks. Independent of any particular experimental results, Luce [25] has noted that the concept of a uniqueness point may be limited given the structure of words. In a computational analysis, Luce found that only 39% of common words are unique before the last phoneme.

4.1 Delayed commitment

The temporal nature of speech would seem to lend itself more naturally to a processing system where immediate commitments to lexical hypotheses are advantageous. However, as discussed, the evidence for immediate commitments based on notions such as the uniqueness point is mixed. Moreover, despite the time locked nature of fluent speech, there are a number of considerations that suggest an advantage for delayed commitment. Listening conditions are typically far from optimal and it may be more efficient to delay final commitment to a single lexical hypothesis as opposed to making many erroneous commitments which must be revised. In some instances, an erroneous commitment may be easily revised in that commitments to other levels of linguistic structure may be minimally effected. In contrast, other commitments (a noun-verb ambiguity) may require time consuming structural revisions. In other instances, important morphological information (such as the morphological head) may serve to disambiguate the preceding syllable. In the following section, we discuss the evidence supporting delayed commitments.

Grosjean [26], [27] has published a series of studies that have explicitly examined the issue of sequential word recognition. These experiments used a gating paradigm where successively larger fragments of a word are presented until the entire word is heard. On each trial, the subject indicated the identity of the item. Grosjean [27] capitalized on an earlier finding [26] in which low frequency one syllable words showed relatively low accuracy identifications, in some cases reaching only 50% accuracy. Grosjean [27] found that words presented in sentence context were not correctly identified until midway into the following words. Subject ratings of identification confidence did not reach 100% until approximately 350 ms after acoustic offset. Connine, Blasko and Hall [28] explicitly examined resolution of acoustic-phonetic ambiguity when the ambiguous string was compatible with two lexical items (dent/tent). In these experiments, subjects labeled the initial phoneme of the ambiguity in sentences where semantic biasing information occurred either three or six syllables downstream from the ambiguity. Labeling responses were influenced by the semantic biasing information when it appeared in the three syllable temporal window but not in the six syllable delay. The results suggest temporal constraints on lexical commitments such that within three to six syllables (approximately 1 sec) a lexical commitment to a single lexical hypothesis is made.

Similar retroactive effects have been demonstrated for single word presentation. Cluff and Luce [29] presented two syllable words (spondees) such as PANCAKE and JACKKNIFE in noise for identification. Based on computational analyses, the spondees were composed of easy syllables, a high frequency word with few neighbors or hard syllables, low frequency word with many neighbors. In the crucial condition, the spondees consisted of a hard syllable followed by an easy syllable (PANCAKE). Cluff and Luce found that the hard syllables were identified much more accurately (83%) in this condition compared to hard-hard sequences (74%). Further, the hard syllables in the hard-easy context were identified as accurately as easy syllables (83%). Cluff and Luce interpreted these data as supporting a retroactive resolution of the hard syllable from the easily recognized easy syllable.

More recently, Zwitserlood and Schriefers [30] have investigated the contribution of processing time to lexical activation. They noted that in continuous speech, the passage of time typically results in more acoustic-phonetic information relevant to a word and an increased opportunity to process earlier parts of the input. The natural confound of additional time and information was separated in an experiment where an intact word or the initial syllable was presented in a cross modal priming task. In one condition, the visual target was presented at the offset of the intact word or the initial syllable. In a critical condition, the initial syllable was followed by a delayed visual target where the delay was equal in duration to the missing word portion. They found that priming effects in the delayed condition were comparable to those in the intact word condition. These results suggest that the impact of a stretch of acoustic-phonetic information is not static and that it does not have immediate impact on the activation level of a lexical entry.

5. Multiple lexical hypotheses, lexical competition and graded activation

A number of important properties of lexical activation have emerged as critical model characteristics. These properties include multiple activation of lexical forms, neighborhood effects and graded lexical activation. Multiple activation has been fairly well established in a number of paradigms. Marslen-Wilson [2] investigated activation of multiple lexical representations using a cross modal priming paradigm (see also [31]). Pairs of words were selected that shared onset information but deviated (became lexically unique) before their offset (e.g. CAPTAIN, CAPTIVE). Semantic associates for both of the alternatives were presented prior to unique specification of a word (e.g. after the 't' - CAPT). The results showed that both word candidates were active. Probing at word offset showed activation only for the word that was actually heard. These results demonstrate clearly simultaneous activation of multiple candidates. Other results have shown that multiple activation does not depend crucially on an intact onset. Connine, Blasko and Wang [32] used perceptual-lexical ambiguities where an ambiguous segment resulted in more than one lexical interpretation (e.g. d/tent where the initial phoneme is ambiguous). Cross modal semantic priming showed that both interpretations were activated despite the ambiguous onset. In a related set of experiments, Andruski, Blumstein and Burton [33] showed that the 'goodness' of a word-initial phoneme determines the degree of semantic priming.

The preceding section showed evidence that activation of a target lexical item occurs in the context of simultaneously activated lexical candidates. In effect, words must compete for recognition amid unique constellations of lexical hypotheses, all of which have varying degrees of frequency and phonological overlap (e.g., phonological neighbors). At issue is the consequence of multiple activation for recognition, specifically whether recognition of a word is impaired when it overlaps with other words. One possible theoretical instantiation of the effects of lexical neighborhood is in terms of lateral inhibition among activated lexical representations (e.g., [6]; Norris, [7]; [34]). A mechanism of lateral inhibition allows activated word forms to inhibit one another in proportion to their level of activation. Other models propose passive, activation-based competition in which recognition occurs when the activation of a candidate is sufficiently greater than its closest competitor [35]. Accordingly, the presence of a competitor has no direct influence on a target word [36]. Instead, competitor effects are relegated to a post lexical decision stage. Distinguishing between the different competition mechanisms is a matter of debate, since they predict similar outcomes. The models distinguish themselves only in their predicted patterns of activation. Passive competition models (e.g., Cohort) predict that recognition of a target word may be delayed in the presence of a competitor, although the pattern of activation remains the same to when a competitor is absent. Alternatively, an inhibitory mechanism predicts that activation of a target word will be reduced in the presence of a competitor due to mutual inhibition.

The empirical evidence for lexical competition spans a variety of paradigms, which taken together, suggest that competitor effects are relatively subtle and short-lived. For instance, Marslen-Wilson [4] found competition effects for spoken words only when the acoustic signal was truncated prior to the point at which a word was unique and presented to subjects in a gating task and a cross modal priming paradigm. No competitor effects were found for words where a response was made after the entire word was heard (lexical decision task and auditory repetition task). Once the lexical identity of the input was unambiguous, competition effects were no longer evident. Thus, activation of competitors decays rapidly once the identity of the target is established with relative certainty. Extending this notion, evidence suggests that competition effects do not dissipate when ambiguity remains unresolved [36]. Using cross modal priming, Marslen-Wilson et al. (Experiment 2), manipulated voice onset time to create ambiguous initial phonemes in words which yielded a single competitor (b/plank - WOOD) or no competitor (d/task - JOB). Virtually identical effects were found for the unambiguous source words. The ambiguous no-competitor prime generated a priming effect similar to its source word, however the ambiguous word-competitor failed to show a priming effect. Such results reflect the signature of lexical competition and arguably, mutual inhibition. The competing word forms inhibit one another comparably and form a type of lexical deadlock, precluding sufficient activation to reach a level of word meaning.

Pre-activation of a close competitor has also been shown to influence identification of subsequently presented target word. Goldinger, Luce & Pisoni [37]; (see also [38]; [39]) found reduced identification performance when target words presented in noise were preceded by phonetic primes presented in the clear (e.g., BULL

- VEER). Similar interference effects were obtained in prime-target pairs sharing initial overlap (e.g., BASE-BAKE; [38]; [40]). However, Radeau et al, showed facilitation for pairs sharing final overlap (e.g., BASE-CASE). In the initial overlap case, the target presumably maps onto the pre-activated competitor until the final segment distinguishes its unique lexical status. In the final overlap case, the initial segment distinguishes the target from the prime, and the therefore the remaining pre-activated phonological overlap serves to facilitate activation of the corresponding lexical form. In sum, position of phonological overlap may dictate either lexical interference or pre-lexical facilitation.

Lexical competition effects are not restricted to single competitors. A neighbor may be defined as a word that differs by one phoneme in any position, and therefore different words have lexical neighborhoods ranging in the number and frequency of potential competitors. Words with many, highly frequent neighbors are typically more difficult to identify when presented in noise than words with few, low-frequency neighbors [37;41]. Similar neighborhood density effects have been found for nonwords presented in the clear in a lexical decision task [41]. However, as Bard [42] describes, neighborhood effects could arguably be driven by a single or a few strong competitors since the chance of encountering a strong competitor is greater in dense neighborhoods.

A final body of research investigating the consequences of lexical competition takes a somewhat different approach by assessing activation of embedded words. As discussed earlier, multiple word forms are activated when aligned from onset [31] or misaligned from onset (e.g, activation of BONE in TROMBONE; [43]. The issue however is whether these embedded words enter into competitive relations with the target word. Cluff and Luce [29] have shown that recognition of bisyllabic items comprised of separate words is affected by the neighborhood density and frequency of its constituent words. Similarly, Vroomen and de Gelder [44] used cross modal priming (in Dutch) to assess activation of an initial syllable word which overlapped with the onset of a second syllable. Results showed a systematic reduction in priming the more competitors that began with the onset of the second syllable. In a word spotting task Norris, McQueen, & Cutler [45] showed similar effects. In addition, performance was facilitated when the initial word did not overlap with the onset of the second syllable and the second syllable had many competitors. In the overlap case, the greater number of potential candidates beginning with the second syllable interferes with recognition of the initial word. In the non-overlap case the pattern is reversed since no part of the initial word competes with alternative lexical hypotheses starting at the second syllable and because the the final sequence is itself part of a dense neighborhood. Additional evidence of competition was obtained by McQueen, Norris, & Cutler [46]. In a word-spotting task, subjects had more difficulty detecting words misaligned from onset in nonsense strings that consisted of real word onsets. For example, subjects had more difficulty spotting MESS in DOMESS (derived from domestic) than SACK in SACRIF (derived from sacrifice). In the former case, activation of MESS is presumably inhibited by the relatively greater activation of DOMESTIC, which benefited from earlier activation. Activation is more comparable For words aligned from onset since a dominant inhibitor has yet to be established. In keeping with assumptions of SHORTLIST and TRACE, this suggests that the strength of lexical inhibition is proportional to the degree of activation afforded to the competitor.

As we have discussed, an item may become physically unique at a specific point in the signal and consequently may begin to be identified with relative confidence. However, a plethora of data shows that processing continues after a uniqueness point and that commitment to a single lexical hypothesis is frequently delayed. The data from lexical competition studies shows that regardless of when an item becomes unique in the signal, its recognition may depend upon the processing characteristics of its lexical relations. Accordingly, accurate recognition within an items lexical environment may necessarily entail delayed commitment since recovery from inaccurate commitment may evoke a relatively high cost to processing.

An assumption that is related to multiple activation that shared by a number of models is that the acoustic-phonetic similarity of the input to a stored representation determines the degree to which a lexical representation is activated. We have conducted a series of experiments that directly addressed the consequences of manipulating acoustic-phonetic similarity [47] via linguistic feature overlap [48]. These experiments demonstrated that a nonword derived from a base word (e.g. SEVEN) by changing the initial phoneme by one or two phonetic features (MINIMAL NONWORD; e.g. ZEVEN) activates the base word. Priming for minimal nonwords was found for both high and low frequency base words. However, a nonword in which the initial phoneme was altered by five or more phonetic features (MAXIMAL NONWORD; e.g. MEVEN) did not activate the base word (see also, [49]). More recently, we have used a phoneme monitoring paradigm where reaction time to detect a target phoneme is measured [50]. Both minimal and maximal derived nonwords were used and in addition, a final set of nonwords (control nonwords) were constructed by changing the consonants and vowels excluding the final syllable. The control nonwords served as a baseline against which to assess any facilitation found for the derived nonwords. Connine et al found that phoneme detection reaction times increased as acoustic-phonetic similarity of the input relative to the base word decreased. In a related set of experiments, Deelman and Connine (1997) have demonstrated that the models differ in the consequence of mismatching information when there is no completely matching lexical entry.

6. Language architecture: Lexical and segmental levels

One major dimension along which models differ is in terms of the direction of information flow among levels. The SHORTLIST model and the Cohort model assume that information is strictly uni-directional from feature/phoneme units to lexical representations. In contrast, a distinguishing characteristic of TRACE is its interactive architecture: activated lexical representations feedback to component phoneme levels to boost activation levels at the lower level. The architectural relationship between lexical and acoustic-phonetic levels is currently a matter of some controversy. Connine and Clifton [51] have argued that the lexical/speech interface is a likely candidate for interactive effects since interaction among levels requires a comparable vocabulary of representation. One task used to support an interactive relationship between lexical and speech levels is the phoneme identification task [51];[52]; see also [53]. These studies have juxtaposed effects of lexical knowledge and effects of non-linguistic variables (monetary payoff). Connine and Clifton showed that listeners labelled lexically biased speech continua (e.g. dype-type and dice-tice) with more identifications that formed words (more 'd' responses in the dice-tice vs. dype-type continuum). The influence of lexical status was confined to those stimuli that are perceptually ambiguous. A reaction time analysis of word and nonword responses for perceptually ambiguous stimuli showed an advantage for word responses (see also [54]). A comparable labelling shift was found when listeners received a larger cash reward for a 'd' or 't' response but no reaction time advantage was found for the biased response. Connine and Clifton argued that these results supported predictions from TRACE where perceptual processes compute a representation from the speech signal in concert with feedback from a lexical representation. Pitt [52] extended these results by subjecting the labelling results for lexical status and monetary payoff to a signal detection theory analysis. He found that only the lexical status effects yielded differences in d' and argued that they reflect a true perceptual effect on speech perception.

A second line of research investigating the interplay of perceptual and lexical processes, the phonemic restoration illusion, has also supported an interactive mechanism between the lexicon and speech perception [55]; [56]). The phonemic restoration illusion occurs when listeners hear a word as being intact although one of the phonemic segments has been replaced with noise of the same duration. Samuel [56] developed the phenomenon into an empirical methodology by constructing two versions of words, and having listeners discriminate between them. In one version, noise replaced a given phonemic segment while in the other version noise was superimposed over the phonemic segment. Subjects heard both versions of each word and made forced-choice decisions as to whether the noise was added or replaced. Responses were converted into hits, misses, and false alarms and thus provided the error rates necessary to conduct traditional signal detection analyses. Consequently, subjects' discriminability (d') between replaced and added versions could be separated from a post-perceptual bias (beta) or tendency to generate a particular response. The stronger the illusion, the less a listener can discriminate between versions, which is reflected in lower d' values. Using this methodology, [56] found that subjects had little difficulty discriminating between versions of items containing little or no lexical information (critical segments excised from real words and phonological nonwords). However, subjects had comparatively great difficulty discriminating between versions of real words. Samuel [57] found that when the critical phoneme and the noise were similar (e.g., /s/ and white noise), discriminability was relatively poor compared to cases in which the critical phoneme and noise differed on acoustical dimensions. Taken together, these results suggests that the restoration illusion is a consequence of an interplay between lexical activation and acoustic/phonetic similarity of the critical segment (noise or noise + phoneme) and its target (phonemic) representation. In a recent extension of this work, Samuel [58] has focused on a robust phenomena in the speech literature, adaptation. Adaptation effects are demonstrated when the boundary separating two phonemes along a speech continua shifts as the result of repeated exposure to an adaptation stimulus (typically, an endpoint stimulus from the continuum). Typically, the adaptor produces fewer labelling responses of that category thereby shifting the category boundary away from the adaptor. Samuel [58] demonstrated that lexically restored phonemes can produce adaptation effects on speech perception comparable to an intact stimulus.

An alternative view of lexical effects on speech perception has been offered by Cutler, Mehler, Norris and Segui [59]. This research focussed on developing a task analysis of the phoneme monitoring paradigm to account for facilitated phoneme detection when the target is in a word carrier. Cutler and Norris [60] argued that the phoneme monitoring task is accomplished via a race between a lexical and pre-lexical level of representation. When a target appears in a nonword, only the less efficient pre-lexical representation is available for a response. A word carrier permits both lexical and pre-lexical level of representation for the response but all other things being equal, the more stable lexical representation is available first (on average). Cutler et al demonstrated a lexical effect in phoneme monitoring in that listeners responded faster to word initial targets than initial nonword targets. However, the effect disappeared when target lists were comprised solely of monosyllabic items. The authors theorized that the homogeneity of the stimuli in the monosyllabic lists caused listeners to shift attention to prelexical levels of representation. Since the lists were purposefully mundane, subjects were not concentrating on the lexical meaning of the speech. Alternatively, the more varied lists caused listeners to shift attention to a lexical level, therefore producing more lexically based responses.

The architectural relationship between perceptual processes and word recognition continues to be a hotly debated area. It may be, however, that an interactive architecture will ultimately be required for any computational system to succeed. One indication of this is that this architecture is an emergent property of recent simulations of the Shortlist model (cf. [61]). In this version of the model, an inspection of the hidden units revealed both phoneme and lexical nodes that were directly sharing information. As such, the architecture of Shortlist converges on the same architecture assumed by highly interactive models such as TRACE.

7. Language architecture: Lexical and sentential

The bulk of the research considered in the previous sections focussed on lexical processing. The question addressed in this section focuses on the contribution of sentence level representations to word recognition. The issue of the architectural arrangement of lexical and sentential representations parallels the previously addressed question of lexical and perceptual processes. What kind of sentential information contributes to lexical processes and when during lexical computations do those effects occur? It should be stressed that the issue is not whether sentence context influences lexical processes- clearly there are many examples where that is the case. The question is when during processing do those effects occur. The way in which the question has been framed is in terms of direct (interactive) as opposed to selectional (modular) effects of sentence context on lexical processing. Two of the paradigms described in Section 6., phoneme restoration and phoneme identification have been used to investigate this issue. Samuel [56] has shown that sentence context effects only the B (beta) component in a signal detection analysis of the phoneme restoration phenomenon. This suggests a bias effect of context. Connine [62] demonstrated a sentence context effect in labelling a word-word speech continuum (e.g. bath-path) but no reaction time advantage was found for sentence consistent responses. Finally, Zwitserlood [63] has shown that activation a word and its competitors is not initially influenced by a biasing sentence context. Context effects emerged late in a word and were argued to reflect selection. In summary, the available evidence suggests that sentence context is used to select among activate lexical hypotheses. Unlike perceptual computations that can benefit directly from lexical knowledge, lexical processing is not directly informed from sentence context.

8. Contribution of attention

Our discussion of the architectural aspects of the word recognition system included various perspectives on the nature of information flow through the system. Research conducted on word recognition has focussed on the nature of how information is utilized, or shared, among different levels of representation (cf., [64]). Since the majority of this research is focused on the flow of acoustic/phonetic information through a given processing architecture, considerable insights can be gained by evaluating studies which focus listeners' attention to various levels of processing. In this section we explore how attention allocated to a given perceptual unit (e.g., phonemes, syllables, or words) can provide insight about how elements across various levels of representation are processed. As discussed earlier, there is data supporting the notion that different levels of representation operate in tandem as well as the perspective that such levels are processed in relative isolation from one another. What follows is a description of some of the more prominent work conducted on the role of attention, using paradigms such as phonemic restoration [65]; [66] and phoneme monitoring [67]; [59]. Much of this work involves the allocation, or focusing, of attention to a specific level of processing. However, at the other end of the attentional spectrum, considerable insight can be gained from investigating processing under conditions of decreased attentional focus through increasing cognitive load. Therefore, we also discuss recent work investigating the significance of acoustic/phonetic cues under conditions of reduced attention [68]. The results previously described suggest that the restoration illusion involves contact with a lexical representation and its influence upon lower level discriminations. At issue is the question of whether attentional focus can be allocated to different processing levels. If so, and more interestingly, can it be determined that processes at each level function independently from one another, in so much that such lower level processes are immune from activity at higher levels? For instance, perhaps the illusion is the result of subjects primarily attending to the lexical level. This would direct attention away from the lower phonemic level - the very level presumably most crucial for making the necessary discrimination. If discrimination differs as a function of the level at which attention is directed, then perhaps these levels can function in relative autonomy from one another. Alternatively, evidence of lexical influence, regardless of allocation of attention, would argue for a more interactive processing architecture.

Using the signal detection procedure described above, Samuel and Ressler [65] investigated this issue by directing listeners' attention in two ways. First, training was provided by having subjects perform added/replaced discriminations with feedback. Second, three types of visual cues were presented prior to each trial. One cue was a visual prime of the word that would be heard on that trial. A second cue type contained the word prime with an asterisk over the phoneme where the noise would appear. A third cue showed only the target phoneme and its position. Finally, a control condition simply presented a series of dashes to subjects prior to each trial. Samuel and Ressler found that only the cue containing the word plus the replaced phoneme location showed higher discriminability scores (indicating less robust phoneme restoration). Knowing in advance just the word or the phoneme did not increase discrimination performance. This suggests that listeners can focus attention to the crucial phonemic unit and discriminate more precisely, but only once contact is made with the lexical item.

Samuel [69] tested whether contact with a lexical representation is necessary or sufficient in mediating attentional focus to a phonemic level. As in Samuel and Ressler [65], Samuel [69] used the combined cue of a word prime and positional target, but created two new types of cues. One cue consisted of the syllable in which the crucial phoneme appeared along with the relative position of the syllable in the word (e.g., -/VER/- preceding "overdue"). A second cue included the syllable prime and the target phoneme location. The results showed facilitated discrimination performance when the combined word prime/location cue was presented. Neither syllable-based cues differed reliably from baseline. These results support the previous claim that directing attentional focus to phonemic levels of processing can be mediated through contact with the lexicon. In fact, the evidence supports the stronger claim that directing attention in this manner requires a mandatory "consult" with a lexical representation.

As previously described, lexical facilitation in phoneme monitoring performance has been shown to depend upon task-generated expectancies and strategies guiding listeners' attention [60]. This shifting of attention between lexical and prelexical (phonemic) levels of representation using phoneme monitoring has been corroborated in a more recent study by Pitt and Samuel [67]. An additional question concerned whether relative costs would be incurred when targets appeared outside the focal point of attention (e.g., unexpected temporal locations). Pitt and Samuel manipulated target location expectations by creating lists loaded for a particular target position. Results showed faster and more accurate performance for targets appearing at expected locations, compared to a baseline control in which target positions were equally distributed. This indicates that subjects were able to develop a listening strategy that optimized attentional allocation to pre-specified segments (phonemes) at given temporal locations. Performance was comparable between words and nonwords for the early target locations, however, a lexical advantage appeared in the last target position. This suggests that listeners switch attention to a lexical representation when it is available.

The studies described thus far involved the focusing of attention to specified levels of processing. This approach was motivated largely by the choice to use aspects of attention as methodological tools to help uncover the underlying structure of the word recognition system. A separate but related issue that remains to be discussed is what reduced attention may bring to bear on the underlying processing architecture. A recent study by Gordon, Eberhardt, and Rueckl [68] provides insight into how various acoustic/phonetic cues are used based on attentional resources. Gordon et al. subjects perform a speech labeling task on a voicing continuum under conditions of reduced or focused attention. In a reduced attention condition, subjects performed an arithmetic task while labeling the stimuli; focused attention required only the labeling response. Two sets of stimuli were used that contained a fundamental frequency typical of voiceless (lower F0) or of voiced (higher F0) consonants. Previous research has shown that F0 influences labelling responses primarily in the ambiguous category boundary region. Gordon et al found that F0 showed similar effects in the reduced and focused attention conditions but also found that F0 influenced responses at the continua endpoints. In other words, the importance of VOT for phonetic perception was reduced relative to F0 onset frequency when attention was not focused on the signal. These results suggest a fundamental change in the way in which lower level phonetic cues are processed as a function of attentional variables.

The role of attention in spoken word recognition and speech processing is twofold. As the majority of this section has described, attentional variables have been used primarily as methodological tools. Focusing attention to specific aspects of the signal and/or specific levels of processing has provided insight to the processing architecture and consequently fueled the debate between autonomous and interactive conceptualizations. Alternatively, research investigating consequences of differential allocation of attention seems to go further to speak to both processing considerations and the nature of attention in its own right. In sum, attentional variables may play a more integral role in speech processing than one is lead to believe given its relative absence in current models of spoken word recognition. Given this, current theoretical directions may be guided in potentially interesting ways by considering how attentional variables may interact in the process of extracting meaning from the speech signal.

Acknowledgements. Preparation of the chapter was supported by NIH Grant R01 DC02134 to the first author. Additional support was provided by the Center for Cognitive and Psycholinguistic Science.

References

[1] Marslen-Wilson, W. & Welsh, A. (1978). Processing interactions and lexical access during word recognition in continuous speech. Cognitive Psychology,10, 29-63.

[2] Marslen-Wilson, W. (1987). Functional parallelism in spoken word recognition. Cognition, 25, 71-102.

[3] Marslen-Wilson, W. (1989). Access and integration: Projecting sound onto meaning. In W.D. Marslen-Wilson (Ed.), Lexical representation and process. MIT Press.

[4] Marslen-Wilson, W. (1990). Activation, competition and frequency in lexical access. In G.T.M. Altmann (Ed.), Cognitive Models of Speech Processing: Psycholinguistic and Computational Perspectives. MIT Press.

[5] Marslen-Wilson, W. & Zwitserlood, P. (1988). Accessing spoken words: On the importance of word onsets. Journal of Experimental Psychology: Human Perception and Performance, 15, 576-585.

[6] McClelland, J. & Elman, J. (1986). The TRACE model of speech perception. Cognitive Psychology, 18, 1-86.

[7] Norris, D. (1994). Shortlist: A connectionist model of continuous speech recognition. Cognition, 52, 189-234.

[8] Klatt, D. (1979). Speech perception: A model of acoustic-phonetic analysis & lexical access. Journal of Phonetics, 7, 279-312.

[9] Pisoni, D.B. & Luce, P.A. (1987). Acoustic-phonetic representations in word recognition. Cognition, 25, 21-52.

[10] Marslen-Wilson, W., & Warren, P. (1994). Levels of perceptual representation and process in lexical access: Words, phonemes, and features. Psychological Review, 101(4), 653-675.

[11] Nygaard, L.C. & Pisoni, D.B. (1995). Speech perception: New directions in research and theory. In J.L. Miller & P.D. Eimas (Eds.), Speech, Language, & Communication. Academic Press.

[12] Frazier, L. (1995). Issues of representation in Psycholinguistics. In J.L. Miller & P.D. Eimas (Eds.), Speech, Language, & Communication. Academic Press.

[13] Lahiri., A. & Marslen-Wilson, W. (1991). The mental representation of lexical form: A phonological approach to the recognition lexicon. Cognition, 38, 245-294.

[14] Gaskell, M.G. & Marslen-Wilson, W.D. (1996). Phonological variation and interference in lexical access. Journal of Experimental Psychology: Human Perception and Performance, 22, 144-158.

[15] Archangeli, D. (1988). Aspects of underspecification theory. Phonology, 5, 183-207.

[16] Connine, C.M. & Deelman, T. Pre-compilation of phonological variation. In preparation.

[17] Gow, D. & Gordon, P.C. (1995). Lexical and prelexical influences on word segmentation: Evidence from priming. Journal of Experimental Psychology: Human Perception and Performance, 21, 344-359.

[18] Cutler, A. & Norris, D. (1988). The role of strong syllables in segmentation for lexical access. Journal of Experimental Psychology: Human Perception & Performance, 14, 113-121.

[19] Cutler, A. & Carter, D.M. (1987). The predominance of strong initial syllables in the English vocabulary. Computer Speech and Language, 2, 133-142.

[20] Marslen-Wilson, W. (1984). Function and process in spoken word recognition. In H. Bouma & D.G. Bouwhuis (Eds.), Attention and performance X: Control of language processes. Hillsdale, NJ: LEA.

[21] Marslen-Wilson, W. & Tyler, L.K., (1980). The temporal structure of spoken language understanding. Cognition, 8, 1-71.

[22] Goodman, J. & Huttenlocher, J. (1988). Do we know how people identify spoken words? Journal of Memory and Language, 27, 684-698.

[23] Taft, M. & Hambly, G. (1986). Exploring the cohort model of spoken word recognition. Cognition, 22, 259-282.

[24] Radeau, M., Mousty, P. & Bertelson, P. (1989). The effect of uniqueness point in spoken-word recognition. Psychological Research, 51, 123-128.

[25] Luce. P.A. (1986). A computational analysis of uniqueness points in auditory word recognition. Perception and Psychophysics, 39, 155-158.

[26] Grojean, F. (1980). Spoken word recognition processes and the gating paradigm. Perception and Psychophysics, 28, 267-283.

[27] Grosjean, F. (1985). The recognition of words after their acoustic offset: Evidence and implications. Perception & Psychophysics, 38, 299-310.

[28] Connine, C.M., Blasko, D.G., & Hall, M. (1991). Effects of Subsequent sentence context in auditory word recognition: Temporal and linguistic constraints. Journal of Memory and Language, 30, 234-250.

[29] Cluff, M.S. & Luce, P.A. (1990). Similarity neighborhoods in spoken two-syllable words: Retroactive effects on multiple activation. Journal of Experimental Psychology: Human Perception and Performance, 16, 551-563.

[30] Zwitserlood, P. & Schriefers, H. (1995). Effects of sensory information and processing time in spoken word recognition. Language and Cognitive Processes, 10, 121-136.

[31] Zwitserlood, P. (1989). The locus of effects of sentential-semantic context in spoken word recognition. Cognition, 32, 25-64.

[32] Connine, C.M., Blasko, D.G., & Wang, J. (1994). Vertical similarity in spoken word recognition: Multiple lexical activation, individual differences, and the role of sentence context. Perception & Psychophysics, 56, 624-636.

[33] Andruski, J.E., Blumstein, S.E., & Burton, S. (1994). The effect of subphonetic differences on lexical access. Cognition, 52, 163-187.

[34] McQueen, J., Cutler, A., Briscoe, T., & Norris, D. (1995). Models of continuous speech recognition and the contents of the vocabulary. Language & Cognitive Processes, 10, 309-331.

[35] Marslen-Wilson, W. (1993). Issues of process and representation in lexical access. In G.T.M. Altman & R. Shillcock (Eds.), Cognitive models of speech processing: The second Sperlonga meeting (pp. 187-210).

[36] Marslen-Wilson, W., Moss, H., & van Halen, S. (1996). Perceptual distance and competition in lexical access. Journal of Experimental Psychology: Human perception and Performance, 22, 1376-1392.

[37] Goldinger, S.D., Luce, P.A., Pisoni, D.B. (1989) Priming lexical neighbors of spoken words: Effects of competition and inhibition. Journal of Memory and Language, 28, 501-518.

[38] Goldinger, S.D., Luce, P.A., Pisoni, D.B., & Macario, J.K. (1992). Form-based priming in spoken word recognition: The roles of competition and bias. Journal of Experimental Psychology: Learning, Memory, & Cognition, 18, 1211-1238.

[39] Slowiaczek, L.M. & Hamburger, M. (1992). Prelexical facilitation and lexical interference in auditory word recognition. Journal of Experimental Psychology: Learning, Memory, & Cognition, 18, 1239-1250.

[40] Radeau, M., Morais, Morais, J., & Segui, J. (1995). Phonological priming between monosyllabic spoken words. Journal of Experimental Psychology: Human perception and Performance, 21, 1297-1311.

[41] Luce, P.A., Pisoni, D.B. & Goldinger, S. (1990). Similarity neighborhoods of spoken words. In G.T.M. Altmann (Ed.), Cognitive Models of Speech Processing: Psycholinguistic and Computational Perspectives. MIT Press.

[42] Bard, E.G. (1990). Competition, lateral inhibition and frequency. In G.T.M. Altman (Ed.), Cognitive models of speech processing (pp. 185-210). Cambridge, MA: MIT Press.

[43] Shillcock, R.C. (1990). Lexical hypotheses in continuous speech. In G.T.M. Altman (Ed.), Cognitive models of speech processing (pp. 24-49). Cambridge, MA: MIT Press.

[44] Vroomem, J. & de Gelder, B. (1995). Metrical segmentation and lexical inhibition in spoken word recognition. Journal of Experimental Psychology: Human Perception & Performance, 21, 98-108.

[45] Norris, D., McQueen, J., & Cutler, A. (1995). Competition and segmentation in spoken-word recognition. Journal of Experimental Psychology: Learning, Memory, & Cognition, 21, 1209-1228.

[46] McQueen, J.M., Norris, D.G., & Cutler, A. (1994). Competition in spoken word recognition: Spotting words in other words. Journal of Experimental Psychology: Learning, Memory, & Cognition, 20, 621-638.

[47] Connine, C.M., Blasko, D.M., & Titone, D.A. (1993). Do the beginnings of spoken words have a special status in auditory word recognition? Journal of Memory and Language, 32, 193-210.

[48] Chomsky, N. & Halle, M. (1968). The Sound Pattern of English. MIT Press.

[49] Marslen-Wilson, W. & Zwitserlood, P. (1989). Accessing spoken words: The importance of word onsets. Journal of Experimental Psychology: Human Perception & Performance, 15, 576-585.

[50] Connine C.M., Titone, D.A., Deelman, T., & Blasko, D.M. (in press). Similarity mapping in spoken word recognition: Evidence from phoneme monitoring.

[51] Connine, C.M. & Clifton, C., Jr. (1987). Interactive use of lexical information in speech perception. Journal of Experimental Psychology: Human Perception and Performance, 2, 291-299.

[52] Pitt, M.A. (1995). The locus of the lexical shift in phoneme identification. Journal of Experimental Psychology: Learning, Memory, & Cognition, 21, 1037-1052.

[53] Ganong, F. (1980). Phonetic categorization in auditory word perception. Journal of Experimental Psychology: Human Perception & Performance, 6, 110-125.

[54] Pitt, M. & Samuel, A. (1993). Autonomous vs interactive models in speech perception: Evidence from an identification study and meta-analysis. Journal of Experimental Psychology: Human Perception and Performance.

[55] Warren, R.M. (1970). Perceptual restoration of missing speech sounds. Science, 167, 392-393.

[56] Samuel, A. (1981). Phonemic restoration: Insights from a new methodology. Journal of Experimental Psychology: General, 110, 474-494.

[57] Samuel, A. (1981). The role of bottom up confirmation in the phonemic restoration illusion. Journal of Experimental Psychology: Human Perception and Performance, 7 1124-1131.

[58] Samuel, A.G. (1997). Lexical activation produces potent phonemic percepts. Cognitive Psychology, 32, 97-127.

[59] Cutler, A., Mehler, J., Norris, D. & Segui, J. (1987). The syllable's differing role in the segmentation of French and English. Journal of Memory and Language, 25, 385-400.

[60] Cutler, A. & Norris, D. (1979). Monitoring sentence comprehension. In W.E. Cooper & E.T.C. Walker (Eds.), Language processing: Psycholinguistic studies presented to Merrill Garrett, pp. 113-134. Hillsdale, N.J.: Erlbaum.

[61] Norris, D.G. (1993). Bottom-up connectionist models of interaction. In G.T.M. Altman & R. Shillcock (Eds.), Cognitive models of speech processing: The second Sperlonga meeting (pp. 211-234).

[62] Connine, C.M. (1987). Constraints on interactive processes in auditory word recognition: The role of sentence context. Journal of Memory and Language, 26, 527-538.

[63] Zwitserlood, P. (1989). The locus of effects of sentential-semantic context in spoken word recognition. Cognition, 32, 25-

[64] Boland, J.E. & Cutler, A. (1996). Interaction with autonomy: Multiple output models and the inadequacy of the great divide. Cognition, 58, 309-320.

[65] Samuel, A.G. & Ressler, W.H. (1986). Attention within auditory word perception: Insights from the phonetic restoration illusion. Journal of Experimental Psychology: Human perception and Performance, 12, 70-79.

[66] Samuel, A.G. (1991). A further examination of attentional effects in the phonemic restoration illusion. Quarterly Journal of Experimental Psychology, 43a, 679-699.

[67] Pitt, M.A. & Samuel, A.G. (1990). Attentional allocation during speech perception: How fine is the focus? Journal of Memory and Language, 29, 611-632.

[68] Gordon, P.C., Eberhardt, J.L., & Rueckl, J.G. (1993). Attentional modulation of the phonetic significance of acoustic cues. Cognitive Psychology, 25, 1-42.

[69] Samuel, A.G. (1991). A further examination of attentional effects in the phonemic restoration illusion. Quarterly Journal of Experimental Psychology, 43a, 679-699.

Issues in Using Models for Self Evaluation and Correction of Speech

Marie-Christine Haton

LORIA/Université Henri Poincaré, Nancy 1
BP 239
54506 Vandœuvre-lès-Nancy, France

Summary. The design of computer-based systems for training purposes requires the necessity of taking into account the different worlds in which the actors operate. In this paper, we deal with speech correction involving a therapist ("the orthophonist"), the trainee, and a technical aid performing extraction and visual displays of speech features. We want to point out some common-sense issues that appear as essential for designing such computer-based systems: the matching between those different worlds, the choice of reference patterns to which the subject's utterances will be compared, the orthophonic check and the definition of a norm which could be considered as a target to be reached by the subject, the matching of the norm with the reference patterns, the matching between the vocal utterances and their technical equivalent, the settlement, and then, the management of a speech education program adapted to the trainee.

Key words: Assisted speech training, visual aid, speech impaired people.

1. Introduction

The use of computers to perform advanced speech processing and to give a visual feedback to users has been a research topic for about twenty five years. Such a system was first described by Nickerson and Stevens [1].

Most of the time, the visual feedback aims at compensating some hearing deficiency through computer-based devices. See [2] to [5] for some of them. Such systems have been designed, either for computer-assisted pronunciation training (SPELL, ILAM, LISTEN, VSA, Speech Viewer II), or for the evaluation of voice features (EVA, Dr Speech). Some of them have been developed in the framework of the European TIDE program (MUSA, HARP, DICTUM).

A great advantage of such devices lays in the possibility given to the user to train his/her voice through repetitive drills, under the objective judgement of the machine. Thus, one crucial point is the choice of reference patterns to which the subject's utterances will be compared.

Speech education and/or rehabilitation with computer-assisted techniques are often adapted to children, under the guidance of a teacher. For hearing impaired people having reached a reasonable oral language level, for people impaired by acquired deafness and whose speech may worsen, or for other speech impaired adults, the concept of self training is of prime importance. The computer may intervene as a guarantee, or as a warning tool in case of voice deterioration. More generally, it can concern speech impaired persons, hearing impaired persons and also foreign language teaching.

The speech processing techniques used in those systems have a strong impact on the system performance as well as on their cost.

In this paper, we refer to our experience concerning the SIRENE 2 system, designed and developed in our laboratory, in co-operation with Thomson-ASM which designed the speech analysis front-end, and with URAPEDA (regional centre for hearing impaired rehabilitation), for evaluation and return on experience.

The system has the following characteristics: use of advanced phonetic decoding for evaluating the vocal productions of the trainee and providing pronunciation hints; close and constant interaction with user groups made up of motivated hearing-impaired persons, who intervened in the iterative process of system development and testing; integration of a voice analysis tool for helping speech therapists and physicians in voice diagnosis.

The author gratefully acknowledges M.P. Chouvet, M.M. Dutel, together with students Céline and Marjolaine, G. Mauguin and the young adults at URAPEDA, and P. Alinat, for their contributions to the work described in this paper.

An extension of the SIRENE 2 system has just started in the framework of the Telematics ISAEUS project. It will be done on a multilingual basis, involving the French, German and Spanish languages. The aim of this project is to design a multilingual, low cost, PC-based system with software on CD-ROM, using a generic frame for the speech exercises and a user-friendly interface.

2. Using models

Three actors in their own worlds are concerned: the orthophonist (O), the subject (S) and the technical aid (T). That aid is twofold: an off-line system used in delayed time for speech analysis before or after the reeducation session itself, and an on-line system used in real time during reeducation. Figure 1 gives a functional view of the computer-assisted speech education process, from the first evaluation made by the therapist in his orthophonic world (OW) to the execution of exercises in the subject world (SW).

DTTW and RTTW correspond respectively to the delayed-time and the real-time technical worlds.

Speech and hearing deficiencies can assume diverse, complex forms. The orthophonist's expertise is to identify deficiencies and to devise reeducation programs. In order for a computer based system to be of any use in that situation, it is necessary to model the subject's voice characteristics as well as the progress he is expected to make. The success of the whole system largely depends on the quality of that modeling.

The orthophonic check is of prime importance. It may be performed with the help of a technical aid. We believe, however, that the computer can only assist human expertise at this point, for at least two reasons.

The first reason is that it is very difficult to distinguish normal voices from pathological voices, and, *a fortiori*, to distinguish between pathologies, on the basis of numerical parameters only (although such parameters as shimmer, jitter and the like can provide useful information.)

The second reason is that the purpose of speech reeducation is more to make the trainee acquire kinaesthetic reference marks to reach good abilities including fluency, good quality and intelligibility rather than reach an ideal performance.

The diagnosis phase provides an orthophonic qualification OQ of the subject abilities that will allow the specialist to define a norm ON. This norm will be considered as a target to be reached by the trainee in the immediate future. Both OQ and ON intervene in the establishment of a subject training exercises base SEB from which the current exercise to be performed will be selected. While the exercise is executed, subject vocal utterances SVU are produced and analysed, providing TVU, the technical equivalent of these vocal utterances. TVU is used as a base for automatic guidance through the drills in SEB.

Fig. 1. A functional view of the computer-assisted speech education process

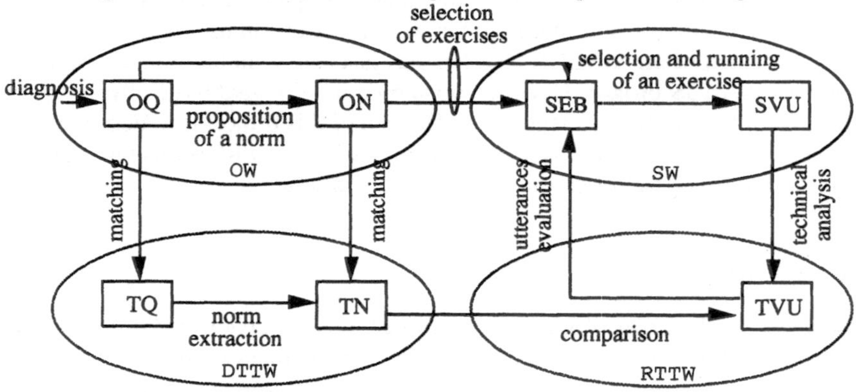

Much effort has to be dedicated to the matching between OQ and ON and their technical equivalent, TQ and TN. How to process speech signal to keep the substance of the original vocal message, how to transpose an ideal norm to be reached in a reeducation program and how to manage this program, here are major points to address for the best use of a technical device for speech education. It is worth noticing that the matter of foreign language teaching (FLT) can be discussed in similar terms.

3. Norm building

To build a norm adapted to the trainee consists in some sense of building reference patterns in the technical world. That constitutes a general problem in automatic speech engineering, we want either to recognise speech utterances, or to train a subject to acquire vocal and articulatory skills...

We will call reference patterns, speech units that can be phonemes, syllables, parts of sentences or complete sentences, encoded in a numerical format, with useful information about their use, and stored in files. They may have several origins:

- they may be parameter variation intervals coming from statistics about the language concerned,
- they could have been pronounced by the specialist or by a given speaker. In this case, they will be used by the subject in situation of visual confrontation with his/her own vocal productions;
- they may arise from voluntary archiving from vocal productions of the subject. In this case, they will play two roles: on one hand, visual confrontation with usual vocal productions and, on the other hand, numerical evaluation of the evolution of subject performances.

In the last case, we have to take care of the danger of using references coming from the subject under training. Actually, concerning people impaired by acquired deafness, for instance, the risk is big not to take into account a defect to which the subject got used, while it could be important to reactivate in a kinaesthetic point of view some elements previously acquired. It is illusive to think that his own reference is a good one. It is the same problem for oralised deaf persons.

Consequently, a preliminary work has to be done before deciding which individual norm will be used.

4. Matching between the subject's world and the technical world

Speech processing offers many powerful methods. Here, our goals are: perform the best evaluation of the trainee's performance, and deliver the best advice.

For this purpose, we have to consider different levels of abstraction. At first, acoustic analysis, providing relevant and precise acoustic features of speech, must be performed. Our analysis system is based on a set of linear filters, the central frequencies and impulse responses (amplitude and phase) of which are deduced from the human cochlea [6] and a strong algorithm, inspired from [7] for the determination of the nature of the speech excitation (voicing or friction).

Then, combination of features, and additional measurements like shimmer, jitter, etc., must lead to more abstract concepts. Articulatory place and manner for phoneme pronunciation are typical of this level. So, the better the first level is, the better the evaluation and the advice will be.

We could consider that a more global approach could lead to the upper level of abstraction, the one of segment identification. Hidden Markov Models [8] or Artificial Neural Networks [9] may have a role to play. Indeed, they do not directly take into account the phonetic features but they can be useful in the case of well identified substitutions between phonemes, for example, as a rough measure of distance between the utterance and the target.

At last, the "perceptual fading" technique may be of some help for FLT or for trainees having reasonable listening abilities. It requires the use of synthesised patterns in which some acoustic cues have been artificially enhanced. Progressive fading of these cues helps the trainee to reach the designed target.

5. Settlement of the speech education program

Designing a speech education program strongly depends on the norm previously established for the concerned trainee. It may vary also according to the situation: child or adult, speech therapy or foreign language teaching. Nevertheless, our work with speech therapists has led us to consider a general canvas to be adapted to those different situations.

As said before, a complete check of speech abilities should lead to the best progression. For that purpose, traditional questionnaires and observation grids may be accompanied by objective features given by the technical aid.

The first step in speech rehabilitation is to make the user adopting an homogeneous breath, without obstacle, tension nor plosion, and controlling his speech intensity and duration. Then, the "breath-voice" coordination intervenes to facilitate a continuous transition between breath and voice in good conditions. At that moment, the subject must be conscious of the vocal cords vibration and drills dedicated to voicing can be envisaged.

The visual feedback will allow to decide which level of fundamental frequency is best suited to the subject, and to train this feature, to make it stable, and then to modulate it in a manner compatible with natural speech intonation.

The following step aims at helping the trainee to master the different facets of voice. This is done through exercises concerning rhythm (for example, in French, syllabic rhythm and "oxytony", lengthening of the stressed syllable, demarcative function of rhythm...), intonation (basic levels, coincidence between minimal intonative units and rhythmic groups, the different modalities of the discourse, etc.), and the speech rate which has an influence upon the other parameters.

Concerning hearing-impaired people, this series of exercises is followed by the exercises for improving phoneme articulation, minimal pairs differentiation and, in a last step, whole sentences. FLT may lead the master to build another kind of progression, depending on the gaps between the mother tongue and the language learnt.

Three phases may be proposed for each drill: sensibilisation, correction, training, as the speech therapist does. To each of them correspond specific advice, levels of difficulty, displays...

As an example, let's consider vowel articulation training. The sensibilisation phase is done through the visual presentation of the vocal and nasal tracts configuration and of the dynamic articulation of the sound. The correction phase involves the real-time presentation of the pronounced vowels in the plane of the two first formants. The goal of the training phase being to consolidate the automatisms acquired in the previous phase, others exercises are proposed, for differentiating minimal pairs for instance.

Observations during supervised training may also bring interesting observations: understanding of instructions, frequency of use of the help function, understanding of the matching between the parameters and their symbolic representation, the way the target is reached, etc.

6. Management of the education program

A question which arises each time a computer system intervenes as an aid for the user is the role to give to this new kind of tutor. In our case, the question is to identify, in the phase of defect correction, if the user can navigate alone inside the drills proposed by the system, or if the system must impose rules of guidance. Three options could be envisaged:

1. the user is invited to follow the specialist advice while being free to access any drill and any display implemented in the system,
2. the specialist prescribes a training program adapted to the subject,
3. the system performs a totally or semi-automatic guidance.

Option 1 seems to us to be adapted to efficient and critical evaluation of the system, performed in a private context, outside institutions. Obviously, option 2 will be adapted to training performed inside institutions. These choices may evolve toward semi-automatic guidance through performance evaluation and updating of the history of the training sessions.

Option 3 takes place in the framework of intelligent computer-aided instruction (ICAI). The aim is to model the specialist approach on four levels: domain knowledge, pedagogical expertise, the way of taking into account the subject's curriculum or a generic representative model of the subjects, and, finally, the subject's ability of choosing the best media for communication. This approach supposes that the system is able to correctly and precisely evaluate trainees performance, to predict what will be its evolution and to update a database including relevant parameters from their utterances. Implementing a semi-automatic guidance appears to be the most reasonable compromise in the present state of technical advances.

7. Conclusion

We have presented in this paper some issues related to the design of a system for speech training, both for impaired people and foreign language teaching.

We have wanted to stress on the necessity to pay a great attention to the modeling of the "worlds" in which the actors operate (therapists, trainees, technical aids for diagnosis and training), and to the matching between those different worlds. It is a necessary condition for designing reliable systems allowing the trainee to be autonomous and motivated.

Most of these ideas have been implemented in our SIRENE 2 system, and validated with young deaf adults. We are presently extending this system on a multilingual basis, in the framework of the European ISAEUS project.

References

[1] Nickerson, R. & Stevens, K.N. (1973) *"Teaching speech to the deaf: can a computer help?"* IEEE Trans. Audio Electroacoustics, AU21, 445-455.

[2] Javkin, H. *et al.* (1993) *"A motivation-sustaining articulatory/acoustic speech training system for profoundly deaf children"*. Proc. IEEE ICASSP'93, I, 145-148.
[3] Rooney, E. *et al.* (1994) *"HARP: an autonomous speech rehabilitation system for hearing-impaired people"*, Proc. ICSLP'94.
[4] Haton, M.C. & Haton, J.P. (1979) *"SIRENE, a system for speech training of deaf people"*, Proc. ICASSP'79.
[5] Elsendoorn, B.A.G. & Coninx, F., editors, [1993] *Interactive Learning Technology for the Deaf*, NATO ASI Series, Springer-Verlag.
[6] Alinat, P. & Pierrel J.M. (1994) *"ROARS: Robust Analysis Speech Recognition System"*. In *Research Reports ESPRIT*, vol. 1, Springer Verlag.
[7] Gold, B. (1962) *"Description of a computer program for pitch detection"*, Proc. Int. Cong. of Acoustics, Copenhagen.
[8] Rabiner, L. & Juang, B.H. (1993) *Fundamentals of Speech Recognition*, Prentice Hall.
[9] Haton, J.P. (1996) *"Neural Networks for Speech Recognition"*. In *Fuzzy Logic and Neural Networks Handbook*, C.H. Chen ed., McGraw-Hill.

The Use of the Maximum Likelihood Criterion in Language Modelling

Hermann Ney

Lehrstuhl für Informatik VI
RWTH Aachen – University of Technology
D-52056 Aachen, Germany

Summary. This paper gives an overview over the use of the maximum likelihood criterion in stochastic language modelling. This criterion and its associated estimation techniques provide a unifying framework for various approaches that seem very much unrelated and different at first glance, such as smoothing and cross-validation, decision trees (CART), word classes obtained by clustering, word trigger pairs and maximum entropy models.

1. Introduction

During the past two decades, a large number of approaches to language modelling have been developed in the context of automatic speech recognition. Among these approaches are smoothing and discounting, decision trees (CART), word classes obtained by clustering, word trigger pairs and maximum entropy models. For these approaches, various criteria and methods are used to train the corresponding language models, e. g.

- the Turing-Good formula for smoothing and discounting [18, 25],
- the EM algorithm for linear interpolation [14, 27],
- the so-called entropy criterion for decision trees (CART) [1, 5],
- the mutual information criterion for both word class clustering [6] and for word trigger pairs [22],
- the entropy criterion for the maximum entropy models [22].

So in this context, the natural question comes up: Are these methods unrelated to each other or is there a principle underlying all these methods, and if yes, what principle? This paper tries to show that virtually all of the language model approaches can be unified using the principle:

structural assumptions + maximum likelihood criterion.

The primary application we consider is large vocabulary speech recognition with applications like text dictation and automatic dialogue systems. Some of the techniques presented are maybe useful in other applications, too, like systems for voice commands and guided dialogues, where a finite state network might be sufficient as language model. For most non-experts and maybe even for the experts in speech recognition, it is still a surprise that the statistical language models, in particular the trigram models, perform as well as they do. Grammar based language models are by far not (yet?) competitive. Therefore, the description focuses on the statistical approaches in language modelling and related issues such as the sparse data problem and smoothing. This paper is only able to touch upon some of the approaches in language modelling. For other approaches and overviews, see [14, 15, 17]. For related topics such as the use of stochastic methods for language acquisition and language understanding, see [13] and [29], respectively.

The paper is organized as follows. We first review the definition of the widely used perplexity criterion and its relation to the maximum likelihood criterion. We then specify and study in detail the *structural assumptions* we need to derive each of the following approaches:

- handling of sparse data
  by the Turing-Good estimates and by absolute discounting,
- partitioning-based models
  like decision trees and clustering-based word classes,
- word trigger pairs,
- maximum entropy models.

2. Perplexity and Maximum Likelihood

The task of a language model is to express the restrictions imposed on the way in which words can be combined to form sentences. In other words, the idea is to capture the inherent redundancy that is present in the language, or to be more exact, in the language subset handled by the system. This redundancy results from the syntactic, semantic and pragmatic constraints of the language and may be modeled by probabilistic or nonprobabilistic ('yes/no') methods.

For large vocabulary recognition tasks, we have to allow any type of word sequence which is difficult to describe deterministically. The task of a stochastic language model is to provide estimates of the prior probabilities $Pr(w_1 \ldots w_N)$ for each word sequence (or sentence) $w_1 \ldots w_n \ldots w_N$. Using the definition of conditional probabilities, we obtain the decomposition:

$$Pr(w_1 \ldots w_N) = \prod_{n=1}^{N} Pr(w_n | w_1 \ldots w_{n-1}) \quad .$$

This equation requires a suitable interpretation of the variable N, the number of words. When considering a single sentence, the number of words is a random variable itself, and we need an additional distribution over the sentence lengths. In practice, the problem is circumvented by applying the above equation to a whole set of sentences and extending the vocabulary by a special symbol (or 'word') that marks the end of a sentence and the beginning of the next sentence.

Strictly speaking, to evaluate the quality of a stochastic language model, we would have to run a whole recognition experiment. However, to a first approximation, we can separate the two types of probability distributions in Bayes' decision rule [2] and confine ourselves to the probability that the language model produces for a sequence of (test or training) words w_n, $n = 1, \ldots, N$. To normalize this prior probability with respect to the number N of words, we take the N-th root and take the inverse to obtain the so-called (corpus or test set) perplexity PP:

$$PP := [Pr(w_1 \ldots w_N)]^{-1/N} \quad .$$

This equation shows that the perplexity is the geometric average of the reciprocal probability over all N words. By the above decomposition of the joint probability and by taking the logarithm, we obtain:

$$\log PP \;=\; -\frac{1}{N} \sum_{n=1}^{N} \log Pr(w_n|w_1 \ldots w_{n-1}) \quad .$$

So far, we have considered the *true* probability distributions. Since these are not known in practice, we have to make suitable assumptions instead and introduce models with certain structures and free parameters. To pave the way for the following sections, we change the notations as follows:

$$(w_1 \ldots w_{n-1}) \;\rightarrow\; h_n \quad ,$$
$$Pr(w_n|w_1 \ldots w_{n-1}) \;\rightarrow\; p_\vartheta(w_n|h_n) \quad .$$

For each position n, the *history* h_n stands for any subset of the full predecessor sequence $(w_1 \ldots w_{n-1})$, e. g. for a trigram $h_n = (w_{n-2}, w_{n-1})$. The *probability model* $p_\vartheta(w_n|h_n)$ replaces the true but unknown probability. The symbol ϑ stands for the set of unknown parameters of the model $p_\vartheta(w_n|h_n)$.

To train or, in statistical terminology, *estimate* these unknown parameters ϑ, it is natural to use as training criterion the minimum of the perplexity. That means, in the above perplexity definition, we replace the true probabilities $Pr(w_n|w_1 \ldots w_{n-1})$ by the model probabilities $p_\vartheta(w_n|h_n)$. Thus, apart from the constant factor $(-1/N)$, the logarithm of the corpus perplexity is identical to the following quantity $F(\vartheta)$:

$$F(\vartheta) \;:=\; \sum_{n=1}^{N} \log p_\vartheta(w_n|h_n) \quad .$$

In statistics [23], this quantity is known as the logarithm of the likelihood function (in short: log-likelihood), and the unknown parameters are *estimated* by maximizing the log-likelihood function $\vartheta \rightarrow F(\vartheta)$:

$$\hat{\vartheta} \;:=\; \arg\max_{\vartheta} F(\vartheta) \quad .$$

Thus for a given corpus and an assumed model, maximizing the likelihood is equivalent to minimizing the perplexity. Therefore, the whole toolbox of maximum likelihood estimation methods can be made available for language modelling although, traditionally in statistics, *unconditional* models are much more common than *conditional* models.

Apart from the mathematical point of view, it is important and instructive to realize the practical problem considered. The likelihood criterion can be interpreted as the quantitative evaluation of a guessing game:

One after the other, we fix each position n in a running text of words $w_1 \ldots w_n \ldots w_N$. Given the knowledge of the history $h_n := w_1 \ldots w_{n-1}$, the language model must *'predict'* the word in position n. *'Prediction'* means: since each word of the vocabulary is possible, the language model assigns to each possible word w a probability $p_\vartheta(w|h_n) > 0$ under the normalization constraint

$$\sum_w p_\vartheta(w|h_n) \;=\; 1 \quad,$$

where the summation is performed over all words of the vocabulary. The *global* score is then obtained by summing the *local* scores which are given by the logarithm of the probabilities $p_\vartheta(w_n|h_n)$ of the actually *observed* words w_n.

It is often convenient to rewrite the log-likelihood functions in terms of the corpus counts $N(h, w)$ defined by:

$$
\begin{aligned}
N(h, w) \quad &:= \quad \sum_{n=1}^{N} \delta(h, h_n)\, \delta(w, w_n) \\
&= \quad \sum_{n:(h,w)=(h_n,w_n)} 1
\end{aligned}
$$

with the usual definition of the Kronecker delta. Thus we can replace the sum over the position index n by a sum over the space of joint events (h, w):

$$
\begin{aligned}
F(\vartheta) \quad &= \quad \sum_{n=1}^{N} \log\, p_\vartheta(w_n|h_n) \\
&= \quad \sum_{h,w} N(h, w) \log\, p_\vartheta(w|h) \quad.
\end{aligned}
$$

For each of the language models considered in this paper, we will specify the following two ingredients:

- *the structure of the model:*
  What is the structure of the model $p_\vartheta(w|h)$ and what are the unknown parameters ϑ ? E. g. in modelfree approaches like the popular bigram, trigram or general m-gram models, the conditional probabilities $p(w|h)$ themselves play the role of the parameters ϑ. The same is true for *modelfree* discounting models. Often, in particular in *modelfree* cases, we will drop the index ϑ in the probability model $p_\vartheta(w|h)$. In other discounting models like absolute discounting, the so-called discounting parameter is the free model parameter ϑ. In CART and clustering, the partition in equivalence classes plays the role of the unknown parameter.

- *the algorithm for maximum likelihood estimation:*

What is a suitable method for finding the optimal parameters:

$$\max_{\vartheta} F(\vartheta) \quad .$$

Often, there is no closed-form solution, and iterative procedures like the EM or GIS algorithm [7, 8] or suboptimal approximations like decision trees (CART) and clustering algorithms for word classes [6, 19] have to be used.

3. Smoothing and Discounting for Sparse Data

To illustrate the sparse data problem in language modelling, we consider a specific example, namely a bigram model. For a bigram model, the free parameters ϑ to be trained are the conditional bigram probabilities themselves. Conventional methods like the maximum likelihood method typically result in the relative frequencies as estimates for the bigram probabilities:

$$p(w|v) = \frac{N(v, w)}{N(v)} \quad .$$

Here, (v, w) is the word bigram under consideration, and $N(v, w)$ and $N(v)$ are the numbers of observed word bigrams (v, w) and words v, respectively. Now assuming a vocabulary size of $W = 20000$ words, there are $W^2 = 400$ million possible word bigrams, but the training corpus rarely consists of more than 10 million words. As a result, the conventional probability estimate for each unseen event is zero, and no more than 2.5% of all bigrams can be observed in training. Using these conventional probability estimates, the recognition system would be unable to recognize any word sequence that contains an unseen word bigram. Therefore, it is crucial to design and train the language models in such a way that probability estimates of zero are avoided. The corresponding methods are referred to as smoothing or discounting. In this section, we will consider two methods: modelfree discounting and absolute discounting.

3.1 Modelfree Discounting and Turing-Good Estimates

To overcome the shortcomings of conventional probability estimates for sparse data, we subtract probability mass from the relative frequencies of seen events and re-assign it to the unseen events. This is achieved by the leaving-one-out method which can be considered to be an extension of the cross-validation method [10, 11]. To illustrate this concept without confusing details, we simplify the problem by considering *joint* probabilities. Given a training text $w_1 \ldots w_n \ldots w_N$, we obtain the set of *joint* events (h, w) by isolating the word w_n and its history h_n in each position $n = 1, \ldots, N$. In general, the history h_n may comprise *all* predecessor words $w_1 \ldots w_{n-1}$. In the case of the so-called m-gram language models, the history is limited to the immediate $(m - 1)$ predecessor words. Here, as usual, we have combined all sentences into a single long word sequence by interpreting the sentence end symbol as a special word of vocabulary. Thus we have a set of *joint* events

$$(h_n, w_n), \quad n = 1, \ldots, N \quad .$$

We consider the process of removing an event, i.e. the observation w_n and its history h_n from the N observations and use it as holdout part. To describe this process, we denote the original count by $r = N(h, w)$ and define the *count* dependent quantities (for a count $r = 0, 1, \ldots, R$):

n_r: number of any event type, i.e. a pair (h, w), seen exactly r times; the formal definition is:

$$n_r \quad := \sum_{h,w:N(h,w)=r} 1 \quad ;$$

p_r: probability of any event type seen exactly r times.

Note that the quantity n_r counts how many event types (h, w) were observed exactly r times. Therefore, it will be called count-count. We have the following constraints:

$$\sum_{r=0}^{R} n_r p_r = 1 \quad \text{and} \quad \sum_{r=0}^{R} n_r r = N \quad .$$

For the sake of clarity, we first write down, in this notational scheme, the *conventional* log-likelihood function for the unknown probabilities $p_r, r = 0, \ldots, R$:

$$F(p_0, \ldots, p_R) = \sum_{r=0}^{R} r n_r \log p_r \quad .$$

When optimizing this function over the unknown parameters $p_r, r = 0, \ldots, R$, we have to observe the normalization constraint $\sum_{r=0}^{R} n_r p_r = 1$. To this purpose, we use the method of Lagrange multipliers for the constraint. Then, by taking the partial derivatives with respect to the unknown parameters, we obtain the usual estimates, i.e. the relative frequencies:

$$p_r = \frac{r}{N} \quad .$$

We now formulate the log-likelihood function in the leaving-one-out framework. After removing one observation as holdout observation, there are only $(r - 1)$ observations of the same event type left in the $(N - 1)$ training observations. Therefore, we have to use the probability p_{r-1} rather than p_r. Each observation $(h_n, w_n), n = 1, \ldots, N$, is used as holdout observation. Observing that there are exactly $r \cdot n_r$ observations for each event type with count r, we obtain the *leaving-one-out* log-likelihood function:

$$F(p_0, \ldots, p_{R-1}) = \sum_{r=1}^{R} r n_r \log p_{r-1} \quad,$$

where the probability p_R is not included in the set of free parameters. The reason is that there is no count $(R + 1)$ in the corpus. As estimate, we typically use the relative frequency $p_R = R/N$. Using the method of Lagrange multipliers for the normalization constraint as before, a straightforward optimization of this criterion over the unknown probabilities $p_r, r = 0, \ldots, R - 1$ results in the formula:

$$p_r = \frac{1 - n_R p_R}{N} \cdot \frac{(r + 1) \, n_{r+1}}{n_r} \quad.$$

Apart from the factor $(1 - n_R p_R)$ which is close to 1, this estimate for p_r is identical to the so-called Turing-Good estimate [12, 25]. It is instructive to compute the total probability mass of unseen events, i. e. we set $r = 0$ in the above equation:

$$n_0 p_0 = (1 - n_R p_R) \cdot \frac{n_1}{N} \quad,$$

which would be zero for the conventional maximum likelihood approach. We see that this probability mass is basically given by n_1/N, i.e. the fraction of event types seen exactly once. To understand the terminology 'discounting', we consider the difference between the relative frequencies r/N and the probability estimates p_r. Ideally, this difference should be larger than zero for each count $r > 0$:

$$\frac{r}{N} - p_r \; > \; 0 \quad.$$

In other words, the original relative frequencies (or counts) are *discounted* so that probability mass is available for the unseen events.

To transfer this approach from the *joint* probabilities $p(h, w)$ to the *conditional* probabilities $p(w|h)$, we introduce the count estimates r^* by the definition:

$$\begin{aligned} r^* \; &:= \; N \cdot p_r \\ &= \; (1 - n_R p_R) \cdot \frac{(r + 1) \, n_{r+1}}{n_r} \quad. \end{aligned}$$

For each pair (h, w), the original count $r = N(h, w)$ in the relative frequency estimate $N(h, w)/N(h)$ is replaced by the associated count estimate $N^*(h, w)$:

$$p(w|h) = \frac{N^*(h, w)}{N(h)} \quad.$$

This has been an informal derivation. For the exact formal derivation, we have to consider the exact log-likelihood function at the level of the *conditional* probabilities [28].

Now this is not the full truth yet. The truth is that the count-counts n_r cannot be used as they are because they are *noisy*. The situation is comparable to estimating a discrete probability distribution by a histogram. The remedy often used is to smooth the count-counts [12] or to limit the method to the small counts, say up to $r = 6$ [18]. A radically different remedy will be used in the following: instead of a *model-free discounting* approach, which we have used so far, we define a *model-based discounting* approach, namely absolute discounting.

3.2 Absolute Discounting

The basic idea of the absolute discounting [26] model is to leave the high counts virtually unchanged. The justification is as follows. The count $r = N(h, w)$ with which a certain event (h, w) has been observed in the training data is likely to change in another set of training data of the same size N. However, we can expect the difference to be small, typically we would expect to see values like $r - 1, r, r + 1$. To take this type of variability into account, we introduce an *average* and therefore *non-integer* count offset which is assumed to be independent of the count value r. Expressed in terms of the counts r and their estimates r^*, we assume (for counts $r > 0$) the difference to be a constant that is *count independent*:

$$r - r^* \cong b = const(r) \quad ,$$

where b denotes the count independent parameter to be estimated. In typical experimentally observed counts (see the examples in [12] and Table 2 in [28]), it has been found that this model of absolute discounting actually provides a good fit to the Turing-Good counts r^* with just one free parameter.

Observing the normalization constraint, we define the model $p_b(w|h)$ of absolute discounting ($0 < b < 1$):

$$p_b(w|h) = \begin{cases} \dfrac{N(h, w) - b}{N(h)} & \text{if } N(h, w) > 0 \\[2em] b \cdot \dfrac{W - n_0(h)}{N(h)} \cdot \dfrac{\beta(w|\overline{h})}{\displaystyle\sum_{w' : N(h, w')=0} \beta(w'|\overline{h})} & \text{if } N(h, w) = 0 \end{cases} \quad .$$

Here, there are two new quantities, the count-count $n_0(h)$ and the distribution $\beta(w|\overline{h})$. For each history h, the count-count $n_0(h)$ is the number of words that were *not* observed as successor words. The distribution $\beta(w|\overline{h})$ is the distribution of the more general history $\overline{h}$ to which we *back off* when there are no observations (h, w), i. e. $N(h, w) = 0$. E. g. for a trigram model, whose history h comprises the two predecessor words, the more general history $\overline{h}$ is typically defined to comprise only the immediate predecessor word.

For this discounting model, we have to consider the leaving-one-out log-likelihood criterion as a function of the discounting parameter b. Ignoring terms not depending on b, we obtain the leaving-one-out log-likelihood function:

$$F(b) = n_1 \cdot \log b + \sum_{r \geq 2} r n_r \cdot \log [r - 1 - b] \quad .$$

Taking the partial derivative with respect to b, we obtain the following equation after separating the term with $r = 2$:

$$\frac{n_1}{b} - \frac{2n_2}{1 - b} = \sum_{r \geq 3} \frac{r n_r}{r - 1 - b} \quad .$$

For this equation, there is no closed-form solution. Instead, we rewrite the above equation:

$$b = \frac{n_1}{n_1 + 2n_2 + \sum_{r \geq 3} \frac{(1-b)\, r}{r-1-b} \cdot n_r} \quad .$$

This equation can be used as an iteration formula [27] and also to derive an upper and a lower bound for b. To obtain a lower bound, we assume $b > 0.5$, which is satisfied in virtually all experimental conditions, and obtain the bounds:

$$\frac{n_1}{n_1 + 2n_2 + \sum_{r \geq 3} n_r} < b < \frac{n_1}{n_1 + 2n_2} \quad .$$

Since in general leaving-one-out tends to underestimate the effect of unseen events, we typically use the upper bound as the leaving-one-out estimate for b. In theory, we could define a more refined model by making the discounting parameter dependent on the history h. However, the experimental results show [27] that there is no advantage by this method.

Instead of absolute discounting, we can also consider other model-based smoothing approaches like linear discounting and linear interpolation [14, 27]. In the case of linear interpolation, the maximum likelihood solution (in connection with leaving-one-out or some other form of cross-validation) is typically computed by the so-called EM algorithm (EM=*expectation-maximization*) [14, 27].

4. Partitioning-Based Models

In this section, we will consider partitioning-based structures used in language modelling. The partitioning is applied to each of the two variables in the language model $p(w|h)$, i. e. either to the word histories h or the words w. For the special case of a bigram language model, we will show how the same partitioning can be used for both variables. These partitioning-based structures can be viewed as another attempt to cope with the problem of sparse data. Another interpretation is that they are able to capture syntactic-semantic dependencies as typically defined by linguistic experts [9]. In all these methods, the criterion for finding the unknown partitionings is the likelihood criterion [6, 14, 16, 27].

4.1 Equivalence Classes of Histories and Decision Trees

First, we consider the level of word histories. The equivalence classes of the possible word histories h will be called states and denoted by a so-called

$$\text{state mapping } S : \ h \to s = S(h) \quad .$$

For these states s, we define new counts based on the elementary counts $N(h, w)$:

$$N(s,w) := \sum_{h:S(h)=s} N(h,w) \qquad N(s) := \sum_{w} N(s,w) \quad .$$

For a given or hypothesized history mapping S, we have the probability model:

$$Pr(w|h) = p(w|S(h))$$

with the *unknown* model parameters $p(w|s = S(h))$ that must satisfy the normalization constraint. By using equivalence states, we can reduce the number of free parameters: For a vocabulary of W words and a history mapping with $|S|$ states, we have only $|S| \cdot (W-1)$ free parameters in comparison with $W \cdot (W-1)$ for a bigram model and $W^2 \cdot (W-1)$ for a trigram model. However, in practice due to the sparse training data, the number of really independent parameters is much more determined by the size N of the training corpus.

We consider the log-likelihood function which now depends on both the mapping S and the parameters $p(w|s)$:

$$\begin{aligned} F(S, \{p(w|s)\}) &= \sum_{h,w} N(h,w) \, \log \, p(w|S(h)) \\ &= \sum_{s,w} N(s,w) \, \log \, p(w|s) \quad . \end{aligned}$$

For a hypothesized mapping S, we maximize the log-likelihood function with respect to $p(w|s)$ and, as usual in such cases, obtain the relative frequencies as estimates:

$$p(w|s) = \frac{N(s,w)}{N(s)} \quad .$$

By using these estimates, the log-likelihood function now depends only on the mapping S:

$$F(S) = \sum_{s,w} N(s,w) \, \log \frac{N(s,w)}{N(s)} \quad .$$

This log-likelihood function can be interpreted as an entropy-like quantity. The important result here is that we have arrived at this criterion by making use only of a structural model and the maximum likelihood criterion. No additional principle has been made use of. To find the optimal mapping S, we can use a K-means-style clustering algorithm [19, 27]. As long as we work with a fixed number of history states and this number is sufficiently small, there is probably no need to recourse to cross-validation and the conventional likelihood criterion is sufficient. For sparse data, of course, we could use one of the leaving-one-out extensions of the likelihood criterion, e. g. the Turing-Good method or absolute discounting.

So far we have considered a 'flat' mapping S with no additional hierarchical structure. We now consider a different approach based on decision trees. The decision tree method is often referred to as CART which stands for "Classification

And Regression Trees" as summarized in a book with the same title [5]. The CART method uses a binary decision tree to model a conditional distribution. Considering a general conditional distribution $p(y|x)$, we have to distinguish the two types of random variables:

- x: The *independent* variable, which stands for an observed or measured value and can take on either continuous (or discretized, also referred to as ordinal) or categorical values.
- y: The *dependent* variable, which is to be *predicted* by the CART method. Depending on the task to which the CART method is applied, the variable y may be a continuous value (as in classical regression) or the class index in a classification task (as in pattern recognition).

Given an observation x, we use the decision tree as follows. Starting at the root of the tree, we ask a yes-no question about the given observation x. Depending on the result, we follow either the left or right branch and ask again questions. This process is repeated until we reach a terminal node of the tree. In the case of a probability estimation task, with which we are faced in language modelling, there is a whole probability distribution $p(y|t)$ assigned to each terminal node t. In a classification task, this distribution can simply be reduced to the most likely class index y. Thus the characteristic property of the CART method is to use a binary decision tree for defining equivalence classes of the independent variable x. As a result, decision trees have the appealing property of being able to visualize the decision-making process and have the chance of being interpreted by a human expert.

For the purpose of language modelling, we have to use the *probability estimation* task in CART. To construct a decision tree, we are given a set of training data (h_n, w_n), $n = 1, \ldots, N$. For each node, a question has to be specified which is achieved by defining subsets over the set of observed histories h. Thus we can identify nodes and questions. So for a node t with left branch t_L and right branch t_R, we have:

$$t, t_L, t_R \subset \{h\} \quad,$$

for which a binary question amounts to a binary split or partition:

$$t = t_L \cup t_R, \quad t_L \cap t_R = \emptyset \quad.$$

For these splits, typically not all possible questions are considered, but only certain types [5]. Categorical values as given by the histories h are split by a binary partition of the possible values. Each candidate split is measured by the improvement in a so-called impurity function, which in our case will be based on the likelihood criterion. For each observation (h_n, w_n) in the training data, we have to consider the probability of the conditional distribution $p(w_n|t)$, where t is the candidate split (or node) we are evaluating with $h_n \in t$. For each observation (h_n, w_n), we assume that somehow we can compute a suitable score

$$g(w_n|t) \quad \text{with} \quad h_n \in t \quad.$$

Assuming that this per-observation score is additive, we obtain the natural definition of a score or impurity function $G(t)$ for the node t:

$$G(t) := \sum_{n:h_n \in t} g(w_n|t) \quad .$$

A popular impurity function is the so-called entropy criterion. As we will show, this criterion is, from the viewpoint of this paper, equivalent to the *conditional likelihood*. To each node t, we assign a probability distribution $p(w|t)$ and define as impurity function:

$$G(t) \quad := \quad - \sum_{n:h_n \in t} \log p(w_n|t) \quad .$$

As before in the case of the mapping $h \rightarrow S(h)$, the variable w_n is a *discrete* variable in a *modelfree* approach so that we obtain as maximum likelihood estimates the relative frequency of the observations sent down to node t. Plugging in these estimates as usual, we obtain the splitting criterion:

$$G(t) \quad = \quad -N(t) \cdot \sum_w p(w|t) \, \log \, p(w|t) \quad ,$$

where $N(t)$ is the number of training observations sent down to the node t. The important result then is that this splitting criterion is nothing else but the familiar maximum likelihood criterion, where the binary tree is used to form equivalence classes for the word histories h that play the role of the unconditional CART variable.

When splitting a node t into t_L and t_R, we have a change in the impurity function. The whole tree is grown by selecting the most effective split of each node. Having found the best split, we then select the best split for each of the successor nodes, and this process is repeated. Typically there is no stopping rule. Instead, a very large tree is constructed and then pruned from the bottom [5]. In [1], the CART method has been used to build a full language model for large vocabulary speech recognition. Other successful applications of the CART and related tree-based methods to speech and language include language understanding [21] and the definition of generalized triphones in acoustic modelling [37].

4.2 Two-Sided Partitionings and Word Classes

In addition to the history mapping h, we introduce for the words w the so-called

$$\text{class (or category) mapping } G : \ w \rightarrow g = G(w) \quad .$$

In this case of *two-sided* mappings or partitionings S and G, we have to distinguish two types of probability distributions:

- $p(g|s)$: probability that, given state $s = S(h)$ with history h, class g is observed in the *next* position.

- $p_0(w|g)$: probability that, given class g in the current position, word w is observed in the *current* position. For each word w, there is *exactly one* class $G(w)$:

$$p_0(w|g) \begin{cases} > 0 & \text{if } g = G(w) \\ = 0 & \text{if } g \neq G(w) \end{cases}$$

Therefore, we can use the somewhat sloppy notation $p_0(w|G(w))$.

For the probability $Pr(w|h)$, we then have the decomposition:

$$Pr(w|h) = p_0(w|G(w)) \cdot p(G(w)|S(h)) \quad .$$

As for the history states s, we introduce new counts depending on the mappings S and G. Using the above defined count $N(s, w)$ and the unigram count $N(w)$, we define:

$$N(s, g) := \sum_{w:G(w)=g} N(s, w) \qquad N(g) := \sum_{w:G(w)=g} N(w) \quad .$$

The log-likelihood function now depends on both mappings G and S and the unknown model parameters and can be written as [27]:

$$F(S, G, \{p(g|s)\}, \{p_0(w|g)\}) =$$
$$= \sum_w N(w) \log p_0(w|G(w)) + \sum_{s,g} N(s, g) \log p(g|s) \quad .$$

Plugging in the relative frequencies for $p_0(w|g)$ and $p(g|s)$ as maximum likelihood estimates, we obtain:

$$F(S, G) = \sum_w N(w) \log N(w) + \sum_{s,g} N(s, g) \log \frac{N(s, g)}{N(s)N(g)} \quad .$$

By identifying the counts with relative frequencies, we can interpret this criterion as the mutual information between the random variable s and the random variable g. Again the point here is that this criterion is obtained for free by using only structural assumptions and the maximum likelihood criterion.

A two-sided *symmetric* mapping can be defined by using the word class mapping $G : w \rightarrow G(w)$ both for the current word and its predecessor words. For a word bigram (v, w), we use $(g_v, g_w) = (G(v), G(w))$ to denote the corresponding word class bigram. For a bigram model, we then have:

$$Pr(w|v) = p_0(w|g_w) \cdot p(g_w|g_v) \quad .$$

The log-likelihood function for such a symmetric mapping is easily derived using the equations derived above:

$$F_{bi}(G) = \sum_w N(w) \log N(w) + \sum_{g_v, g_w} N(g_v, g_w) \log \frac{N(g_v, g_w)}{N(g_v)N(g_w)} \quad ,$$

where the new counts $N(g_v, g_v), N(g_w), N(g_v)$ are defined in the usual way. In a similar way, denoting a word trigram by (u, v, w) and the associated class trigram by (g_u, g_v, g_w), we obtain for the trigram model:

$$F_{tri}(G) =$$
$$= \sum_{w} N(w) \log N(w) + \sum_{g_u, g_v, g_w} N(g_u, g_v, g_w) \log \frac{N(g_u, g_v, g_w)}{N(g_u, g_v) N(g_w)}$$

with some new, but evident count definitions.

So far, the assumption for the two-sided partitioning has been that somehow the mapping $G : w \rightarrow G(w)$ for the word classes is known. Now we outline a procedure by which such mappings can be determined automatically. The task is to find a mapping $G : w \rightarrow g = G(w)$ that assigns each word to one of $|G|$ different word classes and results in a maximum of the log-likelihood function $F_{bi}(G)$ or $F_{tri}(G)$. Since these classes will be found by a statistical clustering procedure, they are also referred to as word clusters. The log-likelihood on the training data, i.e. $F_{bi}(G)$ or $F_{tri}(G)$, is used as optimization criterion. In the spirit of decision-directed learning [10, pp. 210], the basic concept of the algorithm is to improve the value of the optimization criterion by making local optimizations. This means, in moving a word from one class to another, the goal is to improve the log-likelihood criterion. More details about this *exchange* clustering algorithm are given in [19, 24].

5. Word Trigger Pairs

In this section, we study the use of so-called word trigger pairs (in short: word triggers) [3, 22, 34] to improve an existing language model, which is typically a smoothed trigram model in combination with a cache component [20, 27].

To illustrate what is meant by word triggers, we give a few examples:

$$airline \quad \ldots \ldots \quad flights$$
$$concerto \quad \ldots \ldots \quad orchestra$$
$$asks \quad \ldots \ldots \quad replies$$
$$neither \quad \ldots \ldots \quad nor$$
$$we \quad \ldots \ldots \quad ourselves \quad .$$

Thus word trigger pairs can be viewed as long-distance word bigrams. In this view, we are faced with the problem of finding suitable word trigger pairs. This will be achieved by analyzing a large text corpus (i.e. several millions of running words) and selecting those trigger pairs that are able to improve the baseline language model. A related approach to capturing long-distance dependencies is based on stochastic variants of link grammars [30, 36].

The baseline language model used in the following is denoted by $p(w|h)$, where w is the word whose probability is to be predicted for the given history h. For applications where the topic-dependence of the language model is important, e. g. text dictation, the history h may reach back several sentences as for the cache model [20] so that the history length M covers several hundred words, say $M = 400$. To denote a specific word trigger pair, we use the symbol $a \rightarrow b$, where a is the *triggering* word and b is the *triggered* word. In order to combine a trigger pair $a \rightarrow b$ with the baseline language model $p(w|h)$, we define the extended model $p_{ab}(w|h)$:

$$p_{ab}(w|h) \;=\; \begin{cases} q(b|a) & \text{if } a \in h \text{ and } w = b \\[2mm] [1 - q(b|a)] \cdot \dfrac{p(w|h)}{\sum\limits_{w' \neq b} p(w'|h)} & \text{if } a \in h \text{ and } w \neq b \\[4mm] q(b|\bar{a}) & \text{if } a \notin h \text{ and } w = b \\[2mm] [1 - q(b|\bar{a})] \cdot \dfrac{p(w|h)}{\sum\limits_{w' \neq b} p(w'|h)} & \text{if } a \notin h \text{ and } w \neq b, \end{cases}$$

where $q(b|a)$ and $q(b|\bar{a})$ are two interaction parameters of the word trigger pair $a \rightarrow b$. For symmetry reasons, we have introduced an interaction parameter $q(b|\bar{a})$ for the negative case, i. e. when the word a was *not* seen in the history. The unknown parameters $q(b|a)$ and $q(b|\bar{a})$ will be estimated by maximum likelihood. In the above definition, the interaction parameters of the word trigger pair are applied at the highest level, i.e. they always have priority over the baseline language model. This potential shortcoming will be corrected later.

In lieu of the log-likelihood criterion directly, we consider the difference between the log-likelihood of the extended model $p_{ab}(w|h)$ and the log-likelihood of the baseline model $p(w|h)$ on a corpus $w_1 \ldots w_n \ldots w_N$. This difference $\Delta F_{ab}[q(b|a), q(b|\bar{a})]$ is a function of the unknown parameters $q(b|a)$ and $q(b|\bar{a})$:

$$\Delta F_{ab}[q(b|a), q(b|\bar{a})] \;=\; \sum_{n=1}^{N} \log \frac{p_{ab}(w_n|h_n)}{p(w_n|h_n)} \;=\;$$

$$= \sum_{h} \Bigg[N(a; h, b) \log \frac{q(b|a)}{p(b|h)} \;+\; N(a; h, \bar{b}) \log \frac{1 - q(b|a)}{1 - p(b|h)}$$

$$+ \quad N(\bar{a}; h, b) \log \frac{q(b|\bar{a})}{p(b|h)} \;+\; N(\bar{a}; h, \bar{b}) \log \frac{1 - q(b|\bar{a})}{1 - p(b|h)} \Bigg],$$

where the counts $N(.; ., .)$ are defined in the usual way. E.g. the count $N(a; h, b)$ is the number of occurrences that the word b was observed for the history h and the word a appeared at least once in the history. To further rewrite the likelihood improvement and carry out the maximum likelihood estimation for the unknown trigger parameters $q(b|a)$ and $q(b|\bar{a})$, we introduce a set of history independent counts for the co-occurrence of $(a; b), (a; \bar{b}), (\bar{a}; b), (\bar{a}; \bar{b})$:

$$N(a; b) := \sum_{n: a \in h_n, b = w_n} 1 \qquad N(a; \bar{b}) := \sum_{n: a \in h_n, b \neq w_n} 1$$

$$N(\bar{a}; b) := \sum_{n: a \notin h_n, b = w_n} 1 \qquad N(\bar{a}; \bar{b}) := \sum_{n: a \notin h_n, b \neq w_n} 1 \quad .$$

Using these counts and ignoring irrelevant terms, we obtain the log-likelihood improvement as a function of the unknown trigger parameters $q(b|a)$ and $q(b|\bar{a})$:

$$\Delta \tilde{F}_{ab}[q(b|a), q(b|\bar{a})] = N(a; b) \log q(b|a) + N(a; \bar{b}) \log [1 - q(b|a)]$$
$$+ N(\bar{a}; b) \log q(b|\bar{a}) + N(\bar{a}; \bar{b}) \log [1 - q(b|\bar{a})] \quad .$$

The maximum likelihood estimates of the interaction parameters $q(b|a)$ and $q(b|\bar{a})$ are obtained by taking the corresponding derivatives and equating them to zero:

$$q(b|a) = \frac{N(a, b)}{N(a, b) + N(a, \bar{b})} \quad ,$$

$$q(b|\bar{a}) = \frac{N(\bar{a}, b)}{N(\bar{a}, b) + N(\bar{a}, \bar{b})} \quad .$$

Again, by identifying the relative frequencies with the probability estimates in the equation for $\Delta F_{ab}[q(b|a), q(b|\bar{a})]$, we have an expression that can be interpreted in terms of the mutual information between the event set $\{a, \bar{a}\}$ and the event set $\{b, \bar{b}\}$. The situation is complicated by the fact that we have to use *conditional* probabilities $p(w|h)$ as baseline probabilities. In [22, 32], the mutual information criterion was used in connection with a unigram language model $p(w|h)$ as baseline model.

The above model $p_{ab}(w|h)$ has been chosen for mathematical convenience. To arrive at a practical trigger model, a number of modifications are suitable [35]:

- The negative interaction was introduced to illustrate the equivalence to the usual information criterion. For practical likelihood improvement, the negative interaction is not essential.
- As indicated by the results of several groups [22, 32, 34], the word trigger pairs do not help much to predict the next word w for a given history h if this word w is already well predicted by the baseline language model $p(w|h)$, e. g. due to a suitable word m-gram observed in training or due to the cache effect. Therefore, we want to allow the trigger interaction $a \rightarrow b$ only if the probability $p(b|h)$ of the reference model is not sufficiently high. To this purpose, we replace the backing-off model by the corresponding model for linear interpolation with no negative interaction:

$$p_{ab}(w|h) = \begin{cases} [1 - q(b|a)] \, p(w|h) + \delta(w, b) \, q(b|a) & \text{if } a \in h \\ \\ p(w|h) & \text{if } a \notin h \end{cases}$$

with the usual definition of the Kronecker delta. Note that this interpolation model allows a smooth transition from no trigger effect, i. e. $q(b|a) \rightarrow 0$, to a strong trigger effect, i. e. $q(b|a) \rightarrow 1$.

- To define a multi-trigger language model with a large number of trigger pairs, we have to take into account the *interaction* of several trigger pairs. Therefore the *simultaneous* training of all trigger pairs is desirable. Using a conventional baseline model $p(w|h)$, e. g. a smoothed trigram language model, we define the multi-trigger model $p_T(w|h_n)$ (for $h_n = w_{n-M} \ldots w_{n-1}$):

$$p_T(w|h_n) = (1 - \lambda) \cdot p(w|h_n) + \frac{\lambda}{M} \cdot \sum_{m=1}^{M} \alpha(w|w_{n-m})$$

with the trigger parameters $\alpha(w|v) \geq 0$ and the normalization constraint for each v:

$$\sum_w \alpha(w|v) = 1 \quad .$$

The free parameters of this model are the interaction parameters $\alpha(w|v)$ and the interpolation parameter λ. There is no closed-form solution any more for the maximum likelihood criterion, but the EM algorithm provides an efficient iterative solution to this task [35].

6. Maximum Entropy Approach

The advantage of the maximum entropy approach is that it provides a well defined basis for incorporating different types of dependencies into a language model [32]. These different types of dependencies are referred to as *features*. The starting point for the maximum entropy approach is to consider a given set of suitable features, e. g.:

- the usual word m-grams like bigrams and trigrams;
- distant word bigrams and trigrams [32];
- word trigger pairs or similar long-distance word pairs.

Thus each observation of a word m-gram or a word trigger pair in training may be used to define a feature and might be included in the set of selected features if we think this feature is relevant to predicting the next word w given the history h. Often, the criterion is simply to include a feature if its observation count is high enough [32]. Even then, the number of features can be in the order of several millions.

Formally, for a word w and its history h, i.e. the last M predecessor words with, say $M = 100$, we define a feature function for each feature i:

$$f_i(h, w) \in \{0, 1\} \quad .$$

The maximum entropy principle tells us that the most general distribution that satisfies these constraints as expressed by the corresponding feature frequencies has the following functional form [4, pp. 83-87]:

$$p_\Lambda(w|h) = \frac{\exp\left[\sum_i \lambda_i f_i(h, w)\right]}{\sum_{w'} \exp\left[\sum_i \lambda_i f_i(h, w')\right]} \quad ,$$

where for each feature i we have a parameter λ_i and where we define: $\Lambda = \{\lambda_i\}$.

The important result of the maximum entropy principle is that the resulting model has a log-linear or exponential functional form. In the statistical terminology, the distribution of the counts is given by a multinomial distribution, and therefore the underlying sampling approach is referred to as a multinomial one [4, pp. 62-64]. An important difference to the typical applications given in statistical textbooks, however, is again that we are considering *conditional* probabilities.

We consider the log-likelihood function $F(\Lambda)$ for a training corpus with counts $N(h, w)$:

$$F(\Lambda) \quad := \quad \sum_{h,w} N(h, w) \log p_\Lambda(w|h) \quad .$$

To find the optimal set of parameters λ_i for the maximum likelihood criterion, we take the partial derivatives with respect to each of the parameters λ_i and set them to zero:

$$\frac{\partial F}{\partial \lambda_i} \quad = \quad \sum_{h,w} N(h, w) \frac{\partial}{\partial \lambda_i} \log p_\Lambda(w|h) \; = \; 0 \quad .$$

After some elementary manipulations, we obtain:

$$\frac{\partial F}{\partial \lambda_i} = - Q_i(\Lambda) + N_i = 0$$

with the Λ dependent auxiliary function $Q_i(\Lambda)$:

$$Q_i(\Lambda) := \sum_{h,w} N(h)\, p_\Lambda(w|h)\, f_i(h, w)$$

and with the Λ independent feature counts N_i:

$$N_i := \sum_{h,w} N(h, w)\, f_i(h, w) \quad .$$

The GIS algorithm (GIS=*generalized iterative scaling*) is a well known method in statistics for finding the numerical solution of the maximum likelihood equations [7, 32]. It is based on the condition that the number of features for each pair (h, w) with $N(h) > 0$, i. e. with a history h seen in training, is constant with respect to (h, w), i.e.

$$\sum_i f_i(h, w) = F_0 = const(h, w) \quad .$$

(Note that the constant F_0 is not related to the symbol $F(.)$ used for the log-likelihood function.) To satisfy this condition, we can define a complementary constraint in such a way that exactly the above condition is satisfied. This complementary constraint may also be used to handle unseen events, i.e. features not seen in training [33]. Each iteration of the GIS algorithm computes the following parameter update:

$$\lambda'_i = \lambda_i + \Delta\lambda_i$$
$$\text{with} \quad \Delta\lambda_i := \frac{1}{F_0} \log \frac{N_i}{Q_i(\Lambda)} \quad .$$

Thus again, the maximum entropy is an approach that amounts to a special structural assumption about the language model. These parameters are then optimized using the maximum likelihood criterion, which can be accomplished by the GIS algorithm. For more details, see the recent paper [31].

7. Conclusions

The principal goal of statistics as used in the context of language modelling and speech recognition is to learn from observations and make predictions about new observations. This point of view puts more emphasis on the prediction of new observations than on the retrospective interpretation of given observations, which is maybe more along the lines of mainstream statistics as it is traditionally found in textbooks. In our applications, the statistical models are simplifications of complex dependencies in the real world of speech and language. Therefore in most cases, it is a mistake to assume that any such model is a true representation of the underlying processes for speech and language. What we require instead, however, is that the model is useful for predicting new observations. In this paper, we have shown how the maximum likelihood criterion can be used exactly for this purpose in the following approaches:

- handling of sparse data
  by the Turing-Good estimates and by absolute discounting,
- partitioning-based models
  like decision trees and clustering-based word classes,
- word trigger pairs,
- maximum entropy models.

References

[1] L. R. Bahl, P. F. Brown, P. V. de Souza, R. L. Mercer: "A Tree Based Statistical Language Model for Natural Language Speech Recognition", IEEE Trans. on Acoustics, Speech and Signal Processing, Vol. 37, pp. 1001-1008, July 1989.

[2] L. R. Bahl, F. Jelinek, R. L. Mercer: "A Maximum Likelihood Approach to Continuous Speech Recognition", IEEE Trans. on Pattern Analysis and Machine Intelligence, Vol. 5, pp. 179-190, March 1983.

[3] L. R. Bahl, F. Jelinek, R. L. Mercer, A. Nadas: "Next Word Statistical Predictor", IBM Tech. Disclosure Bulletin, Vol. 27, No. 7A, pp. 3941-42, Dec. 84.

[4] Y. M. M. Bishop, S. E. Fienberg, P. W. Holland: 'Discrete Multivariate Analysis', MIT Press, Cambridge, MA, 1975.

[5] L. Breiman, J. H. Friedman, R. A. Olshen, C. J. Stone: 'Classification And Regression Trees', Wadsworth, Belmont, CA, 1984.

[6] P. F. Brown, V. Della Pietra, P. de Souza, R. L. Mercer: "Class-Based n-gram Models of Natural Language", Computational Linguistics, Vol. 18, No. 4, pp. 467-479, 1992.

[7] J. N. Darroch, D. Ratcliff: "Generalized Iterative Scaling for Log-Linear Models", Annals of Mathematical Statistics, Vol. 43, pp. 1470–1480, 1972.

[8] A. P. Dempster, N. M. Laird, D. B. Rubin: "Maximum Likelihood from Incomplete Data via the EM Algorithm", J. Royal Statist. Soc. Ser. B (methodological), Vol. 39, pp. 1-38, 1977.

[9] A. M. Derouault, B. Merialdo: "Natural Language Modeling for Phoneme-to-Text Transcription", IEEE Trans. on Pattern Analysis and Machine Intelligence, Vol. 8, pp. 742-749, Nov. 1986.

[10] R. O. Duda, P. E. Hart: 'Pattern Classification and Scene Analysis', John Wiley & Sons, New York, 1973.

[11] B. Efron, R. J. Tibshirani: 'An Introduction to the Bootstrap', Chapman & Hall, New York, 1993.

[12] I. J. Good: "The Population Frequencies of Species and the Estimation of Population Parameters", Biometrika, Vol. 40, pp. 237-264, Dec. 1953.

[13] A. L. Gorin, S. E. Levinson, A. N. Gertner, E. R. Goldman: "Adaptive Acquisition of Language", Computer, Speech and Language, Vol. 5, No. 2, pp. 101-132, April 1991.

[14] F. Jelinek: "Self-Organized Language Modeling for Speech Recognition", pp. 450-506, in A. Waibel, K.-F. Lee (eds.): 'Readings in Speech Recognition', Morgan Kaufmann Publishers, San Mateo, CA, 1991.

[15] F. Jelinek, J. Lafferty, R. L. Mercer: "Basic Methods of Probabilistic Context Free Grammars", pp. 347-360, in P. Laface, R. de Mori (eds.): 'Speech Recognition and Understanding', Springer, Berlin, 1992.

[16] F. Jelinek, R. L. Mercer, S. Roukos: "Classifying Words for Improved Statistical Language Models", IEEE Int. Conf. on Acoustics, Speech and Signal Processing, Albuquerque, NM, pp. 621-624, April 1990.

[17] F. Jelinek, R. L. Mercer, S. Roukos: "Principles of Lexical Language Modeling for Speech Recognition", pp. 651-699, in S. Furui, M. M. Sondhi (eds.): 'Advances in Speech Signal Processing', Marcel Dekker, New York, 1991.

[18] S. M. Katz: "Estimation of Probabilities from Sparse Data for the Language Model Component of a Speech Recognizer", IEEE Trans. on Acoustics, Speech and Signal Processing, Vol. 35, pp. 400-401, March 1987.

[19] R. Kneser, H. Ney: "Improved Clustering Techniques for Class-Based Statistical Language Modelling", Third European Conference on Speech Communication and Technology, Berlin, pp. 973-976, Sep. 1993.

[20] R. Kuhn, R. de Mori: "A Cache-Based Natural Language Model for Speech Recognition", IEEE Trans. on Pattern Analysis and Machine Intelligence, Vol. 12, pp. 570-583, June 1990.

[21] R. Kuhn, R. de Mori: "Recent Results in Automatic Learning Rules for Semantic Interpretation", Int. Conf. on Spoken Language Processing, Yokohama, Japan, pp. 75-78, Sep. 1994.

[22] R. Lau, R. Rosenfeld, S. Roukos: "Trigger-Based Language Models: A Maximum Entropy Approach", IEEE Int. Conf. on Acoustics, Speech and Signal Processing, Minneapolis, MN, Vol. II, pp. 45-48, April 1993.

[23] E. L. Lehmann: 'Theory of Point Estimation', J. Wiley, New York, 1983.

[24] S. Martin, J. Liermann, H. Ney: "Algorithms for Bigram and Trigram Word Clustering", Fourth European Conference on Speech Communication and Technology, Madrid, pp. 1253-1256, Sep. 1995.

[25] A. Nadas: "On Turing's Formula for Word Probabilities", IEEE Trans. on Acoustics, Speech and Signal Processing, Vol. 33, pp. 1414-1416, Dec. 1985.

[26] H. Ney, U. Essen: "Estimating Small Probabilities by Leaving-One-Out", Third European Conference on Speech Communication and Technology, Berlin, pp. 2239-2242, Sep. 1993.

[27] H. Ney, U. Essen, R. Kneser: "On Structuring Probabilistic Dependencies in Language Modelling", Computer Speech and Language, Vol. 8, pp. 1-38, 1994.

[28] H. Ney, S. Martin, F. Wessel: "Statistical Language Modelling by Leaving-One-Out", in G. Bloothooft, S. Young (eds.): 'Corpus-Based Methods in Speech and Language', Kluwer Academic Publishers, Dordrecht, pp. 174-207, 1997.

[29] R. Pieraccini, E. Levin, E. Vidal: "Learning how to Understand Language", Third European Conference on Speech Communication and Technology, Berlin, pp. 1407-1412, Sep. 1993.

[30] S. Della Pietra, V. Della Pietra, J. Gillett, J. Lafferty, H. Printz, L. Ures: "Inference and Estimation of a Long-Range Trigram Model", Second Int. Colloquium 'Grammatical Inference and Applications', Alicante, Spain, Sep. 1994, pp. 78-92, Springer-Verlag, Berlin, 1994.

[31] S. Della Pietra, V. Della Pietra, J. Lafferty: "Inducing Features of Random Fields", IEEE Trans. on Pattern Analysis and Machine Intelligence, Vol. 19, No. 4, pp. 380-393, April 1997.

[32] R. Rosenfeld: "Adaptive Statistical Language Modeling: A Maximum Entropy Approach", Ph.D. Thesis, School of Computer Science, Carnegie Mellon University, Pittsburgh, PA, CMU-CS-94-138, 1994.

[33] M. Simons, H. Ney, S. Martin: "Distant Bigram Language Modelling Using Maximum Entropy", Proc. IEEE Int. Conf. on Acoustics, Speech and Signal Processing, Munich, Vol. 2, pp. 787-790, April 1997.

[34] C. Tillmann, H. Ney: "Selection Criteria for Word Triggers in Language Modeling", Fourth Int. Colloquium on Grammatical Inference, Montpellier, pp.95-106, Springer, Lecture Notes in Artificial Intelligence 1147, Sep. 1996.

[35] C. Tillmann, H. Ney: "Word Triggers and the EM Algorithm", ACL Special Interest Group Workshop on Computational Natural Language Learning (Assoc. for Comput. Linguistics), Madrid, pp. 117-124, July 1997.

[36] J. Yamron, J. Cant, A. Demetds, T. Dietzel, Y. Ito: "The Automatic Component of the LINGSTAT Machine-Aided Translation System", ARPA Human Language Technology Workshop, Plainsboro, NJ, Morgan Kaufmann Publishers, San Mateo, CA, pp. 158-163, March 1994.

[37] S. J. Young, J. J. Odell, P. C. Woodland: "Tree-Based State Tying for High Accuracy Acoustic Modelling", ARPA Human Language Technology Workshop, Plainsboro, NJ, Morgan Kaufmann Publishers, San Mateo, CA, pp. 286-291, March 1994.

Language Model Adaptation

Renato DeMori[1] and Marcello Federico[2]

[1] University of Avignon, BP 1228, 84 911 AVIGNON, CEDEX 9, France
[2] IRST-Istituto per la Ricerca Scientifica e Tecnologica, Trento, Italy
    email: demori@univ-avignon.fr, federico@itc.it

Summary. This paper reviews methods for language model adaptation. Paradigms and basic methods are first introduced. Basic theory is presented for maximum a-posteriori estimation, mixture based adaptation, and minimum discrimination information. Models to cope with long distance dependencies are also introduced. Applications and results from the recent literature are finally surveyed.

1. Introduction

Current continuous speech recognition systems perform recognition based on a statistical decision criterion. Let W be a word sequence and O be the sequence of acoustic descriptors produced by an enunciation of W. The pair (O, W) can be seen as a jointly distributed random pair. Suppose the true joint distribution $Pr(O, W)$ could be modeled by:

$$Pr(W)Pr(O \mid W) \tag{1}$$

where $Pr(O \mid W)$ is computed with acoustic models and $Pr(W)$ is obtained with language models. Both models are fully specified by a set of parameters. If the models provide a correct formulation of the reality and their parameters are known, then the expected minimum recognition error rate [13] is achieved by selecting a word sequence W' with the following decision rule:

$$W' = \arg\max_{W} Pr(W)Pr(O \mid W). \tag{2}$$

Hypothesis generation using (2) is performed by a search (or decoding) algorithm that exploits a suitable decomposition of the involved probabilities. The acoustic model provides acoustic matching probabilities, i.e. the probability of any subsequence of O matching a given word or phoneme of the language. These probabilities are usually computed with hidden Markov models [34]. The language model (LM) instead provides linguistic probabilities (or scores) for sequences of words that constitute hypotheses of the search process.

The assumptions for the use of rule (2) cannot be validated in real situations. In fact, questions can be raised about the type of models, their statistical distributions and specification parameters, the type and size of the training set, and training conditions regarding microphone, channel, environment, task-dependent phonetic and linguistic facts which are very often different from testing conditions. Due to the fact that conditions may slightly vary during the recognition process, model parameters should evolve accordingly. For all these reasons, the models and parameters obtained by the best known training procedures are far from being ideal.

So far as the LM is concerned, it can be easily observed that moving from one application domain to another induces changes in the lexicon and in the statistical features of the language. In fact, even inside a specific application domain (e.g. radiological reporting) two user groups (hospitals) may employ statistically different languages [15].

Estimating the parameters of a language model on a text sample that is not representative can severely affect recognition performance. This problem can be tackled by adapting the model to topics, people, situations, etc. For example, a language model can be first estimated on a large user-independent text corpus and is then gradually adapted as user-dependent linguistic data become available. Topic adaptation can be obtained by dynamically optimizing the combination of different topic specific language models. Moreover, long distance word dependencies can be captured by modelling word repetition effects [24] and word co-occurrences [27]. Adaptation algorithms may be based on linear combinations of probabilities obtained with training and adaptation data. Another possibility is to refine the probabilities of a LM by looking at the closest model, according to an information theoretic distance, that satisfies a set of constraints derived from the adaptation data. Minimum discrimination information and maximum a-posteriori adaptation methods have been developed for this purpose.

Different types of LMs can be used at different stages of the recognition process, starting with simple models to be used by fast algorithms operating on a very large search space and following with more sophisticated models to be used for making decisions on a restricted lattice of word hypotheses. Language model adaptation can be applied at each level as well.

This tutorial is organized as follows. Section 2. gives some general background on statistical LMs. Section 3. introduces the issue of LM adaptation. Section 4. gives an overview of the basic statistical adaptation methods. Section 5. reports approaches and experimental results from the literature.

2. Background on Language Models

The purpose of LMs is to compute the probability $Pr(W_1^T)$ of a sequence of words $W_1^T = w_1 \dots, w_t, \dots, w_T$. The probability $Pr(W_1^T)$ can be expressed as

$$Pr(W_1^T) = Pr(w_1) \prod_{t=2}^{T} Pr(w_t \mid h_t) \qquad (3)$$

where $h_t = w_1, \dots, w_{t-1}$ can be considered the *history* of word w_t. The probabilities $Pr(w_t \mid h_t)$ may be difficult to estimate as the sequence of words h_t grows. A simplification can be introduced by defining equivalence classes on the histories h_t in order to considerably reduce their number. An n-gram LM approximates the dependence of each word (regardless of t) to the $n - 1$ words preceding it:

$$h_t \approx w_{t-n+1} \dots w_{t-1}. \qquad (4)$$

The n-gram approximation is based on the formal assumption that language is generated by a *time-invariant* Markov process. This greatly reduces the statistics to be collected in order to compute $Pr(W_1^T)$. Clearly, such an approximation causes a reduction in precision. Nevertheless, even a 3-gram (trigram) model may require a large amount of data (text corpus) for reliably estimating its parameters.

In order to further reduce the number of parameters, class based n-gram models can be used. Let $G(w_t) = g_t$ denote the class of the word occurring at time t and let G be the set of all classes. A trigram class based model is in general approximated with the formula:

$$Pr(w_t \mid g_{t-2}\, g_{t-1}) \approx \sum_{g \in G} Pr(w_t \mid g)\, Pr(g \mid g_{t-2}\, g_{t-1}). \qquad (5)$$

Classes may correspond to parts-of-speech (POS), semantic types, or automatically computed word clusters. In general, LMs are evaluated with respect to their impact on the recognition accuracy. However, LMs can be also evaluated in isolation by considering, for instance, their capability of predicting words in a text. The most used performance measure is the so-called *perplexity*. Perplexity is based on the following *log-prob* quantity:

$$LP = -\frac{1}{T} \log_2 \hat{P}r(W_1^T) \qquad (6)$$

where $W_1^T = w_1 \ldots w_T$ is a sufficiently long *test sequence* and $\hat{P}r(W_1^T)$ is the probability of W_1^T computed with a given stochastic LM. Hence, the LM perplexity is defined as:

$$PP = 2^{LP}. \qquad (7)$$

For test sequence it is meant a text independent of that used to estimate the LM. Assuming that the considered language source is *ergodic*, i.e. its behavior can be statistically described by analyzing a sufficiently long word sequence, it can be shown that LP converges in probability to the source *cross-entropy* measured by the LM:

$$H(Pr, \hat{P}r) = \lim_{T \to \infty} -\frac{1}{T} \sum_{W_1^T} Pr(W_1^T) \log_2 \hat{P}r(W_1^T) \qquad (8)$$

which is an upper bound of the source *entropy*:

$$H(Pr) = \lim_{T \to \infty} -\frac{1}{T} \sum_{W_1^T} Pr(W_1^T) \log_2 Pr(W_1^T). \qquad (9)$$

According to basic information theory principles, perplexity indicates that the prediction task of the LM is about as difficult as guessing a word among PP equally likely words. Hence, the smaller the log-prob and the perplexity statistics are, the better the LM is. The perplexity measure in (7) can be seen as a function with arguments a LM and a text sequence. According to this point of view, the here used measure is often called *test-set perplexity*, to distinguish it from the *train-set perplexity*, which is computed on the same text used to estimate the LM. While the test-set perplexity evaluates the generalization capability of the LM to predict words inside new texts, the train-set perplexity measures how much the LM fits or explains the training data.

Unfortunately, even if perplexity is usually a good indicator of the quality of a LM, its correlation with the recognition accuracy is not perfect. In fact, recognition accuracy is surely influenced by the acoustic similarity of words, which is not taken into account by perplexity.

Another important issue with LMs is the choice of the vocabulary size. In fact, due to the exponential growth of the n-gram population with respect to the vocabulary size, an increase of the vocabulary requires much additional training data and computational resources for estimating the n-gram probabilities. The trade-off is to some extent optimized by limiting the vocabulary to the most frequent k words occurring in the corpus. This choice also provides the lowest rate of out-of-vocabulary (OOV) words, i.e. the probability of finding never-observed 1-grams (unigrams). Hence, LM estimation and test-set perplexity are usually performed in *closed vocabulary* modality, i.e. by only involving n-grams made of in-vocabulary words.

3. Adaptation paradigms

Adaptation methods have been proposed that try to cope with different degrees of changes in the language: e.g. domain changes, user changes, topic shifts, etc.

State-of-the-art LMs for ASR are still trained on text corpora that represent specific *domains*, such as medical reporting, commercial correspondence, etc. The reason is that domains usually use restricted languages (sublanguages), from the point of view of the lexicon, syntax, semantics, and/or discourse structure. This favorably reflects on the computational requirements and performance of the ASR system. However, even a simple 2-gram (bigram) LM may require large amounts of training text samples.

In developing an application for a new domain, adaptation techniques allow the amount of training material to be reduced by exploiting samples of possible *close* domains.

Within a domain, variations in the language can be often due to intra-user differences or topic shifts. Both phenomena can significantly affect the a-priori probability of the word sequences that can be uttered. In Figure 1 the word frequencies of two corpora of radiological reports are compared. The texts were produced in two different hospitals. The shape of the cloud around the diagonal axis shows that larger (relative) differences among frequencies can be observed especially for less frequent words, which are in fact the majority.

Topic shifts also affect the probability of words. For instance, inside a radiography report, words like *heart* and *lungs* are much more likely to occur after the word *chest* than *leg*.

User preferences and topic shifts can be coped with by means of adaptation techniques that either try to capture long-distance dependencies or to adjust the n-gram statistics dynamically.

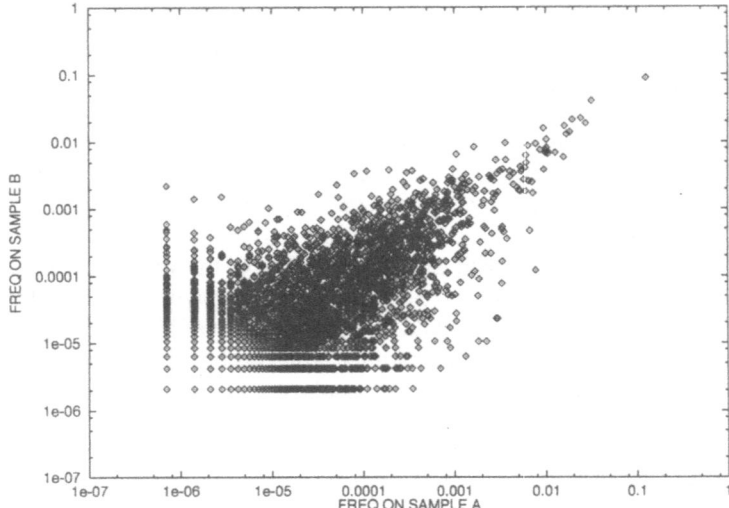

Fig. 1. Unigram relative frequencies measured on two samples (A and B) containing radio-logical reports. Relative frequencies are plotted for all words appearing in both samples. The column and row shaped plots, close to the origin, correspond to words occurring once, twice, etc. in sample B and A, respectively.

From the application and evaluation point of view, different paradigms can be applied. In fact, LM adaptation can be seen as a way to improve the performance of an *out-of-the-box* recognizer, that has only been trained on some general text corpus. Using the system, new texts are produced which reflect the user preferences. These texts are exploited to adapt the LM of the recognizer and therefore to improve recognition performance. Hence, adaptation can be *supervised* by the user who checks and corrects, if necessary, every sentence that has been recognized. Otherwise, adaptation can be *unsupervised*, i.e. it just relies on the recognizer's output. Adaptation can be performed in *incremental mode*, that is the LM is adapted sentence by sentence, or in *batch mode*, i.e. after a significant portion of texts has become available. Incremental adaptation is usually unsupervised, while batch adaptation might either be supervised or unsupervised. In general, topic adaptation requires incremental adaptation, as short term language changes have to be modeled. Domain adaptation intrinsically needs more data, hence batch adaptation is preferable. User adaptation can be performed in both modes. If for instance an a-priori sample of the user is available, adaptation can be performed in batch mode. Moreover, incremental adaptation to the user can be carried out while the system is being used.

3.1 LM adaptation in dialogue systems

Spoken language systems provide a context in which the discourse focus or topic can usually be formalized in an abstract way. In general, interactive or dialogue based systems perform some semantic interpretation of the input utterances, that is usually based on *case-frames*. Semantic interpretation exploits relations between concepts and words or phrases, that may be manually derived or automatically inferred [32]. Semantic relations can be also embodied into LMs in order to provide better linguistic constraints and to gather more information from the speech decoding process. Discourse topic and dialogue state can be more or less reliably inferred from semantic interpretations and other context information. LM adaptation can be performed on the basis of dialogue predictions, which can be made, given the past history, by means of deterministic rules [33] or stochastic models [5]. For instance, a set of topic/dialogue-state specific LMs can be estimated and dynamically combined according to the dialogue prediction. On the other side, LMs could be completely built during run-time, according to the dialogue state. In this case, an efficient method for handling the dynamic construction of LMs is required, as this activity impacts on the ASR system. In Section 5.5, an example of an interactive system is given, in which the implementation of dynamic LMs provides a good trade-off between flexibility and efficiency.

4. Basic statistical methods

Language model adaptation can be seen as an estimation problem in which a parametric model must be determined, given an adaptation sample S, usually very small, and some a-priori knowledge. Typically, such a knowledge is extracted from a large text sample (corpus) S'. Let a training or adaptation sample S be made of words belonging to a finite vocabulary V. Given a fixed history h, a sample S_h can be extracted from S by taking all the n-grams in S that begin with h. This sample $S = hw_{h_1}, \ldots, hw_{h_m}$ can be seen as a realization of m independently and identically distributed (i.i.d) random variables, drawn according to the law represented by the probability distribution $Pr(w \mid h)$. For the sake of simplicity, the history h will be omitted in the notation that follows. Moreover, $Pr(w \mid h)$ is assumed to belong to a parametric family $Pr(w; \theta)$, whose *likelihood* function of a sample S is the multinomial distribution:

$$Pr(S; \theta) = \frac{m!}{\prod_{w \in V} c(w)!} \prod_{w \in V} \theta_w^{c(w)} \quad \text{where} \quad c(w) = \sum_{i=1}^{m} \delta(w_i = w) \qquad (10)$$

with $\delta(e) = 1$ if e is true and 0 otherwise. The multinomial distribution (10) belongs to the exponential family, and the count vector $[c(w)]_{w \in V}$ represents a *sufficient statistic* of S [29].

4.1 Maximum a-posteriori estimation

In the Bayesian estimation [29, 13], the parameter vector θ is considered to be a random variable for which an a-priori distribution $Pr(\theta)$ is assumed. Given a training sample S, and applying the well known Bayes' rule, the a-posteriori distribution of θ is:

$$Pr(\theta \mid S) = \frac{Pr(S \mid \theta) \, Pr(\theta)}{Pr(S)}. \tag{11}$$

The a-posteriori distribution combines the a-priori evidence with the empirical evidence provided by the sample. The maximum a-posteriori (MAP) criterion attempts to find the value of θ that maximizes the a-posteriori probability. By eliminating the constant factor $Pr(S)$ in (11), the MAP criterion becomes:

$$\theta^{MAP} = arg \max_{\theta \in \Theta} Pr(\theta \mid S) = arg \max_{\theta \in \Theta} Pr(S \mid \theta) Pr(\theta). \tag{12}$$

By assuming a non-informative uniform prior distribution, the MAP estimate becomes equivalent to the maximum likelihood one. A sufficient condition for the practical computation of the posterior distribution (11) is that S admits sufficient statistics. In fact, this implies the existence of *reproducing priors*, i.e. distributions $Pr(\theta)$ for which the posterior distribution belongs to the same family as the prior. Interesting reproducing priors, called natural conjugates [37] or kernel densities [13], can be computed if a special sample S', called a-priori sample, is available. In particular, the natural conjugate prior of the multinomial distribution is the Dirichlet distribution [37]. The resulting MAP estimate is:

$$\theta^{MAP} = arg \max_{\theta \in \Theta} \prod_{w \in V} \theta_w^{c(w)+c'(w)} = [\frac{c(w) + c'(w)}{m + m'}]_{w \in V} \tag{13}$$

where $c'(\cdot)$ is the counting function on S'. The same estimate can also be expressed as a convex combination of frequencies of the two samples, with weights proportional to the sample sizes, i.e.:

$$Pr^{MAP}(w) = (\frac{m}{m + m'})f(w) + (\frac{m'}{m + m'})f'(w) \tag{14}$$

where f and f' are the relative frequencies on S and S', respectively. When a large adaptation sample is used (i.e. $m \to \infty$), the MAP estimate asymptotically converges to the maximum likelihood estimate of θ_v (i.e. $f(v)$).

4.2 Linear interpolation

Linear interpolation is a widely used scheme for smoothing n-gram distributions. For instance, the original version of the linear interpolated trigram LM [19] was defined as a linear combination involving lower order empirical distributions as follows:

$$Pr(z \mid xy) = \lambda_1(xy)f(z \mid xy) + \lambda_2(xy)f(z \mid y) + \lambda_3(xy)f(z) \tag{15}$$

where $\forall xy \in V^2$, $\lambda_i(xy) \geq 0$ $(i = 1, 2, 3)$, and $\sum_i \lambda_i(xy) = 1$. In general, the interpolation scheme can be generalized to any linear combination of known distributions:

$$Pr(w; \lambda) = \sum_{i=1}^{l} \lambda_i Pr_i(w) \qquad (16)$$

with λ defined on the l-dimension simplex. It is known that the weight of the above mixture can be estimated from a sample S with the EM (*expectation maximization*) algorithm [12], which numerically approximates their maximum likelihood estimates. The EM algorithm is based on the following iterative formula:

$$\lambda_i^{(n+1)} = \sum_{t=1}^{m} \frac{\lambda_i^{(n)} Pr_i(w_t)}{\sum_{j=1}^{l} \lambda_j^{(n)} Pr_j(w_t)}. \qquad (17)$$

Iterations of (17) are started by initializing the parameter vector uniformly and are stopped as soon as $\lambda^{(n)}$ converges. More robust estimates can be computed by combining the above EM algorithm with cross-validation techniques [16, 30].

The linear interpolation smoothing scheme can be easily extended to the LM adaptation case.

4.2.1 MAP adaptation. The MAP estimate for an n-gram distribution $Pr(w \mid h)$ can be written as:

$$Pr^{MAP}(w \mid h) = \begin{cases} \frac{c(hw)+c'(hw)}{c(h)+c'(h)} & \text{if } c(h) + c'(h) > 0 \\ \\ 0 & \text{otherwise} \end{cases} \qquad (18)$$

This distribution can be smoothed by recursively applying the linear interpolation scheme to the lower order adapted distribution, i.e.:

$$Pr(w \mid h) = \lambda(h) Pr^{MAP}(w \mid h) + (1 - \lambda(h)) Pr(w \mid h') \qquad (19)$$

where h' denotes the lower order $n-1$-gram history. In all recursively defined LMs, the less specific distribution $Pr(w \mid h')$ is either defined according to the scheme of $Pr(w \mid h)$, or is the uniform distribution over V, when the unigram level is reached, i.e. $h \in V$. Of course, parameter estimation of such recursive LMs has to be performed in a bottom-up way, starting from the unigram level.

In order to reduce the number of parameters $\lambda(h)$ to be estimated, histories h are usually grouped into equivalence classes, or *buckets*. Several criteria to form these buckets have been presented in the literature (see [6] for a review). A simple bucketing method is based on the frequency of history h. We assume that the bucketing function maps n-gram histories into natural numbers, i.e. $[\cdot] : V^n \to N$:

$$[h] = \begin{cases} 0 & \text{if } c(h) < k_1 \\ c(h) & \text{if } k_1 \le c(h) < k_2 \\ k_2 + ord(h) & \text{if } k_2 \le c(h) \qquad k_1, k_2 \in N \end{cases} \qquad (20)$$

where ord is any lexical ordering function defined on V^n, and k_1 k_2 are suitable thresholds. The above scheme adapts the number of parameters to the amount of data. Thresholds are set so that all the rarely observed histories share one parameter, intermediate-frequency histories use few parameters, and high-frequency histories use many parameters, i.e. one for each individual history.

4.2.2 Mixture adaptation. The fact that MAP itself can be seen as a linear interpolation, suggests that the previous model could be modified so that the interpolation weights between the posterior and prior frequencies are estimated separately leading to the following model:

$$Pr(w \mid h) = \lambda_1([h])f(w \mid h) + \lambda_2([h])f'(w \mid h) + \lambda_3([h])Pr(w \mid h') \quad (21)$$

where $[h]$ indicates the bucket associated to h, and the already stated conditions on the interpolation parameters are assumed. Finally, to avoid using sample S to estimate both $f(w \mid h)$ and $\lambda([h])$, a cross-validatory EM algorithm should be used.

4.3 Sublanguages mixture adaptation

Mixture model can be clearly extended to combine frequency distributions from several a-priori samples. This corresponds to generalizing the adaptation problem as follows. Let any sublanguage be represented by some text sample. Given l a-priori text samples $S_1, \ldots, S_l$ of sublanguage $L_1, \ldots, L_l$, and a relatively smaller adaptation sample S_{l+1}, of sublanguage L_{l+1}, the problem of LM adaptation is to produce a LM for L_{l+1}.

Two adaptive mixture based LMs can easily be derived. The first is in the spirit of the previous MAP mixture model, and combines sample frequencies:

$$Pr(w \mid h) = \sum_{i=1}^{l} \lambda_l([h])f_i(w \mid h) + \lambda_{l+1}([h])f_{l+1}(w \mid h) + \lambda_{l+2}([h])Pr(w \mid h').$$
$$(22)$$

The second is based on the combination of already estimated LMs of the sublanguages $L_1, \ldots, L_l$, i.e.:

$$Pr(w \mid h) = \sum_{i=1}^{l} \lambda_l Pr_i(w \mid h). \quad (23)$$

In both models, adaptation of the interpolation parameters is carried out on the adaptation sample S_{l+1}. The two models can also be smoothed by adding a further component to each mixture, that is estimated on the collection of text samples as a whole.

4.4 Backing-off

According to the *backing-off* [21] scheme, the n-gram probability is smoothed by selecting the most significant available approximation. For instance, a simple trigram backing-off model [19] is:

$$Pr(z \mid xy) = \begin{cases} s_1 f(z \mid xy) & \text{if } c(xyz) > 0 \\ s_2 f(z \mid y) & \text{if } c(xyz) = 0 \text{ and } c(yz) > 0 \\ s_3 f(z) & \text{otherwise} \end{cases} \quad (24)$$

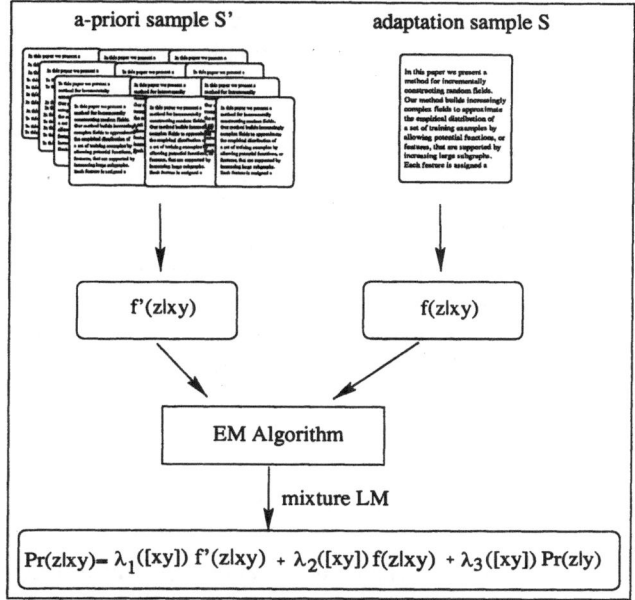

Fig. 2. Mixture based adaptation for a trigram LM

where s_1, s_2, s_3 are suitable normalization constants. Backing-off is an alternative to linear interpolation which is also used for LM adaptation. For instance, in [28] the MAP estimate is integrated into a backing-off LM. A different backing-off model, called *fill-up* model, has been proposed in [2]. The fill-up model switches between the prior and training sample frequencies according to some reliability criterion, i.e.:

$$Pr(w \mid h) = \begin{cases} f^*(w \mid h) & \text{if } c(h, w) > 0 \\[2mm] \alpha(h)f'^*(w \mid h) & \text{if } c(h, w) = 0 \text{ and } c(h) > 0 \text{ and } c'(h, w) > 0 \\[2mm] \beta(h)Pr(w \mid h') & \text{otherwise} \end{cases}$$

(25)

where $\alpha(h)$ and $\beta(h)$ are suitable normalizing constants, and f^* and f'^* are discounted frequencies [2] on S and S'. Discounting is related to the zero-frequency estimation problem [38]. A probability for all the words never observed after the history h is estimated by discounting the empirical distribution $f(w \mid h)$. Hence, a discounted frequency $f^*(w \mid h)$ is estimated, such that:

$$0 \le f^*(w \mid h) \le f(w \mid h),$$

(26)

and the zero-frequency probability $\lambda(h)$ is defined as:

$$\lambda(h) = 1.0 - \sum_{w \in V} f^*(w \mid h).$$

(27)

Redistribution of the probability $\lambda(h)$ is performed proportionally to the less specific distribution $Pr(w \mid h')$. A survey on discounting techniques can be found in [16, 30].

4.5 Maximum Entropy

The maximum entropy (ME) principle has roots in classical statistical mechanics and was extended to statistical applications in [18]. For the sake of simplicity, let any discrete distribution be identified with its parameter vector: $\theta = [\theta_w]_{w \in V}$. The *entropy*[8] of θ is defined as:

$$H(\theta) = -\sum_{w \in V} \theta_w \log_2 \theta_w \qquad (28)$$

where $0 \cdot \log_2 0 = 0$ by definition. Now, let one consider d linear constraints:

$$\sum_{w \in V} \alpha_i(w)\theta_w = p_i \qquad (i = 1, \ldots, d) \qquad (29)$$

where $p_1, \ldots, p_d$ are the only available observations on θ, measuring some *macro properties*, defined by d known constraint functions $\alpha_1(\cdot), \ldots, \alpha_d(\cdot)$. For instance, the constraint functions could specify a partition of V for which the marginal probabilities have been observed. The ME principle suggests estimating θ so that the constraints are satisfied and the entropy H is maximized. This principle can be justified as follows. Let all possible random samples of θ of size m ($m \to \infty$) be grouped into equivalence classes of the count vector $c = [c(w)]_{w \in V}$. Each equivalence class, $\hat{c}$, corresponds to a possible *macro-state* of the sample population V^m, whose elements (*micro-states*) all provide the same empirical estimate $\hat{\theta} = \frac{\hat{c}}{m}$. The ME principle selects the macro-state that satisfies the constraints (29) and has the largest number of micro-states. In fact, the size of a macro-state $\hat{c}$ is:

$$\frac{m!}{\prod_{w \in V} \hat{c}(w)!} \qquad \text{where} \quad \sum_{w \in V} \hat{c}(w) = m; \qquad (30)$$

by applying Stirling's approximation $n! \approx (\frac{n}{e})^n$ it follows that:

$$\frac{m!}{\prod_{w \in V} \hat{c}(w)!} \approx \prod_{w \in V} (\frac{m}{\hat{c}(w)})^{\hat{c}(w)} = 2^{-m \sum_{w \in V} \frac{\hat{c}(w)}{m} \log_2 \frac{\hat{c}(w)}{m}} = 2^{mH(\hat{\theta})}. \qquad (31)$$

It appears that maximizing the entropy of θ under the constraints (29) is equivalent to looking for the largest macro-state satisfying the constraints.

LM estimation through the ME principle is quite recent [11]. In general, if the objective is estimating the n-gram conditional distribution $Pr(w \mid h)$, constraints are defined in terms of characteristic functions of all the n-grams observed over the training sample. In a trigram LM, for example, constraints are introduced for the observed unigrams, bigrams and trigrams. Observations usually correspond to empirical frequencies. A discounting method is also applied to frequencies in order to derive observations even for the never observed n-grams. For instance, bigram constraints have the form:

$$\sum_{xyz \in V^3} \alpha_{\hat{y}\hat{z}}(xyz) \Pr(z \mid xy) = f^*(\hat{z} \mid \hat{y}) \tag{32}$$

where $\hat{y}\hat{z}$ is a specific bigram whose discounted frequency is $f^*(\hat{z} \mid \hat{y})$, and:

$$\alpha_{\hat{y}\hat{z}}(xyz) = \begin{cases} 1 & \text{if } yz = \hat{y}\hat{z} \\ 0 & \text{otherwise} \end{cases} \tag{33}$$

The zero-frequency constraints for the words never observed after word $\hat{y}$ is:

$$\sum_{xyz \in V^3} \alpha_{\hat{y}_-}(xyz) \Pr(z \mid xy) = \lambda(\hat{y}) \tag{34}$$

where

$$\alpha_{\hat{y}_-}(xyz) = \begin{cases} 1 & \text{if } y = \hat{y} \text{ and } f^*(z \mid y) = 0 \\ 0 & \text{otherwise} \end{cases} \tag{35}$$

Constraints for other n-grams are set in a similar way. Hence, the number of constraints set for a trigram LM are of the order of the observed trigrams, bigrams and unigrams. As this number can be very large, even for ordinary tasks, computing a solution for this constrained maximization problem is in general not a trivial task. A general algorithm for ME estimation will be discussed in Section 4.7

4.6 Minimum Discrimination Information

Maximum entropy estimation can be seen as a special case of minimum discrimination information (MDI) estimation. Given a set of constraints (29) on the distribution θ, and an a-priori distribution θ', the MDI estimate of θ is a distribution that satisfies the constraints and minimizes the discriminatory information function or Kullback-Leibler distance:

$$D(\theta, \theta') = \sum_{w \in V} \theta_w \log \frac{\theta_w}{\theta'_w}. \tag{36}$$

In fact, it can be shown [8] that $D(\theta, \theta') \geq 0$, with equality holding if and only if $\theta = \theta'$. In other words, the MDI distribution is the distribution satisfying the constraint that is *closest* to the prior distribution, with respect to the distance D. In particular, if a uniform prior is considered, MDI estimation reduces to ME estimation. MDI estimates as well as ME estimates can be obtained with an algorithm, called *generalized iterative scaling* (GIS) algorithm [9].

Under the MDI estimation paradigm, LM adaptation can be performed as follows [11] (see Figure 3). A prior distribution $Pr'(w \mid h)$ is estimated on a large prior sample S'. Statistics observed on the adaptation sample S are used to set constraints on the objective n-gram distribution $Pr(w \mid h)$ in the way shown above. $Pr(w \mid h)$ is finally estimated through the GIS algorithm.

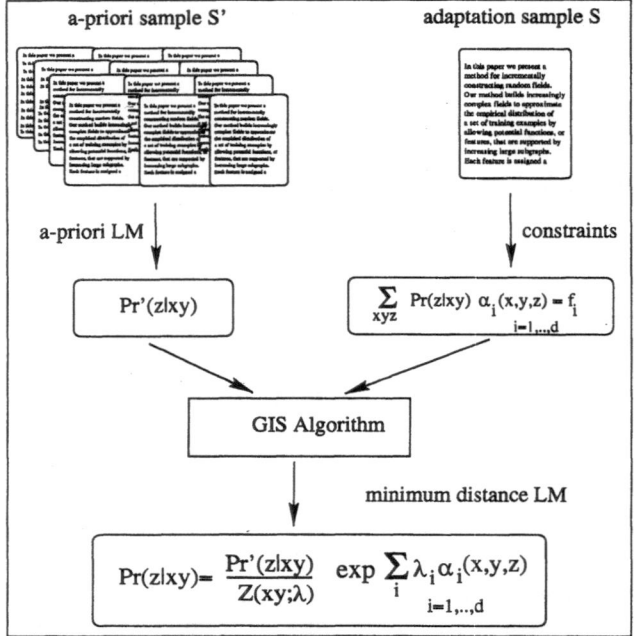

Fig. 3. MDI based adaptation for a trigram LM

4.7 Generalized iterative scaling

Let us now focus on the MDI estimation of the n-gram distribution $Pr(w \mid h)$. It can be shown [9] that every MDI distribution must have the exponential form:

$$Pr(w \mid h) = Z(h; \lambda)^{-1} Pr'(w \mid h) \exp \left(\sum_{i=1}^{d} \lambda_i \alpha_i(hw) \right) \qquad (37)$$

where $Pr'(w \mid h)$ is the a-priori distribution, $\lambda_1, \ldots, \lambda_d$ are parameters determined so that d linear constraints:

$$\sum_{hw \in V^n} \alpha_i(hw) Pr(w \mid h) = p_i \qquad (i = 1, \ldots, d) \qquad (38)$$

are satisfied, and $Z(h; \lambda)$ is the normalization term:

$$Z(h; \lambda) = \sum_{x \in V} Pr'(x \mid h) \exp \left(\sum_{i=1}^{d} \lambda_i \alpha_i(hx) \right) \qquad (39)$$

It can be shown that an algorithm exists, called GIS [9], that converges to a unique solution of the form (37), given that the constraints (38) are consistent. In particular, when the observations p_i $(i = 1, \ldots, d)$ are based on empirical frequencies observed on a sample S, consistency is guaranteed as the empirical distribution

$f(w \mid h)$ is itself a distribution satisfying the constraints. Moreover, it can be shown [9] that MDI estimation becomes equivalent to ML estimation of the parametric model (37) on the sample S.

The GIS algorithm requires some normalization condition on the constraint functions, i.e. $\alpha_i(\cdot)$ $(i = 1, \ldots, d)$ must be real valued functions such that:

$$\sum_{i=1}^{d} \alpha_i(hw) = k \quad \forall hw \in V^n. \tag{40}$$

This condition is easily satisfied when constraints are set with characteristic functions as shown in Section 4.5. In fact, for such constraints it results that $k = n$, as for an arbitrary n-gram hw there is exactly one unigram constraint function with value 1 on hw, exactly one bigram constraint function with value 1 on hw, etc.

Given a training set S of length m, and the n-gram constraints (38) of the type shown in Section 4.5, the GIS algorithm works as follows:

Algorithm:

(0) Set $n = 0$ and $Pr^{(n)}(w \mid h) = Pr'(w \mid h)$

(1) For $i = 1, \ldots, d$, let

$$\lambda_i^{(n)} = \frac{1}{k} \left(\log p_i - \log \sum_{t=1}^{m} \sum_{w \in V} \alpha_i(h_t w) Pr^{(n)}(w \mid h_t) \right)$$

(2) For $t = 1, \ldots, m$ and $\forall w \in V$, set

$$Q^{(n)}(h_t, w) = Pr^{(n)}(w \mid h_t) \exp \left(\sum_{i=1}^{d} \lambda_i^{(n)} \alpha_i(h_t w) \right),$$

$$Pr^{(n+1)}(w \mid h_t) = \frac{Q^{(n)}(h_t, w)}{\sum_{x \in V} Q^{(n)}(h_t, x)}$$

(3) Set $n = n + 1$. If $\lambda^{(n)}$ has converged, set

$$Pr(w \mid h) = Z(h; \lambda^{(n)})^{-1} Pr'(w \mid h) \exp \left(\sum_{i=1}^{d} \lambda_i^{(n)} \alpha_i(hw) \right)$$

and terminate. Otherwise go to step 1.

It can be noticed that the complexity of each iteration is $O(mrd)$, where r is the size of V. Improvements on the original GIS algorithm were recently presented in [10].

4.8 Cache model and word triggers

Modeling long-distance dependencies was the first approach to LM adaptation. Since trigrams do not use anything but the very immediate history, they are unable to adapt to the style or topic of the document and are therefore considered to be a static model. Attempts to predict the word at time t by exploiting correlations at longer distances have been pursued by coupling a static n-gram model with a dynamic component. In general, correlations are searched within a time-window of some hundreds of words. This correspond to approximate the history h_t by:

$$h_t \approx h_t^N = w_{t-N} \ldots w_{t-1} \quad \text{e.g. N=200} \tag{41}$$

Two dynamic models have been mostly successful in the last years: the cache model and the word trigger model.

The cache model is proposed as a dynamic component which tracks short-term fluctuations in word frequencies. The first adaptable statistical LM was based on this principle following the simple hypothesis that a word used in the recent past is much more likely to be used soon than its overall frequency in an n-gram LM would suggest. Hence, assuming a text window of size N, a cache memory can be implemented that maintains the history up to the last N words. Typically, unigram frequencies $f_t^N(w)$ are dynamically computed on the cache memory at time t, and are integrated into the static LM.

The word trigger model exploits dependencies between words that are significantly correlated inside the training corpus. Word triggers can be identified by looking at significant word co-occurrences up to a maximum distance N. The following statistics are usually collected over a large text sample for each word pair x, y:

$$c_N(x, y) = \sum_{t=N+1}^{T} \delta(w_t = y)\delta(x \in h_t^N)$$

$$c_N(\neg x, y) = \sum_{t=N+1}^{T} \delta(w_t = y)\delta(x \notin h_t^N)$$

$$c_N(x, \neg y) = \sum_{t=N+1}^{T} \delta(w_t \neq y)\delta(x \in h_t^N)$$

$$c_N(\neg x, \neg y) = \sum_{t=N+1}^{T} \delta(w_t \neq y)\delta(x \notin h_t^N) \tag{42}$$

where, for instance, $c_N(\neg x, y)$ counts the number of times word x did *not* occur in the histories of word y.

Methods for selecting trigger pairs are usually based on mutual information, χ^2 tests, or likelihood ratio tests. In particular, the latter method results more robust when rarely occurring word pairs must be tested [14]. Experimentally, it results that many trigger pairs are self triggers, i.e. highly correlated word repetitions. It is easy to see that this kind of information is very much related to the unigram cache model statistics shown above.

Statistics about word triggers can be dynamically computed given the current history and integrated within the static LM. For instance, for every trigger pair x, y, such that x occurs in the history h_t, the following empirical frequencies can be employed:

$$f_x(y \mid h_t) = \frac{c_N(x, y)}{c_N(x, \neg y) + c_N(x, y)}$$

$$f_x(\neg y \mid h_t) = \frac{c_N(x, \neg y)}{c_N(x, \neg y) + c_N(x, y)}$$

Trigger pairs and cache memory statistics can be integrated into a static n-gram LM in different ways. The interpolation or backing-off schemes can be used when the cache and the word trigger models are expressed through frequency type statistics. The ME or MDI framework can be employed as well if the cache and trigger word statistics are expressed in terms of distribution constraints.

5. Practical applications of adaptation paradigms

Various types of adaptation have been incorporated into real systems and tested with corpora. Performance of adaptation methods has been evaluated by using training, adaptation, and test sets and by measuring the amount of perplexity reduction on the test set after adaptation.

In some cases, speech recognition tests have also been performed. It appears, in general, that a reduction in perplexity does not necessarily result in an improvement of recognition performance measured as word error rate (WER) reduction. Nevertheless, there is substantial evidence that a considerable reduction in perplexity generally leads to a minor, but substantial reduction in WER. This is the reason why adaptation is used in many commercial systems. Before starting the review of adaptation approaches, an evaluation method for LM adaptation is presented.

5.1 The 1993 ARPA evaluation method

This benchmark refers to the ARPA CSR evaluation in 1993 [31]. The evaluation concerned within-domain sublanguage adaptation. The corpus used was read-speech prompted from issues of the 1990 Wall Street Journal (WSJ). Articles with a minimum of 20 sentences were selected. For an initial test, the vocabulary was closed and adaptation was incremental and supervised. This means that the system knew all the words that could occur and the correct transcription of each previous utterance. The training material was restricted to WSJ texts before 1990. As contrast tests, recognition without LM adaptation and recognition with unsupervised adaptation were also performed. Results were expressed with four different word error rates, respectively for all sentences preceded by 0-5, 6-10, 11-15, and 16 or more sentences in a given article. The ARPA evaluation [31] showed that adaptation reduced the WER to 18.2% as opposed to 21.1% for the same utterances with an unchanged LM. A similar improvement was also achieved by using unsupervised adaptation.

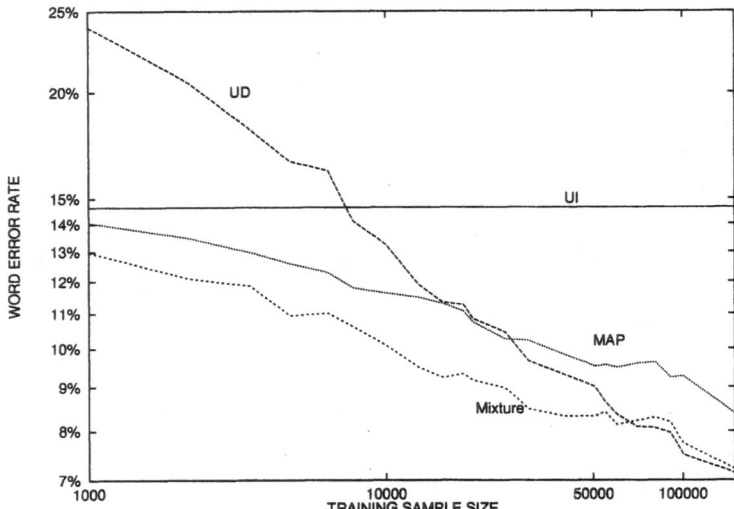

Fig. 4. Continuous speech recognition tests comparing the Mixture interpolation and the MAP interpolation adaptation methods. The UD curve refers to the LM estimated on the adaptation sample S only; the UI line refers to the bigram LM estimated on the user-independent sample S' only.

5.2 Mixture based adaptation

5.2.1 MAP versus mixture adaptation. An experimental comparison between the MAP (Section 4.2.1 and the mixture adaptation (Section 4.2.2) methods was performed in [15]. The task was the adaptation of a general user-independent LM to a new user group. The domain was radiological reporting in Italian. For this task, supervised and batch adaptation was considered. The used data consisted of a user-independent (UI) corpus of 1.9M words, a user-dependent (UD) adaptation sample of 600K words, and a UD test sample of 150 reports dictated by two male speakers. The vocabulary size was of 11,000 words. Speech recognition experiments on bigram LMs were carried out with the speaker-independent real-time continuous-speech recognizer developed at IRST [16].

Performance of the tested LMs with respect to the amount of adaptation data are reported in Figure 4. The best performing LM was the mixture based one. Convergence with the UD LM (trained only on adaptation data), occurs when about 50,000 words of adaptation data are used. After that point, the two LMs seem to behave similarly. By comparing the UI LM (that is trained only in UI data) with the mixture LM, the WER is reduced by 11% after only 1,000 words of adaptation data, and by more than 25% after 5,000 words. With respect to the UD LM, the mixture LM shows relative improvements ranging from 24% to 46% if less than 10,000 training words are used.

5.2.2 Mixtures of sublanguages. In [23] a LM is adapted by linearly interpolating l bigram LMs estimated on different text topics, i.e.:

$$Pr(w \mid y) = \sum_{i=1}^{l} \lambda_i Pr_i(w \mid y). \tag{43}$$

The parameter vector λ is incrementally adapted with the EM algorithm after every word, and only the last N observed words are used for adaptation. Perplexity evaluation on an English corpus showed a 10% reduction in perplexity with respect to an ordinary bigram LM trained on all the texts.

In [7], a mixture of trigram sublanguage models is incrementally adapted by means of the EM algorithm. Best results were obtained by smoothing the mixture with the static global trigram model, i.e.:

$$Pr(w \mid xy) = \sum_{i=1}^{l} \lambda_i Pr_i(w \mid xy) + \lambda_{l+1} Pr(w \mid xy). \tag{44}$$

Improvements in perplexity by 24% are reported with respect to the static global trigram model.

In [17], sentence based LMs are proposed that can be applied to re-score N-best lists or lattices. Sentence level mixtures are computed by interpolating l different topic-specific n-gram LMs. Each topic specific LM is in fact smoothed with a global LM, estimated over all the texts. Hence, the probability of a sentence W_1^T is computed by:

$$Pr(W_1^T) = \sum_{i=1}^{l} \lambda_l \prod_{t=1}^{T} [\theta_k Pr_i(w_t \mid w_{t-1}) + (1 - \theta_k) Pr(w_t \mid w_{t-1})] \tag{45}$$

where weights θ and λ are estimated separately and statically. In fact, little improvement was obtained by making the the the mixture weights λ adaptable. By comparing a static trigram model with a 5-component mixture model, a 17% perplexity reduction was measured.

5.2.3 Text clustering. Documents can be manually assigned to domains or they can be automatically clustered into domains by suitable algorithms.

In [7], a k-means clustering algorithm is proposed that partitions training data texts into clusters. The distance between a text and a cluster is defined as being the perplexity of an unigram LM constructed from the text within the cluster with respect to the text.

In [17], an agglomerative clustering algorithm is proposed that at each step merges the most similar clusters. A similarity measure is used that is based on the combination of inverse document frequencies:

$$S_{ij} = \sum_{w \in A_i \cap A_j} \frac{N_{ij}}{|A^w|} \times \frac{1}{|A_i||A_j|} \quad \text{where: } N_{ij} = \sqrt{\frac{N_i + N_j}{N_i \times N_j}}. \tag{46}$$

A_i denotes the set of words of cluster i, $|A_i|$ its size, $|A^w|$ is the number of articles containing word w, and N_i the number of texts in cluster i.

A clustering algorithm is proposed in [22], that at each step tries to improve the current partition, by moving a text from one cluster to another. For each text A of the corpus W, and for each word w in A, the following log-likelihood is evaluated:

$$LL = \sum_{A \in W} \sum_{w \in A} N(w, A) log \frac{N(w, D(A))}{N(D(A))} \qquad (47)$$

where $D(A)$ is a cluster for text A and $N(\cdot, \cdot)$ stands for the counter of occurrences. Clustering is based on the search for the mapping $D(A)$ that maximizes LL. A similar clustering algorithm and quality measure are also presented in [4].

5.3 Adaptation with a cache model

In [24, 25], an LM based on trigram Parts of Speech (POS) is considered in which a cache memory with room for $N = 200$ words is associated to each of the 153 POS considered. As a word is recognized, it is tagged and inserted into the appropriate cache. If a word has occurred often in the recent past as a member of a given POS class, then it will appear many times in the cache providing a term for modifying the probability of observing that word given the POS class. Hence, each POS has a time-varying empirical distribution $f_t^N(w \mid g)$ associated to. A linear interpolation of static and dynamic distributions of words given a class lead to a remarkable perplexity reduction on a corpus of English texts.

This concept has been extended in various ways. If a word is in a cache, it becomes a *trigger* for increasing the expectation of that word or of related words [36]. Bigram and trigram probabilities can be computed from the words stored in one large cache where all recognized words are inserted as soon as they are recognized [20].

In [26], an LM is proposed, based on the probabilities $Pr(w \mid h_t, D)$ where D is a boolean variable that takes value 1 if word w has occurred in the cache memory h_t^N, and 0 otherwise. Unigram and bigram empirical probabilities are considered. In [26], however, it is not pointed out how distributions can be derived from such probabilities, since D depends on the event to be measured.

In [20], unigram, bigram, and trigram empirical distributions are dynamically estimated on a cache memory of size $N = 1000$. Dynamic frequencies are linearly interpolated to obtain a dynamic trigram model $Pr_t(w_t \mid w_{t-2}, w_{t-1})$. The dynamic LM is then linearly interpolated with a static trigram LM. Dictation experiments showed that the introduction of the cache memory model resulted in a 5% WER reduction with short documents and a 24% WER reduction for long documents.

In [23], experiments showed that adding a cache component to a mixture of sublanguage models reduced the perplexity by 20%.

The linear combination of ME based LMs and cache-based LMs is discussed in [36] showing a 10% perplexity reduction with the addition of a cache component.

In [22], a semantic cache is proposed in addition to the already proposed bigram cache model. Words are clustered into semantic classes and a cache memory is used for each class. Perplexity evaluation on the Wall Street Journal corpus shows a reduction from 95 to 79 with a bigram cache model with a further reduction to 76 with a semantic model.

In [7], a unigram cache is proposed in which word recurrence probabilities decay exponentially over time. Adding the cache component to a static trigram LM provided a 14% perplexity reduction. The introduction of the decay contributed by 2%. The combination of the cache model with an adaptive mixture of sublanguage model provided a 30% perplexity reduction with respect to the static model.

In [17], a static sentence level mixture model is augmented by including a cache component. This technique provided a perplexity reduction of 18%.

5.4 ME and MDI adaptation

In [35], a trigram LM is adapted with respect to the discourse topic, by using the MDI method as shown in Figure 3. Unigram and bigram constraints are derived from a small topic specific sample. A trigram LM derived from general texts is used as the prior distribution. The GIS algorithm is used to find the trigram LM that requires the least perturbation of the general prior model and satisfies the constraints derived from the adaptation sample. Experiments performed on the Switchboard corpus showed an improvement of word accuracy by $3\% - 5\%$ which was comparable to mixture interpolation.

In [27], *word triggers* are proposed for modeling long distance dependencies. Triggers are selected through mutual information. The LM is trained with the ME principle by combining evidence (under the form of constraints) of word triggers and of a static trigram LM. The LM, estimated by means of the GIS algorithm, showed a 12% reduction in perplexity with respect to a trigram model.

5.5 LM adaptation in interactive systems

An example of LM adaptation with respect to the dialogue situation is provided by the SpeeData project [1]. The aim of SpeeData is the development of a multilingual spoken data-entry system. In the considered application domain, the data-entry system is organized in several forms, and manages data-base fields showing variety of formats:

- numbers: dates, portions, amounts of money, etc.;
- fixed texts: limited lists of expressions;
- proper names: owners, company names, locations, etc.;
- free texts: descriptions of properties, rights, etc.

The system accepts multi-lingual continuous speech. The user can fill in more data fields with single utterances, select single fields for input, apply editing commands, etc. Applications like this require a system that dynamically adapts the LM to the status of the interaction. In fact, the active LM mainly depends on the form currently shown on the screen, on the field that can be selected, and on the already filled fields. Moreover, each data-field owns its specific LM which may either correspond to a simple word list, a hand designed grammar, or a statistically trained LM. This

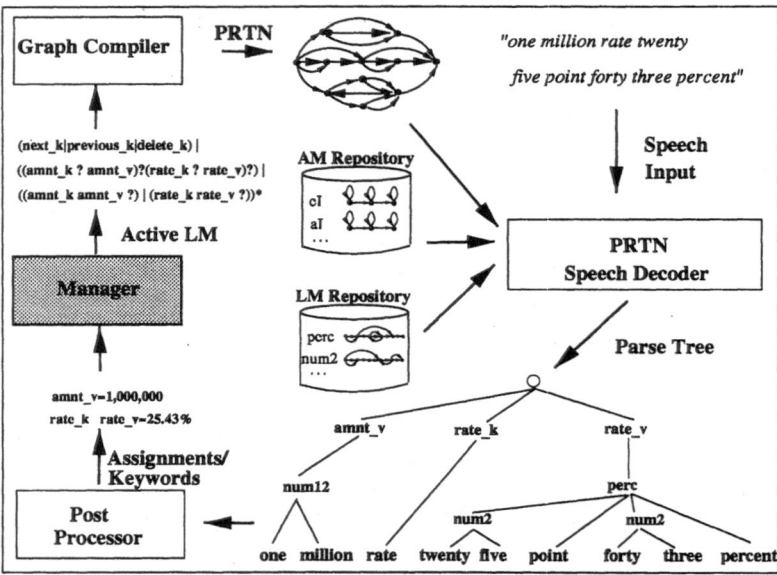

Fig. 5. LM dynamic construction within the SpeeData data-entry system

suggests that the active LM be organized in a hierarchical way. Starting from the bottom, there are *primitive* LMs corresponding to simple data-types like numbers or proper names. Then, there are LMs of data-types that are defined in terms of primitive LMs, e.g. dates or free texts. In fact, free texts are modelled with class based n-gram LMs, where classes correspond to other LMs, e.g. proper names, dates, or even other free texts. All the so defined data-type LMs constitute a repository of static LMs. At any given phase of the interaction, the active LM is dynamically constructed by constraining possible combinations of control commands, keywords, and data-type LMs. All LMs for the data-entry task are represented through *Probabilistic Recursive Transition Networks* (PRTNs). More implementation details can be found in [3].

The information flow of the system is shown in Figure 5. The dialogue manager produces a state dependent *active* LM, in terms of a regular expression. This syntactical expression is run-timely compiled into a PRTN and passed to the speech decoder. This network can refer to other PRTNs which have been pre-loaded by the speech decoder from a LM repository. The incoming speech utterance is decoded into a parse tree. This data structure is then explored by a post processor that transforms it into a sequence of data assignments and/or keywords. Finally, this list is returned to the manager that performs the corresponding actions and updates the dialogue state accordingly.

6. Conclusion

Adaptation of the statistical parameters of a LM is an important step for improving system robustness. Different methods are available that are suitable for application depending on the type of system.

Dynamic LMs can be implemented by switching between pre-determined, fixed models, depending on the truth of some preconditions or by adapting in time the model parameters.

The problem with LM adaptation is that it is difficult to predict modification of probabilities for words which do not appear in the adaptation data. As there is not a well established metric in the word space, more data are usually needed for LM adaptation compared to the adaptation of acoustic models.

Linear combination models use interpolation of probabilities obtained with different LMs. This is very effective and probably the most popular adaptation method, but it has the limitation of imposing a linear smoothing. Among the non-linear combination methods, backing-off models appear to be promising. Criteria for switching between LMs can be statistical or empirical.

MDI or ME based methods have the advantage of being intuitively appealing, extremely general, and that the GIS algorithm is guaranteed to converge for consistent constraints. Nevertheless, this method is computationally expensive, may have a slow convergence and it may be difficult to maintain consistency if constraints are derived with methods that do not use empirical frequencies.

LM sources of knowledge can be n-gram frequencies from different topics, word-trigger statistics, cache memory n-grams, etc. All sources can be used with linear and non-linear methods. In general, simple methods based on linear combination of models, some of them have time-varying statistical parameters, seem to be suitable for most application areas.

Acknowledgement. The authors would like to thank M. Cettolo, M. Matassoni, and M. Gilbert for reviewing this manuscript. Part of the presented work is supported by the European Commission, Telematics Application Programme, project LE-1999.

References

[1] U. Ackermann, B. Angelini, F. Brugnara, M. Federico, D. Giuliani, R. Gretter, and H. Niemann. SpeeData: a prototype for multilingual spoken data-entry. In *Proceedings of the 5th European Conference on Speech Communication and Technology*, Rhodes, Greece, 1997.

[2] S. Besling and H. Meier. Language model speaker adaptation. In *Proceedings of the 4th European Conference on Speech Communication and Technology*, volume 3, pages 1755–1758, Madrid, Spain, 1995.

[3] F. Brugnara and M. Federico. Dynamic language models for interactive speech applications. In *Proceedings of the 5th European Conference on Speech Communication and Technology*, Rhodes, Greece, 1997.

[4] D. Carter. Improving language models by clustering training sentences. In *Proceedings of the 4th Conference on Applied Natural Language Processing*, pages 59–64, Stuttgart, Germany, 1994.

[5] M. Cettolo and A. Corazza. History integration into semantic classification. In *Computational Models of Speech Pattern Processing*, (this volume) in *NATO ASI Series F*, pages 356–361, Springer–Verlag, Berlin, 1998.

[6] S. F. Chen. *Building Probabilistic Models for Natural Language*. PhD thesis, Center for Research in Computing Technology - Harvard University, Cambridge, MA, 1996.

[7] P. Clarkson and A. Robinson. Language model adaptation using mixtures and an exponentially decaying cache. In *Proceedings of the IEEE International Conference on Acoustics, Speech and Signal Processing*, volume 1, pages 799–802, Munich, Germany, 1997.

[8] T. M. Cover and J. A. Thomas. *Elements of Information Theory*. Wiley Series in Telecommunications. John Wiley & Sons, 1991.

[9] J. N. Darroch and D. Ratcliff. Generalized iterative scaling for log-linear models. *The Annals of Mathematical Statistics*, 43(5):1470–1480, 1972.

[10] S. A. Della Pietra, V. J. Della Pietra, and J. Lafferty. Inducing features of random fields. *IEEE Trans. Pattern Anal. Machine Intell.*, PAMI-19(4):380–393, 1997.

[11] S. A. Della Pietra, V. J. Della Pietra, R. Mercer, and S. Roukos. Adaptive language model estimation using minimum discrimination estimation. In *Proceedings of the IEEE International Conference on Acoustics, Speech and Signal Processing*, volume I, pages 633–636, San Francisco, CA, 1992.

[12] A. P. Dempster, N. M. Laird, and D. B. Rubin. Maximum-likelihood from incomplete data via the EM algorithm. *Journal of the Royal Statistical Society, B*, 39:1–38, 1977.

[13] R. O. Duda and P. E. Hart. *Pattern Classification and Scene Analysis*. John Wiley & Sons, New York, NY, 1973.

[14] T. Dunning. Accurate methods for statistics of surprise and coincidence. *Computational Linguistics*, 19:61–74, 1993.

[15] M. Federico. Bayesian estimation methods of n-gram language model adaptation. In *Proceedings of the International Conference of Spoken Language Processing*, Philadelphia, PA, 1996.

[16] M. Federico, M. Cettolo, F. Brugnara, and G. Antoniol. Language modeling for efficient beam-search. *Computer Speech and Language*, 9:353–379, 1995.

[17] R. Iyer and M. Ostendorf. Modeling long distance dependence in language: Topic mixtures vs dynamic cache models. In *Proceedings of the International Conference of Spoken Language Processing*, Philadelphia, PA, 1996.

[18] E. T. Jaynes. Information theory and statistical mechanics. *Physics Reviews*, 106:620–630, 1957.

[19] F. Jelinek. Self-organized language modeling for speech recognition. In A. Weibel and K. Lee, editors, *Readings in Speech Recognition*, pages 450–505. Morgan Kaufmann, Los Altos, CA, 1990.

[20] F. Jelinek, B. Merialdo, S. Roukos, and M. Strauss. A dynamic language model for speech recognition. In *Proceedings of the DARPA Speech and Natural Language Workshop*, Asilomar, CA, February 1991.

[21] S. M. Katz. Estimation of probabilities from sparse data for the language model component of a speech recognizer. *IEEE Trans. Acoust., Speech and Signal Proc.*, ASSP-35(3):400–401, 1987.

[22] G. Kneser and J. Peters. Semantic clustering for adaptive language modeling. In *Proceedings of the IEEE International Conference on Acoustics, Speech and Signal Processing*, volume 1, pages 779–782, Munich, Germany, 1997.

[23] R. Kneser and V. Steinbiss. On the dynamic adaptation of stochastic language models. In *Proceedings of the IEEE International Conference on Acoustics, Speech and Signal Processing*, volume II, pages 586–588, Minneapolis, MN, 1993.

[24] R. Kuhn. Speech recognition and the frequency of recently used words: a modified Markov model for natural language. In *Proceedings of COLING*, pages 348–350, Budapest, Hungary, 1988.

[25] R. Kuhn and R. D. Mori. A cache-based natural language model for speech recognition. *IEEE Trans. Pattern Anal. Machine Intell.*, PAMI-12(6):570–582, 1990.

[26] J. Kupiec. Probabilistic models of short and long distance word dependencies in running text. In *Proceedings of the DARPA Workshop on Speech and Natural Language*, pages 290–295. Morgan Kaufmann, February 1989.

[27] R. Lau, R. Rosenfeld, and S. Roukos. Trigger-based language models: a maximum entropy approach. In *Proceedings of the IEEE International Conference on Acoustics, Speech and Signal Processing*, volume 2, pages 45–48, Minneapolis, MN, 1993.

[28] H. Masataki, Y. Sagisaka, K. Hisaki, and T. Kawahara. Task adaptation using MAP estimation in n-gram language modelling. In *Proceedings of the IEEE International Conference on Acoustics, Speech and Signal Processing*, volume 1, pages 783–786, Munich, Germany, 1997.

[29] A. M. Mood, F. A. Graybill, and D. C. Boes. *Introduction to the Theory of Statistics*. McGraw-Hill, Singapore, 1974.

[30] H. Ney, U. Essen, and R. Kneser. On structuring probabilistic dependences in stochastic language modelling. *Computer Speech and Language*, 8:1–38, 1994.

[31] D. S. Pallet, J. G. Fiscus, W. M. Fisher, J. S. Garofolo, B. A. Lund, and M. A. Pryzbocki. 1993 Benchmark Tests for the ARPA Spoken Language Program. In *Proceedings of the ARPA Human Language Technology Workshop*, pages 51–73, Plainsboro, NJ, 1994.

[32] R. Pieraccini and E. Vidal. Learning how to understand language. In *Proceedings of the 3rd European Conference on Speech Communication and Technology*, pages 1407–1412, Berlin, Germany, September 1993.

[33] C. Popovici and P. Baggia. Specialized language models using dialogue predictions. In *Proceedings of the IEEE International Conference on Acoustics, Speech and Signal Processing*, volume 1, pages 815–818, Munich, Germany, 1997.

[34] L. R. Rabiner. A tutorial on hidden Markov models and selected applications in speech recognition. In A. Weibel and K. Lee, editors, *Readings in Speech Recognition*, pages 267–296. Morgan Kaufmann, Los Altos, CA, 1990.

[35] P. S. Rao, M. D. Monkowski, and S. Roukos. Language adaptation via minimum discrimination information. In *Proceedings of the IEEE International Conference on Acoustics, Speech and Signal Processing*, volume 1, pages 161–164, Detroit, MI, 1995.

[36] R. Rosenfeld. A maximum entropy approach to adaptive statistical language modeling. *Computer Speech and Language*, 10:187–228, 1996.

[37] J. Spragins. A note on the iterative application of Bayes' rule. *IEEE Trans. Inform. Theory*, IT-11:544–549, 1965.

[38] I. H. Witten and T. C. Bell. The zero-frequency problem: Estimating the probabilities of novel events in adaptive text compression. *IEEE Trans. Inform. Theory*, IT-37(4):1085–1094, 1991.

Using Natural-Language Knowledge Sources in Speech Recognition

Robert C. Moore

Microsoft Research
One Microsoft Way
Redmond, WA 98052

Summary. High accuracy speech recognition requires a language model, to specify what word sequences are possible or at least likely. Standard n-gram language models for speech recognition ignore linguistic structures, but more linguistically sophisticated language models are possible. Unification grammars are widely used in natural language processing, and these can be compiled into non-left-recursive context-free grammars that can then be used in real-time speech recognizers by dynamically expanding them into state-transition networks. A hybrid language model incorporating both a unification grammar and n-gram statistics has been shown to increase speech recognition accuracy. Probabilistic context-free grammars and probabilistic unification grammars are also possible.

1. Introduction

At the current state of the art, high-accuracy speech recognition with moderate to large vocabularies (hundreds to tens of thousands of words) requires a language model. That is, in addition to incorporating a model of the correspondence between acoustic signals and words, a recognition system needs to incorporate a model of what sequences of words are possible, or at least likely. Typically, the language models used in speech recognition systems pay little attention to the linguistic structure of utterances. Perhaps the most common language models used for speech recognition incorporate only n-gram statistics, looking at a sliding two-word (bigram) or three-word (trigram) window to judge the likelihood of a recognition hypothesis. Models of this type, however, miss longer-distance constraints on language, resulting in speech recognition output that appears inappropriate or even nonsensical.

For example, using a trigram model, a speech recognizer might misrecognize *the father of the bride* as *the father of the bribe*, even though to a fluent speaker of English, the former phrase is far more natural than the latter. Looking at only a three-word window, *of the bribe* seems not particularly less likely than *of the bride*, since it would occur very naturally as part of a phrase such as *the amount of the bribe*. A larger window would be required to consider *father* and *bride* simultaneously, but as the size of the window increases n-gram models suffer from a sparse-data problem. Consider the phrase *place the rubidium in the container*. The word *rubidium* is

Part of the research described here was conducted while the author was at the Artificial Intelligence Center of SRI International, and was supported by the Defense Advanced Research Projects Agency under Contract N66001–94–C–6046 with the Naval Command, Control and Ocean Surveillance Center.

sufficiently rare in English that using its occurrence to estimate the likelihood of any following words would be difficult, yet the verb *place* by itself makes the subsequent phrase *in the container* highly probable. A simple n-gram model will not capture this effect, however, because of the rarity of the intervening noun.

Examples such as these have led researchers to try to incorporate language models that reflect more linguistic structure into recognition, especially in applications that involve interpretation of speech, rather than simply recognition. In such systems, considerable effort often goes into analyzing the linguistic structure of utterances in the process of applying natural-language processing techniques to extract their meaning. It seems natural to try to use this same information to improve the language model used in recognition.

2. Issues in Language Modeling for Speech Recognition

Before looking in detail at some of the techniques that can be used to incorporate linguistically motivated knowledge sources in recognition, there are some general issues that need to be considered. First, it is important to distinguish between a language model and the search architecture that applies the model. The model itself addresses the question of how information from a particular knowledge source (such as a natural-language grammar) can be mathematically combined with information from other knowledge sources (including acoustic models) to optimize recognition accuracy. Search architectures address the issue of how particular models can be applied computationally. Early work on applying natural-language processing techniques in speech recognition often addressed only the search problem, using rather naive models. It was only later realized that these naive models could actually degrade recognition accuracy rather than improve it, rendering the search problem for such models irrelevant.

An additional set of issues surrounds the question of whether to try to model the language in a particular application qualitatively or probabilistically. A purely qualitative language model simply provides a specification of the permitted word strings, and the search attempts to find the permitted word string that best matches the acoustics. In practical recognizers, such models are often specified in terms of a finite-state grammar, which can be directly incorporated into the acoustic search performed by a hidden-Markov-model (HMM) speech recognizer.

Probabilistic, or statistical language models (such as the n-gram models discussed above) are normally applied using the following instance of Bayes' rule:

$$p(W|A) = \frac{p(A|W) \cdot p(W)}{p(A)}$$

This equation characterizes the maximum likelihood approach to speech recognition. The goal is to find the word sequence W that is most likely given the acoustic evidence A. Bayes' rule expresses this in terms of the probability of A given W, which is provided by the acoustic model, the prior probability of W, which is provided by the language model, and the prior probability of A. Since the probability of A is constant for any particular recognition task, the term $p(A)$ can be ignored, and the problem comes down to finding the word sequence W that maximizes the product of the estimates for $p(A|W)$ and $p(W)$.

Finally, hybrid combinations of the two approaches are possible. For example, an n-gram statistical language model might be combined with a numerical penalty for hypotheses that are judged ungrammatical by some more complex qualitative language model. Or, as we shall see later, a qualitative grammar can be used to structure the hypotheses to which a statistical language model is applied.

Qualitative and statistical language models each have advantages and disadvantages. Statistical language models usually have thousands to millions of parameters that have to be estimated empirically from thousands to millions of words of training data. To be effective, this data has to be from a source that is similar to the language used in the application. A language model trained on newspaper articles would provide a poor fit to the language used in a speech interface to a military command and control system. For novel applications, there may be no reasonable source of language model training data until the system is built and in use. Moreover, even when application-specific training data is available, there is never enough data to estimate all the parameters of the language model directly. Because of the sparse-data problem alluded to above, some parameters of the model always have to be estimated by a combination of lower-order parameters of the model. For example, in a trigram language model, there are always some trigrams that are never seen in the training data. Their probabilities have to be estimated from bigram and unigram data, to avoid assigning them a probability of 0.

Qualitative language models are usually specified in the form of a grammar and lexicon written by hand, with at least some degree of tailoring to the application. This means that human expertise about the application can be substituted for nonexistent training data. The disadvantage is that purely qualitative language models are less robust than statistical models. Statistical models are usually designed to avoid assigning any string within the vocabulary of the system a probability of 0. Hence, any string that might be uttered that is within the vocabulary has a chance of being correctly recognized if the acoustic evidence in favor of it is strong enough. Qualitative language models, on the other hand, simply classify strings as possible or impossible. Since in reality no human expert can anticipate all the meaningful utterances that could arise in a particular application, invariably some strings that are actually uttered will be "out of grammar," and thus be misrecognized, no matter how strong the acoustic evidence for them might be. Moreover, purely qualitative language models are unable to discriminate between two utterance hypotheses that are both classified as possible, even if one is far more likely than the other.

Finally, use of purely qualitative language models makes it impossible to take advantage of robust interpretation strategies that have been developed for natural-language processing. Strategies going by such names as "template matching" or "fragment combining" make it possible to successfully interpret strings, even if they cannot be completely analyzed by the grammar used by the system. If the recognizer is constrained, however, only to produce fully grammatical hypotheses, these robustness strategies will never have the opportunity to be applied, even in cases where they would be successful if applied to the out-of-grammar string actually uttered.

3. Formal Models for Natural Language

The difficulty of incorporating natural-language knowledge sources into speech recognition is largely a function of the complexity of the model. Various types of model have been used in natural-language processing systems, usually associated with a particular type of formal grammar. Of these, the ones that seem most worthy of attention are finite-state grammars, context-free grammars, and various extensions of context-free grammars. Although we present these models in nonprobabilistic form, probabilistic versions of all of them are possible, as we will discuss later.

3.1 Finite-State Grammars

Finite-state grammars can be expressed in a variety of notations. One that is often used to specify language models for speech recognition systems is a set of definitions of grammatical categories in terms of regular expressions over words or other grammatical categories. To ensure that the resulting language is finite-state, no category definition is allowed to be either directly or indirectly recursive. For example, the definition of a simple noun phrase might be written as

```
NP → Det Adj* N
```

which would mean that a noun phrase consists of a determiner (*a*, *the*, and so forth), followed by any number of adjectives, followed by a noun. This definition would classify strings such as *the girl*, *the little girl*, and *the pretty little girl* as noun phrases. The nonrecursivity constraint requires that the definitions of the categories Det, Adj, and N cannot make use of the category NP either directly or indirectly.

Finite-state grammars have the advantageous property that they can be expressed in terms of a finite-state-transition network, which means, as noted above, that they can be incorporated directly into the search performed by an HMM speech recognizer. Unfortunately, finite-state grammars are not expressive enough to describe the structure of natural languages. Suppose we wanted to extend the definition of a noun phrase to include the fact that it can contain a series of prepositional phrases following the noun. A natural way to do this would be to modify the definition of NP and add a definition of PP (for prepositional phrase) as follows:

```
NP → Det Adj* N PP*
PP → P NP
```

This pair of definitions would classify as noun phrases strings such as *the little girl in a red dress on a bicycle*. These definitions no longer constitute a finite-state grammar, however, because the categories NP and PP are defined by mutual recursion.

It has long been been argued (e.g., by Chomsky [1957, p. 21]) that finite-state grammars are inadequate in principle to characterize the word strings of natural languages. The point here, however, is that even if it were possible to define natural-language constructions by finite-state grammars, in many cases it is extremely unnatural to do so. Indeed, in the case under discussion, the same set of strings can be classified as noun phrases by the definition

```
NP → Det Adj* N (P Det Adj* N)*
```

but this denies the grammar writer the very useful category of prepositional phrase, and is so unnatural that it is somewhat difficult to see that it really does define the same set of noun phrases as the definition in terms of prepositional phrases.

Finite-state language models would be more useful if it were possible to use a more general form of specification such as a context-free grammar, but transform it into an equivalent finite-state grammar if one exists. Unfortunately, this transformation problem is undecidable in the general case. (See Theorem 8.15 of Hopcroft and Ullman [1979, p. 205].)

3.2 Context-Free Grammars

Context-free grammars can be defined in several ways, including simply removing the nonrecursivity constraint in the definition we used for finite-state grammars. More usually, they are defined as a set of rules that have a single atomic grammatical category on the left-hand side, and a sequence of atomic categories and words on the right-hand side, such as

```
S  → NP VP
NP → Det N
NP → NP PP
PP → P NP
VP → V NP
```

As long as multiple rules are allowed for a single category, eliminating the use of regular expressions on the right-hand sides of rules does not reduce their expressive power in terms of the sets of strings that they define.

The ability to define grammatical categories by mutual recursion makes context-free grammars much more suitable for defining linguistically based language models, since they can at least model the gross surface structure of natural-language expressions. They suffer, however, in their ability to model the fine detail of constraints on natural language. The rule S → NP VP, for instance, is intended to say that a sentence can consist of a noun phrase followed by a verb phrase, but this omits many important details. Generally, the noun phrase and the verb phrase agree in person and number; the verb is tensed; and an important feature of the sentence, whether it is interrogative or declarative, depends on whether the noun phrase is interrogative (i.e., whether it begins with a "question word" such as *who* or *what*).

Additional information such as this can be expressed in a context-free grammar, but only at the cost of greatly expanding the number of categories and rules. To take the issue of agreement between noun phrases and verb phrases in person and number, English has three persons (first for *I*, *we*, and so forth; second for *you*; and third for others) and two numbers (singular and plural). To express the constraint that in a sentence the subject noun phrase and the verb phrase generally agree in person and number, we would have to have six different categories each for noun phrases and verb phrases (one for each combination of person and number), and six rules of the form S → NP_first_singular VP_first_singular, and so forth. If we wish to add the information that whether the sentence is interrogative or declarative depends on the form of the noun phrase, we would have to add even more distinctions among categories and more rules. This verbosity of context-free grammars in describing the details of natural languages has led researchers in natural-language processing to augment context-free grammars in various ways to allow grammars to be more concise.

3.3 Augmented Context-Free Grammars

Over the years, many kinds of augmentation for context-free grammars have been proposed to make them more useful for specifying grammars of natural languages. In the 1970s, augmented transition networks, or ATNs (Woods, 1970), were widely used. In this formalism, context-free grammars were represented in the form of recursive transition networks, and the augmentations were defined by procedures attached to the arcs of the networks. The disadvantage of procedural augmentations to context-free grammars is that they have to be tailored to the particular processing algorithms used. Thus, the procedural annotations to a grammar appropriate for top-down parsing would be different from those appropriate for bottom-up parsing, because the information to be processed would become available in a different order.

Since the mid-1980s, the dominant approach to supplementing the expressive power of context-free grammars has shifted from ATNs or other procedural formalisms to declarative formalisms, particularly those based on various styles of "unification grammar." These formalisms have the advantage that their meaning is well-defined independent of any processing strategy or algorithm (although in practice there are interactions between the way a grammar is written and the processing algorithm that affect the efficiency, and even the termination, of the processing strategy).

The following example shows a rule written in the unification-based formalism used in the Core Language Engine (Alshawi, 1992) developed by SRI International's research group in Cambridge, England. This formalism is also used in the Gemini system (Dowding et al., 1993, 1994) developed by SRI in Menlo Park, California:

```
S:[tensed=yes] →
        NP:[person=P,num=N] VP:[tensed=yes,person=P,num=N]
```

What unification grammar adds to context-free grammar is the notion of feature constraints. In the notation presented here, grammatical categories are specified in terms of a major category symbol (such as S, NP, or VP), plus a set of feature constraints expressed by equations of the form feature=value. The power of the formalism comes from the ability to constrain a feature not to a specific value, but to a variable that also appears as the value of some other feature, requiring the two features to have the same value, without having to specify what value that is. Thus, in the sample rule, the noun phrase and verb phrase must agree in person and number—that is, the person and num features of the noun phrase must have the same respective values as the person and num features of the verb phrase. The term "unification grammar" comes from the fact that when the grammar is applied, the feature constraints are unified. This means that if the rule above is applied to a noun phrase of the category NP: [person=first,num=singular], the variable P will take on the value first, and the variable N will take on the value singular, which will in turn force the verb phrase to be of the category VP: [tensed=yes,person=first,num=singular].

3.4 Expressive Power of Grammar Formalisms and the Requirements of Natural Language

The example above shows how a unification grammar can specify a context-free language more concisely than a context-free grammar, but unification grammars can also define languages that are beyond the power of context-free grammars. In fact, it is easy to show that unification grammars can model any Turing machine. The construction is quite straightforward. Let there be a major category symbol corresponding to each Turing machine state, with each major category having three features: one representing the symbol currently being scanned by the head, one which encodes the tape to the left of the head, and one which encodes the tape to the right of the head. With this representation it is easy to write a grammar rule corresponding to each possible move of the Turing machine, for example,

```
S1: [current=a,left=X,right=(Y,Z)]  →
        S2: [current=Y,left=(b,X),right=Z]
```

This rule represents the following Turing machine move: When in state S1, if the current symbol is a, replace it by b, move to the right, and go to state S2. When a final state is reached, additional rules can simulate moving to the beginning of the tape and outputting its contents.

The power of unification grammar to express non-context-free languages relies on the use of features that have an infinite value space. In the example above, setting the value of the left feature of the category S2 to the value (b,X), where X can be any value of the left feature of the category S1, implies that the value space for the feature left includes arbitrarily deeply nested ordered pairs. This is an infinite space, which indeed it must be, if it is to represent all possible Turing machine tapes to the left of the Turing machine head.

If we limit the features used in unification grammars to finite value spaces, however, then the formalism becomes equivalent to context-free grammars in the languages it can define. It is easy to show that an arbitrary unification grammar is equivalent to a grammar consisting of all possible fully instantiated substitution instances of the rules of the original grammar. If the features all have finite value spaces, then there will be finitely many such substitution instances, and these substitution instances can simply be viewed as the nonterminal symbols of a corresponding context-free grammar.

An important question is whether the additional expressive power that is gained by allowing unification grammars to have infinitely valued features is actually needed to describe natural languages—that is, whether all natural languages are in fact formally describable by context-free grammars. Since the earliest development of formal language theory, linguists have attempted to construct arguments to show that natural languages are not context-free, but Pullum and Gazdar (1982) have argued rather persuasively that these early arguments failed to achieve their goal.

Most of the recent attention on this topic has focused on whether there are natural languages that have unbounded cross-serial dependencies. This refers to expressions of the form $\ldots a_1, a_2, \ldots a_n, \ldots b_1, b_2, \ldots b_n, \ldots$ for arbitrary n, such that the form of b_i depends on the form of a_i. Languages characterized by such expressions are known to be non-context-free. Pullum and Gazdar consider data suggesting that Dutch contains unbounded cross-serial dependencies, but argue that while the "correct" linguistic description of Dutch may indeed involve such dependencies, the data does not show that the strings of Dutch constitute a non-context-free language, essentially because the surface effects of the cross-serial dependencies in Dutch are not strong enough to rule out other ways of generating the same strings by a context-free grammar. However, subsequent data collected by Shieber (1985) regarding Swiss-German and by Culy (1985) concerning the African language Bambara have been generally accepted as demonstrating cross-serial dependencies in natural languages that cannot be modeled by context-free grammars.

Although the arguments of Shieber and Culy may show that in principle there are natural-language phenomena that cannot be modeled by context-free grammars, it is not clear that this is of much practical consequence to the choice of grammar formalism for language modeling in speech recognition. First, only a handful of languages are currently accepted as having constructions that are inherently non-context-free. Second, if these phenomena are modeled by allowing infinitely valued features in unification grammar, an approximating context-free grammar can be constructed by partitioning the value space into finitely many equivalence classes. (If all values for a feature are put into a single equivalence class, this amounts to simply ignoring the feature in question.) This method results in a grammar that accepts every string accepted by the more detailed grammar, plus some additional strings that violate the constraints on the infinitely valued features. For purposes of language modeling for recognition, this seems preferable in terms of robustness to the alternative of an approximating context-free grammar that rejects certain strings permitted by the more detailed grammar.

4. Search Architectures for Natural-Language-Based Language Models

If a language model more complex than a finite-state grammar is to be used in speech recognition, then the question arises of how to incorporate the language model into the recognition search. Historically, three approaches to this problem have been predominant: word lattice parsing, N-best filtering or rescoring, and dynamic generation of partial grammar networks.

4.1 Word Lattice Parsing

Word lattice parsing (Chow and Roukos, 1989; Boisen et al., 1989) is probably the oldest approach to integration of complex language models into recognition. In this approach, the recognizer produces a set of word hypotheses, with an acoustic score for each potential pair of start and end points for each possible word. A natural-language parser or other complex language model is then used to find the path through the word lattice having the best combined acoustic and language model score. In the case of a qualitative language model, this is usually the fully grammatical string having the best acoustic score.

Older implementations of word lattice parsing were not particularly efficient, because they had to deal with a large degree of word boundary uncertainty. Normally, a word lattice of adequate size for accurate recognition will contain dozens of instances of the same word with slightly different start and end points. One approach to this problem (Chow and Roukos, 1989) is to associate with each word or phrase a set of triples of start points, end points, and recognition scores. Each possible parsing step is then performed only once, but a dynamic programming procedure must also be performed to compute the best score for the resulting phrase for each possible combination of start and end points for the phrase.

Recently, a much more efficient approach to processing of word lattices has been developed by Mohri (1997). In this approach, word lattices are treated as weighted finite-state transducers, which can be determinized and minimized using automata theoretic methods, resulting in lattices that have no word boundary ambiguity for any sequence spanning the lattice. Moreover, Mohri's empirical results show that this can be done very efficiently, with the resulting lattices being much smaller than the input lattices,.

4.2 N-best Filtering or Rescoring

N-best filtering or rescoring (Chow and Schwartz, 1989; Schwartz and Austin, 1990) is a very simple search architecture in which the recognizer enumerates its N-best full recognition hypotheses, which are reprocessed using other complex knowledge sources, such as natural-language-based language models. Qualitative language models can be implemented by N-best filtering, in which the recognizer simply produces an ordered list of hypotheses, and the language model chooses the first one on the list that is accepted by its grammar. Probabilistic or other numerical language models can be implemented by N-best rescoring, in which the recognition score for each of the N-best recognition hypotheses is combined with a score from the complex language model, and the hypothesis with the best overall score is selected.

The advantage of the N-best approach is its simplicity. The disadvantage is that it seems inefficient for large values of N. The computational cost of the best method known for exact enumeration of the N-best recognition hypotheses (Chow and Schwartz, 1989) increases linearly with N, but an approximate method exists (Schwartz and Austin, 1990) that increases the computational cost of recognition only by a small constant factor independent of N. The additional knowledge sources, however, then have to be applied independently to each hypothesis, so that for large values of N, this phase will take a long time.

4.3 Dynamic Generation of Partial Grammar Networks

In an HMM speech recognizer, a finite-state grammar can be represented as a set of state-word-state transitions. Any type of linguistic constraints can, in fact, be represented as such a set, but for language models beyond finite-state grammars in complexity, the set of transitions will be infinite. A network representation can be used, however, if the network is generated dynamically to include only the transitions needed for the recognition search for a particular utterance.

In the method of Moore, Pereira, and Murveit (1989; Murveit and Moore, 1990), when a word is successfully recognized beginning in a given grammar state, the recognizer sends the word and the state it started in to the natural-language parser, which returns the successor state. To the parser, such a state encodes a parser configuration. When the parser receives a state-word pair from the recognizer, it looks up the configuration corresponding to the state, advances that configuration by the word, creates a name for the new configuration, and passes back that name to the recognizer as the name of the successor state. If it is impossible, according to the grammar, for the word to occur in the initial parser configuration, then the parser sends back an error message to the recognizer, and the corresponding recognition hypothesis is pruned out. Word boundary uncertainty in the recognizer means that the same word starting in the same state can end at many different points in the signal, but the recognizer has to communicate with the parser only once for each state-word pair. Because of this, the parser does not have to consider either acoustic scores or particular start and end points for possible words, those factors being confined to the recognizer.

The method of Moore, Pereira, and Murveit is designed to work with any kind of natural-language-based language model for which left-to-right parsing is possible, and was implemented for the kind of unification grammar described in Section 3.3. For context-free grammars, a simpler method of dynamic generation of partial grammar networks has been pioneered by Dupont and Snyers (1989; Dupont 1993). In this method, an finite-state network is constructed from the grammar rules for each nonterminal symbol, but the network may incorporate transitions for nonterminals as well as words. Whenever a nonterminal is encountered in the course of the recognition search, the network for that nonterminal is copied in place of the nonterminal.

This method will handle any context-free grammar as long as the grammar includes no direct or indirect left recursion (which would cause an infinite sequence of expansions of nonterminal symbols to occur). Techniques similar to this enable commercially developed recognizers from Microsoft (Huang et al., 1995) and Nuance Communications (1996) to accept non-left-recursive context-free grammars.

5. Compiling Unification Grammars into Context-Free Grammars

In Section 3.4 we pointed out that unification grammars with finitely valued features are equivalent to context-free grammars in expressive power. In this section we describe a method for efficiently compiling such unification grammars into context-free grammars, while eliminating left recursion, and give empirical results on the efficiency of the resulting grammars when used with the Nuance speech recognition system.

5.1 Instantiating Unification Grammars

Given a unification grammar with only finitely valued features, it would be trivial to transform all unification rules into context-free rules, by generating all possible full feature instantiations of every rule and making up an atomic name for each combination of category and feature values that occurs in these fully instantiated rules. This can easily increase the total number of rules to a size that would be too large to deal with, however. We therefore instantiate the rules in a more careful way that avoids unnecessarily instantiating features or generating unnecessary combinations of feature values.

A prime example of the problem we need to circumvent is a rule that occurs in a grammar developed by SRI for CommandTalk (Moore et al., 1997), a spoken-language interface to a military simulation system:

```
coordinate_nums:[] → digit:[] digit:[] digit:[] digit:[]
```

This rule says that a set of coordinate numbers can be a sequence of four digits. In the CommandTalk grammar, the digit category has features (e.g., singular vs. plural, 0 vs. non-0) that would generate at least 60 combinations if all instantiations were considered. So, if we naively generated all possible complete instantiations of this rule, we would get at least 60^4 rules. Even worse, we need other rules to permit as many as eight digits to form a set of coordinate numbers, which would give rise to 60^8 rules.

To avoid this unnecessary expansion in the number of rules, we define the set of atomic categories by considering, for each daughter category of each rule, all instantiations of just the subset of features on the daughter that are constrained by the rule. Thus, if a rule does not constrain a feature on a particular daughter category, an atomic category will be created for that daughter that is under-specified for the value of that feature. Since our example rule puts no constraints on any of the features of the digit category, by generating an atomic category that is under-specified for all features, we need only a single rule in the derived grammar.

In addition to using categories that are under-specified for values of certain features, we are careful to generate only instances of rules that could actually arise in applying the given grammar and lexicon. The algorithm works as follows:

1. Generate an initial set of rules incorporating "don't care" values for unconstrained features:
   - Treat each lexical entry as a rule whose mother is the category of the lexical entry and whose daughters are a word or word sequence.
   - Transform each rule of the original grammar and lexicon by placing every expression and subexpression, including variables, that occurs in a feature value inside the wrapper val, for example, transform foo(a,X) into val(foo(val(a),val(X))). This prevents "don't care" values from unifying with expressions corresponding to variables in the original grammar, while still permitting unification of any pair of expressions corresponding to a unifiable pair of expressions in the original grammar.
   - Specify an atomic "don't care" value for all the unconstrained features on daughter categories in rules, removing the val wrapper around any expression containing the "don't care" value.
   - For each category modified to include "don't care" values for some of its features, make a corresponding "don't care" variant of every rule whose mother matches the original category.

2. Generate an intermediate set of rules by computing all instances of the initial rules that can arise by recursively applying grammar rules to the lexicon, as in a bottom-up chart parser:
   - Process each rule as an active edge, none of whose daughters has been matched.
   - When processing an active edge, if the next unmatched daughter is a word, move the daughter from the list of unmatched daughters to the list of matched daughters, and process the resulting active edge, if an identical edge has not already been processed.
   - When processing an active edge, if the next unmatched daughter is a category, for each possible unification of the category against an inactive edge, move the resulting unified category from the list of unmatched daughters to the list of matched daughters, and process the resulting active edge, if an identical edge has not already been processed.
   - When processing an active edge, if there are no more unmatched daughters, create an intermediate rule from the mother and daughters of the active edge, and process the mother as an inactive edge, if an identical edge has not already been processed.
   - When processing an inactive edge, for each possible unification of the inactive edge against the next unmatched daughter of an active edge, move the resulting unified category from the list of unmatched daughters to the list of matched daughters, and process the resulting active edge, if an identical edge has not already been processed.

3. Generate a fully instantiated set of rules by traversing the intermediate rules top-down from the "start" categories of the grammar, instantiating any remaining variables in features to a dummy value.
4. Transform the fully instantiated rules into a context-free grammar, by constructing a unique atomic name for each instantiated category.

Step 1 of this algorithm handles the "don't care" feature values, as we discussed previously. The key to the efficiency of this algorithm, however, is step 2, since the only combinations of feature values that we ever consider are those created in step 2 by unifying active and inactive edges, rather than generating all possible combinations of feature values and then discarding those that are never used. Step 3 ensures that we include in the final grammar only rule instances that can be used in a complete analysis of an utterance, by filtering the rules top-down. Step 4 converts the grammar to context-free form, by replacing the complex category structures with atomic category names.

This compilation algorithm was tested on a version of the CommandTalk grammar that had 892 phrasal rules and 1688 lexical rules. The compilation took about 5 minutes on a Sun SPARCstation 10 with a 90-MHz HyperSPARC processor. The resulting context-free grammar had 2886 phrasal rules and 5719 lexical rules; because of the care the algorithm takes not to introduce unnecessary rules or categories, the grammar size increased only by about a factor of 3. We estimated that the size of the equivalent context-free grammar without introducing "don't care" feature values would have been approximately 10^{18} rules, almost all of which would be rules for sequences of coordinate numbers. Without the rules for coordinate numbers, we estimated that the size of the context-free grammar without "don't care" feature values would have been about 10^6 rules.

5.2 Removing Left Recursion from Context-Free Grammars

The standard way to remove left recursion from a context-free grammar is to convert the grammar to Greibach normal form, or GNF (Hopcroft and Ullman, 1979, pp. 94–99). In a GNF context-free grammar, the left-most daughter in each rule is a word, which obviously makes left recursion impossible. To eliminate left recursion from grammars generated for the Nuance recognizer, we produce grammars in a generalized version of Greibach normal form, or GGNF, taking advantage of the fact that the Nuance grammar formalism allows regular expressions on the right-hand side of grammar rules. In a GGNF grammar each category is defined by a single rule whose right-hand side is a GGNF regular expression over words and categories. In a GGNF regular expression the left-most subexpression is either a word, a GGNF regular expression, or an alternation of GGNF regular expressions. A GGNF regular expression has the property that any string of words and categories matching the regular expression must begin with a word. Thus, any GGNF rule is equivalent to a set of GNF rules. For additional compactness in the grammar, all alternations and sequences are flattened, so that no alternation has an alternation as an immediate subexpression and no sequence has a sequence as an immediate subexpression, and all alternative sequences are prefix merged, so that no alternation has as immediate subexpressions two sequences (including sequences of length 1) that have the same first element.[1]

We conducted tests to compare the speed performance of a GGNF grammar to an equivalent finite-state grammar when used with the Nuance recognizer. In general, unification grammars cannot be compiled to finite-state grammars, even when restricted to finitely valued features. However, we had imposed additional restrictions on the CommandTalk grammar to make this possible, since earlier versions of the Nuance recognizer lacked the capability to accept context-free grammars that were not finite-state. The comparison was made with the CommandTalk grammar and a test set of about 900 within-grammar utterances, on a Silicon Graphics Indigo 2 with a 150-MHz R4400 processor. Recognition was 1.265 × real time with the finite-state grammar, and 0.856 × real time with the GGNF grammar. This represents a 32% speedup using the GGNF form of the grammar. (Word recognition accuracy was 92.5% with both grammars.)

In terms of recognition speed, the GGNF grammar was clearly superior to the finite-state grammar, but the time required to compile the GGNF grammar into a Nuance recognition package was much greater. The finite-state grammar took 42 seconds to compile, while the GGNF grammar took 83 minutes. Since this compilation happens offline, for some purposes the penalty in compilation speed might not matter, but we felt that it was great enough to be an impediment to the grammar development cycle. We found another variation in grammar form and compilation settings, however, that offered almost as fast recognition, and much faster compilation.

The variant of GGNF in question is distributed-alternation GGNF, or DAGGNF. In a DAGGNF grammar an alternation can occur in a sequence only as the last element. This is achieved by distributing (in the algebraic sense) alternations over sequences, and prefix merging the result. The purpose of DAGGNF is to make the prefix merging already imposed in GGNF apply to more cases. A DAGGNF grammar can be significantly larger than the corresponding GGNF grammar, so Nuance compilation would take even longer, except for the fact that it seems to virtually duplicate a major phase of optimization in Nuance compilation, which we can therefore turn off, using a compile-time flag provided by Nuance.

Performing a Nuance compile, with the optimization turned off, of the DAGGNF form of the version of the CommandTalk grammar reported above reduced Nuance compilation time from 83 minutes to 4:57 minutes. Nuance recognition speed under the test conditions reported above decreased from 0.856 × real time to 0.890 × real time, an increase of 4.0%, but still 30% faster than with the finite-state form of the grammar.

---

[1] Prefix merging a sequence of length 1 with longer sequences requires using a notation for optional expressions, the merged sequence consisting of the common element followed by optional extensions.

6. Robust Natural-Language-Based Language Models

The techniques described in the preceding section have proven effective for defining qualitative language models based on unification grammars, for the recognition of within-grammar utterances. As we have previously discussed, qualitative language models are incapable of correctly recognizing out-of-grammar utterances, since they effectively assign such utterances a probability of 0. To use natural-language-based language models in a more robust way, it is necessary to convert their results into some sort of numerical score that can be combined with the numerical scores assigned by other knowledge sources, including acoustic models, in such a way that failure to find a completely grammatical analysis of a hypothesis does not totally rule it out.

There are several simple ways in which this could be done. Probably the simplest is to give a fixed numerical penalty to any hypothesis that is not accepted by the grammar. More information can be extracted from a natural-language grammar, however, either by computing the "edit distance" (total number of substitutions, insertions, and deletions) of a hypothesis from the closest grammatical string, or by counting the minimal number of grammatical fragments needed to cover the hypothesis. These methods have the advantage that they have very few parameters to estimate, perhaps only a single weighting factor to combine the language model score with the other knowledge sources. This means that the only data required for parameter estimation would be a small tuning set, in contrast to the large training sets required for fully statistical language models that have tens of thousands (or more) of parameters to estimate.

Experience suggests, however, that these minor variations on essentially qualitative methods do not offer much improvement over conventional n-gram language models, if sufficient training data is available to use statistical methods. In the 1993 DARPA benchmark speech recognition evaluation on the Air Travel Information Service (ATIS) task, SRI used a knowledge source based on the minimal number of grammatical fragments found by a unification-based grammar (Moore et al., 1994). Adding this knowledge source to a recognition system that also incorporated a word-class-based trigram language model reduced the word error rate only from 5.4% to 5.2%.

6.1 Combining Linguistics and Statistics in a Language Model

After the initial attempt in 1993 to use a unification-based natural-language grammar as a knowledge source for language modeling in speech recognition, the following year SRI undertook to develop a more integrated approach to combining linguistic and statistical factors in a single language model (Moore et al., 1995). Again, the approach was based on the idea that, even when the grammar fails to provide a complete analysis of an utterance, the number of grammatical (and semantically meaningful) phrases needed to span the utterance is likely to be smaller than for incorrect competing hypotheses. This is illustrated by an example from the December 1993 ARPA ATIS benchmark test set:

hypothesis: [*list flights*][*of fare code*][*a*][*q*]
reference: [*list flights*][*of fare code of q*]

These two word strings represent the SRI DECIPHER recognizer's first hypothesis for the utterance and the reference transcription of the utterance, each bracketed according to the best analysis that the SRI Gemini ATIS grammar was able to find as a sequence of semantically meaningful phrases. Because of a missing sortal combination, Gemini did not allow the preposition *of* to relate a noun phrase headed by *flights* to a noun phrase headed by *fare code*, so it was not possible to find a single complete analysis for either word string. Gemini was, however, able to find a single phrase spanning *of fare code of q*, but required three phrases to span *of fare code a q*, so it still strongly preferred the reference transcription of the utterance over the recognizer's first hypothesis.

SRI's original attempt to use this information in language modeling simply took this raw count of number of fragments, plus some other ad hoc factors, to compute a language model score. In the model that was subsequently developed, the Gemini grammar was still used to analyze a recognition hypothesis as a sequence of semantically meaningful fragments, but then n-gram statistics were used to estimate the probability of the hypothesis under that analysis. The resulting language model was a kind of multilevel n-gram model. The top level was a trigram model of the probability of a hypothesis as a sequence of types of fragments. That is, the model estimated the probability of an utterance being a sequence of, for instance, a sentence followed by a modifier phrase followed by a nominal phrase. The fragment types used were sentence, nominal phrase, modifier phrase, filler (e.g., *please*), and "skipped". This last category consisted of the sequences of words that were left over in determining the best coverage of the utterance in terms of well-formed phrases of major semantic classes in the ATIS domain. A trigram model was used for this level to model well the important case of an utterance consisting of the *begin_utterance* token, a single sentence, and the *end_utterance* token.

The next level of the model was a word-class-based four-gram model of each fragment. Initially, separate models were estimated for each type of fragment, but this method suffered from splitting the training data into smaller pools. The method of modeling fragments that was finally adopted was a single four-gram model, but one that treats each fragment as a sequence starting with a token such as *begin_sentence* or *begin_nominal_phrase*. The result is that the probability of the first few words of each fragment is conditioned on what type of fragment it is, but once there are several words of context, the probability estimates are conditioned on that rather than on the type of fragment. The four-gram model was smoothed by linear combination with lower-order models, the weights being estimated by deleted interpolation. For the word-class models, the classes were generated semiautomatically from the Gemini lexicon. The probability estimates described above were combined into a simple joint probability estimate for a hypothesis under an analysis as a sequence of fragments.

This Gemini-based language model was used in the December 1994 ATIS speech recognition benchmark evaluation. The results for SRI's recognizer for all test utterances both with and without this knowledge source are given in Table 1.

The baseline DECIPHER recognizer incorporated a word-class-based trigram language model. As the table shows, the improvement in recognition by rescoring DECIPHER output with the Gemini-based language model was about 15% on both word error and utterance error. These differences were measured to be statistically significant at the 95% confidence level on all four significance tests used in the evaluation.

Table 1. December 1994 ATIS benchmark test results

	Word Error	Utterance Error
Baseline DECIPHER	2.5%	15.5%
DECIPHER+Gemini	2.1%	13.1%
Improvement	14.6%	15.1%

It is notable that this improvement in recognition accuracy, by adding a natural-language-based knowledge source, was achieved relative to a state-of-the-art baseline recognizer. (The baseline SRI recognizer was otherwise the top-performing system in this evaluation.) Repeatedly in the past, improvements have been obtained with natural-language-based knowledge sources in recognition, only to have the improvements disappear as baseline recognition accuracy has improved. The results described here represent the only significant improvement we are aware of obtained by using a natural-language-based knowledge source in conjunction with a current state-of-the-art recognizer in a blind test.

6.2 Fully Statistical Natural-Language Grammars

The model we have just described combines linguistic information and statistical information in a principled way, with a substantial resulting gain in recognition accuracy. However, only a single level of linguistic structure—the utterance as a sequence of linguistic fragments—is statistically modeled. One might hope for additional improvement in recognition accuracy by modeling more of the linguistic structure statistically. Recently, there has been much interest in statistical natural-language grammars, but so far, there appear to be no significant results in using them to improve speech recognition. Hence the work and ideas described here should be regarded as promising, but unproven, in regard to language modeling for speech recognition.

6.2.1 Probabilistic Context-Free Grammar. The most obvious candidate for fully statistical natural-language grammars are probabilistic context-free grammars. In this form of statistical language model, each context-free grammar rule has associated with it a conditional probability for the right-hand side given the left-hand side. For example, assigning the rule S → NP VP a conditional probability of 0.5 means that, if there is a phrase of category S, there is a probability of 0.5 that it consists of a phrase of category NP followed by a phrase of category VP. If the top-level category of the grammar is taken to have a probability of 1, then this immediately defines a probability for each analysis tree allowed by the grammar, by multiplying all the conditional probabilities associated with the rules used in the analysis.

For such an approach to produce a good enough language model to compete with conventional n-gram models, however, some form of lexical or semantic conditioning must be included. (This observation was perhaps first made in print by Church [1989].) For example, without information about likely combinations of lexical subjects, verbs, and objects, a statistical grammar incorporating only syntactic constraints would rate *the pizza ate the boy* about as likely as *the boy ate the pizza*. The most common response to this observation in work on statistical natural-language grammars has been to include the lexical head of each phrase as part of the information used to estimate probabilities in a statistical context-free grammar. (Informally, the lexical head of a phrase is the main word of the phrase.) In *the boy ate the pizza*, the subject would be categorized not just as a noun phrase (or perhaps a singular noun phrase), but as a noun phrase whose lexical head is *boy*, and the verb phrase would be categorized as a verb phrase whose lexical head is *ate*.

Taking into account the lexical head of each category mentioned in a rule means that, in effect, conditional probabilities must be estimated for all possible combinations of lexical heads for the nonterminals that appear in a given context-free rule. So, the S → NP VP rule would have separate estimates for each combination of noun as the head of the NP and verb as the head of the VP. This causes a very severe sparse-data problem, since a substantial proportion of the combinations of a particular context-free rule and particular lexical heads for each nonterminal in the rule would occur no more than once in even a very large training corpus. This problem is addressed by making various independence assumptions and using back-off smoothing methods.

Models of this general type have been extensively explored in the past few years (Charniak, 1995; Collins, 1996, 1997; Grishman and Sterling, 1993; Lafferty et al., 1992; Magerman, 1995). Although not all of this work has been explicitly framed as probabilistic context-free grammar, all of it has involved atomic grammatical categories annotated with lexical head information. There seem to be no reported results using such models for speech recognition, however. Instead, the chosen task for evaluating these models tends to be sentence parsing, where the main figure of merit is accuracy in structurally bracketing the sentences. The current state-of-the-art performance seems to be that of Collins (1997), whose method produced completely correct bracketings for 64% of the sentences of 100 words or less using a standard partitioning of the Penn Treebank corpus (Marcus et al., 1993) into training and test sets.

6.2.2 Probabilistic Unification Grammar. None of the work cited above on probabilistic context-free grammar addresses the degree of detail routinely modeled in unification grammars for natural language. Much of the work adopts the linguistic categories used in the Penn Treebank, which has 35 lexical categories (not counting 13 categories for symbols and punctuation, which would not arise in spoken language) and 14 phrasal categories. Contrast this with the report of Black et al. (1993), who found that parsing a 15,000-sentence corpus with their unification grammar of English used 23,431 distinct unification categories produced by different combinations of categories and feature values. Thus, if unification grammars are also to be annotated with lexical-head information, as seems necessary to be competitive with simple n-gram models, the sparse-data problem will be correspondingly worse than with typical probabilistic context-free grammars.

Perhaps as a result of this problem, there seem to be no published reports of models that attempt to incorporate all the constraints of a complex unification grammar into a statistical model. Briscoe and Carroll (1993) use a unification grammar in a statistical model, but estimate the probabilities in their model based only on the context-free approximation formed by ignoring all the feature constraints. The feature constraints are imposed afterwards as a filter, but they are not incorporated into the statistical model per se. The statistical model of Black et al. (1993) is more complex, but somewhat similar in its treatment of unification constraints. For conditioning purposes Black et al. replace their complex feature-based categories by 50 atomic syntactic categories and 50 atomic semantic categories so there are at most 2500 possible linguistic categories available, compared to the 23,431 categories they report as actually arising in parsing their corpus. Goodman (1997) describes a method based on independence assumptions and back-off smoothing that could, in principle, be applied to arbitrary unification grammars, but he tests it only on a simplified model with many fewer features than a realistic unification grammar of a natural language would need, and he gives no reason to believe that the method would scale up particularly well.

There is a rather different approach to the sparse-data problem for probabilistic unification grammar, that, while it is as yet untested, might provide a solution. Instead of focusing on the set of possible categories, which may be enormous, we focus instead on the set of rules that actually compose the grammar. The size of the rule set is typically fairly small; there are seldom many more than 1,000 phrasal syntactic rules in a hand-written unification grammar, either for a general grammar for a natural language, or a task-specific grammar for a particular application. The question we need to answer is: on what should we condition the probability of a particular rule being correct in a given context? It seems clear that the most important single factor that determines whether a particular rule correctly applies in a given partial derivation is whether the unification constraints allow it to apply. If the unification constraints are not satisfied, then it is impossible for the rule to apply, so its probability of being the correct rule would be 0. This suggests that a primary conditioning factor in estimating the probability of a rule should be whether the unification constraints on the rule are satisfied. At its crudest, this would mean counting or estimating the number of times in the derivation of a training corpus that a given rule was correct and the number of times the constraints on the application of the rule were met, and taking the ratio between these counts as an estimate of the conditional probability of the rule.

For example, suppose, in the analysis of the sentence *which man did John see*, we have a partial derivation that correctly predicts that the sentence is a *wh*-question whose initial noun phrase must unify with the category NP: [number=X,wh=yes]. That is, the initial noun phrase must be an interrogative noun phrase. Now suppose we have the following three rules to consider:

```
NP:[number=X,wh=no] → Name:[number=X]
NP:[number=X,wh=Y]  → Pron:[number=X,wh=Y]
NP:[number=X,wh=Y]  → Det:[number=X,wh=Y] N:[number=X]
```

The first rule says that a non-*wh*-noun-phrase can be a proper name. The unification constraints on this rule are not satisfied, so the partial derivation we are considering is counted neither as a case in which the rule could apply nor as a case in which it is correct. The second rule says that a noun phrase can be a pronoun, and that the noun phrase is a *wh*-noun-phrase if the pronoun is a *wh*-pronoun (e.g., *who*). Since this rule is not excluded by the constraint wh=yes on the predicted noun phrase, this counts as an instance in which the rule could apply. However, since the noun phrase we actually have at this point in the data, *which man*, is not a pronoun, this is not a case in which the rule is correct. The third rule says that a noun phrase can be a determiner followed by a noun, and that the noun phrase is a *wh*-noun-phrase if the determiner is a *wh*-determiner (e.g., *which*). In this case, the rule can apply because its unification constraints are met, and it is correct, since the noun phrase in question is of the form determiner followed by noun.

Estimating the conditional probability of a unification rule as the ratio of the frequency with which it is correct to the frequency with which its unification constraints are met takes all unification constraints into account and seems to get at perhaps the most important factor in determining the likelihood of the rule being correct. It has a major problem, however. A valid probability distribution must sum to 1 for all possible hypotheses in any context. For the method we have just proposed, however, this will not always be the case, since a unification-based rule can have different competitors in different contexts, because of differing unification constraints. In the current example, if the noun-goes-to-pronoun rule and the noun-goes-to-determiner-noun rule are the only possible ways of expanding an interrogative noun phrase, then in this context the probability of those two rules must sum to 1. However, in another context when we are expanding a noun phrase that is not constrained to be interrogative, we would want to assign a non-0 probability to the noun-phrase-goes-to-proper-name rule.

This problem can be addressed by expanding the model slightly to take into account what other rules are applicable. We can partition the rules in a unification grammar into mutually exclusive sets of possible competing rules, taking into account only the major category category symbols that appear in the rules. Call a set of rules an "alternative rule set." Note that the notion of "alternative rule set" will be relative to a particular order of derivation. For top-down derivation, an alternative rule set would consist of all the rules having the same major category symbol for the left-hand side of the rule, just as in probabilistic context-free grammars. For bottom-up derivations, however, the alternative rule sets might be defined by the major category symbol of the first or last category on the right-hand side of the rule. The treatment that follows, however, applies to any derivation process and any method of partitioning the rules into mutually exclusive subsets such that each rule falls into the same subset as all its possible competitors with respect to that derivation process.

Each alternative rule set gives us a property of a context that lets us easily define a probability distribution for which rule is correct in that context. Suppose $\{R_1, \ldots, R_k\}$ is the alternative rule set that contains all the rules that can apply in some particular context. Let $a(R_i)$ be a function that is 1 if rule R_i could apply in the context and is 0 otherwise, and let $c(R_i)$ mean that R_i is the correct rule in the context. What we need to estimate is $p(c(R_i)|a(R_1) = x_1, \ldots, a(R_k) = x_k)$ for $1 \leq i \leq k$, where x_i is 1 or 0, depending on whether rule R_i is applicable. In general, there will be too many possible combinations of values for $a(R_1), \ldots, a(R_k)$ to estimate this directly, but we can approximate it by using Bayes' rule,

$$p(c(R_i)|a(R_1) = x_1, \ldots, a(R_k) = x_k) =$$

$$\frac{p(a(R_1) = x_1, \ldots, a(R_k) = x_k|c(R_i)) \cdot p(c(R_i))}{p(a(R_1) = x_1, \ldots, a(R_k) = x_k)}$$

and making the following conditional independence assumption:

$$p(a(R_1) = x_1, \ldots, a(R_k) = x_k|c(R_i)) \approx$$

$$p(a(R_1) = x_1|c(R_i)) \cdot \ldots \cdot p(a(R_k) = x_k|c(R_i))$$

To make the approximation sum to 1 in all cases, we estimate the denominator of the right-hand of Bayes' rule simply to be the normalization constant found by summing the estimates for all the individual cases.

With this decomposition, the only parameters we need to estimate directly are those of the form $p(c(R_i))$ and $p(a(R_i) = 1|c(R_j))$, where R_i and R_j are members of the same alternative rule set. (Note that $p(a(R_i) = 0|c(R_j)) = 1 - p(a(R_i) = 1|c(R_j))$.) Moreover, many of the $p(a(R_i) = 1|c(R_j))$ parameters will be constrained by the grammar to be 0, because even though R_i and R_j are in the same alternative rule set, the grammar is structured in such a way that the unification constraints on both rules could never be satisfied simultaneously in any partial derivation. In a grammar with 1,000 rules, it would be rare for any single rule to have more than perhaps 25 true possible alternatives, so 25,000 is probably an upper bound on the total number of parameters we would have to directly estimate for syntactic probabilities.

This model looks quite tractable, but so far it incorporates only syntactic constraints, not lexical constraints. These are not normally expressed directly in unification-based grammar rules, but are represented using some additional formal mechanism, such as sortal restrictions on semantics representations (Alshawi, 1992, Chapter 9). This suggests an elegant factoring of syntactic and lexical constraints, based on viewing grammars, not as merely correlating strings with syntactic structures, but as mapping between semantic representations and syntactic representations. The kind of unification-based grammar formalism we have been discussing is easily extended to associate possible semantic representations, or logical forms, with each syntactic structure (Alshawi, 1992, Chapter 5; Moore, 1995, Chapter 10), so that a combined syntactic/semantic language model could be defined as

$$p(W) = \sum_L p(W|L) \cdot p(L)$$

In this formula, W is a word string and L is a logical form, so to apply this model, we would find all logical forms that could be expressed by a given word string, and sum the products of the prior probability of each logical form (which would incorporate lexical semantic constraints) and the probability of the word string given the logical form. For a particular word string and logical form, we can compute an estimate of the probability of the word given the logical form exactly as we proposed above, by summing over all grammatical analyses that associate the word string and the logical form. Nothing in the model needs to be changed to extend it to incorporate constraints between syntax and logical form, except that the notion of a rule being applicable or correct must take into account constraints supplied by the logical form at a particular stage of the derivation.

Since logical forms are typically represented as nested functional expressions, if they are annotated with sortal information, as they are in the Core Language Engine and Gemini, it is fairly straightforward to recursively estimate their prior probability by (1) conditioning the probability estimates for the functor of a term on the semantic sort of the term, and (2) conditioning the probability estimates for the sorts of arguments of a functor on the functor and the semantic sorts of sibling arguments. It is often the case that the functors used in logical form representations are largely limited to two arguments, in which case the number of parameters needed to estimate to probability of the semantic representations would be on the order of the number needed to estimate a trigram language model.

Thus, the combined syntactic and semantic model we have sketched should require on the order of the same number of parameters to estimate as an n-gram language model, but have the advantage of incorporating both long-distance syntactic constraints and lexical constraints that are conditioned on semantically meaningful grammatical relationships rather than simple word adjacencies.

7. Summary

We have explored in detail two different kinds of language model for speech recognition that incorporate constraints from natural-language grammars. For situations in which training data for statistical models is not available, we have described methods for using complex natural-language grammars by compiling them into context-free grammars that can be efficiently incorporated into the search architecture of a commercially available real-time speech recognizer. If adequate training data for statistical models is available, we acknowledge that conventional n-gram models are difficult to improve on, but we have described one experiment in which a hybrid language model incorporating both n-gram statistics and constraints from a qualitative natural-language grammar was able to substantially improve the performance of a state-of-the-art conventional HMM-based recognizer. Finally, we have presented more speculative work on statistical context-free grammars, and described how this work might be extended to incorporate all the constraints of a complex unification-based grammar.

References

Alshawi, H. (ed.) (1992) *The Core Language Engine*, The MIT Press, Cambridge, Massachusetts.

Black, E., F. Jelinek, J. Lafferty, D. M. Magerman, R. Mercer, and S. Roukos (1993) "Towards History-based Grammars: Using Richer Models for Probabilistic Parsing," in *Proceedings of the 31st Annual Meeting of the Association for Computational Linguistics*, Columbus, Ohio, pp. 31–37.

Boisen, S., Y.-L. Chow, A. Haas, R. Ingria, S. Roukos, and D. Stallard (1989) "The BBN Spoken Language System," in *Proceedings Speech and Natural Language Workshop February 1989*, Philadelphia, Pennsylvania, pp. 106–111 (Morgan Kaufmann Publishers, Inc., San Mateo, California).

Briscoe, T., and J. Carroll (1993) "Generalized Probabilistic LR Parsing of Natural Language (Corpora) with Unification-Based Grammars," *Computational Linguistics*, Vol. 19, No. 1, pp. 25–59.

Charniak, E. (1995) "Parsing with Context-Free Grammars and Word Statistics," Technical Report CS–95–28, Department of Computer Science, Brown University, Providence, Rhode Island.

Chomsky, N. (1957) *Syntactic Structures*, Mouton & Co., The Hague, Holland.

Chow, Y.-L., and S. Roukos (1989) "Speech Understanding Using a Unification Grammar," in *Proceedings of the IEEE Conference on Acoustics, Speech, and Signal Processing*, Glasgow, Scotland, pp. 727–730.

Chow, Y.-L., and R. Schwartz (1989) "The N-Best Algorithm: An Efficient Procedure for Finding Top N Sentence Hypotheses," in *Proceedings Speech and Natural Language Workshop October 1989*, Cape Cod, Massachusetts, pp. 199–202 (Morgan Kaufmann Publishers, Inc., San Mateo, California).

Church, K. W. (1989) "Syntactic Parsing May Not Help Speech Recognition Very Much," in *Spoken Language Systems Working Notes*, AAAI Spring Symposium Series, Stanford, California, pp. 6–9.

Culy, C. (1985) "The Complexity of the Vocabulary of Bambara," *Linguistics and Philosophy*, Vol. 8, No. 3, pp. 345–351.

Collins, M. J. (1996) "A New Statistical Parser Based on Bigram Lexical Dependencies," in *Proceedings of the 34th Annual Meeting of the Association for Computational Linguistics*, Santa Cruz, California, pp. 184–191.

Collins, M. J. (1997) "Three Generative, Lexicalized Models for Statistical Parsing," in *Proceedings of the 35th Annual Meeting of the Association for Computational Linguistics*, Madrid, Spain, pp. 16–23.

Dowding, J., J. M. Gawron, D. Appelt, J. Bear, L. Cherny, R. Moore, and D. Moran (1993) "Gemini: A Natural Language System for Spoken-Language Understanding," in *Proceedings of the 31st Annual Meeting of the Association for Computational Linguistics*, Columbus, Ohio, pp. 54–61.

Dowding, J., R. Moore, F. Andry, and D. Moran (1994) "Interleaving Syntax and Semantics in an Efficient Bottom-Up Parser," in *Proceedings of the 32nd Annual Meeting of the Association for Computational Linguistics*, Las Cruces, New Mexico, pp. 110–116.

Dupont, P. (1993) "Dynamic Use of Syntactical Knowledge in Continuous Speech Recognition," in *Proceedings Third European Conference on Speech Communication and Technology*, Berlin, Germany, pp. 1959–1962.

Dupont, P., and D. Snyers (1989) "Efficient Dynamic Expansion of Context-Free Grammar in Speech Recognition," in *Overview of Research in Speech Recognition at PRLB in 1988*, Philips Research Laboratory, Brussels, Belgium, pp. 32–68.

Goodman J. (1997) "Probabilistic Feature Grammars," in *Proceedings of the International Workshop on Parsing Technologies*, Boston, Massachusetts.

Grishman, R., and J. Sterling (1993) "Smoothing of Automatically Generated Selectional Constraints," in *Proceedings Human Language Technology Workshop*, Plainsboro, New Jersey, pp. 254–259 (Morgan Kaufmann Publishers, Inc., San Francisco, California).

Hopcroft, J., and J. Ullman (1979) *Introduction to Automata Theory, Languages, and Computation*, Addison Wesley Publishing Company, Reading, Massachusetts.

Huang, X., A. Acero, F. Alleva, M.-Y. Hwang, L. Jiang, and M. Mahajan (1995) "Microsoft Windows Highly Intelligent Speech Recognizer: Whisper," in *Proceedings 1995 International Conference on Acoustics, Speech, and Signal Processing*, Detroit, Michigan, pp. 93-96.

Lafferty, J., D. Sleator, and D. Temperley (1992) "Grammatical Trigrams: A Probabilistic Model of Link Grammar," *Probabilistic Approaches to Natural Language Working Notes*, AAAI Fall Symposium Series, Cambridge, Massachusetts, pp. 89–97.

Magerman, D. M. (1995) "Statistical Decision-Tree Models for Parsing," in *Proceedings of the 33rd Annual Meeting of the Association for Computational Linguistics*, Cambridge, Massachusetts, pp. 276–283.

Marcus, M. P., B. Santorini, and M. A. Marcinkiewicz (1993) "Building a Large Annotated Corpus of English: The Penn Treebank," *Computational Linguistics*, Vol. 19, No. 2, pp. 313–330.

Mohri, M. (1997) "Finite-State Transducers in Language and Speech Processing," *Computational Linguistics*, Vol. 23, No. 2, pp. 269–311.

Moore, R. (1995) *Logic and Representation*, CSLI Publications, Center for the Study of Language and Information, Stanford University, Stanford, California.

Moore, R., M. Cohen, V. Abrash, D. Appelt, H. Bratt, J. Butzberger, L. Cherny, J. Dowding, H. Franco, J. M. Gawron, and D. Moran (1994) "SRI's Recent Progress on the ATIS Task," in *Proceedings of the Spoken Language Technology Workshop*, Plainsboro, New Jersey, pp. 72–75 (Morgan Kaufmann Publishers, Inc., San Francisco, California).

Moore, R., D. Appelt, J. Dowding, J. M. Gawron, and D. Moran (1995) "Combining Linguistic and Statistical Knowledge Sources in Natural-Language Processing for ATIS," in *Proceedings of the Spoken Language Systems Technology Workshop*, Austin, Texas, pp. 261–264 (Morgan Kaufmann Publishers, Inc., San Francisco, California).

Moore R., J. Dowding, H. Bratt, J. M. Gawron, Y. Gorfu, and A. Cheyer (1997) "CommandTalk: A Spoken-Language Interface for Battlefield Simulations," in *Proceedings of the Fifth Conference on Applied Natural Language Processing*, Association for Computational Linguistics, Washington, DC, pp. 1–7.

Moore, R., F. Pereira, and H. Murveit (1989) "Integrating Speech and Natural-Language Processing," in *Proceedings Speech and Natural Language Workshop February 1989*, Philadelphia, Pennsylvania, pp. 243–247 (Morgan Kaufmann Publishers, Inc., San Mateo, California).

Murveit, H., and R. Moore (1990) "Integrating Natural Language Constraints into HMM-based Speech Recognition," in *Proceedings 1990 International Conference on Acoustics, Speech, and Signal Processing*, Albuquerque, New Mexico, pp. 573–576.

Nuance Communications (1996) *Nuance Speech Recognition System, Version 5, Developer's Manual*, Menlo Park, California.

Pullum, G. K., and G. Gazdar (1982) "Natural Languages and Context-Free Languages," *Linguistics and Philosophy*, Vol. 4, No. 4, pp. 471–504.

Schwartz, R., and S. Austin (1990) "Efficient, High-Performance Algorithms for N-Best Search," in *Proceedings Speech and Natural Language Workshop June 1990*, Hidden Valley, Pennsylvania, pp. 6–11 (Morgan Kaufmann Publishers, Inc., San Mateo, California).

Shieber, S. M. (1985) "Evidence Against the Context-Freeness of Natural Language," *Linguistics and Philosophy*, Vol. 8, No. 3, pp. 333–343.

Woods, W. A. (1970) "Transition Network Grammars for Natural Language Analysis," *Communications of the ACM*, Vol. 13, No. 10, pp. 591–606.

How May I Help You?

A.L. Gorin, G. Riccardi and J.H. Wright

AT&T Research, Florham Park, New Jersey
email: {algor, dsp3, jwright}@research.att.com

Summary. We are interested in providing automated services via natural spoken dialog systems. By *natural*, we mean that the machine understands and acts upon what people actually say, in contrast to what one would like them to say. There are many issues that arise when such systems are targeted for *large populations of non-expert users*. In this paper, we focus on the task of automatically routing telephone calls based on a user's fluently spoken response to the open-ended prompt of *"How may I help you?"* . We first describe a database generated from 10,000 spoken transactions between customers and human agents. We then describe methods for *automatically acquiring language models for both recognition and understanding* from such data. Experimental results evaluating call-classification from speech are reported for that database. These methods have been embedded within a spoken dialog system, with subsequent processing for information retrieval and form-filling.

Key words: Spoken language understanding; spoken dialog system; speech recognition; stochastic language modeling; salient phrase acquisition; topic classification.

1. Introduction

There are a wide variety of interactive voice systems in the world, some residing in laboratories, many actually deployed. Most of these systems, however, either explicitly prompt the user at each stage of the dialog, or assume that the person has already learned the permissible vocabulary and grammar at each point. While such an assumption is conceivable for frequent expert users, it is dubious at best for a general population on even moderate complexity tasks. In this work, we describe progress towards an experimental system which shifts the burden from human to machine, making it the device's responsibility to respond appropriately to what people actually say.

The problem of automatically understanding fluent speech is difficult, at best. There is, however, the promise of solution within constrained task domains. In particular, we focus on a system whose initial goal is to understand its input sufficiently to route the caller to an appropriate destination in a telecommunications environment. Such a call router need not solve the user's problem, but only transfer the call to someone or something which can. For example, if the input is *"Can I reverse the charges on this call?"* , then the caller should be connected to an existing automated subsystem which completes collect calls. Another example might be *"How do I dial direct to Tokyo?"*, whence the call should be connected to a human agent who can

Reprinted from *Speech Communication (23)1-2, Gorin, Riccardi and Wright "How may I help you?", pp 113–127, 1997,* with kind permission from Elsevier Science - NL, Sara Bugerhartstraat 25, 1055 KV Amsterdam, The Netherlands

provide dialing instructions. Such a call router should be contrasted with traditional telephone switching, wherein a user must know the phone number of their desired destination, or in recent years navigate a menu system to self-select the desired service. In the method described here, the call is instead routed based on the *meaning* of the user's speech.

This paper proceeds as follows. In Section 2, an experimental spoken dialog system is described for call-routing plus subsequent automatic processing of information retrieval and form-filling functions. The dialog is based upon a feedback control model, where at each stage the user can provide both information plus feedback as to the appropriateness of the machine's response [G95]. In Section 3, a database is described of 10K fluently spoken transactions between customers and human agents for this task. In particular, we describe the language variability in the first customer utterance, responding to the prompt of *"How may I help you?"* in a telecommunications environment.

In Section 4, we describe the spoken language understanding (SLU) algorithms which we exploit for call classification. A central notion in this work is that it is not necessary to recognize and understand every nuance of the speech, but only those fragments which are salient for the task [G95]. This leads to a methodology where understanding is based upon recognition of such salient fragments and combinations thereof.

There are three main components in our SLU methodology. First is to automatically acquire salient grammar fragments from the data, modeling those parts of the language which are meaningful for the task plus their statistical associations to the machine actions. Second is to recognize these fragments in fluent speech, searching the output of a large vocabulary speech recognizer. The statistical language model which constrains this recognizer embeds automatically-acquired fragments in a stochastic finite state machine, providing an efficient approximation to an n-gram model with variable length units [R96] . Third, we exploit these multiple recognized fragments to classify the call-type of an utterance. Since the SLU is embedded within a dialog system, the classifier provides both the best (rank 1) and secondary (rank 2, etc.) decisions. Finally, in Section 5, we report on experimental results for call-classification from the above-mentioned speech database, training on 8K utterances and testing on 1K.

2. A Spoken Dialog System

The goal of a call-router is to recognize and understand the user's speech sufficiently to determine the call-type. Dialog is necessary since, in many situations, the call type cannot be determined from a single input. This can be due to an ambiguous request or to imperfect performance of the spoken language understanding (SLU) algorithms.

One important component of dialog is *confirmation*, wherein the machine proposes its understanding of the user's input, receiving reinforcement feedback as to the appropriateness of its response. In previous work, we have described a dialog model for integrating the meaning of initial and secondary user inputs based on a feedback control system [G95] [G95b]. In case the machine's interpretation is in error, the subsequent user input can provide both disconfirmation plus *clarifying* information. Finally, once the call-type has been successfully negotiated and agreed upon, there is often a subsequent form-filling dialog to *complete* the information necessary to provide a service.

To illustrate these issues, we now present several transcripts of spoken human/machine dialogs with this system. The first dialog demonstrates the basic call-routing function where the user explicitly confirms the machine's interpretation, (where M denotes machine and U denotes a human user).

M: How may I help you?
U: *Can you tell me how much it is to Tokyo?*
M: You want to know the cost of a call?
U: *Yes, that's right.*
M: Please hold on for rate information.

The second dialog illustrates an information retrieval stage after call-type classification. Observe that in addition to the call-type, the SLU algorithm must also extract the place-name *Chicago*. Further observe the use of an implicit confirmation request, wherein the user's silence is interpreted as affirmation.

M: How may I help you?
U: *What's the area code for Chicago?*
M: You want an area code.
U: *(silence)*
M: The area code for Chicago is 312.

In some cases, the input is ambiguous or misunderstood, so the device must then engage in a clarifying dialog. There are several possible strategies for such clarification. For the ambiguous user input, *"Can I charge this call please?"*, the SLU algorithm (cf. Section 4) generates a most likely classification of *'third number billing'* with secondary decisions of *'card billing'* or *'collect'*. The dialog illustrated below asks for clarification via proposing the top choice. Alternatively, the machine could provide a short menu, e.g. *"Do you want A or B?"*, or in this case where there is no clear winner amongst the three choices, ask *'How do you want to bill the call?'*.

M: How may I help you?
U: *Can I charge this call please?*
M: You want to charge this to another number?
U: *No, to my Universal card.*
M: You want to charge this call to a credit card.
U: *(silence)*
M: Please speak your card number.

In the above example, the device interprets the user's clarifying response to be a combination of disconfirmation (*"no"*) and additional call-type information (*"to my Universal card"*) as per our feedback control model [G95b]. The following example illustrates a call-type classification followed by a form-filling subdialog.

> M: How may I help you?
> *U: I tried calling my Aunt and got a wrong number.*
> M: You need a billing credit.
> *U: (silence)*
> M: Please speak the number that you dialed.
> *U: 908 582 2778*
> M: Was the call billed to the phone that you're calling from?
> *U: Yes it was.*
> M: You will be given credit, thank you for calling.

There are many open issues in optimizing the user interface of such a system, in particular those of selecting the best prompts to stimulate maximally consistent and informative responses from large populations of non-expert users [A96][B96]. In the next section, we discuss the variations in response to a particular open-ended prompt, in the situation where users are talking to a human agent.

3. Database

In order to enable experimental evaluation, we generated a database of 10K spoken transactions between customers and human agents. First, both channels of the dialog were recorded from the agents' headset jacks onto a digital audio tape (DAT). At the end of each transaction, a control key was manually depressed (by the human agent) to generate a touch-tone code, serving both as a segmentation marker and a call-type label. These recordings were then automatically segmented, filtered and downsampled to generate a stereo speech file for each transaction.

We then focused on the first customer utterance, responding to the greeting prompt of *"How may I help you?"*. These utterances were endpointed, orthographically transcribed and then labeled as to call-type and quality of the speech and channel. We remark on the distinction between the call-action labels provided by the agents and by the labelers. The agent's touch-tone tag comprised an on-the-spot single label for the entire transaction. The labelers, however, based their decision on the *first customer utterance* only, plus were allowed to select more than one call-label per utterance. We observed that 84% of the utterances were labeled with a single call-type, 16% with two (e.g. COLLECT and PERSON-TO-PERSON), then a small remainder (0.6%) with 3 labels. It is possible for the agent-generated call-type to not match any of the labeler's, since sometimes the first utterance is ambiguous, with things becoming clear only after some dialog. An issue for future study is the correlation between these labeling methods, plus an analysis of the reasons for their mismatches. Since the experiments of Section 5 are based on the first utterances only, those are the labels which are used for training and testing.

Several samples of (first) utterances follow, where digits are replaced with the symbol *x*.

Examples

I need to make a long distance phone call and charge it to my home phone number

yes how much is it to call the number I just dialed

yes where is area code x x x

yes what time is it in area code x x x right now I'm trying to gauge the time difference

I just I'm trying to get a number from information

Although people's spoken language varies widely, most of the time they are asking for one of a moderate number of services. We selected a subset of 14 services plus an *OTHER* class to subsume the remainder. This distribution is highly skewed, as illustrated in the rank-frequency plot in Figure 1.

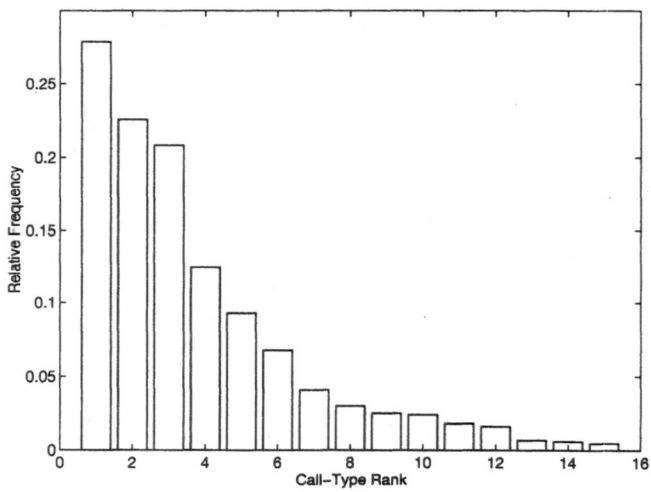

Fig. 1. Rank frequency distribution of call-types

We now discuss the vocabulary in this database. Of the 10K utterances, 8K are used for training language models for recognition and understanding, 1K for testing and the remaining 1K reserved for future development. Figure 2 shows the increase in vocabulary size accumulated over the 8K utterances, with a final value of ~ 3600 words. Even after 8K utterances, the slope of the curve is still significantly positive. We examined the tail of the lexicon, i.e., the last 100 vocabulary words accumulated in the training utterances. Approximately half were proper nouns (either people or places), but the other half were 'regular words' (e.g. *authorized, realized, necessary,* ...). The out-of-vocabulary rate at the token-level in the test sentences is 1.6%. At the sentence-level, this yields an OOV rate of 30% (which is also observed in the slope of vocabulary growth in Figure 2). Thus, approximately one out of three utterances contains a word *not* in the training data. As will be detailed in Section 4, the test-set perplexity using a statistical trigram model is ~ 16.

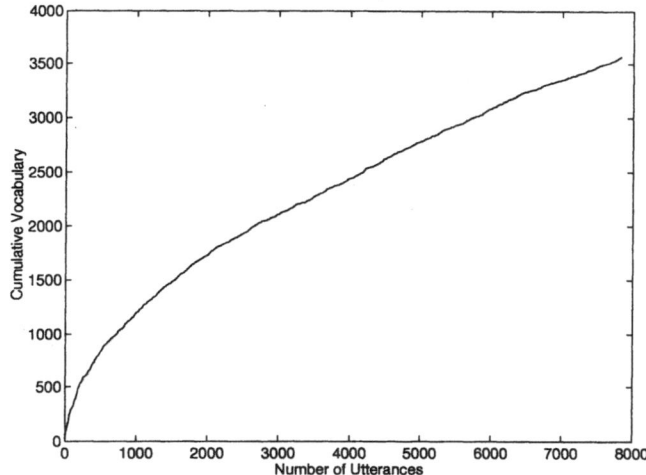

Fig. 2. Vocabulary growth in database

Utterance length varies greatly, from a minimum of one word (e.g. *"Hello?"*) to 183, with an average of 18 words/utterance. We remark on the definition of 'first customer utterance'. The labelers were instructed that the customer's first utterance was completed when the human agent began responding. Back-channel affirmations from the agent such as *'uh-huh'* were transcribed and marked, but did *not* indicate the end of the customer's utterance. An issue for future research is to understand to what degree such utterances will shorten when people are talking to a machine rather than a human. The distribution of these lengths for the 10K transcriptions is shown in Figure 3. In that same figure, the cumulative distribution is also shown. Observe that almost all of the sentences have length less than 60. Observe also that the median is approximately equal to the mean (18), although the distribution is highly skewed. Recall that these utterances are the initial user response to the greeting prompt.

Similarly, one can histogram the duration of these initial utterances, as shown in Figure 4. The average duration of an utterance is 5.9 sec, so that the speaking rate in this database is approximately 3 words per second.

4. Algorithms

In this section, we describe the algorithms underlying this system and experiments. A key notion is that for any particular task, it is *not* necessary to recognize and understand every word and nuance in an utterance. That is, to extract semantic information from spoken language, it suffices to focus on the salient fragments and combinations thereof. There are three major issues that we address:

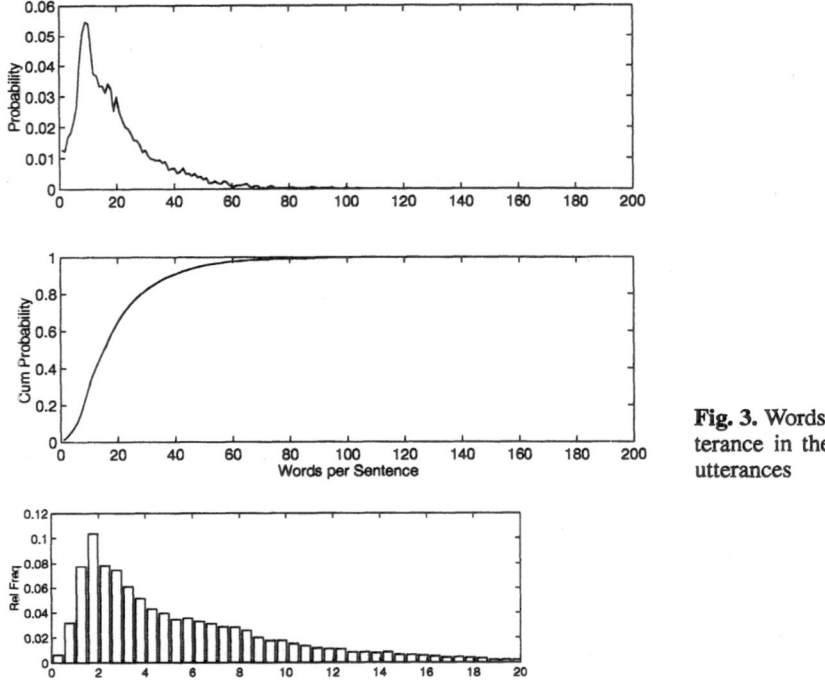

Fig. 3. Words per utterance in the initial utterances

Fig. 4. Duration distribution of initial utterances

- How do we acquire the salient grammar fragments for this task?
- How can we recognize these fragments in fluent speech?
- How do we map multiple recognized fragments to a machine action?

Our technical approach is to avoid hand-crafted models throughout, focusing on machine learning methods which automatically learn the structure and parameters of statistical models for each stage from data. In subsection 4.1 we describe algorithms to automatically acquire salient words, phrases and grammar fragments for a task. We recognize these in fluent speech via searching the output of a large-vocabulary recognizer (LVR), whose language model is a stochastic finite state machine (SFSM) with embedded automatically-acquired phrases. The LVR and training algorithm for the recognizer language model are described in subsection 4.2. We then formulate a call-classification from those multiple recognized salient fragments as described in subsection 4.3.

4.1 Salient Fragment Acquisition

We are interested in constructing machines which learn to understand and act upon fluently spoken input. For any particular task, certain linguistic events are critical to recognize correctly, others not so. We have quantified this notion via *salience* [G95], which measures the information content of an event for a task. In previous experiments, salient words have been exploited to learn the mapping from unconstrained input to machine action for a variety of tasks [G93][G94a][G94b][H94][M93][S93]. In this work, we build upon the ideas introduced in [G96b] to automatically acquire salient phrase and grammar fragments for a task, exploiting both linguistic and extra-linguistic information in the inference process. In particular, the input to this inference algorithm is a database of transcribed utterances labeled with associated machine actions. It is these associated actions which comprise the extra-linguistic information in this task. While there is a large literature on automated training of stochastic language models, such efforts have traditionally exploited only the language itself, with the goal of within-language prediction to improve ASR [J90]. Learning from language alone is actually a much harder problem than people are faced with, who acquire language during the course of interacting with a complex environment. This algorithm, following that intuition, exploits both language and extra-linguistic information to infer structure.

Communication and Salience

We briefly review the intuitions underlying salience, following [G95]. Consider devices whose purpose is to understand and act upon fluently spoken input. The goal of communication in such systems is to induce the machine to perform some action or to undergo some internal transformation. The communication is judged to be successful if the machine responds appropriately. We have explored this paradigm in some detail [G95], in particular contrasting it with traditional communication theory, where the goal is to reproduce a signal at some distant point.

Following this paradigm, we have constructed several devices which acquire the capability of understanding language via building statistical associations between input stimuli and appropriate machine responses [G93][G94a][M93][S93][H94]. The meaning of an input stimulus (e.g. a word) can be defined [G95] via its statistical associations to a device's input/output periphery. This has the attractive property of grounding meaning in a device's experiences and interactions with its environment. Viewing this set of associations as a vector enables one to define a semantic distortion between events as the distance between their association vectors. The salience of an event is then defined as its distance from a null-event.

In the case that associations are defined via mutual information between events, then this semantic distortion can be shown to be equivalent to the relative entropy between the *a posteriori* distributions of the output actions (conditioned upon the two input events). The salience of an event is then the unique non-negative measure of how much information that event provides about the random variable of appropriate machine responses. The reader is referred to the tutorial paper in [G95] for a detailed discussion of these ideas. We remark that there are related but slightly different salience measures that have been discussed in [G97].

Salient Phrase Fragments

In previous work, we introduced the notion of a salient word, demonstrating that a rudimentary mapping from input to machine action can be constructed based on only that subset. For example, a salience analysis of the database for this task yields the values in table 1

Table 1. Some salient words

Word	Salience	Word	Salience
difference	4.04	dialed	1.29
cost	3.39	area	1.28
rate	3.37	time	1.23
much	3.24	person	1.23
emergency	2.23	charge	1.22
misdialed	1.43	home	1.13
wrong	1.37	information	1.11
code	1.36	credit	1.11

We now search the space of phrase fragments, guided by two criteria. First, within the language channel, a word pair $v_1 v_2$ is considered as a candidate unit if it has high mutual information,

$$I(v_1, v_2) = \log_2 \left[P(v_2|v_1)/P(v_2) \right] \tag{1}$$

This measure can be composed to recursively construct longer units, by computing $I(f, v)$ where f is a word-pair or larger fragment. We remark that this is an approximation to the mutual information of the full n-tuple [C91]. We then introduce an additional between-channel criterion, which is that a fragment should have high information content for the call-action channel. Following [G95], where f is a fragment and $\{c_k\}$ is the set of call-actions, denote its salience by

$$S(f) = \sum_k P(c_k|f)I(f, c_k) \tag{2}$$

This salience measure is a mutual information averaged over the call-actions. It has been shown to be the unique non-negative measure of how much information an event in one channel provides for the random variable of the second channel [B68]. We perform a breadth-first search on the set of phrases, up to length four (an implementation artifact), pruning it by these two criteria: one defined wholly within the language channel, the other defined via the fragment's extra-linguistic associations. The within-channel associations are computed via mutual information and/or the ρ measure of section 4.2. The extra-linguistic associations are computed via the salience of (2). The following table illustrates some salient and background phrase fragments generated by this algorithm. Three attributes of each fragment are provided. First, the mutual information between the final word in the fragment and the preceding subfragment, denoted MI. Second, the peak of the *a posteriori* distribution $P(c_k|f)$, denoted P_{max}. Third, the call-type for which that peak occurs, denoted *Call-Type*. When the peak is between 0.5 and 0.9, then the fragment is only moderately indicative of that call-type and so is provided within parentheses. When the peak is low (<0.5), then it is a background fragment not strongly associated with any single call-type, so none is provided.

Table 2. Salient and background phrase fragments

MI	Phrase Fragments	P_{max}	Call-Type
7.4	made a long distance	0.93	Billing Credit
7.3	long distance	0.55	(Billing Credit)
7.1	I would like	0.24	
6.9	area code	0.65	(Area Code)
6.3	could you tell me	0.37	
5.6	the area code for	0.92	Area Code
5.3	I'm trying	0.33	
5.0	a wrong number	0.98	Billing Credit
4.9	a long distance call	0.62	(Billing Credit)
4.8	the wrong number	0.98	Billing Credit
4.4	I'm trying to	0.33	
4.3	long distance call	0.62	(Billing Credit)
4.3	I just made a	0.93	Billing Credit
4.1	I'd like to	0.18	

For example, consider the fragment *'long distance'*, which has a strong co-occurrence pattern within the language channel, thus a high mutual information (MI=7.3). However, it is not a very meaningful phrase in the sense that the most likely call-type (given that phrase in an utterance) is a billing credit query, but only with probability 0.55. Consider on the other hand an extension of that phrase, *'made a long distance'*, which both has high mutual information (MI=7.4) and strongly connotes a billing credit query with probability 0.93. A similar discussion can be made for the fragments *'area code'* and *'the area code for'*. There are several background fragments in the list, which have strong co-occurrence patterns but are not indicative of any particular call-type, such as *'I would like'* and *'could you tell me'*. Such fragments are useful for creating improved models for speech recognition, as addressed in subsection 4.2.

Salient Grammar Fragments

We now consider a method for combining salient phrase fragments into a grammar fragment. For example, in Table 2, consider the two salient phrases *'a wrong number'* and *'the wrong number'*. Clearly, these should not be treated independently, but rather combined into a single unit. The key idea is that there are two similarity measures, one in the language channel, the other extra-linguistic. Within-channel, there are various measures to compute similarity of word-strings (e.g. a Levenshtein distance). We impose the extra-linguistic constraint, however, that in order for two strings to be clustered, then their meaning must be similar.

For sake of exposition, we restrict attention to a single call-type, focusing on salient fragments for billing credit queries only, based on the transcriptions. Table 3 illustrates the growth of a salient grammar fragment for this call-type. The first pass of the algorithm determines the salient words for billing credits, for which the top choice is *'wrong'*. The others are *'dialed'*, *'credit'*, *'disconnected'*, *'misdialed'* and *'cut'*.

The word *'wrong'* is strongly indicative of billing credit (denoted Cr), with $P(Cr|wrong)=0.92$. The coverage is low, however, with only 48% of those queries containing that word. The local context of this salient word is then evaluated for those elements which sharpen the semantics, i.e. increase the classification rate. The top choice for expanding local context is then *'wrong number'*, which sharpens the *a posteriori* probability to 0.98. Similarly, other left and right contexts are added, leading to the grammar fragment

$F(wrong) = (a|the|was)$ wrong $(number|eos|call)$,

where *eos* is the end-of-sentence marker, | indicates disjunction (or) and concatenation indicates conjunction in order. The grammar fragment with the kernel *'wrong'* is then denoted $F(wrong)$. At this point, the semantics is quite sharp, with the *a posteriori* probability being 0.97, although the coverage has dropped to 0.42. This process is then repeated to construct fragments surrounding the other salient words for this call-type, denoted $F(dialed)$, etc. As this expression becomes too long to fit in the table, we indicate the fragment from the previous row by '—'. By incrementally adding these fragments, the coverage is increased to 0.64 while maintaining a high classification rate of 0.95.

Table 3. Growth of a salient grammar fragment for distinguishing billing credit queries

Prob correct $P(Cr\|G)$	Coverage $P(G\|Cr)$	Fragment G
0.92	0.48	wrong
0.98	0.41	wrong number
0.95	0.45	wrong (number|eos|call)
0.97	0.42	(a|the|was) wrong (number|eos|call)
0.95	0.50	F(wrong) | F(dialed)
0.95	0.57	F(wrong) | F(dialed) | F(credit)
0.95	0.59	— | F(disconnected)
0.95	0.64	— | F(misdialed) | F(cut off)

Again for the sake of exposition, let's consider the two-class problem of distinguishing billing credit queries from the others, still restricting attention to transcriptions only. (In Section 5, we will report on a full multi-class experiment from speech). For any particular salience threshold, a particular set of grammar fragments will be generated. A most rudimentary decision rule would be based simply whether one of these fragments matches a substring of the recognizer output. For example, the following are some illustrative correct detections of a billing credit query, based on such a matching scheme. The substring which matches a grammar fragment is highlighted by capitalization plus connection with underscores. Digit sequences are denoted '*xxx*'.

Correct Detections

i placed a call and i GOT_A_WRONG_NUMBER earlier this afternoon.

yes i MISDIALED a number.

I_WAS_CUT_OFF when trying to call this number.

I_WAS_DIALING 1 xxx xxx xxxx and i got someone else

yes I_JUST_DIALED AN_INCORRECT_NUMBER

yes I would like TO_GET_CREDIT_FOR a number I called

There are two types of errors that occur in such a classifier. First is a *false detection*, i.e. classifying a call as a billing credit when it was not. Second is a *missed detection*, i.e. a billing credit query that was classified as other. The operational costs of such errors can be quite different. For example, a missed detection in a call-router leads to a missed opportunity for automation, while a false detection leads to an incorrect routing. Several examples of such errors are shown below.

False Detections

yes i have a number here and i don't know if it's A_WRONG_NUMBER

I was trying to get xxx xxx xxxx and it said it WAS_DISCONNECTED

Missed Detections

I am trying to call wooster and the number I have rings to a different number

I'm going to blame this one on my wife I misread her handwriting

I'm dialing xxx xxx xxxx and I keep getting bells and things like that

4.2 Recognizing Fragments in Speech

In this subsection, we describe our methodology for recognizing salient fragments in fluent speech. Traditionally, the problem of spotting words or fragments in speech has been approached via constructing models of the those fragments plus a background model to subsume their complement. When there are a small number of fragments, it was sufficient to describe the background via a low-level filler model [W90]. As the problem size increases, however, such methods do not scale well. Intuition tells us that the best background model is the rest of the language, leading one to apply large vocabulary recognition and then search the ASR output for the salient fragments. For example, experiments along these lines for keyword spotting using LVR were reported in [P93] [M94].

The ASR engine in our experiments is a research version of AT&T's Watson speech recognizer [S97]. We use an off-the-shelf acoustic model trained on a separate database of telephone-quality read-speech based on the methods in [L94] with shared de-correlation matrices across distributions. The lexicon is based on the 8K training set of Section 2, with a single phoneme-based dictionary pronunciation of each word [R95a]. The language model, pronunciation models and full-context acoustic phone models are composed on-the-fly via the methods of [R95].

The recognizer is constrained by a stochastic language model which approximates an *n*-gram model on variable-length phrase units. These phrase units are automatically acquired from the database based on their utility for minimizing the entropy of the training corpus. At this point, these phrases are acquired separately and according to a different criterion than the salient fragments of the previous subsection. It is a subject for future research to integrate these two methods, in order to optimize the recognizer to maximize the understanding rate.

Language Modeling

For language modeling to constrain the recognizer, we automatically train a stochastic finite state grammar represented via a *Variable Ngram Stochastic Automaton* (VNSA) [R96]. A VNSA is a non-deterministic automaton that allows for parsing any possible sequence of words drawn from a given vocabulary. Moreover, it implements a backoff mechanism to compute the probability of unseen word-tuples. The stochastic automaton is automatically generated from the training corpus according to the algorithm presented in [R96]. The order of a VNSA network is the maximum number of words that can be used as left context. I.e., if the order is n and w_j denotes the jth word in an utterance, then it utilizes the conditional probabilities $Prob(w_i|w_{i-n+1}, \ldots, w_{i-1})$. VNSAs have been used to approximate standard n-gram language models yielding similar performance to standard bigram and trigram models [R96]. Since they are represented as stochastic finite state machines, their incorporation into a one-pass Viterbi speech decoder is straightforward and efficient. Furthermore, they can be exploited in a cascade of transducer compositions for speech processing to include intra and inter-word phonotactic constraints [P97].

Automatically Acquired Phrases

Traditionally, n-gram language models for speech recognition assume *words* as the basic lexical unit. However, there are several motivations for choosing longer units for language modeling. First, not all languages have a predefined word unit (e.g. Chinese). Second, many word tuples (phrases) are strongly recurrent in the language and can be thought as a single lexical entry, e.g. *'area code'*, *'I would like to'* or *'New Jersey'*. Third, for any model of a fixed order, we can selectively enhance the conditional probabilities by using variable length units to capture long spanning dependencies. In previous work [R96], the effectiveness of incorporating *manually* selected phrases in a VNSA has been shown.

In this paper, building upon [R97], we describe an algorithm for automatically generating and selecting such variable length units based on minimization of the language perplexity $PP(T)$ on a training corpus T. We remark that while there has been other research into automatically acquiring entropy-reducing phrases [G95a] [M95], this work differs significantly in the language model components and optimization parameters.

The phrase acquisition method is an iterative process which converges to a local minimum of $PP(T)$, as illustrated in Figure 5. In particular, given a fixed model order n and a training corpus T, the algorithm proceeds as follows.

Re-estimation Algorithm for the ASR Language Model

Parameters: Let K be the number of candidates generated at each iteration, and M be the number of iterations.

Initialization: Let T_{11} be the initial training corpus T, and let λ_{11} be the language model of order n trained from that corpus.

Iterate for $m = 1$ *to* M,

> *Generate* a ranked set of K candidate phrases from symbol pairs in the lexicon of training set T_{m1}, denoting these via $(x\_y)_k$. The ranking is via the correlation measure ρ described below.
>
> For each candidate phrase, $k = 1$ *to* K
>
> > *Filter* the current training corpus $T_{m,k-1}$ by replacing each occurrence of the phrase with the phrase unit $(x\_y)_k$. Denote this new filtered set by T_{mk}.
> >
> > *Train* a new language model (still of order n) from T_{mk}, denoted λ_{mk}
> >
> > *Test* whether adding this candidate phrase decreases perplexity, i.e. whether $PP(\lambda_{mk}, T_{mk}) < PP(\lambda_{m,k-1}, T_{m,k-1})$. If so, then continue, else reject this candidate phrase via setting $T_{mk} = T_{m,k-1}$.
> >
> > *next k*
>
> *next m*
>
> *Train* a final language model from the filtered corpus T_{MK} plus the original T, with lexicon comprising all original words plus the acquired phrases.

The algorithm is initialized with the training corpus T, with the initial language model λ_{11} corresponding to a stochastic n-gram model on words. For each iteration, the first step is to generate and rank candidate symbol-pairs (x,y) based on a *correlation coefficient*

$$\rho(x,y) = P(x,y)/[P(x) + P(y)] \qquad (3)$$

where $P(x)$ denotes the probability of the event x and $P(x,y)$ denotes the probability of the symbols x and y occurring sequentially. At the first iteration, x and y are both words, in subsequent iterations they are potentially larger units. Observe that $0 \leq \rho(x,y) \leq 0.5$. We remark that this correlation measure has advantages over mutual information with respect to ease of scaling and thresholding [R97].

Thus, a phrase x_y is selected only if $P(x,y) \cong P(x) \cong P(y)$ (i.e. $P(y|x) \cong 1$) and the training set perplexity is decreased by incorporating this larger lexical unit into the model. After the M iterations are completed, there is the final step of retraining the language model from the final filtered corpus T_{MK} plus the original T. This preserves the granularity of the original lexicon, generating alternate paths through the SFSM comprising both the new phrases plus their original word sequences. I.e., if the words '*long*' and '*distance*' only occur together in the corpus leading to the acquisition of the phrase '*long_distance*', this final step preserves the possibility of the words occurring separately in some test utterance.

4.3 Call Classification

We make a decision as to which of the 15 call-types to classify an utterance in a particularly straightforward manner. The speech recognizer described in subsection 4.2 is applied to an utterance, producing a single best word recognition output. This

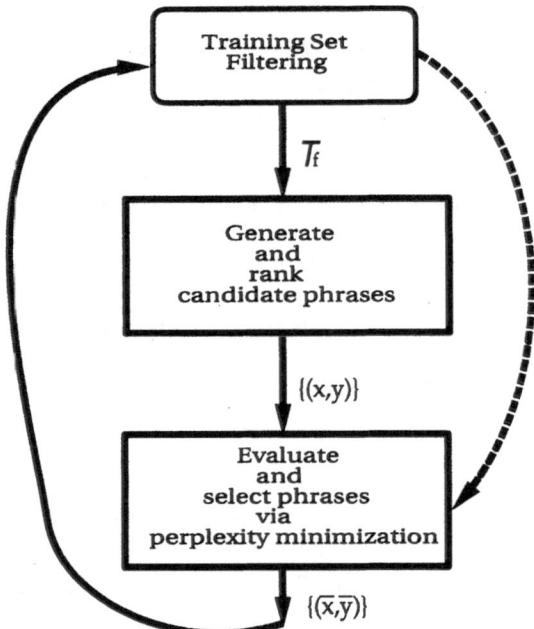

Fig. 5. Phrase selection via entropy minimization

ASR output is then searched for occurrences of the salient phrase fragments described in subsection 4.1. In case of fragment overlap, some parsing is required. The parsing algorithm is a simple one, selecting longer fragments over shorter ones, then proceeding left to right in the utterance. This yields a transduction from the utterance s to a sequence of associated call-types. To each of these fragments f_i is associated the peak value and location of the *a posteriori* distribution,

$$p_i = \max_k P(C_k|f_i) \tag{4}$$

$$k_i = \operatorname{argmax}_k P(C_k|f_i) \tag{5}$$

Thus, for each utterance s we have a sequence $\{f_i, k_i, p_i\}$. The decision rule is to select the call-type of the fragment with maximum p_i, i.e. select $C_{K(s)}$ where

$$i(s) = \operatorname{argmax}_i p_i \tag{6}$$

$$K(s) = k_{i(s)} \tag{7}$$

If this overall peak is less than some threshold, P_T, then the utterance is rejected, i.e. if $p_{i(s)} < P_T$.

Several examples are given below, listing the transcription then ASR output using the phrase-bigram grammar of subsection 4.2 with the detected fragments highlighted via capitalization and bracketed via underscores. The transduction into call-types with associated scores is then given, with the peak fragment indicated via underlining.

Examples 1-4 demonstrate robustness of the salient fragments in the presence of recognition errors. The fifth example illustrates an ASR error which yielded a salient fragment, where *'stick it on my'* was misrecognized as *'speak on my'* (observing that *stick* was not in the training data, thus is an out-of-vocabulary word). The final example involves a user who thought they were talking to a hardware store. In this case, the recognizer performs poorly because of a large number of out-of-vocabulary words. However, the call-classification is indeed correct–leading to transfer to a human agent.

Examples of Call Classification

1. Transcr: yes I just made a wrong telephone number
   ASR+parse: *not help me yes I JUST MADE A long telephone number*
   Transd+decision: {*CREDIT 0.87* }

2. Transcr: hello operator I get somebody to speak spanish
   ASR+parse: *motel room you get somebody speak SPANISH*
   Transd+decision: {*ATT SERVICE 0.64*}

3. Transcr: hi can I have the area code for saint paul minnesota
   ASR+parse: *hi can THE AREA CODE FOR austin for minnesota*
   Transd+decision: {*AREA CODE 1.00*}

4. Transcr: yes I wanted to charge a call to my business phone
   ASR+parse: *yes I WANNA CHARGE call to distance phone*
   Transd+decision: {*THIRD NUMBER 0.75*}

5. Transcr: hi I like to stick it on my calling card please
   ASR+parse: *hi I'D LIKE TO SPEAK ON MY CALLING CARD please*
   Transd+decision: {*PERSON PERSON 0.78*} {*CALLING CARD 0.96*}

6. Transcr: I'm trying to find out a particular staple which fits one of your guns or a your desk staplers
   ASR+parse: *I'm trying TO FIND OUT they've CHECKED this is still call which six one of your thompson area state for ask stay with the*
   Transd+decision: {*OTHER 0.86*}

5. Experiment Results

The database of Section 3 was divided into 8K training and 1K test utterances. The remainder of the 10K database has been reserved for future validation experiments. Salient phrase fragments were automatically generated from the training transcriptions and associated call-types via the methods of subsection 4.1. In particular, the length of these fragments was restricted to four or less and to have training-set frequency of five or greater. An initial filtering was imposed so that the peak of the *a posteriori* distribution for a fragment is 0.6 or greater. We have observed in previous experiments [G95] that the numerical value of salience is influenced by the fragment frequency, as is typical for information-theoretic measures. Figure 6 shows a scatter plot of salience versus within-channel information content $i(f) = -\log_2[P(f)]$. It is thus advantageous to introduce a *frequency-tilted salience threshold* of the form

$$sal(f) \geq \alpha i(f) + \beta, \tag{8}$$

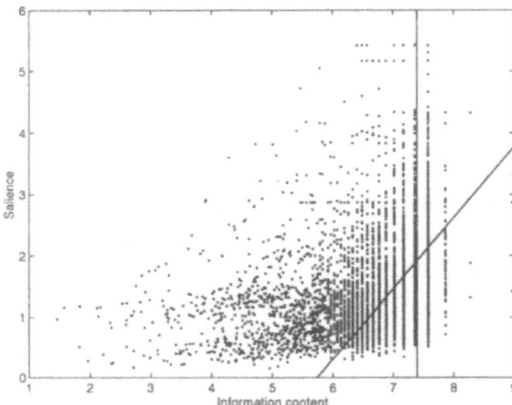

Fig. 6. Salience versus information for phrase fragments

The values of α and β can be varied and evaluated empirically. In the scatter plot of Figure 6, two thresholds also shown: the vertical line for the frequency threshold, the other for the frequency-tilted salience threshold. In this experiment, we select the values of α and β via a statistical significance test. For any particular fragment, we evaluate the null hypothesis that its observed *a posteriori* distribution $P(C_k|f)$ occurs as a random sample of the prior distribution $P(C_k)$. Computed via a multinomial distribution, a significance level of 1% is imposed, yielding the tilted salience threshold shown in Figure 6. This reduces the total number of phrase fragments by about 20%. There are approximately 3K such salient fragments, with length distributed as in Table 4

Table 4. Length distribution of salient phrase fragments

Length of Salient Fragment	1	2	3	4
Relative Frequency	0.04	0.18	0.43	0.35

The speech recognizer is as described in subsection 4.2. The VNSA language model is trained via 20 iterations of the algorithm in 4.2, with 50 candidates per iteration. For the phrase-bigram model, this yields 783 phrases in addition to the original 3.6K word lexicon. The length of these fragments varies between 2 and 16, distributed as shown in Table 5.

Table 5. Length distribution of VNSA phrase fragments

Length of VNSA Fragment	2-3	4-5	6-7	8-16
Relative Frequency	0.88	0.07	0.03	0.02

We first compare word accuracy and perplexity as a function of the language model. Table 6 shows the word accuracy (defined as probability of correct detection minus probability of insertion) as the language model is varied. Recall that phrase units comprise both the original lexicon of words plus variable-length phrases induced by entropy minimization. Bigrams and trigrams are 2nd and 3rd order models respectively on whichever lexicon is specified. Figure 7 shows the test set perplexity as a function of language model units and order. We observe that in both of these within-language performance measures, phrase-bigrams fall between word-bigrams and word-trigrams, but with the computation and memory requirements of word-bigrams.

Table 6. Word accuracy versus language model

Unit Type	Bigram	Trigram
words	49.5 %	52.7%
words + phrases	50.5%	52.7%

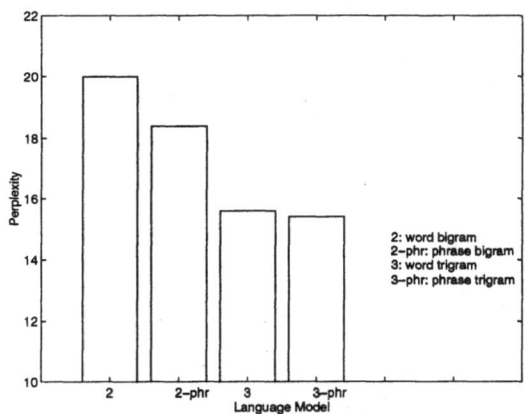

2: word bigram
2-phr: phrase bigram
3: word trigram
3-phr: phrase trigram

Fig. 7. Test set perplexity versus language model

The ASR output is then searched for salient fragments, as described in subsection 4.3. If no fragments are found, then the utterance is rejected and classified as *other*. The number of salient fragments per utterance found in the 1K test set varies between zero and 11, distributed as shown in Table 7.

Performance Evaluation

The salient fragments recognized in an utterance are then rank-ordered, as was described in subsection 4.3. We now measure the performance on the test data in

Table 7. Number of salient fragments recognized in an utterance

Salient Fragments per Utterance	0	1	2	3	4	5	6-11
Relative frequency	0.14	0.25	0.28	0.18	0.07	0.04	0.04

terms of true classification rate and false rejection rate. For each test utterance, the decision between *accept* and *reject* is based on the top-ranked call type. If this is *"other"*, or if the associated probability fails to reach a designated threshold, then the call is rejected. Otherwise, the call is accepted and the accuracy of the attempted classification (at rank 1 and rank 2) is determined using the label set for that call. The desired goal for calls labeled *"other"* is that they be rejected. The *false rejection rate* is the proportion of calls not labeled *"other"* that are rejected. At rank 1, the *true classification rate* is the proportion of accepted calls for which the top-ranked call type is present in the label set.

At rank 2, the true classification rate is essentially the proportion of calls for which either the first or second highest ranked call type is present in the label set. However, for a small number of calls the label *"other"* is paired with another call type, and a rejection at rank 2 is then counted as a correct outcome for such a call. We include such cases in the true classification rate at rank 2 because at that point the call has been accepted for handling by the dialog system and contributes to the measure of success appropriate to it.

With these definitions we can plot the ROC curve of true classification rate against false rejection rate. Let's introduce the following notation and definitions for a particular utterance,

- C is the list of call-type labels (recall that this is typically a single label)
- $\hat{A}_1$ denotes the decision to accept the call at rank 1
- $\hat{R}_1$ and $\hat{R}_2$ denote the decision to reject the call at rank 1 and rank 2 respectively
- $\hat{C}_1$ and $\hat{C}_2$ denote the rank 1 and rank 2 call types from the recognized fragments.

We then measure, over the set of 1K test utterances, the following probabilities.

$$\textit{False rejection rate} = P(\hat{R}_1 | \textit{other} \notin C) \tag{9}$$

$$\textit{True classification rate (rank 1)}$$
$$= P(\hat{C}_1 \in C | \hat{A}_1) \tag{10}$$

$$\textit{True classification rate (rank 2)}$$
$$= P\left((\hat{C}_1 \in C) \cup (\hat{C}_2 \in C) \cup (\hat{R}_2 \cap \textit{other} \in C) \Big| \hat{A}_1 \right) \tag{11}$$

We generate a performance curve by varying the rejection threshold from 0.6 to 0.95. Figure 8 shows the rank 1 performance curves for several different ASR language models. As a baseline for comparison, the performance on transcribed output (i.e. error-free ASR) is also shown. It is interesting to note that call-classification performance is significantly higher than word accuracy – confirming the intuition that some events are crucial to recognize for a task, others not so. It is also worthwhile noting that while the phrase-bigram language model for ASR performs worse than word-trigrams with respect to word accuracy, it performs *better* with respect to call-classification rate. This is because the variable-length phrase-units selectively provide long-spanning local dependencies, with the side-effect of improving recognition of the salient fragments. This result reinforces the intuition that optimizing recognition for understanding is an important research issue. We now compute both

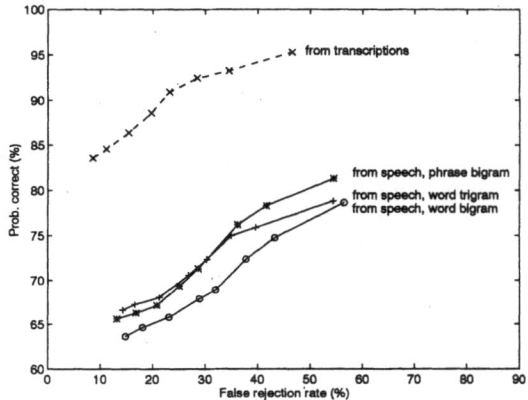

Fig. 8. Call-classification performance for varying ASR language models

rank 1 and rank 2 performance using the phrase-bigram model for ASR, with performance shown in Figure 9.

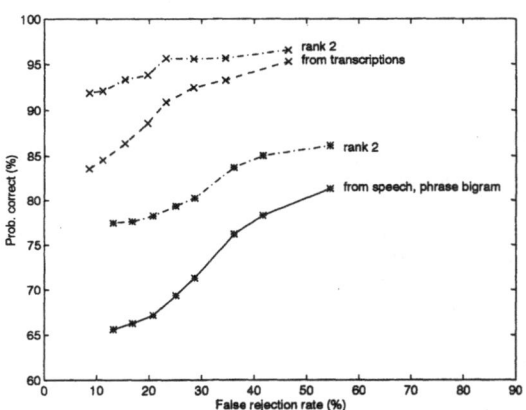

Fig. 9. Rank 1 and rank 2 performance

6. Conclusions

We have described progress towards a natural spoken dialog system for automated services. By *natural*, we mean that the machine understands and acts upon what people actually say, in contrast to what one would like them to say. A first stage in this system is call-classification, i.e. routing a caller depending on the *meaning* of their fluently spoken response to *'How may I help you?'* We have proposed algorithms for *automatically acquiring* language models for both recognition and understanding, experimentally evaluating these methods on a database of 10K utterances. These experiments have shown that understanding rate is significantly greater than recognition rate. This confirms the intuition that it is not necessary to recognize and understand every nuance of the speech, but only those fragments which are *salient* for the task.

Acknowledgements. The authors wish to thank Larry Rabiner, Jay Wilpon, David Roe, Barry Parker and Jim Scherer for their support and encouragement of this research. We also thank Mike Riley and Andrej Ljolje for many hours of useful discussion on the ASR aspects of this effort. We finally thank our colleagues Alicia Abella, Tirso Alonso, Egbert Ammicht and Susan Boyce for their continued collaboration in the creation of a spoken dialog system exploiting the methods of this paper.

References

[A96]   A. Abella, M. Brown, B. Buntschuh, *"Developing Principles for Dialog-Based Interfaces"*, Proc. ECAI Spoken Dialog Systems Workshop, Budapest. August 1996.

[B68]   N.M. Blachman, *"The amount of information that y gives about x"*, IEEE Trans. on Information Theory, vol. IT-14, pp. 27-31, Jan. 1968.

[B96]   S. Boyce and A.L. Gorin, *"User Interface Issues for Natural Spoken Dialog Systems"*, Proc. Intl. Symp. on Spoken Dialog (ISSD), pp. 65-68, Philadelphia, Oct. 1996.

[C91]   T.M. Cover and J.A. Thomas, *"Elements of Information Theory"*, Wiley (1991).

[G93]   A.N. Gertner and A.L. Gorin, *"Adaptive Language Acquisition for an Airline Information Subsystem"*, Artificial Neural Networks for Speech and Vision, pp. 401-428, (ed. R. Mammone), Chapman and Hall, 1993.

[G94a]   A.L. Gorin, S.E. Levinson, and A. Sankar *"An Experiment in Spoken Language Acquisition"*, IEEE Trans. on Speech and Audio, pp. 224-240, vol. 2, no. 1, part II, Jan. 1994.

[G94b]   A.L. Gorin, H. Hanek, R. Rose and L. Miller," *Spoken Language Acquisition for Automated Call Routing,"* Proc. of the Intl. Conf. on Spoken Language Processing, pp. 1483-1485, Sept. 1994, Yokohama, Japan.

[G95]   A.L. Gorin, *"On Automated Language Acquisition"*, 97(6), pp. 3441-3461, Journal of the Acoustical Society of America (June 1995).

[G95a]   E. Giachin, *"Phrase Bigrams for Continuous Speech Recognition"*, Proc ICASSP, pp. 225-228, Detroit, 1995.

[G95b]   A.L. Gorin, *"Spoken Dialog as a Feedback Control System"*, Proc. Of the ESCA Workshop on Spoken Dialog Systems, Denmark June 1995.

[G96a]   A.L. Gorin, B.A. Parker, R.M. Sachs and J.G. Wilpon, *'How may I help you?'*, Proc. of IVTTA, pp. 57-60, Basking Ridge, Sept. 1996.

[G96b]   A.L. Gorin, *"Processing of Semantic Information in Fluently Spoken Language"*, Proc. of Intl. Conf. on Spoken Language Processing (ICSLP), pp. 1001-1004, Philadelphia, Oct. 1996.

[G97]   P.N. Garner and A. Hemsworth, *"A Keyword Selection Strategy for Dialog Move Recognition and Multi-Class Topic Identification"*, Proc. ICASSP, pp 1823-1826, Munich (1997).

[H94]   E.A. Henis, S.E. Levinson and A.L. Gorin, *"Mapping Natural Language and Sensory Information into Manipulatory Actions"*, Proc. of the eighth Yale workshop on adaptive and learning systems, June 1994.

[J90]   F. Jelinek, *"Self-organizing Language Models for Speech Recognition"*, in Readings in Speech Recognition, (edited by A. Waibel and K. Lee), Morgan-Kaufmann, pp. 449-456.

[L94]   Ljolje, A., *"High accuracy phone recognition using context clustering and quasi-triphonic models"*, Computer Speech and Language, 8:129-151, 1994.

[M93]   L.G. Miller and A.L. Gorin, *"Structured Networks for Adaptive Language Acquisition"*, International Journal of Pattern Recognition and Artificial Intelligence, vol. 7, no. 4, pp. 873-898 (1993).

[M94]   J. McDonough and H. Gish, *"Issues in Topic Identification on the Switchboard Corpus"*, Proc. of ICSLP, pp. 2163-2166, Yokohama, 1994.

[M95]   Matsumura T. and Matsunaga S. : *"Non-uniform unit based HMMs for continuous speech recognition"* Speech Communication Vol.17 No.3-4, pp.312-329, 1995

[M96]   Masataki H. and Sagisaka Y. : *"Variable-order N-gram generation by word-class splitting and consecutive word grouping"* Proceedings of ICASSP Vol.I pp.188-191, 1996

[P93]   B. Peskin, *"Topic and Speaker Identification via Large Vocabulary Speech Recognition"*, Proc. of the ARPA Workshop on Human Language Technology (ARPA, Washington D.C.) 1993.

[P97]   F. Pereira and M. Riley, *"Speech recognition by composition of weighted finite automata"*, in Finite-State Devices for Natural Language Processing, (E. Roche and Y. Schabes, editors), MIT Press, Cambridge (1997)

[R95]   M. Riley, F. Pereira and E. Chung, *"Lazy transducer composition: a flexible method for on-the-fly expansion of context-dependent grammar networks"*, Proc. ASR Workshop, Snowbird, 1995

[R95a]   Riley, M.D., Ljolje, A., Hindle, D. and Pereira, F., *"The AT&T 60,000 Word Speech-To-Text System"*, EUROSPEECH-95, pp. 207-210, Madrid, 1995

[R96]   G. Riccardi, R. Pieraccini, E. Bocchieri, *"Stochastic Automata for Language Modeling"*, Computer Speech and Language, 10(4), pp. 265-293, 1996.

[R97]   G. Riccardi, A.L. Gorin, A. Ljolje and M. Riley, *"Spoken Language Understanding for Automated Call Routing"*, Proc. ICASSP, pp. 1143-1146, Munich 1997.

[S93]   A. Sankar and A.L. Gorin, *"Visual Focus of Attention in Adaptive Language Acquisition"*, Artificial Neural Networks for Speech and Vision, pp. 324-356, (ed. R. Mammone), Chapman and Hall, 1993.

[S97]   R. D. Sharp et al., *"The WATSON Speech Recognition Engine"*, Proc. ICASSP, pp. 4065-4068, Munich, 1997.

[W90]   J.G. Wilpon, L.R. Rabiner, C.H. Lee and E.R. Goldman, *"Automatic Recognition of Keywords in Unconstrained Speech using Hidden Markov Models"*, IEEE Trans. on ASSP, pp. 1870-1878, vol. 38, no. 11, November 1990.

Introduction of Rules into a Stochastic Approach for Language Modelling

Thierry Spriet and Marc El-Bèze

LIA, Université d'Avignon, LIA
CERI - BP1228 84911 Avignon cedex 9 France
email: `spriet@univ-avignon.fr`

Summary. Automatic morpho-syntactic tagging is an area where statistical approaches have been more successful than rule-based methods. Nevertheless, available statistical systems appear to be unable to hold long span dependencies and to model unfrequent structures. In fact, part of the weakness of statistical techniques may be compensated by rule-based methods. Furthermore, the application of rules during the probabilistic process inhibits the error propagation. Such an improvement could not be obtained by a post processing analysis. In order to take advantage of features that are complementary with two approaches, a hybrid approach has been followed in the design of an improved tagger called ECSta. In ECSta, as shown in this paper, a stack-decoding algorithm is combined with the Viterbi classical one.

Key words: morpho-syntactic, statistical, rule-based, mixed approach, stack decoding

1. Introduction

Automatic tagging at a morpho-syntactic level is an area of Natural Language Processing where statistical approaches have been more successful than rule-based methods [1]. Nevertheless, available systems based on statistical algorithms appear to be unable to hold long span dependencies and to model unfrequent structures. In fact, part of the weakness of statistical techniques may be compensated by rule-based methods. In order to take advantage of features that are complementary in the two approaches, a hybrid approach has been followed in the design of an improved tagger called ECSta. It is difficult not to say impossible to use rules in the Viterbi [6] search or with the Forward-Backward algorithm. Although ECSta mainly relies on statistical models, it jointly uses some knowledge rules. In doing that, care has been taken to preserve the consistency of the mathematical model. In ECSta, a stack-decoding algorithm is combined with the Viterbi classical one. In this paper, the strategy for the combination of the two components is described. A probabilistic tagger is a process designed to disambiguate a multiple choice of syntactic tags for each lexical component of a word sequence. The structure used by such a system is called hypotheses lattice. The probability of a tag sequence C depends on the history $H(c_i)$ of each tag c_i and is computed as follows :

$$P(C) = \max_x \prod_{i=1}^{N} P(c_i/H(c_i))P(w_i/c_i)$$

Often the tag history is limited to two previous units. Classical algorithm keeps memory of the history represented by the best path leading to c_i (Viterbi) or only the probabilities of the paths leading to c_i (Baum-Welch) [2].

In order to introduce new history-dependent information into $P(C)$ the complete history of each node of the lattice should be kept. Since it is not conceivable to consider all the paths (about a billion for a typical French sentence), a new algorithm, based on the A* strategy is proposed in the following.

2. Stack Decoding Strategy

2.1 The Algorithm

The A* algorithm [4] applied to our case, finds the optimal path in a graph where the nodes are partial syntactic interpretations of the sentence, and each connection corresponds to the tagging of a word or a word compound. It expands a tag sequence with all the tags proposed by the dictionary for the first word following the sequence. This algorithm is admissible and the minimum number of hypotheses is considered. For an A* strategy, a good evaluation function is needed, or, at least, an excellent estimation of this function. The better the estimation is, the fewer hypotheses are explored. The estimation function is described in the next paragraph, and is called $E(C)$. The principle of the algorithm is to partly develop the most competing hypothesis. At each step, $E(C)$ points out the hypothesis which represents the best candidate.

2.2 The Evaluation Function

The evaluation function must estimate an upper bound of the score which can be reached by a completely developed tagged word sequence. We can distinguish three parts in its computation. If $CS_{1,i}$ is a tagged word sequence $c_1..c_i$ in a sentence of N words, where c_k represents the tag of the word k, then :

$$E(CS_{1,i}) = V(c_1, .., c_{i-1})S(c_i)VB(c_1, .., c_i)$$

- $V(c_1, .., c_{i-1})$ is the score obtained by the sequence $c_1, .., c_{i-1}$, and can be provided by a probabilistic model,
- $S(c_i)$ is the consequence of the choice of the tag c_i for the word i,
- $VB(c_1, .., c_i)$ is the estimation of the best score which can be obtained by the development of the sequence.

The function E needs to be continuous and decreasing with the size of the sequence. It means that $E(CS_{1,i-1}) \geq E(CS_{1,i})$. The probabilities provided by a stochastic model respect these properties and lead to a straightforward computation of the two first components. With a tri-class model, we have :

$$V(c_i, .., c_{i-1}) = V(c_1, .., c_{i-2})S(c_{i-1}) \;\; ; \;\; S(c_i) = P(c_i/c_{i-2}, c_{i-1})P(w_i/c_i)$$

and, if $C_{i+1,N}$ is a sequence of tags for $w_{i+1}, .., w_N$,

$$VB(c_1, .., c_i) = argmax_{C_{i+1,N}} P(C_{i+1,N})$$

$$= argmax_{C_{i+1,N}} \prod_{k=1}^{N-1} P(c_{k+1}/c_{k-1}, c_k)P(w_{k+1}/c_{k+1})$$

So that, the score $E(CS_{1,i})$ is similar to a Viterbi scoring if $CS_{1,i}$ is a part of the sequence pointed out by a classical Viterbi algorithm.

VB may be estimated, in a very simple method, through a Viterbi-Back scheme, so that a first pass has to be performed in a right-left direction. In this way, all VB estimates are made available before the search of the best path in the lattice. It is now possible to introduce extra-information into $E(CS_{1,i})$ by adding to the $S(c_i)$ a third term :

$$S(c_i) = P(c_i/c_{i-2}, c_{i-1})P(w_i/c_i)T(c_i)$$

where $T(c_i)$ corresponds to a confidence rate associated to the choice c_i. By default, $T(c_i) = 1$, which gives full confidence to the stochastic model. In practice, $T(c_i)$ is the result of the application of rules applied to the history with unrestricted length.

2.3 Peculiar Advantages of the Algorithm

Notwithstanding the gain directly obtained by these rules, the algorithm performs some context corrections that a post processing would not be able to do. When the evaluation function has penalized a sequence of tags, the algorithm explores competitive paths in the lattice, with the following possibilities :

- find some errors on a tag c_{i-k} by detecting an error on the tag c_i.
- have a different analysis of the right context of c_i, if a rule gives another interpretation for this tag.

A post-processing can only change a tag to another but does not analyze the consequences for the other tags of the sequence. Nevertheless, changing a tag in a sequence changes the probability of the whole sequence and can encourage the model to revise the affectation of neighbouring tags. Figure 1 shows the kind of corrections which can be made by a stack decoding strategy. This example refers to

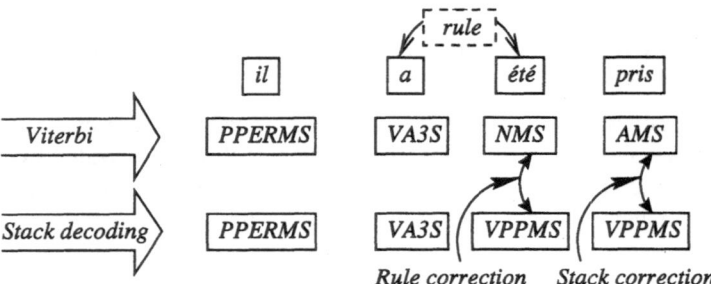

Fig. 1. Double correction

the french sentence "Il a été pris de court.", for which a classical algorithm makes two errors. The first one comes from a training drift , which is explained in the following section. The second one, proves the stack decoding ability to avoid the error propagation by taking advantage from extra-information and to produce another tag sequences analysis. Here, the statistical model chooses the tag *AMS* (*masculine singular adjective*) for *pris* because the sequence *Noun-Adjective* is more frequent than the sequence *VPPMS-VPPMS* (two past participles).

3. Rules

3.1 Correction of Biases

Even a good model may have some biases, considering as reference a classical triclass language model used with a Viterbi-decoder [5]. It was trained on newspaper corpora (Le Monde) with approximately 40 million words for 103 parts of speech. A tagging error rate of nearly 6% was observed on a test corpus. More details can be found in [3]. Our purpose here is to analyse the errors. It is interesting to note that the error rate can be divided into 3 equal parts. The first type of errors is due to a divergence of the model for the following reasons :

- The past participle *été* got the tag *Singular Masculine Noun* (summer) after the auxiliary verb *avoir*. This is impossible in French. In order to cope with this problem, we have implemented the rule : "If there is the tag *auxiliary* for any derivation of the verb *avoir* just before the word *été*, $T(SingularMasculineNoun) = 0$".
- The preposition *en* is wrongly tagged *Adverb* and more often *Personal Pronoun* before a *country name* or a *noun*. As this is also impossible in French, we have introduced a similar kind of rule into the rule set.

The second type of errors is due to a wrong assignment of the OOV (Out-Of-Vocabulary) tag. Proper nouns are involved in about 67% of these errors. Most of the cases can be corrected by few simple rules. The rules need to introduce a new tag (*Proper Noun*) in the lattice. In doing that, the VB scores become inconsistent because this new tag was not taken into consideration in the computation of the probabilities. In practice, if we want to use this kind of rules, we have to produce two tags (*OOV* and *Proper Noun*) for all OOV units in the first step of the algorithm, which computes the Viterbi-Back scores.

For the last type, the frequencies of the errors are so low that it is not specially interesting to write specific rules. We find here spelling mistakes, dictionary errors, under-represented structures or long span dependencies.

3.2 Under-represented Structures and Long Span Dependencies

The two previously mentioned rules involve a binary decision, namely whether the structure is possible or impossible. In this part, we show how to estimate $T(c_i)$ for under-represented structures or long span dependencies which are completely inhibited by the statistical model. Unfortunately we are still obliged to perform a hand made error analysis. The first step is of course to describe as accurately as possible the under-represented structure or the dependencies. If it is possible to exactly describe the context $AC(c_i)$ of the structure such that no other structures can appear in this context, we can use the same kind of rules as previously. Otherwise, we have to train $T(c_i)$ for this context. This means counting $NM(c_i)$ on a test corpora which is the number of mistakes made by the statistical model on $AC(c_i)$. If $TN(c_i)$ is the occurrence number of $AC(c_i)$ we have:

$$T(\overline{c_i}) = \frac{1 - NM(c_i)}{TN(c_i)}; T(ci) = \frac{TN(c_i) + NM(c_i)}{TN(c_i)}$$

where $\overline{c_i}$ represents all the tags proposed by the model except c_i.

It is interesting to point out that we can use all the tags of the sequence to describe $AC(c_i)$ while in the Viterbi algorithm the access is limited to the two previous tags. Thanks to this, it is possible to develop rules for long span dependencies. Rules can be automatically produced with decision trees [7]. The questions used to train the trees must concern a longer history than the one used by the n-gram model. In fact, the constraints which can be expressed with a short history are already taken into account by the statistical model.

4. Multi Level Interactions

4.1 Linguistic and Syntactic

Another application where this mixed approach can be interesting is in linguistically disambiguating homophonic outputs of a speech recognition system. We begin by developing all the nodes in order to associate syntactic information to each word proposed by the acoustic component, so we obtain a hypotheses lattice which can be used by a tagger. In this graph, all the words $w_{i,1}, ..w_{i,x}$ are homophones and may have different tags, each of them giving additional paths so that the number of paths is strongly increasing. It is clearly impossible to develop all the paths for a classical tagging task. A classical approach would search the best "linguistic" sequence (with a n-gram) or the best syntactic sequence (with a n-class) but the interaction of these two levels is not real. Even with a combination of n-grams and n-classes, the calculation of the probabilities is not completely interactive. With the kind of rules we propose, these two levels can be used together. For example, a rule can verify that a word sequence does not appear within a particular tag sequence. In a n-class, this inter-connection of syntactic and linguistic level is only assumed through the probability $P(w_i/c_i)$ [8].

4.2 Phonology

In the framework of the decoding strategy, we develop specific rules based on phonologic constraints for the lexical unit sequence. The possibilities offered by the

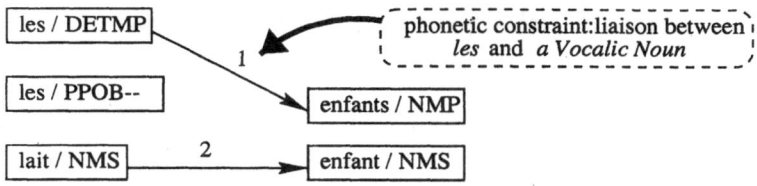

Fig. 2. A phonetic constraint

joint use of phonetic, lexical and syntactic levels in a unique rule is very interesting. It can incorporate peculiarities of a given language as for example the "liaison" phenomenon in the French language. Figure 2 gives an example of this.

In this case, the 3-class tagger prefers the path 1 because of the frequent syntactic structure *DETMP-NMP* (*masculine plural determiner* followed by a *masculine plural noun*). If the acoustic component has not detected the liaison between *les* and *enfants*, ECSta penalises the path 1 and finally chooses the path 2.

5. Conclusion

A combined solution involving probabilistic methods and a rules based approach, has been proposed. A strategy has been described for jointly using a set of rules and a stochastic model. The main interest of this is to introduce under-represented structures and log span dependencies in a stochastic model.

References

[1] Brill E., A simple Rule-Based Part of Speech Tagger, 3nd conf. on Applied Natural Language, Trento, Italy, April 1992, pp.63-66.
[2] Derouault A.M., Merialdo B., Natural Language Modeling for Phoneme-to-text transcription, IEEE trans. on Pattern analysis and machine intelligence, vol. PAMI-8 No 6, Nov. 1986.
[3] El-Bèze M., Spriet T., Intégration de contraintes syntaxiques dans un système d'étiquetage probabiliste, TAL January 1996.
[4] Hart P., Nilsson N., Raphael B., A formal basis for the heuristic determination of minimum cost paths, IEEE Trans. Systems Sci. Cyberne, 1968, pp 100-107.
[5] Jardino M., Adda G., Automatic determination of a stochastic bi-gram class language model, Proc. Grammatical Inference and Applications, 2nd Int. Coll. ICGI94, Spain, Sept. 1994, pp.57-65.
[6] Jelinek F., Continuous speech recognition by statistical methods, Proceeding of the IEEE, vol. 64, April 1976, pp.532-556.
[7] Kuhn R.,De Mori R., The application of semantic classification trees to natural language understanding, IEEE Trans. on pattern analysis and machine intelligence, vol. 17, No. 5, may 1995.
[8] Mérialdo B., Tagging text with a probabilistic model, ICASSP 1991, Toronto, vol. S2, pp 809-812.

History Integration into Semantic Classification

Mauro Cettolo and Anna Corazza

IRST - Istituto per la Ricerca Scientifica e Tecnologica I-38050 Povo, Trento, Italy
email: {`cettolo,corazza`}@itc.it

Summary. In spoken language systems, the classification of coherent linguistic/semantic phrases in terms of semantic classes is an important part of the whole understanding process. Basically, it relies on the plain text of the segment to be classified. Nevertheless, another important source of useful information is the dialogue context. In this paper, a number of different ways to integrate the dialogue history into the semantic classification are presented and tested on a corpus of person-to-person dialogues. Best result gives a 3.6% reduction of the error rate with respect to the performance obtained without using history.

1. Introduction

In machine-mediated person-to-person communication, care should be devoted to preserve communication in a robust way. The task considered here involves bilingual negotiation to fix an appointment [1]. The information that the machine is required to extract from the input utterance mainly consists of its overall purpose (here called Dialogue Act, DA) and of some parameters for which usually a local syntactical analysis is sufficient.

This paper concerns the first of these two problems, namely the choice of the DA to be assigned to a given input. It is given that the input sentence can be automatically segmented into coherent semantic units to be then classified. First results on the segmentation problem are presented in [3].

In addition to the text of the sentence, one important information source that could effectively help classification is represented by the dialogue history, i.e. the last uttered DAs. Different strategies to consider context in classification are here discussed and experimentally compared.

Semantic Classification Trees (SCTs) [1, 4, 5] do not only choose the DA to associate to the input, but they can also evaluate a probability distribution on the DA set. This distribution can be used if DA hypotheses are to be scored on the basis of different information sources (acoustic, syntactical, and so on). Eventually, the most likely DA will be chosen. In the here considered case, SCT probability distribution is integrated with the probability of the dialogue history, i.e. the last DAs occurring in the conversation.

After the evaluation of the baseline without using context information, two different approaches to history integration were considered. The first approach combines in various manners the distribution probability obtained by SCTs, with DA-trigrams. The second approach directly uses SCTs: new possible questions are considered concerning the history, such as "Is the previous DA x?".

The next section introduces SCTs. Section 3. describes the corpus utilized in the experiments, while Section 4. presents the different kinds of history integration considered together with the experimental results obtained for each of them. Results are discussed in Section 5..

2. Classifier

The SCTs used for the experiments presented here are implemented by following [2], with the only exception being the pruning algorithm, presented in [6]. The set of all possible questions that can be associated to each internal node is represented by *keywords*: every word in the vocabulary can be a keyword, with the constraint that it must appear a minimum number of times in the training corpus. Note that this is not a restriction, because rare words are not significant to the task.

Given an input sentence, the SCT is requested to decide to what DA it corresponds (classification), and to evaluate the DA distribution, which will be used for history integration, as described in the following. Each node in the tree is associated to a question regarding the presence of a keyword in the input sentence.

3. Data

The experiments were performed by using a dialogue corpus collected at IRST [1], which is composed of 201 monolingual person-to-person Italian conversations for

Table 1. Statistics describing the corpus

	Training	Test	Whole Corpus	
# dialogue	181	20	201	
# turn	2784	302	3086	
# segment	5619	679	6298	
$|W|$ (non-noise)	28858	3610	32468	
$|V|$ (non-noise)	1353	514	1433	
DA distributions				
request-response	972	99	1071	17.0 %
affirm	927	129	1056	16.8 %
state-constraint	753	88	841	13.4 %
suggest	609	80	689	10.9 %
closing	433	54	487	7.7 %
acknowledge	402	42	444	7.0 %
negate	354	44	398	6.3 %
task-definition	267	44	311	4.9 %
garbage	173	30	203	3.2 %
opening	186	16	202	3.2 %
confirm-appointment	156	19	175	2.8 %
request-suggestion	149	8	157	2.5 %
accept	123	12	135	2.1 %
request-clarification	60	6	66	1.0 %
clarification	34	4	38	0.6 %
reject	21	4	25	0.4 %
total	5619	679	6298	100.0 %

which acoustic signal, word transcriptions and linguistic annotations are available. The two speakers were asked to fix an appointment, observing the restrictions shown on two calendar pages they were given; they did not see each other and could hear the partner only through headphones. The conversations took place in an acoustically isolated room, were naturally uttered by the speakers, without any machine mediation, and involved 61 different persons, 22 females and 39 males.

The dialogues were transcribed by annotating all extra-linguistic phenomena such as mispronunciations, restarts and human noises. The linguistic labeling includes the turn segmentation into semantic units and their labeling in terms of 16 DAs. The whole corpus was then divided into training and test sets. The latter consisted of 20 complete dialogues, for a total of 302 turns and 679 semantic units. The statistics of training and test sets, including the distributions of DAs, are reported in Table 1.

4. Dialogue History Integration

The classification based on SCTs takes into account only the text of the sentence. On the other hand, the dialogue history could be effectively used to disambiguate the correct DA in several situations.

The usefulness of the history in predicting the current DA can be evaluated from the graph presented in Figure 1 where the probability of finding the correct solution in the N-best hypotheses is depicted in two cases: the guess is only based on the frequencies of each DA in the training corpus or on the trigram statistics. It can be noted that the curve referring to this last case is higher than the first one, showing that trigrams might help in improving DA prediction.

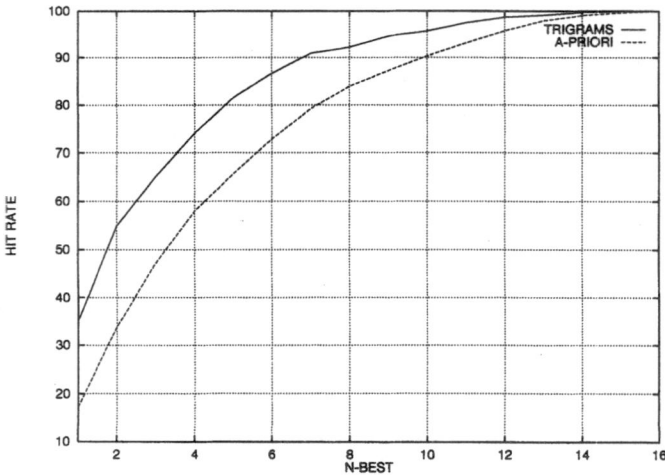

Fig. 1. Probability that the actual DA falls inside N-best, selected by using: (a) trigrams DA prediction, and (b) only a-priori DA probabilities

Summing up, two different conditional probability distributions are defined on the DA set: the first one, $\Pr(DA \mid \text{history})$ is identified by the sequence of previous DAs; the second one, $\Pr(DA \mid \text{words})$, only by the current word sequence. The former is the probability distribution given by trigrams of DAs; the latter was estimated by using SCTs. These distributions were combined by following the strategy depicted in Figure 2.

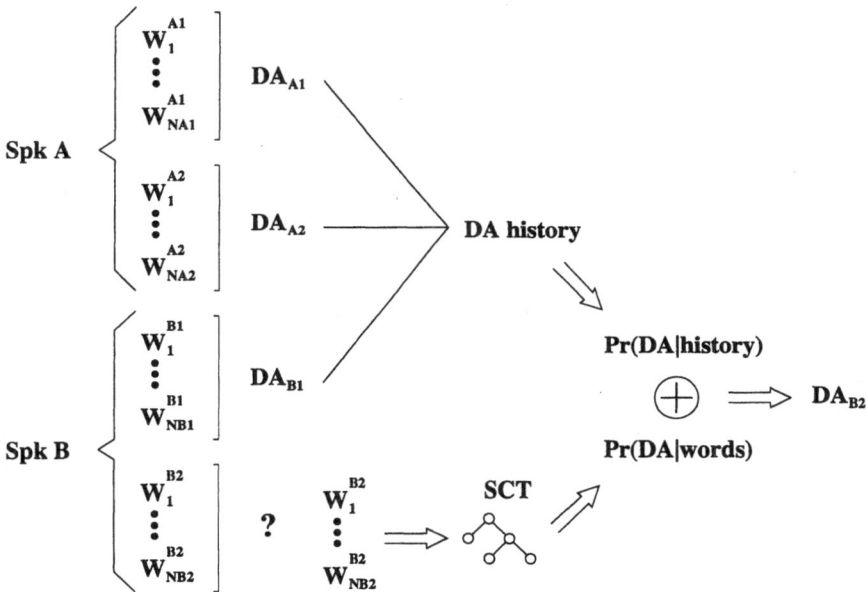

Fig. 2. Scheme for dialogue history integration into the SCT-based segment classification

Three different methods were tested to integrate the two probabilities. All of them assume that the two distributions are independent, which seems reasonable in the case considered.

In the first strategy, a linear combination of the two log-probabilities is used to evaluate the probability of each DA.[1] The DA having highest probability is then output. In the other two, a two-step approach is adopted: a restricted set of the most likely hypotheses is first determined on the basis of $\Pr(DA \mid \text{history})$ and then rescored by $\Pr(DA \mid \text{words})$. These last two strategies only differ for the criterion with which the size of the restricted set is chosen. In one case, referred to as N-best in Figure 3, the size of this set is defined independently from the input. On the contrary, in the second case this size is decided dynamically, and depends on

---

[1] From the probability theory point of view, the product of the two distributions should be divided by the a-priori probability of the classes, but it performed worse than the simple (weighted) product.

classifier	error rate	ΔE (wrt $\star$)
$\Pr(DA \mid \mathrm{w})$	29.0 % $\star$	-
$\Pr(DA \mid \mathrm{h})$	64.9 %	-
$\Pr(DA \mid \mathrm{w})^{\lambda} \cdot$ $\cdot \Pr(DA \mid \mathrm{h})^{1-\lambda}$	28.0 %	-3.6 %
N-best rescoring	29.0 %	0.0 %
δ-best rescoring	28.4 %	-2.0 %

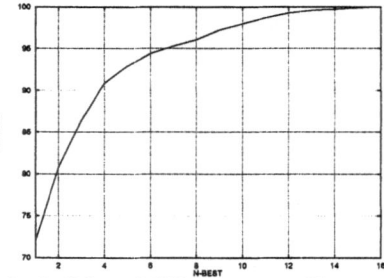

Fig. 3. On the left, classification results by using single KSs and different types of integration are reported. The graph on the right depicts the probability that the actual DA falls inside N-best obtained by integrating the history probability into the classification probability

the input: all and only the hypotheses whose probability difference from the best one is less than a threshold δ are chosen for rescoring. The parameters used in the integration, i.e. the linear combination factor, N and δ, were chosen using a trial-and-error approach.

In a more direct way, the last DAs can be added to SCTs possible questions, so that the classification is no longer based only on the text of the current sentence, but also on the last DAs. Preliminary experiments have also shown that it is convenient to join to each DA its relative position.

The integration between SCT and DA trigrams by the three strategies described above gave the experimental results reported in Figure 3, where the strategy having the best performance is the linear combination of the two (log) distributions. This strategy also turns out to be better than the direct integration in the SCT, where an error rate of 28.6% was obtained, which is slightly worse, even if comparable, with the lowest error rate obtained with trigrams.

The graph in Figure 3 plots the probability that the correct solution falls inside the N-best hypotheses obtained selected with the linear combination strategy.

5. Discussion

In evaluating the results presented, it should be mentioned that they are optimistic, as they are obtained by using the *true* history, while in reality only the *estimated* history is available, which may contain errors. Moreover, in these experiments, spoken words were used, instead of the recognizer output, so that even the baseline result has to be considered a lower bound for the expected error in real conditions.

In the VERBMOBIL project, a problem similar to that presented here is faced. In fact, a "shallow" translation is based on a set of DAs to be assigned to each input sentence. Interestingly, in [9] it is reported that the N-gram based approach outperformed a more complex dialogue model, based on a finite state automaton. The results presented on the prediction power of N-grams can be compared with those depicted in Figure 1. The adopted experimental set up is very similar to that considered here: 17 DAs, 41 dialogues for training, 81 for testing. The obtained accuracy is of 37.47% within the first best, 56.50% within the second best and 69.52% within the third best: in all cases slightly better, but very similar, to ours.

In [7, 8], two different approaches are compared, the first based on polygrams, the other on SCTs that differ from those used here just because they also use reciprocal position of keywords in the sentence. Experiments on the classification task give an error rate of about 40% on 19 DAs in both cases. On the contrary, in [10] only 8 DAs are considered, obtaining an error rate on the test set of 20.6% through a recurrent neural network.

Clearly, all the works presented are not precisely comparable, neither with ours or each other. In fact, the domain is not precisely the same, nor is the DA list. Moreover, the languages are different, and therefore, the corpora used for training and test. Nevertheless, the trend that can be extracted from the results and the conclusions are very similar.

The overall results of the approach are not really exciting and it seems to us that they are not going to drastically improve, at least as long as such model is maintained. Nevertheless, it is interesting to note that when the two or three best hypotheses are considered, the probability of hitting the right one is much higher, as showed in Figure 3. A convenient dialogue strategy can then disambiguate between such hypotheses.

References

[1] B. Angelini, M. Cettolo, A. Corazza, D. Falavigna, and G. Lazzari. Multilingual Person to Person Communication at IRST. In *Proc. of ICASSP*, Munich, Germany, 1997.

[2] L. Breiman, J. Friedman, R. Olshen, and C. Stone. *Classification and Regression Trees*. Wadsworth, Pacific Grove, Cal., 1984.

[3] M. Cettolo and A. Corazza. Automatic Detection of Semantic Boundaries. In *Proc. of the European Conference on Speech Communication and Technology*, Rhodes, Greece, Sept. 1997.

[4] M. Cettolo, A. Corazza, and R. De Mori. A Mixed Approach to Speech Understanding. In *Proc. of ICSLP*, Philadelphia, USA, 1996.

[5] R. De Mori and R. Kuhn. The Application of Semantic Classification Trees to Natural Language Understanding. *IEEE Transactions on Pattern Analysis and Machine Intelligence*, PAMI-17(5):449–460, May 1995.

[6] S. Gelfand, C. Ravishankar, and E. Delp. An Iterative Growing and Pruning Algorithm for Classification Tree Design. *IEEE Transactions on Pattern Analysis and Machine Intelligence*, 13(6):163–174, 1991.

[7] M. Mast, R. Kompe, S. Harbeck, A. Kiessling, H. Niemann, E. Noeth, E. Schukat-Talamazzini, and V. Warnke. Dialog Act Classification with the Help of Prosody. In *Proc. of ICSLP*, Philadelphia, USA, 1996.

[8] M. Mast, E. Noeth, H. Niemann, and E. Schukat-Talamazzini. Automatic Classification of Speech Acts with Semantic Classification Trees and Polygrams. In *Proc. of IJCAI, workshop "New Approaches to Learning for Natural Language Processing"*, pages 71–78, Montreal, Canada, 1995.

[9] N. Reithinger and E. Maier. Utilizing Statistical Dialogue Act Processing in Verbmobil. In *Proc. of Annual Meeting of the ACL*, 1995.

[10] S. Wermter and M. Loechel. Learning Dialog Act Processing. In *Proc. of the 16th International Conference on Computational Linguistics*, pages 740–745, Kopenhagen, Denmark, 1996.

Multilingual Speech Recognition

E. Nöth and S. Harbeck and H. Niemann

Lehrstuhl für Mustererkennung (Informatik 5)
Universität Erlangen–Nürnberg, Martensstr. 3, 91058 Erlangen, Germany
email: noeth@informatik.uni-erlangen.de

Summary. We present two concepts for systems with language identification in the context of multilingual information retrieval dialogs. The first one has an explicit module for language identification. It is based on training a common codebook for all the languages and integrating over the output probabilities of language specific n–gram models trained over the codebook sequences. The system can decide for one language either after a predefined time interval or if the difference between the probabilities of the languages succeeds a certain threshold. This approach allows to recognize languages that the system can not process and give out a prerecorded message in that language. In the second approach, the trained recognizers of the languages to be recognized, the lexicons, and the language models are combined to one multilingual recognizer. Only allowing transitions between the words from one language, each hypothesized word chain contains words from just one language and language identification is an implicit by-product of the speech recognizer. First results for both language identification approaches are presented.

1. Introduction

There has been a growing interest towards language identification with the transition of speech research from laboratory systems to real life applications: consider an automatic speech understanding system for information retrieval over the telephone that is installed in Germany and that is intended to be used by the majority of the population. It will either have to be able to handle German with a wide variety of foreign accents or be able to handle German, Turkish, Greek, Italian, etc. or exclude guest workers as customers. Things get worse if the system is intended for travel information and foreign tourists are its potential customers.

In this paper we present our approach to language identification in the context of the multilingual and multifunctional speech understanding and dialog system SQEL (Spoken Queries in European Languages). The system is being developed in the EC funded Copernicus project COP-1634. Partners are the Universities of Erlangen (Germany), Kosice (Slovak Republic), Ljubljana (Slovenia), and Pilsen (Czech Republic). The system is intended to handle questions about air flight (Slovenian system) and train connections (German, Slovak, and Czech system) in these four languages [1]. We verify our results on the SQEL corpus with two additional databases: the ATIS/EVAR [5] database with German and English sentences and the "Bundessprachenamt" corpus (BSPA) with 13 different languages.

Basis of the system is the EVAR system, the architecture of which is based on the German SUNDIAL demonstrator (ESPRIT project P 2218) [3]. Even though

---

[1] It will not be truly multifunctional in the sense that one can ask in one language questions about several applications and switch between applications during one dialog.

major changes were made – especially in the *Linguistic Analysis* [6] and the *Dialog* module [2] – the general architecture of the SUNDIAL demonstrator was kept for the EVAR system. EVAR can handle continuously spoken German inquiries about the German IC train system over the telephone.

The rest of the paper is organized as follows: In section 2. we will explain the architecture of the national SQEL demonstrators by looking at the current EVAR system. Following this, we will motivate and introduce two different system architectures for the two versions of the integrated multilingual SQEL demonstrator. The main difference is that the first architecture (section 3.1) has an explicit language identification module, whereas in the second architecture (section 3.2), the language identification is a by–product of the speech recognition process. Following this we will explain the principle of the explicit language identification in section 3.3. In section 4. we will present results and conclude with an outlook to future work in section 5..

2. Architecture of the National SQEL Demonstrators

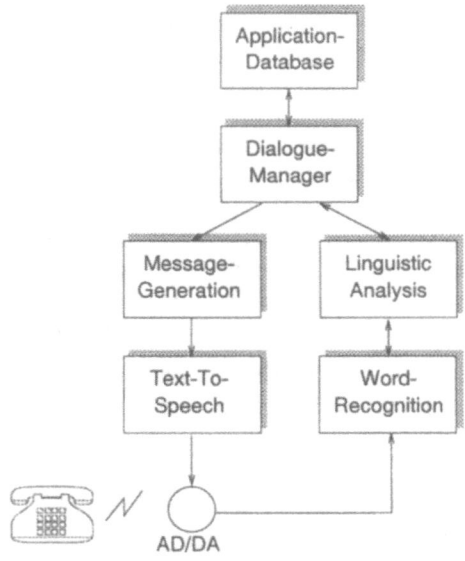

Fig. 1. Architecture of the German SUNDIAL demonstrator

Figure 1 shows a system overview of the German SUNDIAL demonstrator as well as of the EVAR system and the intended national SQEL demonstrators. Each of the four demonstrators will handle one language and one application. Here we describe the EVAR system, since an improved version of it will be the German SQEL demonstrator and since it is the only SQEL demonstrator that is already fully functional. The main components of the system are:

- *Word Recognition:* The acoustic front end processor takes the speech signal and converts it to a sequence of recognized words. Ideally the recognized words are the same as what was actually spoken. Using state–of–the–art technology, the word recognizer module performs the steps signal processing, feature extraction and a search based on hidden Markov models (HMM). Signal processing techniques include channel adaptation and sampling of the speech signal. The well known mel–cepstral features as well as the first derivatives are calculated every 10 msec. Using a codebook of prototypes the first recognition step is a vector quantization. These features are used in a beam search operating on semi–continuous HMMs. Output of the module is the best fitting word chain. A description of the word recognition module can be found in [7, 11].
- *Linguistic Analysis:* The word string is interpreted and a semantic representation of it is produced. A UCG (unification categorial grammar) approach [4, 1] is used, to model the user utterances. Partial descriptions are used, which leads to higher robustness w.r.t. ill-formed word sequences. The method of delivering partial interpretations is the key to enhanced robustness of the parser. A description of the linguistic analysis module can be found in [6].
- *Dialog Manager:* This module takes the semantic representation of the user utterance and performs the interpretation within the current dialog context. It decides upon the next system utterance. Specialized modules within the dialog manager for contextual interpretation, task management, dialog control and message generation are communicating via a message passing method. A description of the dialog manager can be found in [2].
- *Application Database:* The official German InterCity train timetable database is used. Ljubljana will use the Adria Airline database, Pilsen and Kosice will use the Czech and Slovak InterCity train timetable database.
- *Message Generation* and *Text–to–Speech:* In order to have a complete dialog system this module transforms the textual representation of the system utterance into sound. We use a simple concatenation of canned speech signals (All words that the system can say are recorded and stored as individual files).

In the next section we will describe the planned adaptation steps to build an integrated demonstrator that will be able to handle dialogs in all four languages.

3. Language Identification with Different Amounts of Knowledge about the Training Data

Of course, the best language identification module is a multilingual recognizer. In speech recognition this can be implemented in the following way: starting with the speech signal, run several recognizers in parallel. Each recognizer is specialized to one language, i.e. has an acoustic and a language model of one language. Then for each given point in time, one can identify the spoken language, based on the score (probability) for the best matching word chain in each of the recognizers. However, in this case the recognizers have to give comparable judgements. Also, if the system has to recognize N languages then N recognizers have to run in parallel, and $N-1$

recognizers do work that is unnecessary for the system. Another problem with this approach is that you can only recognize these N languages.

Consider the situation that you want the SQEL system to be able to identify more than the four languages and react appropriately if a question is uttered in a language that can not be handled by the system. For instance, if the system identifies that an utterance was uttered in Polish, it can react with a prerecorded Polish utterance like

> The SQEL system detected a Polish utterance. Unfortunately, so far the system can only handle dialogs in Czech, German, Slovak, and Slovenian. Please ask your question again in one of these languages.

Clearly, the language identification module will not have the same quality of training data for additional languages. We might only have Polish speech samples where we know the language, but not what was said. Also, the samples might be from a very different domain, and the other necessary resources (pronunciation lexicon, stochastic language models) might not be available.

Our strategy for integrating the national demonstrators into one system is twofold:

- Build a system with explicit language identification. The only label of the training data for the language identification is the spoken language. The topic or the spoken words of the training utterance will not be known. We will describe the architecture of this system in section 3.1.
- Develop a multilingual recognizer for the N languages. In this case the same amount of labeled training data and resources (pronunciation lexicon, stochastic language models) has to be available for the languages to be identified as for the languages to be recognized. The language identification is done implicitly during the decoding of the utterance. We will describe the architecture of this system in section 3.2.

3.1 A System with Explicit Language Identification

Figure 2 shows a system overview of the intended final SQEL demonstrator with explicit language identification. As can be seen, the major changes affect the word recognition module and the information flow between the modules. Since we plan to use as many software modules as possible from the EVAR system, many of the internal changes can be implemented via switches for language specific resource files. To do this, the modules have to have a control channel in addition to the existing data channel. The control channel will be used to pass messages like identity of the language and current application. The four–way arrows in Figure 2 indicate switches, the double arrows indicate data flow and the single arrows indicate control flow. The *signal processing* can be done independent of the language. The next steps — vector quantization and HMM search — need language dependent data. What is needed are language dependent codebooks, lexicons and stochastic language models. If the module has information about what language was uttered, it can simply

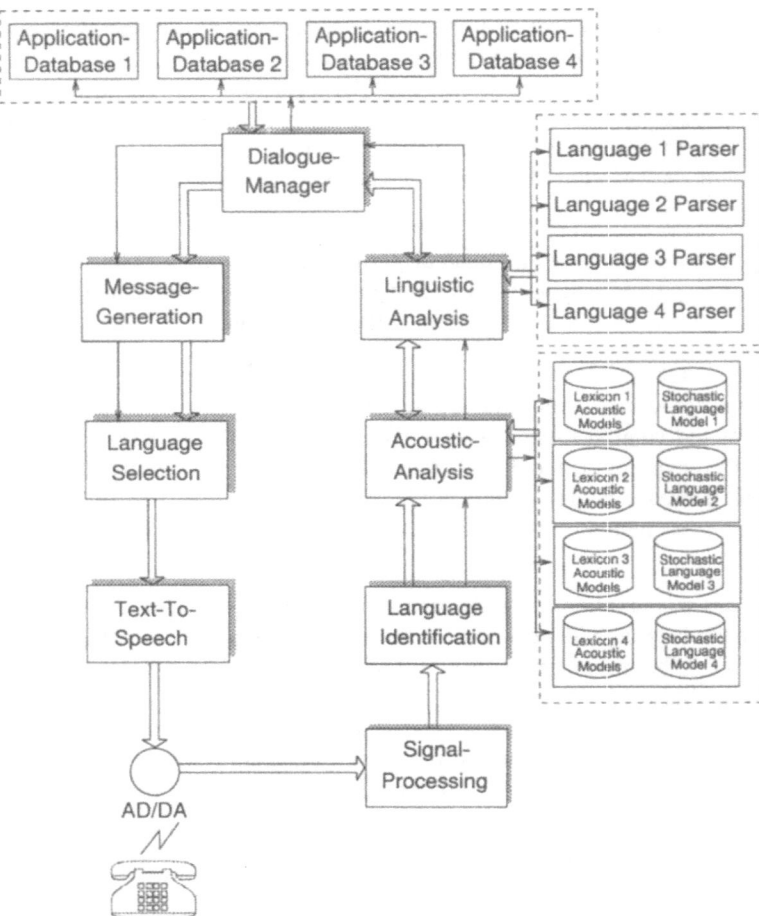

Fig. 2. Architecture of the SQEL demonstrator with an explicit language identification module

switch to the resource files of the right language. Therefore a *language identification* module has to be added to the system that has to identify the language and pass a message to the remaining modules. The module will be activated at the beginning of the dialog. To save computation time, we use the same mel–cepstral features as the recognition module. After a certain time interval a classification step between the N languages is performed. The application requires a decision after only a few seconds, because users tend to make short queries [2]. An overview of algorithms for language identification is given in [9, 13]).

During the analysis of further user utterances the language identification module simply passes on the extracted feature vectors and causes no delay.

Clearly, there is a tradeoff between recognition accuracy and delay time for the task of language identification: The longer the utterance, for which the sequence of feature vectors is computed, the more language specific sounds have been uttered by the caller and the better the automatic language identification will be. On the other hand, the recognition has to wait for the language identification decision, before it can start. As mentioned above, it is not clear yet, how long the utterance has to be for languages as close as Czech and Slovak, in order to be able to classify the language at an acceptable rate. This leads us to an alternative approach presented in the next section.

3.2 A System with Implicit Language Identification

We want to use all knowledge sources that are available as early as possible, i.e. apply n speech recognizers for the language identification process. To reduce the computational load mentioned above, we build a recognizer that contains all words from all languages in its dictionary. By using a stochastic bigram language model that only allows transitions between words within one language, each hypothesized word chain will only contain words of one language.

The basis for our multilingual speech recognition system are monolingual speech recognizers. We use semi-continuous HMMs for acoustic and bigrams for linguistic modeling. The monolingual recognizers are trained in the ISADORA [10] environment which uses polyphones with maximum context as subword units. The construction of the multilingual speech recognizer is as follows:

1. Increase the number of codebook density functions to reflect the language de-pendent codebooks. For example when having two different languages with a codebook of 256 density functions per language, then the multilingual recog-nizer will have 512 density functions.
2. Add special weight coefficients to the HMM output density functions to reflect the increased number of available density functions. The new weight coeffi-cients are set zero, so that every density function belonging to different lan-guages has no influence on the output probability of the HMM.
3. Construct a special bigram model which consists of the monolingual bigrams and does not allow any transitions between the languages as shown in equa-tion 1.

$$P(\text{word}_{language_i}|\text{word}_{language_j}) = 0 \quad \text{for } i \neq j. \tag{1}$$

Figure 3 shows the alternative system architecture. One might argue that this approach will slow down the recognition, just like running N smaller recognizers in parallel, since we quadruple the lexicon. At the beginning of the recognition process every word of the multilingual vocabulary is possible, so that there are a lot of different search paths. After a few seconds the most probable paths will be in the correct language. The acoustic models of the other languages should result in paths with lower scores. The beam search algorithm [8] is used to restrict the search space to paths through the word hypotheses graph which contain more reliable hypotheses. Experiments showed that this suboptimal search strategy has no bad effects on the word recognition rate. So using the beam search strategy in forward decoding only paths of the correct language should be expanded. After a few words it should be as

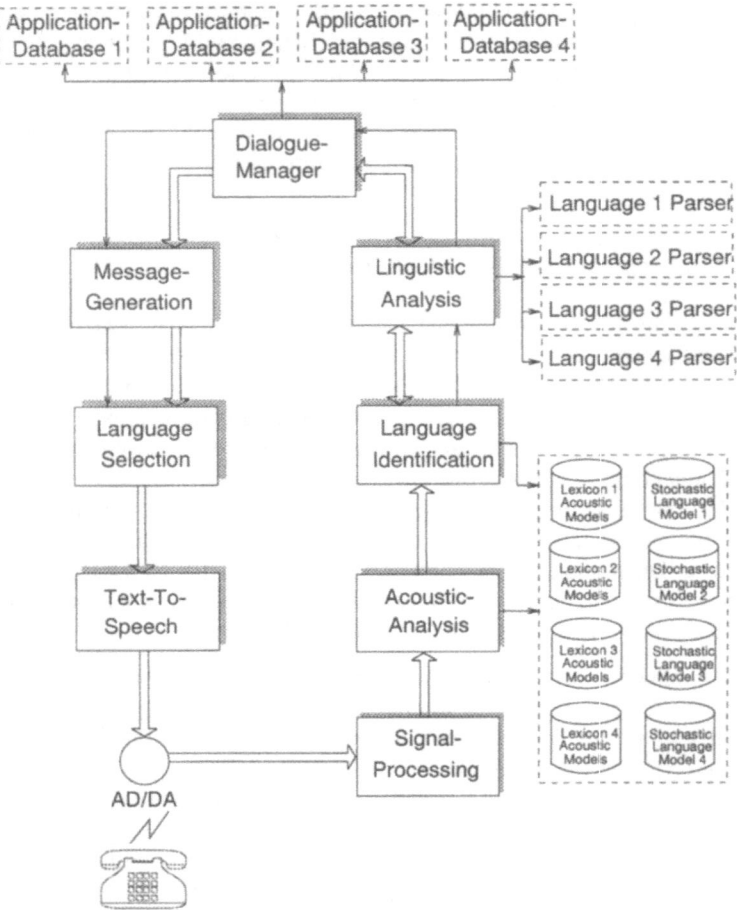

Fig. 3. Architecture of the SQEL demonstrator with implicit language identification

fast as the monolingual speech recognition system. In Figure 4 the number of states inside the beam for the multilingual and the monolingual speech recognizers are compared for one English sentence. At the beginning of the sentence all available languages are possible. Therefore, the number of states is significantly higher than in the monolingual case. After a short time (less than 2 seconds) all states of the wrong language are pruned and the number of states inside the beam is the same as the one for the monolingual recognizer. Therefore the increase in computational load is negligible while for running N recognizers in parallel it increases by more than N (see section 4.).

In the next section we will describe the language identification module that will be used in the first system architecture. The implementation of the implicit language

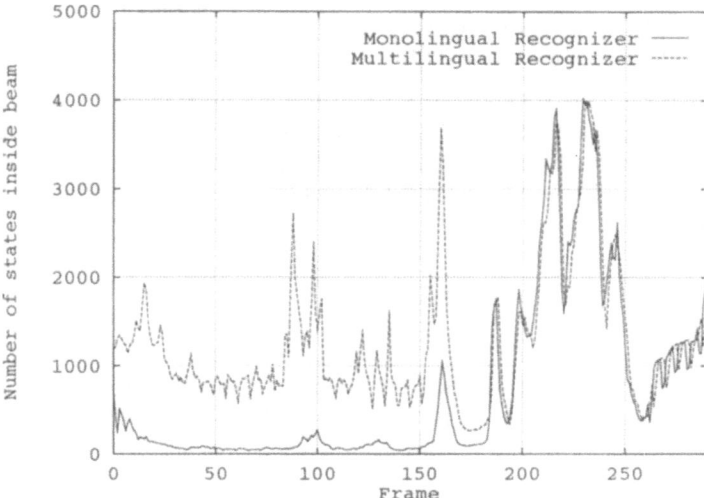

Fig. 4. The number of states inside the English monolingual and the German/English multilingual speech recognizer evaluated on an English sentence. At the beginning of the sentence the multilingual recognizer has much more states inside the beam because all languages are still possible. After 200 frames (2 seconds) all states of German models are pruned and the number of states of both recognizers are the same.

identification for the second system is straight forward and we will not further elaborate on it.

3.3 Language Identification Based on Cepstral Feature Vectors

We want to build a module that only knows the identity of the training utterances, because we want to train additional languages. In order to be as efficient as possible, we want to use as many processing steps of our speech recognition system as possible. The following steps are performed:

- Extract the same mel-cepstral features and derivatives as for the recognition task. Thus after the identification no new feature extraction is necessary.
- Take an appropriate subset of the features. The lower cepstral coefficients are more sound specific, i.e. language specific, whereas the higher coefficients are more speaker specific. In preliminary experiments [12] good results were achieved with using the first six mel–cepstral coefficients from three consecutive frames resulting in a feature vector of length 18.
- Train a vector quantizer with the training data from all the languages together. Output of the vector quantizer is a sequence of indices, i.e. we use a hard vector quantizer.
- Train N n–gram language models over the symbol sequences for the N languages.
- To identify the language, calculate the sequence of vector quantizer symbols and calculate the N n–gram probabilities in parallel. For each language sum up over

the sequence of the negative logarithms of the n–gram probabilities. At any time the algorithm can then decide for the most likely language.

Note that for large values of N a beam search can be used, i.e. after a certain interval, languages that are below a certain threshold, are discarded. Also, the module can decide for the language with the highest probability either after a fixed time interval or if the difference between the best and the second best alternative exceeds a certain threshold.

4. Results

In this chapter we want to present some preliminary results of explicit and implicit language identification experiments based on the SQEL , the ATIS and the BSPA corpus.

Results for Explicit Language Identification

For the explicit language identification we tried to recognize the three Slavic SQEL languages Czech, Slovenian and Slovak. We trained the quantizer and the three n–gram language models with 4 hours of speech (1.7 hours from 42 Slovenian, 1.5 hours from 30 Slovak, and 1 hour from 23 Czech speakers).

Line "w/o LDA" in Table 1 shows recognition rates of the explicit language identification module for 17 minutes from 9 independent test speakers (4 Slovenian, 3 Slovak, and 2 Czech speakers). We achieve even better results when using a transformation of the feature space with the *Linear Discriminant Analysis* (Line "with LDA" in Table 1)

Table 1. Recognition rate for explicit language identification between three languages with and without LDA feature transformation. Forced decision after 2 seconds (or at the end of the utterance, if it is shorter than 2 seconds).

	rec. rate for Czech	rec. rate for Slovenian	rec. rate for Slovak
w/o LDA	91.47 %	98.67 %	93.65 %
with LDA	96.76 %	97.95 %	98.67 %

Considering the small amount of training data, the similarity of these three languages and the time to decide, these results are very encouraging. However, it should be kept in mind, that so far we used high quality speech input and that the speech material is read speech from a restricted domain. Nevertheless, at least the restricted domain is realistic for an application in a human machine dialog system.

To prove our results even on noisy and inhomogeneous data we evaluated our algorithm on a different corpus, which was collected by the Federal Institute for Language Engineering (Bundessprachenamt) in Germany. This corpus was collected via television and radio and contains 13 different languages (German, English, Arabian,

Chinese, Italian, Dutch, Polish, Portuguese, Rumanian, Russian, Swedish, Spanish and Hungarian). Contrary to the SQEL database the recording conditions are very variable, the domain is not restricted and spontaneous and read speech are mixed. To keep the amount of training data constant for each language we use half an hour of speech per language and the rest of data for testing (between 10 minutes and two hours, 30 minutes in the average). Therefore the recognition rates given in Table 2 are class wise averaged. The perplexity of the BSPA corpus was much higher than in the restricted domain of the SQEL corpus, leading to a higher number of different observed n–grams. Nevertheless Table 2 shows that the approach works even under worse conditions, for example a class wise averaged recognition rate of 76 percent is achieved when 60 seconds segments are classified. Nevertheless the results show that when the number of languages or the complexity of the task which is handled by the system increases the necessary time to identify the language is not acceptable for the overall dialogue system. This leads us to the approach with implicit identification which is evaluated in the next section.

Table 2. Class wise recognition rates for explicit language identification between 13 language on the BSPA corpus for different segment lengths

rec. rate for 2s	rec. rate for 10s	rec. rate for 30s	rec. rate for 60s
34.8 %	55.7 %	68.5 %	76.0 %

Results for Implicit Language Identification

We tested our multilingual speech recognizer approach on two different databases. The first database is the Slovak and Slovenian corpus of the SQEL project, the second contains spontaneous corpora for German (EVAR) and English (ATIS) (see also Table 3).

Table 3. Description of training and test sets used for the multilingual speech recognizer

Language	Amount of train- ing data	Number of speakers (calls)	Amount of test data	Number of speakers (calls)	Level of spontaneity
Slovak	4.5 h	30	40 min	4	read
Slovenian	4.5 h	42	40 min	6	read
German	7 h	804	1 h	234	spontaneous
English	7 h	46	2 h	30	spontaneous

In our first experiment we took the baseline system as described in section 3.2 and evaluated it on the SQEL database (see Table 4, row multilingual). The word accuracy of the Slovenian sentences is the same as in the monolingual recognizer, but the accuracy for Slovak sentences decreased by 60 percent. The problem is that the beam search often canceled all states of the correct language after a short time in almost all Slovak sentences. Once there are no Slovak states inside the beam, the recognizer cannot return to the Slovak language. When using two different beams inside the forward decoding, one very big beam for the first part of an utterance and one normal one for the rest of the utterance, we increased the recognition rate of Slovak with the side effect of higher computation time.

Row "multilingual with multilingual silence" in Table 4 and Table 5 show the effect of using a multilingual silence category. Instead of using different silence models for each language all silence models for all languages are in one common category. This method allows transitions between the languages by using a silence model during decoding. The word accuracy using the multilingual word recognizer was almost as good as using the correct monolingual recognizers in both databases. Additionally the computation time was almost as good as using the correct monolingual recognizer, whereas the time of a recognizer evaluated on an out-of-language speech signal increases drastically: if one speaks a sentence in a foreign language into an automatic speech recognition system, the recognition time generally increases significantly, because nothing matches well and thus the dynamically adapted beam width [7, p. 120] goes up. Table 5 shows the computation time for the monolingual and the multilingual recognizers.

Table 4. Recognition rates for the multilingual word recognizer and the monolingual recognizers in the languages Slovak and Slovenian

Recognition rates (word accuracy)		
Monolingual Slovenian	91 %	
Monolingual Slovak		86 %
Multilingual	91 %	29 %
Multilingual with multilingual silence	90 %	87 %

Table 5. Computation time for the multilingual word recognizer and the monolingual recognizers in the languages Slovak and Slovenian

Computation time with multilingual silence		
Recognizer	Slovenian	Slovak
Monolingual Slovenian	20 min	1 h
Monolingual Slovak	1 h	20 min
Multilingual	20 min	20 min

Table 6 shows the same results hold when running a bilingual German/English recognizer. Again the word accuracy stayed practically the same.

Table 6. Recognition rates for the multilingual and the monolingual recognizers on the ATIS/EVAR task

Recognizer	WA on English	WA on German
Monolingual	63 %	71 %
Multilingual	64 %	71 %

5. Conclusions and Future Work

We presented two concepts for systems with language identification in the context of multilingual information retrieval dialogs. The first architecture is a straightforward integration of an explicit language identification module. It has the advantage of being able to recognize languages that can not be processed by the system and allows an appropriate reaction. It has the disadvantage of delaying the recognition process until the spoken language can be identified with a high accuracy. The alternative approach is to combine the monolingual recognizers to one recognizer. By forcing word transitions to stay within one language, the system identifies the language and decodes the utterance simultaneously. Since the beam search eliminates partial hypotheses with bad scores, the size of the search space approaches that of the monolingual recognizers. Thus, the delay caused by increased vocabulary size is small. The approach utilizes the available speech data more efficiently than the explicit language identification, but can not identify additional languages.

For the explicit identification preliminary experiments with the three Slavic SQEL languages were presented that showed that the language can be identified with high accuracy after only two seconds.

For the implicit identification we presented first results with a bilingual recognizer for Slovenian and Slovak and for English and German, indicating that the combined system can achieve the same recognition rates on both languages as the two monolingual recognizers. The time behavior also stayed the same as for the monolingual recognizers whereas the time behavior of the monolingual recognizers on the wrong language showed that our approach is superior to running N recognizers in parallel for language identification purposes.

In the future we plan to extend the number of different languages inside the multilingual recognizer. Because with an increasing number of languages the number of output probabilities is growing, we want to examine an approach of sharing the same codebook or the same acoustic models for subword modeling.

Acknowledgement. This work was partly funded by the European Community in the framework of the SQEL–Project (Spoken Queries in European Languages), Copernicus Project No. 1634. The responsibility for the contents lies with the authors.

References

[1] F. Andry, N. Fraser, S. McGlashan, S. Thornton, and N. Youd. Making DATR Work for Speech: Lexicon Compilation in SUNDIAL. *Computational Linguistics*, 18(3):245–267, Sept. 1992.

[2] W. Eckert. *Gesprochener Mensch–Maschine–Dialog.* Berichte aus der Informatik. Shaker Verlag, Aachen, 1996.

[3] W. Eckert, T. Kuhn, H. Niemann, S. Rieck, A. Scheuer, and E. G. Schukat-Talamazzini. A Spoken Dialogue System for German Intercity Train Timetable Inquiries. In *Proc. European Conf. on Speech Communication and Technology*, pages 1871–1874, Berlin, 1993.

[4] R. Evans and G. G. (eds.). The DATR Papers: February 1990. Technical report, Cognitive Science Research Paper CSRP 139, University of Sussex, Brighton, 1990.

[5] C. H. J. Godfrey and G. Doddington. The ATIS Spoken Language Systems Pilot Corpus. In *Speech and Natural Language Workshop*, pages 96–101. Morgan Kaufmann, Hidden Valley, Pennsylvania, 1990.

[6] G. Hanrieder. *Inkrementelles Parsing gesprochener Sprache mit einer linksassoziativen Unifikationsgrammatik.* PhD thesis, Universität Erlangen–Nürnberg, 1996.

[7] T. Kuhn. *Die Erkennungsphase in einem Dialogsystem,* volume 80 of *Dissertationen zur Künstlichen Intelligenz.* Infix, St. Augustin, 1995.

[8] B. Lowerre and D. Reddy. The Harpy Speech Understanding System. In W. Lea, editor, *Trends in Speech Recognition*, pages 340–360. Prentice–Hall Inc., Englewood Cliffs, New Jersey, 1980.

[9] Y. K. Muthusamy, E. Barnard, and R. A. Cole. Reviewing automatic language identification. *IEEE SIGNAL PROCESSING MAGAZINE*, pages 33 – 41, Oktober 1994.

[10] E. Schukat-Talamazzini, T. Kuhn, and H. Niemann. Speech Recognition for Spoken Dialogue Systems. In H. Niemann, R. De Mori, and G. Hanrieder, editors, *Progress and Prospects of Speech Research and Technology: Proc. of the CRIM / FORWISS Workshop*, PAI 1, pages 110–120, Sankt Augustin, 1994. Infix.

[11] E. G. Schukat-Talamazzini. *Automatische Spracherkennung – Grundlagen, statistische Modelle und effiziente Algorithmen.* Vieweg, Braunschweig, 1995.

[12] V. Warnke. Landessprachenklassifikation. Studienarbeit, Lehrstuhl für Mustererkennung (Informatik 5), Universität Erlangen–Nürnberg, 1995.

[13] M. Zissman. Comparison of four approaches to automatic language identification of telephone speech. *IEEE Trans. on Acoustics, Speech and Signal Processing*, 4:31–44, 1996.

Toward ALISP: A proposal for Automatic Language Independent Speech Processing.

Gérard Chollet[1], Jan Černocký[2], Andrei Constantinescu[1,5], Sabine Deligne[1,3] and Frédéric Bimbot[1,4]

[1] CNRS URA-820, ENST, Signal Department, 46 rue Barrault, 75634 PARIS cedex 13
[2] Technical University of Brno, Institute of Radioelectronics, (CZECH REPUBLIC)
[3] now with IBM, T.J. Watson Research Center (Yorktown Heights, USA)
[4] IRISA (FRANCE)
[5] On the leave from IDIAP to ASCOM HASLER (SWITZERLAND)

Summary. The models used in current automatic speech recognition (or synthesis) systems are generally relying on a representation based on phonetic symbols. The phonetic transcription of a word can be seen as an intermediate representation between the acoustic and the linguistic levels, but the *a priori* choice of phonemes (or phone-like units) can be questioned, as probably non-optimal. Moreover, the phonetic representation has the drawback of being strongly language-dependent, which partly prevents reusability of acoustic resources across languages. In this article, we expose and develop the concept of ALISP (Automatic Language Independent Speech Processing), namely a general methodology which consists in inferring the intermediate representation between the acoustic and the linguistic levels, from speech and linguistic data rather than from a priori knowledge, with as little supervision as possible. We expose the benefits that can be expected from developing the ALISP approach, together with the key issues to be solved. We also present preliminary experiments that can be viewed as first steps towards the ALISP goal.

1. Introduction

Linguists claim that human language is articulated into multiple levels of elementary units, which are themselves organised in systems. Sentences are made of words, which belong to different syntactic classes, words can be decomposed into morphemes of different kinds, morphemes themselves are articulated into phonemes, characterised by various features... The description of the phonological, morphological and lexical systems, for a given language is usually based on an expert analysis of the linguistic material, which requires a certain number of *a priori* hypotheses on the nature of the relevant units. However, most people speak without knowing what is a phoneme or a morpheme, which tends to show that a representation of the speech structure can be inferred from linguistic data, without requiring specific knowledge in linguistics. It is thus interesting to investigate on means that could be used to let a representation emerge from speech data, rather than basing it on the traditional linguistic description.

The concept of ALISP, Automatic Language Independent Speech Processing, covers a general methodology that is directed towards the description of linguistic material with as little supervision and *a priori* knowledge as possible, using instead some general constraints that are felt to be underlying the human communication process. In particular, it is hypothesized that the linguistic code is the result of a

compromise between an economy principle (a law of "least effort") and the necessity to maintain sufficient discrimination between distinctive elements used for
communication. Here, concepts from Information Theory and techniques for Data
Mining may, among others, have a role to play, for expressing the constraints and
for inferring representations that satisfy them.

This paper does not provide any general solution to ALISP. It is mainly aimed at
identifying the key issues it involves. In section 2. we discuss the practical benefits
than can be expected from ALISP. Then, we identify, in section 3., the main difficulties that prevent a straightforward solution to the problem. We describe in sections
4. and 5. a few tools and preliminary experiments that can be viewed as first steps
towards ALISP.

2. Practical benefit of ALISP

Current speech technology is based on a description of speech in terms of phonetic
elements. As a result, it is highly supervised, relying on the massive use of hand-
processed data, and also highly dependent on the target language. A significant part
of the resources deployed to build a system for a given language are not reusable
for another language. For instance, the design of a speech recognizer requires large
amount of *annotated* speech database, and it is usually necessary to resort to a significant quantity of phonetically labelled data, in order to initialise the acoustic models. Moreover the quality of the recognition lexicon depends on the availability of
an exhaustive phonetic dictionary covering most pronunciation variants. Assuming
this resource is available, the use of phonetic symbols makes it difficult to consistently describe the acoustic variability behind the various pronunciation variants. In
a similar way, speech synthesis systems are based on speech units that result from a
fragile compromise between linguistic knowledge, ad hoc definitions and arbitrary
choices. In this case again, the speech segments have to be carefully recorded and
manually segmented from a well-designed speech database. In the context of speech
coding, the lack of a robust symbolic representation of speech makes it difficult to
reach very low bit rates for compression.

One of the concepts behind ALISP is that there exists some structure common
to all languages of the world, at least because they are produced with a similar vocal
apparatus, and that it could be used for a language independent representation of
speech. A first step in that direction is to design a language-independent process for
inferring a language-dependent representation of speech from speech data, with as
little supervision and annotation as possible. While classical approaches to speech
recognition tend to optimize the joint probability $\mathcal{P}(O, W)$, where O and W are the
acoustic observations and the linguistic units, and where the phonetic representation
is *a priori* considered as optimal, the idea behind ALISP is that an intermediate
representation R should be inferred from the data, so as to optimize $\mathcal{P}(O, W|R)$.
Moreover, the intermediate representation could be (at least in a first step) inferred
so as to maximize $\mathcal{P}(O|R)$ under a constraint of parsimony, i.e. without requiring
labelled data.

In the context of ALISP, a key resource is therefore a large database of speech signal in the target language. For this purpose, a large quantity of unlabelled speech data can easily be obtained from radio or TV broadcasts, telephone conversations, CD-ROMs, etc... They are available for many speakers, in many languages and various audio quality. The cost for collecting them is negligible in front of the cost that would be needed to collect and annotate specific data. From a large database, a fully automatic procedure is used in order to derive a representation of the speech signal for that language, that satisfies a compromise between the accuracy of the description of the acoustic observations and the economy of the representation. This intermediate representation has then to be mapped to the word level: only this step may require the use of a set of annotated data and the help of general phonetic knowledge. But the phonetic representation should guide the mapping rather than be the intermediate representation between the acoustic and the word levels. In summary, most of the structure and of the representation of speech model are inferred from the speech signal, whereas the mapping with the higher levels take place in a second step.

3. Issues specific to ALISP

3.1 Selecting features

In the framework of ALISP, where the definition of the speech units is based on the detection of regularities in the speech signal, special attention should be paid to the selection of highly informative feature vectors. It is all the more essential as it will condition the ability to identify patterns in the speech signal, as well as the nature of these. Besides, it is desirable to limit the dimensionality of the observation space in order to keep the search algorithms manageable when trying to detect typical trajectories in this space. As a result, ALISP urges to find concise parameterization schemes, capable of expressing the redundancy of the speech signal with a minimum number of parameters. To this respect, the extraction of segmental features, instead of frame-based features, seems a relevant direction. It allows indeed a more economical feature representation, where each segmental feature vector is a sort of "summary" of the information conveyed in a (well chosen) segment of the original features. We give an example of such a parameterization scheme in section 4.1.

3.2 Modeling speech units

The choice of an optimal model for modeling speech units depends on the particular feature vectors used, but also on the speech units to be represented. Hidden Markov Models (HMMs) are most often used as they proved to be efficient to describe phonemes for instance. However, HMM suffer some well known shortcomings, and the formulation of alternative models is an active field of research. In the framework of ALISP, the question of choosing an optimal model is more complex and becomes a circular problem, because the speech units are not known in advance, and because their derivation is likely to be dependent on the choice of the model. The choice of the unit model does thus also conditions the emergence of "well-behaved" units. Consequently, ALISP should benefit from the formulation of models overcoming some of the HMM limitations. Segment models [17] for instance aim at relaxing the basic HMM assumption according to which state-bound observations are independent and identically distributed. Though theoretically more attractive than conventional HMM, they have not produced a major breakthrough in terms of speech recognition performances. However it might be speculated that the gain from relaxing inadequate HMM constraints will better benefit a framework where it does not only profit the modeling of already pre-defined units, but where it also profits the definition of these units. Work in this direction, where the derivation of acoustic units is combined with segmental modeling, is reported in [3].

3.3 Defining a derivation criterion

The extent to which acoustically derived subword units will evidence more consistency than linguistically defined ones is conditioned to the formulation of a proper derivation criterion. Large vocabulary recognition systems need to rely on subword units, because it allows to extend the vocabulary without having to collect additional speech data. On the other hand, subword units are highly sensitive to coarticulation, so that they are usually defined as context-dependent phones. However it is still unclear which context information should be maintained for each phone, as the extent to which it is affected by contextual variability also depends on the pronunciation variant. Besides, increasing the set of pronunciation variants in all contexts may hurt performances as it also increases the confusability. An issue for ALISP is thus to express these requirements of description accuracy and of non-confusability with an objective criterion. Some clues to this problem may be found in the field of Information Theory, especially in Minimum Description Length approaches [19], where the above compromise can be stated as the problem of finding the best balance between the accuracy of the model and its complexity, measured as its description length. A way to relate the derivation of the units with their ability to serve as recognition units, is to use a Minimum Classification Error criterion, along the lines described in [15].

3.4 Building a lexicon

Major sources of difficulty with building a lexicon based on acoustic units are, first, that there are no one-to-one correspondences between these and the linguistic representation, and second, that the boundaries of the acoustic units do not necessarily coincide with the boundaries of the words. A straightforward way to elude the second difficulty is to constrain the alignment of the acoustic units on the speech data. However it is likely to limit a potential advantage of data-driven units, which is to provide a natural way of modeling cross-word coarticulation. We review possible ways to address these difficulties, depending on whether the acoustic units are mapped to phonemes, or to lexicon entries.

In this first scenario, the lexicon entries are still built based on a phonetic description. Acoustic units need thus to be mapped to phonemes, which can be done by analyzing the confusion matrix obtained from an unconstrained alignment of phonetically labelled data [22], or by introducing and evaluating inter-model distances [1] [16]. A drawback of this approach is to require the labelling of a still undetermined amount of speech data. Also, the need for language dependent phonetic dictionaries remains. On the other hand, it does not require that the boundaries of the acoustic units coincide with the boundaries of the words. Constraints can be placed instead on the concatenation of the decoded phonemes to ensure that the decoded string of phonemes can still be read as a succession of lexicon entries (like the experiments reported in section 5.3 for instance). Alternatively, lexicon entries can be recovered by correcting the phoneme insertions, deletions and substitutions.

In this second scenario, lexicon entries are described as combinations of acoustic units, so that the need of a phonetic dictionary is suppressed. Lexicon entries are transcribed as sequences of acoustic units. Then, a few pronunciation variants are selected for each entry by clustering its various transcriptions, leading to a deterministic type of lexicon. The clustering may be optimized using a discriminative criterion based on the mutual information across the transcriptions [10]. Alternatively, a stochastic automaton modeling the pronunciations variants for each entry can be inferred from the transcriptions, resulting in a statistical lexicon [18]. A limitation of this scenario is that the boundaries of the acoustic units need to coincide with the boundaries of the lexicon entries. This limitation can be at least partially released by a judicious choice of the lexicon entries. These may include some multi-words so that the acoustic units can span over their inner word boundaries. An issue is then to select only the multi-words expected to be most frequently affected by cross-word coarticulation [3], since maintaining all multi-words in the lexicon would increase too much the cost of the search. More generally, the very notion of what a lexicon entry should be might need to be reconsidered.

It is far from obvious which procedure can best interface the acoustically derived representation of speech with its linguistic representation. However, this task is all the more challenging as it is quite likely to condition the extent to which acoustically derived units can help reducing the gap between the performances of word based and subword based speech recognition.

4. Some tools for ALISP

In the experiments described in section 5., we use algorithmic tools, two of which are briefly described in this section.

4.1 Temporal Decomposition

Temporal Decomposition (TD) is a model of spectral evolution introduced in [2], which describes a speech segment as a linear combination of a limited set of vectors called targets. The temporal contribution of each target is expressed by an interpolation function (see Figure 1). As spectral characterizations of acoustic events, the target vectors are expected to show less variability than the original frames while

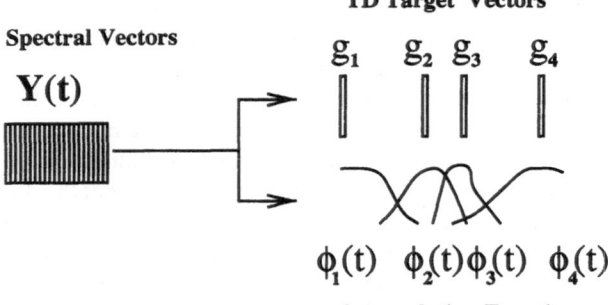

TD Target Vectors

Spectral Vectors

$g_1 \quad g_2 \; g_3 \quad g_4$

$Y(t)$

$\phi_1(t) \quad \phi_2(t)\phi_3(t) \quad \phi_4(t)$

Interpolation Functions

Fig. 1. *A graphic illustration of temporal decomposition (TD)*

requiring a reduced number of parameters. TD can be understood as a non-uniform subsampling scheme that captures the most significant spectral events in the acoustic signal. As it maps overlapping segments of frame based feature vectors to target vectors, it allows the extraction of segmental features.

4.2 The multigram model

The multigram model was originally developed in order to model variable-length regularities within streams of symbols [4] [11] : hence the name *multigram*, as opposed to n-grams, for which the dependencies are supposed to be of fixed length. It can be understood as a production model, where a source emits a string Z of units, called multigrams, drawn from a limited set $\{z_i\}$. Each multigram gives rise to a variable-length sequence of elementary observations. We assume that, in practice, the only observable output of this process is the string of observations O, resulting from the concatenation of all sequences. Given an observed string O, the goal is to retrieve the set of distinct underlying multigrams $\{z_i\}$, and to identify in O the observation sequences originating from a common multigram, which involves finding a segmentation S of O, as illustrated on Figure 1. In the framework of ALISP, the set

Table 1. *Inference of the underlying multigram structure,* (S, Z), *of an observed string of data,* O

Z:	$z_{(i_1)}$	$z_{(i_2)}$	$z_{(i_3)}$ $\cdots$
	$\Uparrow$	$\Uparrow$	$\Uparrow$
S:	$[\, o_{(1)}\; o_{(2)} \,] \oplus$	$[\, o_{(3)} \,] \oplus$	$[\, o_{(4)}\; o_{(5)}\; o_{(6)} \,] \cdots$
		$\Uparrow$	
O:	$o_{(1)}\; o_{(2)}\; o_{(3)}\; o_{(4)}\; o_{(5)}\; o_{(6)} \cdots$		

of underlying multigrams $\{z_i\}_i$ can be thought of as a set of acoustic units derived without any supervision from a given stream of spectral vectors O. It is derived by maximizing jointly both the likelihood of the data and of the set $\{z_i\}$:

$$\{z_i\}^* = \arg\max_{\{z_i\}} \mathcal{L}(O \mid \{z_i\}) \, \mathcal{L}(\{z_i\}) \tag{1}$$

The *a priori* distribution of all possible sets is unknown, but, according to Information Theory, the likelihood $\mathcal{L}(\{z_i\})$ is inversely related to the number of bits required to fully specify the set $\{z_i\}$. As a result, selecting a set according to criterion (1) is equivalent to searching for an optimal trade-off between its adequacy to the observed string O, and its complexity, measured as a number of bits. This principle is related to Minimum Description Length (MDL) approaches [19], the expected advantage of which is to reduce the risk of over-learning, often noticed with ML estimations [20]. Criterion (1) is optimized through an EM (Expectation-Maximization) procedure, where the data likelihood and the complexity of the model are alternately optimized [13].

5. Experiments

In this section, we describe three experiments that relate on ALISP. The first one illustrates the difficulty of reusing acoustic models across languages. The second one presents a method for very low bit rate coding that exploits the structure of speech and the third one reports speech recognition experiments based on acoustic units.

5.1 Cross-Language Recognition

In section 2., it was argued that a starting point towards ALISP was to set up a language-independent procedure to derive language-dependent data-driven speech units. In this section, we further justify this methodology, by reporting experiments which illustrate the difficulty of reusing a set of phoneme models defined and trained for a given language (the source language) to perform recognition of another language (the target language). An awkward part of this approach is to build a lexicon of the target language vocabulary, especially because it may contain phonemes which do not exist in the source language. In our experiments, the phonemes of the target language vocabulary were mapped to the phonemes of the source language by a phonetician.

We followed this protocol to recognize isolated digits, using Swiss French and/or American English as source languages, and French, British English, German, Spanish, and Italian as target languages. A set of 35 HMM was defined to model the Swiss French phonemes, and it was trained on a subset of the Swiss French Poly-Phone [9] database consisting of about 3000 sentences uttered by more than 300 different speakers (male and female). Another set of 38 HMM was defined to model the American English phonemes by clustering the original 61 NTIMIT phone labels, and it was trained on the NTIMIT database. For both sets of models, the feature vectors consist of 12 MFCC, plus delta, delta-delta and energy coefficients. The speech data to recognize for each target language consisted of a subset of the Speechdat telephone database [6], where 300 to 500 speakers are pronouncing one digit. Digit recognition accuracies of the cross-language recognition experiments are reported in Table 2.

Table 2. Word accuracies for isolated digit recognition across different languages

Target Language	Models of the Source Language		
	Swiss French	American English	Mixed set
French	90%	44%	
British	52%	53%	
German	76%	68%	
Spanish	82%	51%	70%
Italian	89%	69%	80%

Not surprisingly, the best accuracy (90 %) is obtained when French digits are recognized with the models trained on the Swiss French phonemes[1]. A similar score (89 %) is obtained by using the same models on the Italian digits, which tends to evidence a high phonemic similarity between French and Italian. On the other hand, Swiss French phoneme models are hardly reusable to recognize British English digits, as it results in the worst accuracy (52 %) with these models. As for the American English phoneme models, they perform worse than the Swiss French phoneme models across all target languages. This might be accounted for by the fact that the recording conditions of the NTIMIT and of the Speechdat database are fairly different. It might also be due to an unlucky clustering of the original 61 NTIMIT phone labels. Especially, the recognition accuracy on the British English digits tends to be lower than those obtained on the other target languages, but it should be mentioned that the quality of the British English speech subset happened to be relatively lower. It is also a well-known fact that American English and British English recognizers perform substantially worse, when confronted with each other's data.

On the whole, the results of our cross-language recognition experiments tend to evidence that the phonemic similarity between two languages greatly varies depending on the considered languages. In most cases, phoneme models trained on one language are re-usable on another language only at the cost of a significant degradation of the performances, which limits their cross-language usability to bootstrap purposes. Even though various languages are described using common phoneme symbols, the acoustic realizations of these phonemes across the languages is subject to a considerable variability.

5.2 Very low bit rate speech coding

The core issue of very low bit rate (100-200 bits/s) speech coding [14] is to minimize the bit rate of the coded data, while still preserving the intelligibility of the

---

[1] A score of 90 % for isolated digits might appear rather low, but it is essentially due to the fact that the test and train sets come from different databases. Indeed, using the same models on a test subset of the PolyPhone data (but still different from the subset used for training) resulted in a 97% [8].

reconstructed speech. We know from Information Theory that, for a given distortion level, the bit rate can be reduced by coding variable-length segments reflecting the structure of the data [7] [21]. As it is based on the inference of typical patterns at the acoustic level, ALISP offers a relevant framework to address the issues of very low bit rate coding. In this paper we report on an extension of our previous work on coding [5]. The whole procedure is as follows:

1. **Selection and modeling of speech units.** The speech training database is converted into a string of symbols by applying Temporal Decomposition (TD) and by quantizing each TD target vector. Prior probabilities of the sequences of symbols occurring in this string are estimated applying a multigram algorithm. Each sequence having a non zero probability is used to define a left-to-right HMM of $2n + 1$ states (no skips allowed), which is trained on the corresponding segments of the original feature vectors. Then, the HMM are aligned on the training speech data, and only the unit models used more than a pre-specified threshold are selected, and assigned a code index.
2. **Speech coding.** The speech data is encoded by concatenating the bit indices of the HMM along the best Viterbi alignment, plus additional bits conveying the information necessary for the reconstruction of the speech.
3. **Speech reconstruction.** A few speech segment examples of each speech unit are stored both in the coder and decoder side. Each time an input speech segment is coded using a speech unit index, the speech unit example best matching this input speech segment is selected, based on a DTW alignment score between the LPCC frames. A few bits specifying the selected example and an energy correction factor are added to the bit code of the unit. In the decoder side, the speech signal is recovered, like in segmental speech synthesis, by concatenating the pre-stored speech unit examples, according to the bit specifications.

The above procedure was applied to a subset of the Swiss French database Poly-Var [9] consisting of 218 phone calls made by a single speaker. It was split into a training set of about 5 hours and a test set of about 1 hour. The speech signal was parameterized into 10 LPCC vectors complemented with a pitch and a log energy coefficient, from which TD target vectors were computed (reducing by 6 the number of the original LPCC frames). The stream of TD vectors was quantized using 64 symbols, and an inventory of 1666 sequences was derived from the resulting string of symbols[2]. In average, a sequence in this inventory corresponds to a speech segment of about 113 ms.. The 1666 HMM defined for each of the sequences in the inventory were trained using 3 observation streams: the original LPCC, the ΔLPCC, and the log-energy, Δlog-energy coefficients.

During the alignment step, the prior probabilities of the speech units are used to weight the acoustic scores, with different factor weights: $\gamma = 0, 5,$ or 10. Increasing the value of γ biases the choice of the speech units, so that speech units with high priors are advantaged, to the expense of the acoustic similarity criterion. It can

---

[2] Sequences of up to 5 symbols were allowed, but all 4 and 5-grams had their probability falling to zero.

Table 3. Expected bit rates [bits/sec] for various weight factors γ. R_e and R_u correspond respectively to bit rates where the length of the code indices are computed based on the unit priors ($-\log_2 p_i$ bits), or assuming a uniform distribution ($\log_2 N$ bits). R_s is the average number of sequences per second

		train. set			test set		
γ	N	R_e	R_u	R_s	R_e	R_u	R_s
0.0	1514	202	206	11.07	207	211	11.40
5.0	1201	126	132	7.19	127	132	7.29
10.0	894	99	106	5.95	98	106	5.95

thus be understood as a tuning parameter allowing to displace the trade off between compression and distortion.

The number of speech units N, and the expected bit rates to encode the train and test sets using these speech unit, are reported in Table 3. The bit rates include the bits conveying the speech reconstruction information (3 bits to specify 1 example out of the 8 examples, and 5 bits specifying the energy correction factor). As expected, the bit rates decrease as γ increases, reflecting the fact that the distribution of the speech unit indices in the encoded data becomes closer to the distribution assumed to optimize the length of the indices. Audio examples can be downloaded from the web at www.fee.vutbr.cz/~cernocky/Icassp98.html. For $\gamma = 0$ and a bit rate of 211.2 bits/s on the test set, the quality of the reconstructed speech is mostly intelligible and sounds relatively natural (without the artifacts characterizing the frame based reconstruction methods). However, the quality varies across the words, tending to be better for the words occurring most frequently in the database, and it deteriorates to a not fully intelligible speech as γ increases. These preliminary results illustrate the gain from inferring sub-word regularities from speech data without supervision, to further reduce speech coding rates. However, it still needs to be improved, especially to reduce the influence of the word distribution in the speech data, and to adapt it to a multi-speaker environment.

5.3 Mono-Speaker Continuous Speech Recognition

In this section, we report on speech recognition experiments, where the speech recognition units are acoustic multigrams. An inventory of variable-length acoustic units is derived from an observed stream O of continuous-valued spectral vectors and their consistency as speech recognition units is assessed, based on comparative experiments with linguistic units.

The speech signal is converted into a string of symbols by applying Temporal Decomposition (TD) to a stream of 16 LPCC vectors (reducing the number of feature vectors by 5) and by quantizing the TD target vectors. Quantization is done by reporting the most likely state sequence of a 32-state Ergodic HMM trained on the target vectors. It is used to derive an initial set of acoustic multigrams, each of them being modeled by an HMM (see Figure 2). An iterative process is then used to select among these initial units. The selected HMM are used to produce the most likely

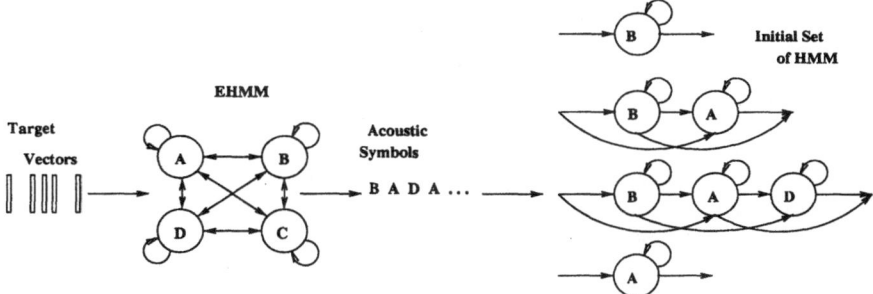

Fig. 2. *Definition of an initial set of HMMs*

transcription of the target vectors into acoustic symbols. It is obtained by reporting the state sequence of the Viterbi alignment. The transcription into acoustic symbols is then automatically aligned with the transcription into phonetic symbols: variable length sequences of acoustic symbols and of phonemes are assigned a probability of co-occurrence, using the joint multigram model introduced in [12].

During the recognition test, the most likely transcription into acoustic symbols of a test stream of TD target vectors is decoded into phonemes based on the probabilistic many-to-many mapping provided by the joint multigram model. Note that the concatenation of the phonetic sequences output during this decoding process may result in a string which does not necessarily correspond to a succession of lexicon entries. For a word based language model to be used within this framework, constraints are placed on the concatenation of the decoded phonemes. In our experiments, a new sequence of phonemes can be concatenated to the partially decoded string, only if the resulting string is still readable as a succession of lexicon entries.

The speech database from which the acoustic multigrams are derived is a set of French sailing weather forecasts, corresponding to about 60 minutes of continuous speech (single male speaker). Another 30 minute set of weather forecasts is used for the recognition experiments. The vocabulary is about 400 words, and the bigram perplexity on the test set is 10. Three reference systems based on recognition units which are either phonemes, triphones, or words are built, also using the TD target vectors as training observations. The results obtained with the reference systems and with the multigram system are in Table 4. When no language model is used, the recognition scores evaluate the reliability of the acoustic modeling within each system, and also, in the multigram case, the reliability of the probabilistic mapping between the sequences of acoustic symbols and of phonemes. The phonetic accuracy obtained with multigram units (70.6 %) is intermediate between the scores obtained with triphones[3] (55.5 %) and with words (81.2 %). The use of a word bigram language model compensates for the relative lack of reliability of the phoneme and triphone models, which, in this case, outperform the models trained on words, with word accuracies of respectively 83.9 % and 86.5 %. The integration of the same

---

[3] at least when, like in our experiments, no model interpolation, nor tied estimation techniques are used to warrant the reliability of the estimates for the models of triphones.

Table 4. Comparison of systems based on multigrams and on linguistic units

| | Acoustic modeling of | | | |
	phonemes	triphones	words	multigrams
Number of HMM at convergence	37	1636	403	613
Average frequency of a model on the training set	1280.0	29.9	34.6	30.4
Average number of states per model	1	1	4.7	3.5
Phonetic accuracy with no language model	47.0 %	55.5 %	81.2 %	70.6 %
Word accuracy with a word bigram model	83.9 %	86.5 %	82.3 %	76.1 %

language model within the multigram system, though also quite profitable, does not allow to exceed a 76.1% word accuracy. The relative lack of efficiency of the linguistic component in the multigram case illustrates the difficulty of integrating top down information with bottom up derived models such as multigram models. The way the multigram level and the linguistic level have been interfaced in our experiments is far from optimal. First of all, both levels are optimised separately: the most likely linguistic output is decoded from a single intermediate representation only (the most likely transcription into acoustic symbols). Besides, both the way sequences of phonemes are mapped to sequences of acoustic symbols, and the way lexicon entries are recovered from the sequences of phonemes lack flexibility. It can happen that a sentence cannot be recovered using only the sequences of phonemes mapped to the decoded acoustic sequences with a non zero probability.

The above approach to speech recognition is a step towards a more data-driven acoustic modeling, where an intermediate representation of the speech signal is derived independently from any phonetic knowledge. However, it is sub-optimal in many ways, and especially, improved procedures are needed to better model the correspondences between the intermediate representation and the word level.

6. Conclusions

The experiments reported in the previous section are clearly very preliminary illustrations of the concepts on which ALISP is based. Nevertheless, they are works converging towards the goal of replacing the conventional phonetic representation by a more optimal data-driven intermediate representation that would better bridge the gap between the acoustic and the linguistic levels. The concept of ALISP is based on the hypothesis that there exist some general principles that govern the structure of the spoken language code and its mapping with the linguistic level. Expressing these constraints is a first step towards the automatic discovery of appropriate units. However, in a context where the structure of speech is learnt from data, some of the tools and procedures originally designed for the use of language dependent units may not be mostly appropriate, and ALISP urges the need to find alternative schemes. As stressed in this paper, changing from an approach to speech processing conceptually grounded in phonetics to ALISP raises numerous issues, but it also offers challenging perspectives. Among these is the possibility to gain maybe some clues of possible means for the human brain to build a representation of this structure with no prior knowledge in linguistics. On a more practical side, ALISP should benefit several areas of speech processing by reducing the need for language dependent knowledge, and thus for supervision.

References

[1] O. Andersen, P. Dalsgaard and W. Barry. Data-driven Identification of Poly- and Mono-phonemes for four European Languages. *Eurospeech* 1993.

[2] B. S. Atal. Efficient coding of LPC parameters by temporal decomposition. In *Proc. IEEE ICASSP*, pages 81–84, 1983.

[3] M. Bacchiani, M. Ostendorf, Y. Sagisaka and K. Paliwal. Design of a speech recognition system based on acoustically derived segmental units. In *Proc. IEEE ICASSP*, pages 443–446, 1996.

[4] `F. Bimbot, R. Pieraccini, E. Levin, and B. Atal. Variable length sequence modeling: Multigrams. *IEEE Signal Processing Letters*, 2(6):111–113, 1995.

[5] J. Cernocky, G. Baudoin, and G. Chollet. Speech Spectrum representation and coding using Multigrams with distance. In *Proc. IEEE ICASSP*, pages 1343–1346, Munich, 1997.

[6] G. Chollet, A. Constantinescu and K. Choukri. Multilingual Speech Databases: Beyond SpeechDat. In *IWSSIP*, Poznan, 1997.

[7] A. Chou and T. Lookabaugh. Variable dimension vector quantization of linear predictive coefficients of speech. In *Proc. IEEE ICASSP*, pages 505–508, Adelaide, 1994.

[8] A. Constantinescu, O. Bornet, G. Calloz and G. Chollet. Validating Different Flexible Vocabulary Approaches on the Swiss French PolyPhone and PolyVar Databases. *Proc. ICSLP*, Philadelphia, 1996.

[9] A. Constantinescu and G. Chollet. Swiss PolyPhone and PolyVar: building databases for speech recognition and speaker verification. In *Speech and image understanding, Proc. of 3rd Slovenian-German and 2nd SDRV Workshop*, pages 27–36, Ljubljana, Slovenia, 1996.

[10] A. Constantinescu and G. Chollet. On Cross-Language Experiments and Data-Driven Units for ALISP? IEEE Workshop on Automatic Speech Recognition and Understanding (ASRU), St. Barbara, 1997.

[11] S. Deligne and F. Bimbot. Language Modeling by Variable Length Sequences: Theoretical Formulation and Evaluation of Multigrams. In *Proc. IEEE ICASSP*, pages 169–172, Detroit, 1995.

[12] S. Deligne, F. Yvon and F. Bimbot. Variable-length sequence matching for phonetic transcription using joint multigrams. In *Proc. of EUROSPEECH 95*, pages 169–172, Madrid, 1995.

[13] S. Deligne and F. Bimbot. Inference of Variable-Length Linguistic and Acoustic Units by Multigrams. *Free Speech Journal (http://www.cse.ogi.edu/CSLU/fsj/)*, 4, 1997.

[14] B. F. C. Jaskie. A survey of low bit rate vocoders. *DSP & Multimedia technology*, pages 26–40, Apr. 1994.

[15] B.H. Juang and S. Katagiri. Discriminative Learning for Minimum Error Classification. In *IEEE Transactions on Signal Processing*, Vol. 40, No. 12, pp. 3043-3053. 1992.

[16] J. Koehler. Multi-lingual phoneme recognition exploiting acoustic-phonetic similarities of sounds. *ICSLP*, 1996.

[17] M. Ostendorf, V.V. Digalakis and O.A. Kimball. From HMM's to Segment Models: a unified view of stochastic modeling for speech recognition. *IEEE Transactions on Speech and Audio Processing*, Vol.4, 360–378, 9/1996.

[18] K. Paliwal Lexicon-building methods for an acoustic subword based speech recognizer In *Proc. IEEE ICASSP 90*, pages 729–732.

[19] J. Rissanen. Modeling by shortest data description. *Automatica*, 14:465–471, 1978.

[20] J. O Ruanaidh and W. J. Fitzgerald. Numerical Bayesian Methods Applied to Signal Processing. Springer-Verlag, 1996.

[21] Y. Shiraki and M. Honda. LPC Speech Coding based on Variable-Length Segment Quantization. *IEEE Trans Acoustic, Speech and Signal Processing*, 36(9):250–257, 1988.

[22] P. Sovka Cross-Language Experiments Verifying the Approach of Converting French Flexible Vocabulary to Czech *ENST Paris, research report of sabbatical stay*, 1/1997

Interactive Translation of Conversational Speech

Alex Waibel

Interactive Systems Laboratories
Carnegie Mellon University 5000 Forbes Avenue, Pittsburgh PA 15217 (USA)
email: waibel@cs.cmu.edu

Summary. We present JANUS-II, a large scale system effort aimed at interactive spoken language translation. JANUS-II now accepts *spontaneous* conversational speech in a limited domain in English, German or Spanish and produces output in German, English, Spanish, Japanese and Korean. The challenges of coarticulated, disfluent, ill-formed speech are manifold, and have required advances in acoustic modeling, dictionary learning, language modeling, semantic parsing and generation, to achieve acceptable performance. A semantic "interlingua" that represents the intended meaning of an input sentence, facilitates the generation of culturally and contextually appropriate translation in the presence of irrelevant or erroneous information. Application of statistical, contextual, prosodic and discourse constraints permits a progressively narrowing search for the most plausible interpretation of an utterance. During translation, JANUS-II produces paraphrases that are used for interactive correction of translation errors. Beyond our continuing efforts to improve robustness and accuracy, we have also begun to study possible forms of deployment. Several system prototypes have been implemented to explore translation needs in different settings: speech translation in one-on-one video conferencing, as portable mobile interpreter, or as passive simultaneous conversation translator. We will discuss their usability and performance.

1. Introduction

Multilinguality will take on spoken form when information services are to extend beyond national boundaries or across language groups. Database access by speech will need to handle multiple languages to service customers from different language groups. Public service operators (emergency, police, telephone operators and others) frequently receive requests from foreigners unable to speak the national language. Already multilingual spoken language services are growing. Telephone companies in the US (AT&T Language Line), Europe and Japan now offer language translation services over the telephone, provided by human operators. Movies and television broadcasts are routinely translated and delivered either by dubbing, subtitles or multilingual transcripts. With the drive of automating information services, therefore, comes a growing need for automated multilingual speech processing. While few commercial multilingual speech services yet exist, intense research activities are underway. The major aims are: (1) Spoken Language Identification, (2) Multilingual Speech Recognition and Understanding for human-machine interaction, (3) Speech Translation for human-to-human communication. Speech translation is the most ambitious of the three, as it requires greater accuracy and detail during the analysis, and potentially needs to track highly disfluent and colloquial conversational speech.

2. Background

In the not too distant past, the possibility of one day being able to carry out a telephone conversation with a speaker, with whom you share no common language, appeared remote. With the state of the art in speech recognition and machine translation still far short of perfection, the combination of the two technologies could not be expected to deliver acceptable performance.

The late '80s and early '90s, however, have seen tremendous advances in speech recognition performance, propelling the state of the art from speaker dependent, single utterance, small vocabulary recognizers (e.g. digits) to speaker independent, continuous speech, large vocabulary dictation systems at around 10% word error rate. Similarly, machine translation has advanced considerably, and a number of text translation products are now commercially available.

2.1 The Problem of Spoken Language Translation

Beyond improving each component, however, it has become increasingly clear, that good speech translation cannot be achieved by mere combination of better speech recognition and machine translation components. Just as continuous speech recognition has become possible without attempting to achieve perfect phoneme recognition performance (In fact phoneme accuracy still ranges between 50% and 70%), the problem must be attacked in its entirety. Closer inspection of actual human spoken dialogs verifies this intuition. Consider an actual spoken dialog between two Spanish speakers trying to agree on a time for an appointment. The following example shows a manually produced careful transliteration of the utterance, the way it was actually spoken by the speaker:

> "...sí sí el viernes diecinueve puedo sí porque sabes me voy de viaje d hoy la verdad así es que este mes es muy viajero me voy el día seis de viaje y estoy hasta el doce así que el día diecinueve me viene muy bien francamente..."

Running this utterance through a commercial text translation system, the following translation results was obtained. (Note, that this would even assume perfect speech recognition):

> yes yes on friday nineteen can yes because know I go me of trip D today the truth such is that this month is very traveler I go me the day six of trip and I am until the twelve as soon as the day nineteen comes me very well outspokenly

What went wrong? The fact is humanly spoken sentences are hardly ever well-formed in the sense that they seldom obey rigid syntactic constraints. They contain disfluencies, hesitations (um, hmm, etc.), repetitions (".... so I, I, I guess, what I was saying."), and false starts ("..how about we meet on Tue.. um.. on Wednesday....."). Yet put in the context of discussion they are still perfectly understandable for a human listener. A successful speech translation system therefore cannot rely on perfect recognition or perfect syntax. Rather, it must search for a semantically plausible interpretation of the speaker's intent while judiciously ignoring linguistically unimportant words or fragments.

The problem described is exacerbated by recognition errors and environmental noises that occur during speech recording, such as coughs, laughter, telephone rings, door slams, etc.. Without proper treatment, these noises may be recognized as one of the words in the vocabulary, potentially causing great damage in the translation process. The dramatic variation in is another problem to be accounted for in human-to-human dialog recognitions. In fast speech, considerably higher error rates are observed due to coarticulation, reduction or elisions between the words.

A spoken dialog does not consist of sentences in the classical sense, nor are we provided with punctuation markers to delimit them. Instead, each utterance is fragmentary and each speaker's turn often contains two or more sentences or concepts (*"... no, Tuesday doesn't work for me...how about...Wednesday morning...Wednesday the twelfth"*). Even if we were given punctuation markers, attempts to translate such fragmentary utterances frequently result in awkward output.

To provide useful spoken language communication across language barriers, we must therefore *interpret* an utterance, or extract its *main* intent, rather than attempt a sentence by sentence translation. This often involves summarization. Thus we wish to "translate" the previous Spanish example above as:

"... I'm available on Friday the nineteenth..."

Only by way of a semantic and pragmatic interpretation within a domain of discourse can we hope to produce culturally appropriate expressions in another language.

2.2 Research Efforts on Speech Translation

Speech translation research today began with systems in the late eighties and early nineties whose main goal was to demonstrate feasibility of the concept. In addition to domain constraints, these early systems had fixed speaking style, grammatical coverage and vocabulary size. Their system architecture was usually strictly sequentially, involving speech recognition, language analysis and generation, and speech synthesis in the target language. Developed at industrial and academic institutions, they represented a modest, yet significant first step toward multilingual communication. Early systems include independent research prototypes developed by ATR [1], AT&T [2], Carnegie Mellon University and the University of Karlsruhe [3], NEC [4], and Siemens AG.

Most were developed through international collaborations that provided the cross-linguistic expertise. Among these international cooperations, the Consortium for Speech Translation Advanced Research, or C-STAR, was formed as a voluntary group of institutions committed to build speech translation systems. It arose from a partnership among ATR Interpreting Telephony Laboratories (now Interpreting Telephony Laboratories) in Kyoto, Japan, Carnegie Mellon University (CMU) in Pittsburgh, USA, Siemens AG in Munich, Germany, and University of Karlsruhe (UKA) in Karlsruhe, Germany. Additional members joined forces as partners or affiliates: ETRI (Korea), IRST (Italy), LIMSI (France), SRI (UK), IIT (India), Lincoln Labs (USA), DFKI (Germany), MIT (USA), and AT&T (USA). C-STAR continues to grow and to operate in a fairly loose and informal organizational style with each of its partners building complete systems or component technologies, thereby maximizing the technical exchange and minimizing costly software/hardware interfacing work between partners. In addition to the activity of consortia such as C-STAR, and the industrial research described above, there are government sponsored initiatives in several countries. One of the largest is Verbmobil, an eight year effort sponsored by the BMFT, the German Ministry for Science and Technology [5] that involves 32 research groups.

3. JANUS-II - A Conversational Speech Translator

JANUS [3] was one of the early systems designed for speech translation. It was developed at Carnegie Mellon University and University of Karlsruhe in the late '80s and early '90s in partnership with ATR (Japan) and Siemens AG (Germany). Since then it has been extended at both sites to more advanced tasks. Results from these efforts now contribute to ongoing spoken language translation efforts in the US (Project Enthusiast) and Germany (Project Verbmobil). While the first version, JANUS-I, processed only syntactically well-formed (read) speech over a smaller (500 word) vocabulary, JANUS-II now operates on spontaneous conversational human-human dialogs in limited domains with vocabularies of around 3000+ words. At present, it accepts English, German, Spanish, Japanese and Korean input and delivers translations into German, English, Spanish, Japanese or Korean. Further languages are under development.

Beyond translation of syntactically well formed speech, or (relatively well behaved) human-machine speech utterances, the research focus for JANUS-II has been on the translation of *spontaneous conversational human-to-human speech*. In the following we introduce a suitable database and task domain and discuss the JANUS-II spoken language translator.

3.1 Task Domains and Data Collection

To systematically explore the translation of spoken language, a database for training, testing and benchmarking had to be provided. For realism in practical situations a task domain had to be chosen that requires translation between humans trying to communicate with each other, as opposed to tasks that aim at information retrieval (human-machine). Some applications of speech translation (See section below.) will have elements of human-machine dialogs, when a computer intervenes in the communication process providing feedback to the users. In other situations, however, simultaneous translation of ongoing human-to-human conversations is desired.

A symmetric negotiation dialog is chosen. As a task domain, many sites have adopted the appointment scheduling domain proposed in the Verbmobil project. To elicit natural conversations that are nonetheless contained and, more importantly, comparable across languages, we have devised sets of calendars with given constraints, that get progressively more complex and generate more conflicts between speakers. Subjects are simply requested to schedule a meeting with each other and do so at their own pace and in whatever fashion they wish to express themselves. The same calendars can be used (in translated form) for monolingual dialog recordings in each language. The dialogs are recorded in an office environment, typically using push-to-talk buttons to activate recording. The recordings are transcribed carefully and double-checked to ensure that *all* acoustic events (including repetitions, false starts, hesitations, human and non-human noises) are transcribed and listed in the transcripts as they occur in the signal. Several sites in Europe, the US and Asia are now collecting and transcribing data in this fashion. More than 2,000 dialogs corresponding to about half a million words have been collected for English. Somewhat smaller databases to date have been collected for German, Spanish, Korean and Japanese by various sites as well.

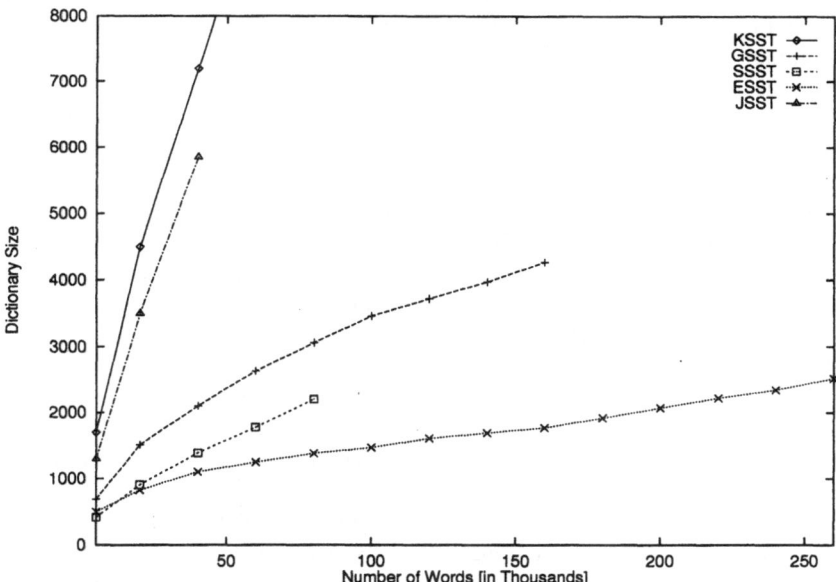

Fig. 1. Vocabulary growth as a function of database size

Figure 1 shows for various languages the growth of the resulting vocabularies as a function of the number of words spoken by subjects. For up to a quarter of million spoken words the domain vocabulary grows to around 3000 words in English. We note that (as always in spontaneous speech) there cannot be full saturation even at that level, as there will always be new words to contend with. Interesting in this figure is also the rapid growth in vocabulary size for Japanese, Korean and even for German. This is the result of using full form 'word' entries in the dictionary, a strategy that is appropriate for English, debatable for German, and inappropriate for Japanese and Korean. German, characterized by large numbers of inflections and noun compounds, and Japanese/Korean by packaging entire phrases into inflected forms generate many more variants from root forms than English and have to be broken down into subunits. In Spanish we have also explored two different data collection strategies: (1) A push-to-talk button scenario on one side, which requires the speaker to hold down a record a button while talking to the system. (2) A cross-talk scenario on the other allowing speakers to speak simultaneously and taking turns whenever they want to. The speech of each dialog partner is recorded on separate channels. Each of these two recording scenarios is evaluated in actual speech translation system tests.

3.2 System Description

The key to the problem of speech translation is finding an approach to dealing with uncertainty and ambiguity at every level of processing. A speaker will produce ill-formed sentences, and noise will surround the desired signal; the speech recognition engine will produce recognition errors; the analysis module will lack in coverage, and without consideration to dialog and domain constraints each utterance will be ambiguous in meaning.

JANUS-II was designed to deal with this problem by applying all sources of knowledge (from acoustic to discourse) successively to narrow the search for the most plausible translation. Two approaches appear possible: (1) to provide feedback (backtracking) from later knowledge sources to earlier knowledge sources, (2) to maintain a list or graph of possibilities at each stage and narrow these possibilities as each subsequent knowledge source is applied. The second approach is selected, mostly for efficiency reasons. It does not require backtracking or repeating earlier processing stages and allows, in principle, for incremental speech translation, that is, continuous recognition and translation, potentially while the speaker is speaking.

Figure 2 shows a system overview. The main system modules are speech recognition, parsing, discourse processing, and generation. Each module is language independent in the sense that it consists of a general processor that can be loaded with language specific knowledge sources.

Speech is accepted through a signal processing front-end and processed by the speech recognition module. Stationary background noises (computer hum, telephone hiss, air conditioner, microphone channel) are removed or normalized by signal enhancement techniques in the signal processing front-end. Nonstationary human and non-human noises such as coughs, lip smacks, breathing, and telephone rings, door slams, etc. are modeled as 'garbage' models and recognized individually as noise-words. To avoid having to create models for each conceivable noise in the world, a clustering algorithm reduces these garbage words to a more manageable number of up to seven prototypical noise garbage categories.

The recognition module then generates acoustic scores for the most promising word hypotheses, given a pronunciation dictionary. It uses Hidden Markov Models (HMM) and HMM-Neural Net hybrid technologies combined with statistical language models [6] in an attempt to produce most robust recognition performance.

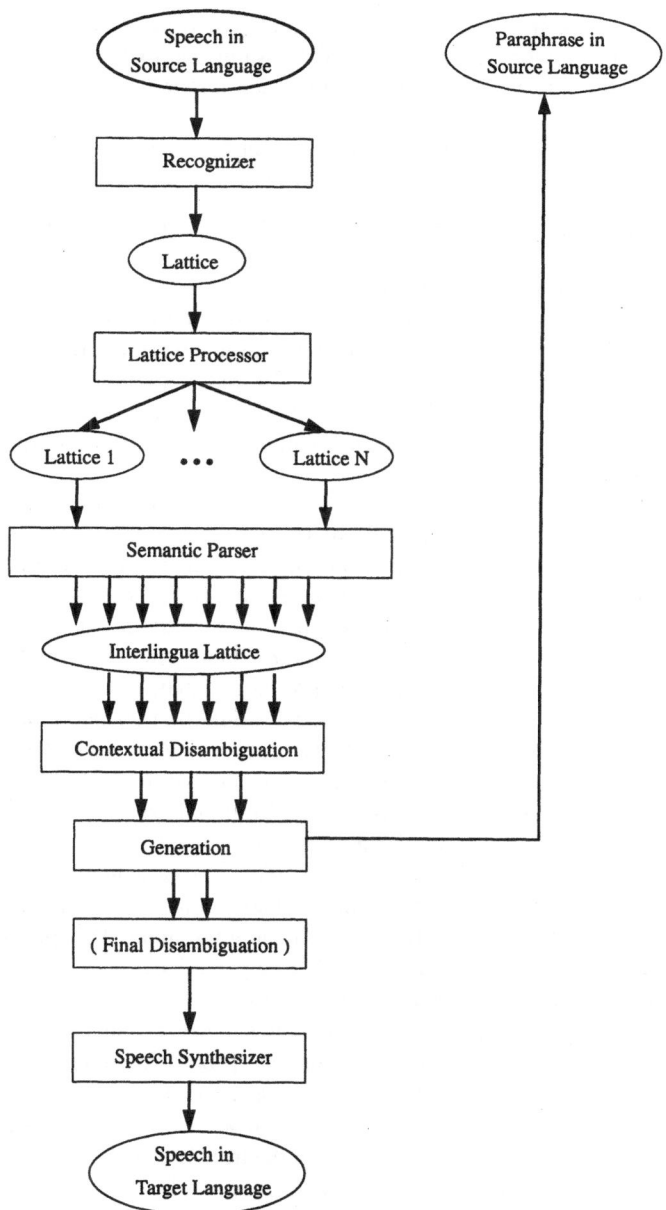

Fig. 2. JANUS-II System Level Overview

In lieu of the best recognition hypothesis, the JANUS-II recognition engine returns a lattice of near-miss hypothesis fragments organized as a graph. This graph is then reduced by a lattice processor that has two functions:

- Eliminate redundant or unproductive alternatives, such as arcs that differ only by different noise-word hypotheses on the assumption that a confusion between such noise alternatives (say, key-click vs. microphone tap) have no bearing on translation accuracy.
- Break a long utterance into usable smaller sublattices according to rough prosodic cues, such as pauses and hesitations for further processing.

The resulting shorter and reduced lattices are then passed on to the language analysis.

Unlike JANUS-I that relied heavily on syntactic analysis, JANUS-II employs almost exclusively a semantic analysis. This is to obtain robust interpretation of the meaning in spite of ill-formedness of expression and recognition error from the input. Several parsing approaches are used: A semantic pattern based chart parser (Phoenix) and GLR*, a stochastic, fragment based extension of an LR parser. Both employ semantic grammars and derive a language independent representation of meaning called "Interlingua".

There are three main advantages for the Interlingua approach: First, it aims to reduce the dependence of the output sentence on the structural form of the input language. What matters is the intent of the input utterance, whatever the way it was expressed. Sentences like *"I don't have time on Tuesday"*, *"Tuesday is shot"*, *"I am on vacation, Tuesday"*, can now all be mapped onto the same intended meaning "I am unavailable on Tuesday", and an appropriate sentence in the output language can be generated. Even culturally dependent expressions can be translated in a culturally appropriate fashion. Thus *"Tuesday's no good"* could be translated into *"Kayoobi-wa chotto tsugo-ga warui"* literally: *"As for Tuesday, the circumstance is a little bit bad"*. The second advantage of the Interlingua approach is the comparative ease by which additional languages can be added. Thus only one output generator has to be written for each new output language, as opposed to adding an analysis and an generation module for each language pair. The ease of generating output in any language constitutes the third opportunity and advantage: generating an output utterance in the input language thereby paraphrasing the input. This permits the user to verify if an input utterance was properly analyzed. This very important feature improves the usability of a speech translation, as the user most likely does not know if an output translation in an unknown language is correct or not.

Semantic representations in natural language processing have, of course, been studied extensively over the years, leading to a number of Interlingua based text translation systems (see [11, 12, 13] for review). We find the use of an interlingua based approach particularly advantageous for the translation of spontaneous speech, as spoken language is syntactically more ill-formed and less reliable, but the semantics typically more contained.

For each recognition hypothesis emerging from the recognizer, a semantic analysis is performed, resulting in a rank ordered list or a lattice of meanings. Naturally, not every recognition hypothesis will result in a different hypothesis, nor will every recognition hypothesis result in a semantically plausible hypothesis, so that a substantial reduction in remaining hypotheses can be achieved. The semantic analysis in the JANUS-II system is provided by one of several parsing schemes, the Phoenix parser, the GLR* parser (see discussion below), and several exploratory connectionist and statistical learning parsers.

After parsing, a discourse processor or contextual disambiguation can be applied to select the most appropriate meaning from the Interlingua lattice, based on the additional consideration of the context or discourse state. There are three different approaches that can be used to perform this selection or reordering: 1.) discourse plan based inference mechanisms, 2.) Interlingua N-grams (conditioning the current meaning on previous dialog states, and 3.) a dialog finite state machine. The proper weighting of each of the disambiguating strategies is obtained by training statistics over a large training database.

Following the parsing stage, generation of an appropriate expression in the output language is performed, followed by speech synthesis in the output language. For synthesis, JANUS-II resorts to commercial synthesis devices and/or builds on the speech synthesis research work of our partners.

3.2.1 Recognition Engine. The baseline JANUS-II recognizer uses two streams of LDA coefficients derived over melscale, power and silence features. It uses a three pass Viterbi Decoder, Continuous Density HMM's, Cross-Word Triphones and speaker adaptation. A channel normalization and explicit noise models are designed to reduce stationary background noise and human and non-human noise events.

In our effort of enhancing the overall system performance, we continue to improve the underlying speech and translation strategies. Particularly, in the light of our need to rearrange and redeploy our recognizer for different languages and different tasks, we wish to automate many aspects of the system design that might otherwise be predetermined once.

Improved results have recently been achieved through the following strategies [6]:

- **Data Driven Codebook Adaptation** - These are methods aimed at automatically optimizing the number of parameters.
- **Dictionary Learning** - Due to the variability, dialect variations, and coarticulation phenomena found in spontaneous speech, pronunciation dictionaries have to be modified and fine-tuned for each language. To eliminate costly manual labor and for better modeling, we resort to data-driven ways of discovering such variants.
- **Morpheme Based Language Models** - For languages characterized by a richer morphology, use of inflections and compounding than English, more suitable units than the 'word' are used for dictionaries and language models.
- **Phrase Based and Class Based Language Models** - Words that belong to word classes (MONDAY, TUESDAY, FRIDAY...) or frequently occurring phrases (e.g., OUT- OF-TOWN, I'M-GONNA-BE, SOMETIMES-IN-THE-NEXT) are discovered automatically by clustering techniques and added to a dictionary as special words, phrases or mini-grammars.
- **Special Subvocabularies** [7] - Special Confusable Subvocabularies (e.g. Continuous Spelling for Names and Acronyms) are processed in a second classification pass using connectionist models.

3.2.2 Robust Parsing Strategies. Two main parsing strategies are used in our work: the Phoenix Spoken Language Parser, and the GLR* robust parser.

- **The Phoenix Spoken Language System** [8] was extended to parse spoken language input into slots in semantic frames and then use these frames to generate output in the target language. Based on transcripts of scheduling dialogs, we have developed a set of fundamental semantic units that represent different concepts of the domain. Typical expressions and sentence patterns in a speaker's utterance are parsed into semantic chunks, which are concatenated without grammatical rules. As it ignores non-matching fragments and focuses on important key phrases, this approach is particularly well suited to parsing spontaneous speech, that is often ungrammatical and subject to recognition errors. Generation based on conceptual frames is terse but delivers the intended meaning.
- **The GLR* Parser** [9] - As a more detailed semantic analysis we also pursue GLR*, a robust extension of the Generalized LR Parser. It attempts to find a maximal subsets of an input utterance that are parsable, skipping over unrecognizable parts. By means of a semantic grammar GLR* parses input sentences into an interlingua, a language independent representation of the meaning of the input sentence. Compared to Phoenix interlingua generated by GLR* offers greater level of detail and more specificity, e.g. different speaker attitudes and levels of politeness. Thus, translation can be more natural, overcoming the telegraphic and terse nature of concept based translation. As GLR* skips over unrecognizable parts, it has to consider a large number of potentially meaningful sentence fragments. To control the combinatorics of this search, stochastic parsing scores and pre-breaking of the incoming lattices are used to reduce the ambiguity. GLR* has greater computational requirements but produces more detailed translation.

3.3 Performance Evaluation

To assess the performance and relative progress in the development of speech translators, several evaluation measures have to be devised. Evaluations can be performed at three levels:

- Speech Recognition Rate - Measured, as usual, by counting substitution, deletion and insertion errors over an unseen test database.
- Semantic Analysis based on Transcripts - This can be measured, if a 'desired' interlingua representations (the reference) has been established over a new test set. The drawback of this approach is that it is subjective and requires considerable manual labor.
- End-to-End Translation Accuracy based on 1.) Transcriptions and 2.) Recognizer Input. Each clause or conceptual fragments (not each turn) represents an event for evaluation to avoid undue weighting of short confirmatory remarks (e.g., *"That's right"*, *"OK"*). Output is then judged by a panel of three judges under the criteria "Good", "Acceptable" and "Bad", where Acceptable means an utterance was translated awkwardly, but still transmits the intended meaning. Utterances that were established as 'out-of-domain' were counted as acceptable, if they produced

an acceptable translation nonetheless or rejected the utterance as 'out-of-domain', and they were counted as bad otherwise.

Figure 3a shows the recognition results obtained over the course of recent development on a Spanish conversational translator for the scheduling domain. As can be seen, the initial recognition accuracy was quite low, which is explained in part by insufficient data in the initial stages of development for a new language. In other parts, however, the results reflect the difficulty of processing human-to-human conversational dialogs. As other research teams have found (see ICASSP'95, for example) on similar tasks (e.g., the Switchboard corpus, where, due to higher perplexity and the additional difficulty of telephone bandwidth, the results of only 50+% word accuracy have so far been achieved), human-to-human dialogs are highly disfluent, heavily coarticulated, vary enormously in speaking rate, and contain many more short poorly articulated words than read or human-machine speech. Indeed, better accuracies (exceeding 80%) can be observed in the scheduling domain, when speakers are not conversing with each other but are cognizant of the fact that they are talking to a computer.

Figure 3a also shows a comparison between speech collected using a push-to-talk switch and free cross-talk dialogs. While both are human-to-human, cross-talk appears to result in even less well behaved speech and thus is more difficult than push-to-talk speech. For other languages (English, German, Japanese), JANUS-II currently delivers similar word accuracies of 70+%. In recent evaluations carried out by the Verbmobil project using five different recognition engines recognition these accuracies up to 70% were found to be the best achievable for conversational German so far.

Figure 3b shows the result of end-to-end speech translation performance over a set aside test set. The results were obtained by scoring the translations produced by three different grammars from three different moments in the development cycle. The same test set was used to test all three grammars (of course, without any development in the interim). Reassuringly, translation accuracy was found to improve with grammars of greater coverage. It can be seen that translation accuracies up to 85% can be achieved based on transcribed spoken language, and up to 74% using the two parsers, Phoenix and GLR*.

Table 1 finally shows an interesting comparison between cross-talk and push-to-talk conditions. It was carried out using the Phoenix parser in both cases over several unseen test sets. Human translators report translating the rapid-fire turn-taking in spontaneous dialogs as unacceptably difficult. Based on these reports, we predicted that cross-talk speech would be much harder to recognize and to translate by machine as well. Since we have to compare results from different test dialogs (with considerable variability in performance) to check this prediction, we note that a precise comparison under equal conditions is not possible. Within our task domain and over multiple tests, however, a surprising trend appears to emerge. While cross-talk speech is indeed generally harder to recognize than push-to-talk, it results in shorter turns that were found to translate as well or better. Thus, translation of uninhibited human conversational dialogs appears to be no more problematic than controlled turn taking. The difficulties human translators experience with rapid cross talk di-

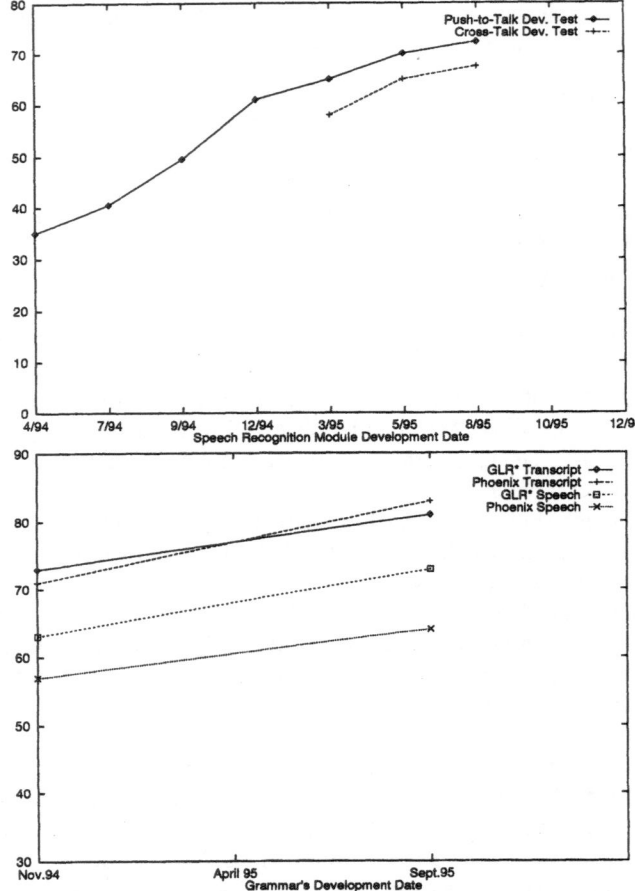

Fig. 3. Performance results for the recognition (a) and translation (b) of Spanish conversational dialogs

alogs might be related to the human cognitive load of tracking two parallel speech channels, rather than any intrinsic translation difficulty of the material.

4. Applications and Forms of Deployment

The need for Spoken Language Interpretation arises in different situations, each posing different challenges and opportunities. We have begun experimenting with three such different application scenarios: 1.) Spoken language interpretation in an interactive video conferencing environment, 2.) Portable Speech Translator, and 3.) Simultaneous Dialog Translation.

Table 1. Comparison between push-to-talk and cross talk dialogs (human-human)

	Speech Recognition Accuracy	Translation of Transcript	Speech-to-Speech Translation
Push-to-Talk Data	71%	74%	52%
Cross-Talk Data	70%	81%	73%

4.1 Interactive Dialog Translation

The prototype video conferencing station with spoken language interpretation facility uses two displays: One facing the user and another, touch sensitive display, embedded in the desk. The user operates his own station by way of the desk screen. A record button activates speech acquisition and displays both the recognition result and a paraphrase of the analyzed utterance. This is accomplished by performing a generation from the (language independent) interlingua back into the user's language. The user can now verify if the paraphrase reflects the intended meaning of the input utterance. If so, he presses a send button, which replaces the paraphrase by the translation into the selected output language and sends it on to the other video conferencing site. At the other site, the translation appears as subtitles under the transmitted video image of our user. It is also synthesized in the target language for speech output. The translation display can also be used to run collaborative virtual environments such as joint white-boards or applications that the conversants make reference to. Translation can be delivered in about two times real time.

The video conferencing station is a cooperative translation environment, where both conversants are trying to be understood and can verify the systems understanding of a spoken utterance. It can therefore benefit from user feedback and can more easily assure correctness. It also offers alternative modes for user input as well as for error recovery: Input can be provided by handwriting or typing in addition to speech. In case of error these alternative modalities can be applied to generate a new paraphrase and translation [10]. In this way, effective communication can be provided despite imperfect recognition and translation. In addition to offering a variety of recovery mechanisms, the translation station also elicits somewhat more benign user speaking style than human-to-human conversational speech.

Work is in progress that exploits this opportunity for error correction. To recover from human and machine error, a number of strategies have been explored [11], including repair by respeaking, spelling, and handwriting as alternative redundant modes of human-computer interaction. Recovery can typically be achieved within one or two tries. The JANUS- II system also offers other simple forms of assistance, such as letting the user simply type over erroneous recognitions.

The interface allows the user to select different output languages by language buttons on the translator screen. The input language is either set by the user at system start-up or can be set automatically by a language identification module as a preprocessor. In effect, the system begins by processing an incoming speech utterance via recognizers of several languages, and the most probable language is selected based on the goodness of match.

The environment still offers many opportunities for further study of the human factors of interactive spoken language translation. The best trade-off between processing speed and accuracy, the role of repair and multimodality in the translation process, how to deal with out-of domain utterances, how to learn and integrate new words or concepts, are all issues for continuing investigation.

4.2 Portable Speech Translation Device

JANETTE is a downsized version of JANUS-II. The system runs on a Laptop PC (a 75 MHz Pentium) with 32 MB of memory. In this configuration the system currently still takes about twice as long per utterance to translate than on our video stations. The system can be carried in a knapsack or a carrying bag. Translation is presented either by an acoustic earpiece, or by a wearable heads-up display. The wearable heads-up display displays the translation in text form on see-through goggles, thereby allowing the user to see subtitles under the face of the person he/she is talking to. This alternate presentation of translation result allows for greater throughput, as the translation can be viewed without interrupting the speaker. While acoustic output may allow for feedback with the system, a simultaneously displayed translation may therefore provide greater communication speed. The human factors of such new devices still await further study in actual field use.

4.3 Passive Simultaneous Dialog Translation

The language interpreting systems described so far offer the opportunity for feedback, verification and correction of translation between two cooperative conversants who want to cooperate with each other. Not every situation affords this possibility, however. In N-party conference situations, foreign TV or radio broadcasts, or simultaneous translation of speeches or conversations, a passive un-cooperative translation situation is encountered. Here, the speaker cannot be involved in the communication process for verification of the translation. Also, in the case of conversational speech, this kind of translation is likely to be particularly difficult as it requires processing of human-to-human speech, greater coarticulation, and potentially more difficult turn taking phenomena. Indeed, the rapid succession of sometimes overlapping turns makes the cognitive planning of a translation particularly difficult for humans attempting to translate conversational dialog.

Our results reported above for cross-talk and push-to-talk dialogs, however, suggest that the same cognitive limitations experience by human translators do not hold for machines: two separate speech translation processes can easily process separate channels of a dialog and produce translations that keep up with the conversants. In our lab, a conversational translator has been installed that slices turns at major breaking points and sends the corresponding speech signals to an array of 5 processors, that incrementally generate translations during the course of a human conversation (here, once again, two subjects negotiating a meeting). Despite the disfluent nature of such an interactive and rapid conversation, translation of conversational dialogs within this domain can be performed accurately more than 70% of the time.

Acknowledgements. The author wishes to express his gratitude to his collaborators, Jaime Carbonell, Wayne Ward, Lori Levin, Alon Lavi, Carol VanEss Dykema, Michael Finke, Donna Gates, Marsal Gavalda, Petra Geutner, Thomas Kemp, Laura Mayfield, Arthur Mc-Nair, Ivica Rogina, Tanja Schultz, Tilo Sloboda, Bernhard Suhm, Monika Woszczyna and Torsten Zeppenfeld.

This Research would not have been possible without the support of the BMBF (project Verbmobil) for our work on the German recognizer, the US Government (project Enthusiast) for Spanish components, and ATR Interpreting Telecommunications Laboratories for English speech translation collaboration. Thanks are also due to the partners and affiliates in C-STAR, who have helped define speech translation today.

References

[1] T. Morimoto, T. Takezawa, F. Yato, S. Sagayama, T. Tashiro, M. Nagata, and A. Kurematsu, "ATR's Speech Translation System: ASURA", EUROSPEECH 1993, pp. 1295.

[2] D.B. Roe, F.C.N. Pereira, R.W. Sproat, and M.D. Riley, "Efficient Grammar Processing for a Spoken Language Translation System", ICASSP 1992, Vol. 1, pp. 213.

[3] A. Waibel, A.M. Jain, A.E. McNair, H. Saito, A.G. Hauptmann, and J. Tebelskis "JANUS: A Speech-To-Speech Translation System Using Connectionist and Symbolic Processing Strategies", ICASSP'91, 1991.

[4] K. Hatazaki, J. Noguchi, A. Okumura, K. Yoshida, T. Watanabe, "INTERTALKER: An Experimental Automatic Interpretation System Using Conceptual Representation", ICSLP 1992.

[5] W. Wahlster "First Results of Verbmobil: Translation Assistance for Spontaneous Dialogues" ATR International Workshop on Speech Translation, November 8-9, 1993.

[6] B. Suhm, P. Geutner, T. Kemp, A. Lavie, L. Mayfield, A.E. McNair, I. Rogina, T. Schultz, T. Sloboda, W. Ward, M. Woszczyna, and A. Waibel, "JANUS: Towards Multilingual Spoken Language Translation" in Proceedings of the ARPA Spoken Language Technology Workshop, Austin, TX, January 1995.

[7] H. Hild, and A. Waibel, "Integrating Spelling into Spoken Dialogue Recognition", EUROSPEECH, Vol. 2, pp. 1977.

[8] W. Ward, "Understanding Spontaneous Speech: The Phoenix System", ICASSP 1991, Vol. 1, pp. 365.

[9] A. Lavie and M. Tomita, "GLR* - An Efficient Noise-Skipping Parsing Algorithm for Context-Free Grammars", Proceedings of Third International Workshop on Parsing Technologies, 1993, p. 123.

[10] M.T. Vo, R. Houghton, J. Yang, U. Bub, U. Maier and A. Waibel, "Multimodal Learning Interfaces", in Proceedings of the ARPA Spoken Language Technology Workshop, Austin, TX, January 1995.

[11] A.E. McNair and A. Waibel, "Improving Recognizer Acceptance through Robust, Natural Speech Repair", ICSLP 1994, Vol. 3, pp. 1299.

[12] Hutchins, W.J. and H. Somers. An Introduction to Machine Translation, Academic Press, San Diego, 1992.

[13] Hovy, E.H., "How MT Works", Special Feature on Machine Translation, Byte Magazine, (167–176), January, 1993.

[14] S. Nirenburg, J.C. Carbonell, M. Tomita, and K. Goodman, "Machine Translation: A Knowledge-Based Approach", Morgan Kaufmann, San Mateo, 1992.

Multimodal Speech Systems

Françoise D. Néel and Wolfgang M. Minker

LIMSI-CNRS, B.P. 133
91403 Orsay Cedex, France
email: `neel, minker@limsi.fr`

Summary. This chapter describes the various knowledge sources required to handle human-machine multimodal interaction efficiently: they constitute the task, user, dialogue, environment and system models. The first part of the chapter discusses the content of these models, emphasising the problems occurring when speech is combined with other modalities. The second part focuses on spoken language characteristics, describes different parsing methods (rule-based and stochastic) using a task model, and briefly presents the integration of the rule-based method in an end-to-end information retrieval system.

Key words: multimodality, spoken language processing, natural language understanding, semantic grammar, Hidden Markov Models, stochastic/rule-based methods, dialogue strategies

1. Introduction

When interacting with each other, humans spontaneously make use of several simultaneous perception and production capabilities (gesture, vision, touch, etc.). The fact that advanced technologies such as speech recognizers and synthezisers, data gloves, touch screens, helmet mounted displays, eye-trackers, etc., become now available, should let us presume that computers will offer more intuitive means of communication than the traditional screen/keyboard peripherals.

The machine, however, is far from possessing the same type of competence as humans do, and especially is unable to skilfully combine those new devices. If humans correctly interpret the information they exchange, even in difficult conditions, it is because they a priori share a more or less common knowledge about the world in general and about the topic of the interaction in particular.

This knowledge proceeds from a complex imbrication of various types of information. Designing a system therefore requires accurate analysis and modelling of those various knowledge sources involved at all levels of the comprehension process. In a multimodal context, the system relies on such knowledge to interpret the user's request more precisely and to select the most appropriate form to present the information to the user [1].

In this chapter, multimodal speech systems are considered in a broad sense and defined as *interfaces* enabling two parties (the user and the application domain) to exchange information. Those systems are limited to task-oriented dialogues in which a precise goal is pursued. Each party can use different modes and modalities such as speech combined with other natural means (gesture, position or gaze capture); each can perform independent actions and evolve in either identical or different environments. The term *mode* is associated to the perception and production processes used by humans to exchange information with the world (vision, hearing, taste, touch, gesture, haptic properties, etc.). The term *modality* refers to the various forms that a mode can take: for example, song, whisper, speech are different modalities of the acoustic mode; a visual scene may be represented by text, fixed or animated images, etc.

The remainder of this chapter is organised as follows: Section 2 presents and discusses the content of the knowledge sources involved in all interaction process. The examples do not pretend to exhaustivity, but have been chosen to better specify the role of speech in multimodal interaction, and to identify the problems occurring when speech is used with other modalities. Section 3 focuses on spoken language understanding and describes task models integrated in information retrieval systems. Different methods (rule-based and stochastic) for modelling the task are discussed with respect to portability to different languages and similar task domains.

2. System Architecture: Knowledge Sources and Controllers

Most interactive speech systems have a *centralised* architecture in which the dialogue controller directly accesses all knowledge bases and acts as a supervisor. Various types of knowledge have already been well identified: knowledge about the user, the task (application domain) and the dialogue. Beyond those, multimodality also requires to model the system itself as well as the environment in which the interaction takes place.

Some aspects of the models are *static*, meaning that they are pre-existent to the dialogue, some are *dynamically* built or modified from the ongoing exchanges between the user and the machine.

According to the application domain, the control can be *distributed* among several specialised controllers which behave as transducers and are capable of independent actions [2] (Figure 1). The roles of the different controllers similar to multiple agents can be defined as follows: the *interaction controller* merges the various types of information provided by different devices into a high-level unified language structure. The *dialogue controller* extracts the semantic content and translates it into a codified language understandable by the *application controller*. Then the dialogue

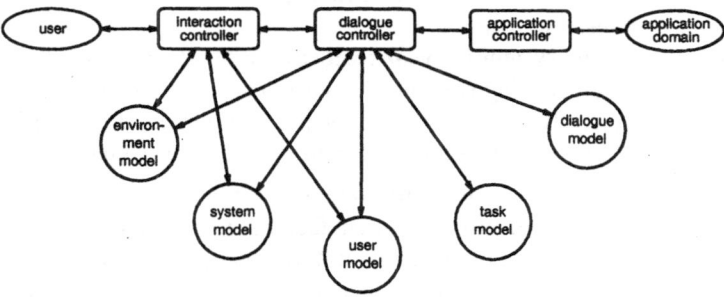

Fig. 1. General architecture of a multimodal interactive system, (from [2])

controller completes the information received from the application and builds a semantic frame for providing a response to the user. The interaction controller chooses the best form of presentation depending on several criteria such as the task priority, the internal states of the user, of the environment, of the dialogue and of the devices. The interaction controller may take decisions without referring to the dialogue controller. It may trigger *reflex actions*, whenever the semantic interpretation is unambiguous or useless (direct feedbacks), and whenever perception affects production: for example, this controller can automatically augment the intensity value of the speech synthesis if there is too much background noise.

2.1 Environment Model

Most interactive speech systems poorly take the environment model into account: its role however appears crucial in professional domains such as avionics and space shuttle control, or in general vehicle applications and information retrieval services using interactive kiosks. The model may consist of physical measures indicating the characteristics of the environment in which the interaction takes place: e.g. speech-to-noise ratio, vibration frequency, light exposure, temperature, etc.

During interaction, the initial values of the *static* model may constantly be updated with information from sensors (*dynamic* part). In multimodal interactive systems, information about the speech-to-noise ratio or the vibration frequency, for example, can lead the system to authorise or forbid vocal interaction.

The environment model may be directly related to the task model when the latter includes a complete representation of the setting (descriptions of landscapes, landmarks, routes), like in navigation aid systems.

2.2 System Model

The system model (also called *interaction* model) mainly consists of a description of each device (speech recognizer, speech synthesizer, data glove, eye-tracker, etc.) in terms of capacities, conditions of use, constraints of combination. Meta-knowledge about what the system limitations are with respect to the task, is also necessary to convey a true image of the system capabilities to the user and not mislead his expectations: for example, a system must be able to guide the user to another more adequate service if necessary.

During interaction, the system is *dynamically* aware of which device is active and which internal action is being performed at each instant: this meta-knowledge allows the system to correctly process input and output information, and in particular to select the best combination of modalities, according to device availability.

• **Importance of the time factor for multimodal integration** In order to interpret the input message correctly, one most important factor is related to time event processing [3]: the events and units produced by different peripheral devices must be merged correctly. Otherwise the message risks to be totally misunderstood, because each device has its own processing time. This aspect is critical when speech recognition is combined with other devices: a speech recognizer needs much more time to recognise a word than a tactile screen to detect the co-ordinates of a pointing device. The system therefore receives an information stream from all the different devices in a sequence which may not correspond to the real sequence in which the stream has been produced by the user (for example, when speech recognition is used with mouse clicks, the word-click sequence is in a complete reverse order). It is necessary to assign to each word of the message the dates which correspond to the exact times of its production by the user, in order to enable the system to reconstitute the correct chronological order. This is particularly obvious when deictics are used: in the somehow modified well-known example *put these boxes here*, each word of the message must be correctly matched with each successive designation gesture.

Another example is the combination of a speech recognition system and an eye-tracker. If the user is specifying some action to be performed while gazing at a spot on the screen, the system should be able to keep the track of the user's gaze direction. When the speech recognition process is over, his gaze direction risks to have changed. In such cases, it is necessary to associate a state history to all devices whose internal state frequently changes.

The time factor is also important for information output: the system must be able to carefully synchronise visual and vocal information. Furthermore, in case of message interruption, synthesis must be resumed at the beginning of a semantic unit (word or phrase) to preserve the meaning, which again implies a temporal event history and date assignment to each word.

Time also carries information by itself: the same sequence of words can be interpreted in two different ways depending on their precise temporal distribution or closeness. Two words pronounced within a short interval are usually linked by a co-reference relation and may be merged into one command; if their utterances are separated by a pause, they may refer to two completely separate commands. The word sequence alone is therefore not sufficient. It is once more necessary to know the exact beginning and ending dates of each word in the sequence as they have been pronounced, and to define time intervals which allow the system to decide whether two events may or may not be merged in order to be correctly interpreted.

Time intervals may be either explicitly specified [4] or automatically trained from multimodal corpora and implicitly processed in neural approaches [5].

• **Basic principles for mode combination** One might expect that the interaction quality improves as the number of modes and modalities increases. In fact, the combination of modes and modalities implicitly respects several criteria which are based on fundamental principles such as equivalence, complementarity, redundancy and specialisation [6] defined as follows:

equivalence: if the same action can be performed using one mode or another, those modes are considered as equivalent. In reality, an exact equivalence between modes is rarely observed: a visual scene is never strictly equivalent to its corresponding linguistic description.

complementarity: modes are considered as complementary, when an action cannot be interpreted if one mode is missing: the instruction *put this here* cannot be understood if a gesture does not accompany the word utterance. The gesture alone cannot be interpreted either. Speech can easily be used in complement, since it can be superimposed to any mode.

redundancy: different modes are defined as redundant when they are simultaneously used to provide the same information: for example, the same message can be produced by speech synthesis and also be displayed on the screen. It appears that when different modes are used in that way, an additional implicit intention is conveyed and the semantic content differs as compared to using only one mode: alarm messages make use of redundancy (to attract the user's attention), for example.

specialisation: a mode is considered to be specialised when it is best adapted to convey one specific type of information. The visual display mode seems most appropriate to spatial information.

2.3 User Model

User modelling enhances system effectiveness and usability, in so far as it allows each individual user to have a customised system. It has given birth to several types of studies in various fields including artificial intelligence, linguistics, human-computer interaction and cognitive sciences, since user modelling covers a large scope of aspects ranging from physical to cognitive levels. Some illustrative features are mentioned below.

● **Physical aspects** The physical aspects refer to the prototypical capacities of user's perception or production. In some cases, they indicate possible permanent handicaps, such as blindness or deafness, which prohibit the use of one mode or modality.

When the user model is *dynamically* refined through interaction, it allows for system adaptation. According to the environment properties, initial physical parameters may be modified: a user may be temporarily deaf due to background noise, or blinded by dazzling lights. In some applications, it appears to be most important to capture the user's position or gaze direction in order to allow the system to either display information on a screen if the user is looking at it, or to synthesise the message if the user is moving around [7].

The model may also include the user's articulatory performance, i.e. his pronunciation (phonological variants, dialectal accent), his speaking style (affected or slackened), etc.

● **Cognitive aspects** Cognitive aspects are multiform and deal with user's beliefs, desires, intentions and commitments concerning both the task and the system. They also deal with the user's own mental ability to adapt. Some aspects affecting the user's behaviour are listed below:

frequency of use: regular users of a system expect short-cuts, whereas episodic users need more precise and step-by-step instructions. During interaction, users' preferences might gradually be detected by the system.

familiarity with computer: the user's behaviour also depends on his familiarity with computer interaction. An expert in computer science does not generally overestimate the system capabilities.

user's knowledge about the application: users might be either expert or novice. Their knowledge about the application domain greatly influences the dialogue structure. A user generally knows which information he is looking for (for example, flight or train departure or arrival times). But he may not be aware of how to get it, because he ignores how the database is structured. He may even not know precisely which information he is searching for: within the framework of a CNET/CNRS-GRECO (French National Research Group on Speech Processing) project [8], spontaneous task-oriented dialogues were recorded in order to analyse users' linguistic behaviour and strategies. In a task concerning a railway information service, users directly asked for precise information, using a quite straightforward schema (departure and arrival stations, train time and category) [9]. In another task concerning a university information and orientation centre, students were more passive, since they often ignored which courses were available and which ones they were authorised to follow.

Professional users and the general public have rather different expectations. Professional users accept a certain number of constraints such as using intrusive devices (helmet or data gloves) or training speech recognizer, whereas the general public requires natural speech synthesis, but may accept isolated spoken word recognition if they are guided by the system.

User modelling appears crucial in car-navigation aid applications, since the already existing on-board systems are limited to presenting information in a most rigid way. To allow for improved interaction, the system should possess strategies similar to those adopted by humans to build common spatial references from description of itineraries and spatial scenes: in particular, the system should offer descriptions which coincide with the user's direct perception of the scenery, while assuming security and comfort. Analyses of both linguistic and gestural behaviours of car-pilots and co-pilots involved in different driving situations reveal the following observations [10]:

- verbal interaction seems adequate, because it does not attract the driver's attention from his main driving task; but gestures (which might be represented by visual designation icons on a screen) appear necessary in order to provide more global and precise instructions, and to complete verbal information. Various forms of presentation (icons, earcons, graphics, etc.) might be combined.
- the choice of the relevant strategy (step by step/incremental or global) used for spatial description depends on the characteristics of the environment (complexity of cross-roads) and on the user's perception (the degree of landmark visibility, familiarity or significance (a well-known restaurant constitutes a better shared reference than the number of streets to by-pass)).
- anticipation is necessary to respect security: information about two close cross-roads must be given in the same message.
- the driver's reactions largely depend on his previous knowledge of the route and of the traffic regulations.

2.4 Task Model

The task model may be viewed as a semantic description of the objects or concepts which are manipulated in the application during interaction. It also includes conceptual relations between objects. The model may be represented by a list of items or basic concepts structured into a frame [11]. The concepts are generally associated with a set of words and of syntactic structures used to express them; they may also refer to gestures and other sign languages. Objects have attributes represented as slots in a frame, and the goal of the task may be considered to be fulfilled when values are assigned to each slot of the frame. Values may be either obtained through dialogue or inferred from relations existing between slots.

The task model does not necessarily include a complete representation of the application domain (information database, etc.), but only those parts which are needed to handle and to improve the interaction. In the air-traffic controller training system PAROLE [12], aircrafts in the observed sector were represented by their concept values concerning level, speed and heading, etc.: as soon as an incompatibility between several parameters was detected in the controller's instruction, an error message could immediately be produced by the system, without accessing the air traffic simulator. In a yellow pages information system developed at CNET/IRISA [13], the geographical database was structured into significant regions: either into administrative entities (such as districts) or into areas centred around an important city. This task representation was used by the dialogue model. A valued network of weakened sub-questions concerning the domain was built on this structure. The dialogue model was able to adopt a co-operative strategy by relaxing constraints in this network and to propose the closest relevant answer to an initial question.

The task model may be *static* (the latter example) or *dynamic* (the former example in which aircraft parameters are updated through interaction).

• **Task hierarchy** In more complex tasks, goals are multiple and can be decomposed into sub-goals. The task model in that case is represented by a hierarchical structure of sub-tasks which can be performed either sequentially or in parallel. Very elementary tasks may be identified such as those concerning internal system actions (*reading a value*, *verifying a value*, etc.). They may also be defined as common to several sub-tasks and even to different application domains, determining to what extent the task model is application-independent.

• **Task priority** When sub-tasks are performed in parallel, a supplementary criterion must be taken into account in order to determine the priority of each. In critical applications such as space shuttle on-board systems, different levels of priority (sometimes called *criticity* in order to distinguish it from the dialogue priority) may be assigned to each task: four values (*present, relevant, urgent, vital*) were proposed in [14]. These criteria could allow the prototype to choose which information to display on which part of the configurable screen (e.g. *vital* information in the centre and *present* information in a corner) and to select the best combination of modes (e.g. redundant earcons and flashing lights for *vital* information, in order to attract the pilot's attention).

2.5 Dialogue Model

Dialogue modelling relies on numerous linguistic and philosophic analyses of pragmatics i.e. of the language usage [15]. Conversational analysis show that humans do not choose their expressions according only to purely linguistic (lexical and syntactic) constraints, but rather to underlying principles, called by Grice [16] [17] *conversational implicatures*, which allow them to communicate more efficiently. Research in this domain mainly concerns language (speech/text) interaction. Until now multimodal interaction has been most often limited to direct *object manipulation* which does not require sophisticated dialogue modelling.

• **Conversational implicatures** Grice identifies four basic maxims of conversation which jointly express a general *co-operative principle*:

quality: a contribution is assumed to be true, otherwise there is a misleading intention.

quantity: a contribution should be just as informative as required. Ellipsis and anaphoras are therefore commonly used in spontaneous language in so far as they refer to what has previously been mentioned or what is expected to be known. Bunches of details, on the contrary, convey different additional meanings.

relevance: a contribution should be relevant to the topic of discussion; unreasonable changes of focus are not expected.

manner: a contribution should be perspicuous and easily understandable (avoiding ambiguity or obscurity).

Deviations are observed whenever humans have only partial information about their partner and about the discourse domain, which is frequently the case. Several exchanges are then necessary to update mutual intentions and knowledge in order to recover from rule violation.

• **Speech acts** Speech utterances can be interpreted as actions: the dialogue partner's cognitive state is expected to be modified by an utterance, in a similar way to what is observed when a user changes the state of an object (e.g. takes an object or breaks it), or more generally changes the state of his environment by performing a physical action. Linguistic actions, called *speech acts* [18] [19], involve three basic functions:

locutionary: corresponding to the surface form of the utterance, and referring to its syntactic structure (question, statement, order);
illocutionary: corresponding to the speaker's intentions;
perlocutionary: corresponding to the effect on the listener.

In the example *you are coming* the locutionary function is a statement (composed of three words). The illocutionary function refers to the speaker's intention which may be an acknowledgement, an order (with an emphatic stress on the first word), a question or a surprise expressed by corresponding prosodic cues. The perlocutionary function corresponds to the way the message is interpreted by the hearer (he may be threatened or just pleased).

The surface form does not always fully coincide with the speaker's intention, since an utterance may have one or more purposes: e.g. the assertion *it is warm here* can also be a disguised request for opening a window. This is what Searle calls an *indirect speech act* [15] [19].

• **Dialogue acts** In human-machine interaction, several threads must be conducted in parallel to ensure task completion. Speech acts are therefore extended to the notion of *dialogue acts* [20] which include at least three tightly embedded simultaneous processes:

informative: concerns all factual information referring to the task. Because users' and machine representations of the task do not exactly coincide, messages need to be exchanged to complete, disambiguate, modify or specify object parameters. These messages are directly linked to the different steps of the task completion.

communicative: ensures that the dialogue proceeds smoothly, and mainly concerns the way how information is exchanged through any communication channel. This implies dialogue about dialogue (metadialogue). This aspect is particularly crucial when communicating via the telephone line: phatic messages, such as *do you hear me*, are necessary to ensure a good transmission. These communicative messages, such as repetitions and confirmations, may occur any time during the interaction.

interactive: refers to the roles assumed or expected by each partner. It concerns sociological (politeness) but also functional aspects (the degree of partners' expertise). A computer may be considered by the user to play the role of a tool, of an assistant/agent, of a mediator or of a peer (an equal partner). If this role cannot be assumed, some mismatch may provoke errors or requests for explanation [21].

• **Dialogue hierarchical structure** A dialogue minimal unit consists of at least one question and the corresponding relevant response (which defines an *exchange*). Should interaction be non-problematic, the dialogue model would be practically unified to the task model. But, most frequently, as already explained, several exchanges are necessary to match the user's intentions and expectations with the task requirements, the system and environment constraints. Several exchanges are therefore required to update the different models accordingly.

The dialogue model is often decomposed into hierarchical scripts or subdialogues which describe both the permissive chaining between possible situations and the authorised shifts of focus from the ongoing discussed topic. Dialogue act identification enables the system to directly control the interaction: after a correct identification of the user's ongoing dialogue acts, the dialogue controller may choose the best strategy in order to recover from deviation or mismatch from the ideal minimal interaction.

• **Examples of dialogue models and strategies** Dialogue models differ from one another with respect to dialogue act categorisation and to strategies. There exists quite a consensus about the hierarchical structures which may consist of three or four different levels: the Esprit SUNDIAL project [22] relied on four embedded levels (*transaction*, *exchanges*, *interventions* and *dialogue acts*); the German national project VERBMOBIL [23] also includes four levels (*dialogue, (greeting, ne-*

gotiation) phases, turns, dialogue acts), the greeting phases integrating dialogue initialisation and closing politeness phenomena. Dialogue acts are more or less task-dependent like *accept, suggest, reject appointment date and location* in VERBMO-BIL, but some are more general like *garbage, deliberate* or *feedback* and concern usual errors, deviations and problematic situations.

Various dialogue strategies have been identified, such as *directive, reactive, co-operative/suggestive, negotiative*. Early speech interactive systems made a large use of *directive* and *reactive* strategies because of the constraints imposed by the speech recognizer performance limitations. The first speaker-independent flight information retrieval prototype developed at CSELT in the middle of the eighties used a complete *directive* strategy in order to oblige the user to answer the system questions with only one word (yes/no, names of cities, times...). The system VODIS [24] developed by Cambridge University in collaboration with British Telecom used *reactive* strategies in which the dialogue mainly included correction and confirmation subdialogues; but in case of recognition failure, the system was also able to switch to a more *directive* strategy. In [13], the valued lattice built on the task domain enables the system to *suggest* the closest answer to the question by gradually relaxing constraints: the number of exchanges can therefore be considerably reduced. The AGS project [25], also developed at CNET in France, proposes a *co-operative* strategy based on users' reasoning processes (represented by first-order logical predicates). The VERBMOBIL project whose objective is to mediate human-human interaction about appointment scheduling tasks, makes use of *negotiating* strategies [26].

2.6 Models Interdependency

Clear distinction between environment, system, user, task and dialogue models are not always easy to identify, since there are several possible interdependencies between these models. Some have already been mentioned: a user may become deaf due to the background noise, the user's perception is influenced by the system descriptions of visual scenes, etc. The structure of the task model enables specific dialogue strategies. The models may be in conflict with each other: e.g. task priority may sometimes prevail over dialogue priority, and vice versa.

Existing systems do not possess complete descriptions of all models: depending on the application domain, the models which play important roles are more refined than others. In multimodal interactive systems, environment and system models appear to be as important as task and dialogue models.

2.7 Role of Speech in Multimodal Applications

One might wonder why, in spite of its complexity, integrate speech with other modalities. There may be several reasons. The spoken language:

- improves task efficiency, since it can easily be superimposed with other modalities, especially with gestures [27] [28].
- helps user's identification or verification in so far as it offers a complement to other identification or verification devices [29].

- can serve as a substitute for a mode which cannot be used because of users' physical handicaps: for example, MEDITOR [30], a text editor for blind young people, uses speech to describe visual information and to improve control over the system.
- avoids explicit changes of context when visual information is predominant: e.g. when creating objects, a designer can directly control them by modifying their colour, shape or orientation and constantly keep an eye-contact with them [31].

3. Information Speech Systems

Since speech represents one of the most complex modalities involving high level processes, this part of the chapter is devoted to semantic representation and interpretation, and task model integration into spoken language systems. The task model is implemented using a case grammar formalism which seems well adapted to spontaneous speech processing. Two different parsing methods (rule-based and stochastic) are discussed in terms of efficiency and portability across languages and similar tasks. The integration of the understanding component with a dialogue model into a spoken language system is described briefly at the end of this section.

3.1 Spontaneous Language Characteristics

Several corpora were recorded in the middle of the eighties in Europe and in the United States in order to better apprehend the way people speak and behave when performing specific tasks. The above-mentioned experimentation carried out at the SNCF (French National Railway Society) telephone train timetable information service [32], aimed at recording spontaneous dialogues in two different conditions: the real users communicated either with a human operator, or with a simulated computer, using a Wizard-of-Oz protocol (which made them believe they were talking to a machine). In both setups, no specific instructions were given to the users. The analyses of the corpora revealed the great variety of formulations and linguistic structures spontaneously used by humans, as compared to the written form. Spoken language is a real-time creation process which does not allow for long reflection or back-tracking. It therefore combines two apparently conflicting phenomena: language redundancy and language economy. A human is permanently updating and correcting his utterances in order to improve their semantic content and to comply to what he intends to say; he simultaneously tends to rely on his partner's awareness of the situation, by referring to the information already exchanged, in a more or less anaphoric or elliptic way and by respecting underlying principles such as Grice's quantity rule [16] [17].

In the example taken from the GRECO transcripts of human-human telephone dialogues [32] *(O:Operator,U:User)*:

O - arrivée 19:43
O - arrival 19:43
U - 19h, oui, c'est d'bonne heure/y euh vous n'avez pas plus tard?
U - 19, yes, it's early/there ah it's not possible later?

O - un peu plus tard alors y en a un à 17h10
O - a little later then there's one at 17:10
U - alors 17h10, oui
C - well 17:10, yes
O - euh ça vous va là comme horaire ou...
O - ah this is convenient for you this schedule or...
U - ben 17h10 euh je pense/ pour 21h40, pa(r)ce que, euh l'hôtel est à l'extérieur de :/ de la
chapelle donc euh, euh oui ou alors je sais pas/ qu'est/qu'est-c'qu'i(l) y a/qu'est-c'qu'il
y a après l'17h10?
*U - well 17:10 ah I think/ for 21:40, b'cause, ah the hotel is outside / The Chapelle, therefore
ah ah yes or then I don't know/ what's/ what's there/ what's there after the 17:10*
O - arrivée 16h59, j'en ai pas entre...
O - arrival at 16:59, there's none between...

the following characteristics can be observed:

at the utterance level: Under time pressure, speakers repeat parts or phrases of the
message, producing many ill-formed sentences, disfluencies, false starts and self-
corrections. They make a systematic use of prosody to convey interrogative meaning to assertive sentences [33]. Under physical effort, they need to make pauses
that may not be significant, and use linguistic cues (*therefore, there, well, then,*
etc.) which play the role of punctuation markers. Spoken language is not only a
linguistic vehicle per se, it also functions as a tool to involve the listener's attention, comprehension and sympathy. This explains the great percentage of statements that are just feedback remarks, appreciation or justifications (*it's early,
because the hotel is outside The Chapelle*). In that respect, it is not easy to detect
which parts of the utterance play a significant and central role, comparable to the
verb role in written sentences.
at the interaction level: As already mentioned, a dialogue is a co-operative action
between two parties, each possessing some representation of the topic which is
being discussed. The underlying challenge is therefore to share (take or give) control, whenever necessary. This provokes frequent interruptions, overlapping utterances to complete or reformulate what has just been said, or sudden stops at the
point where completion is thought to be perfectly predictable. Anaphoric or elliptic phrases are frequently used to provide the most informative elements quickly,
and to avoid already mentioned items to be explicitly repeated. Telephone dialogues are also characterised by constant repetitions of the information provided,
especially when digits and numbers are concerned. Several reasons might be put
forward for such a behaviour [34]:
  – repetitions are a form of implicit confirmation (the user lets the listener know
    what he has grasped, and expects to be contradicted if he has misunderstood);
  – they also allow for a better memorisation of the information (to palliate the lack
    of visual track);
  – they finally ensure a phatic control (the dialogue partner knows that the user is
    still in contact).

Such phenomena are less frequent in the Wizard-of-Oz protocol, but still sub-
sist sufficiently to render difficult any automatic interpretation using traditional approaches. Understanding spontaneous speech therefore requires adapted formalisms

which analyse the sentence structure on a semantic rather than syntactic level, and extract semantic, ignoring global syntactic malformations. One possible knowledge representation is based on the use of semantic case grammars.

3.2 Case Grammar Formalism used for Task Modelling

The original concept of a case frame as described by Fillmore [35] is based on a set of universally applicable cases expressing the relationship between a verb and its constituents. Bruce [36] extended the Fillmore theory to any concept-based system, and defined an appropriate semantic grammar (Figure 2), which is illustrated in Section 3.3.

reference word:	case frame or concept identifier
case frame:	set of cases related to a concept
case:	attribute of a concept
case marker:	surface structure indicator of a case
case system:	complete set of cases of the application

Fig. 2. Semantic case grammar formalism, (from [37])

Case grammars have already been applied in the context of speech understanding in an electronic mail domain [38] and in a prototype designed as an aid to professional language training in the air-traffic control domain [39] [40]: in the latter, this choice appears adequate, as the professional language (called a *phraseology*), though meant to be constrained, offers many characteristics of spontaneous language (inversion, ellipsis, different formulations for a same semantic content, etc.), when used by operational controllers.

• **Application domains** Spoken language information retrieval prototypes using a case grammar formalism have been developed at LIMSI-CNRS for two similar application domains: air and train information retrieval services. The ATIS (Air Travel Information Services) task was defined as a common research task for data collection and evaluation support within the ARPA (Advanced Research Project Agency) Speech and Natural Language program [41] [42]. Prototypes have been developed for two languages, English [43] and French (the French version is called L'ATIS [44]). In the Esprit Project 9075 MASK (Multimodal-Multimedia Automated Service Kiosk), the aim is to offer travellers a train timetable information kiosk in French using multimodal input (speech and touch) and multimedia output (sound, video, text and graphics) [45]. The three systems (ATIS, L'ATIS, MASK) allow users to speak naturally without any constraint of formulation. The vocabulary used is about 2000 words for each.

• **System components** A generic spoken language system for database access mainly includes the following functional components: speech recognition, semantic analysis and dialogue management as well as response generation (Figure 3). The

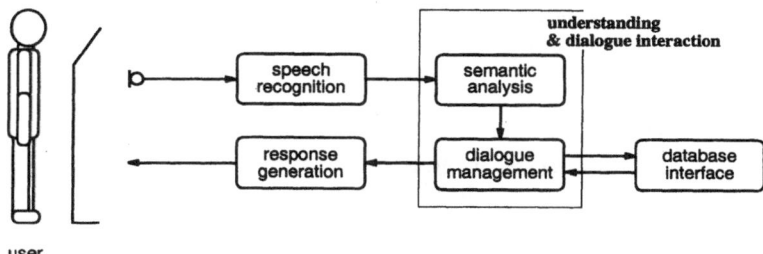

Fig. 3. Overview of a spoken language system for database access, (from [37])

speech recognizer output word string is passed to the semantic analyser which extracts the meaning of the query and builds an appropriate semantic representation which depends on the task model. The database request generator uses this meaning representation eventually completed by the information from the dialogue history, to generate a database query in an adapted coded language. The response generator may present the interaction result under various forms (text, speech, tables or graphics) to the user.

3.3 Different Parsing Methods

Rule-based and stochastic methods for the semantic analysis were compared in two application domains on similar test corpora [43] [46] [47] [57].

For the semantic analysis, the use of the case grammar formalism is exemplified on the following query (taken from ATIS corpus):

> *you get could you give me a ticket <u>price</u> on [uh] [throat clear] a flight <u>first</u> class <u>from</u> <u>San Francisco</u> to <u>Dallas</u> please*

<u>price</u>: **reference word** or **concept** (which identifies the concept **airfare**); other concepts may be **book, flight,** etc.

<u>San Francisco, Dallas, first</u>: respectively corresponding to **from-city** (departure city), **to-city** (arrival city) and **class** (first class) **cases;** other cases are **itinerary, departure time,** ...

<u>from, to,</u> etc.: **case markers.**

The parsing process considers less than 50% of the query to be semantically meaningful. The hesitations and false starts are ignored.

A declarative language containing a list of possible case frames and associated cases is used to describe the case system [44] [46]. The task model is represented by a hierarchical architecture containing several levels ranging from conceptual to intermediate and basic structures: the example (Figure 4) presents the case frame related to the concept **airfare.**

Conceptual level	CASEFRAME airfare {REFERENCE WORDS: prix, tarif, coûte, ... itinerary: @itinerary. times: @times. ...}	
Intermediate level	SUBFRAME itinerary {from-city: (quitte, de) @city. to-city: (à, pour, vers) @city. ...}	Common levels
	SUBFRAME times {rel-departure: (partir) @comparative. departure: @depart-hour-minute. ...}	
	SUBFRAME depart-hour-minute {depart-hour: (partir) @hour (heure). depart-hour: (partir) @noon-midnight. depart-minute: (partir, heure, midi, minuit) @minute. ...}	
Basic level	SUBFRAME city {city: denver, boston, atlanta, ...}	
	SUBFRAME comparative {comparative: avant, après}	
	SUBFRAME hour {hour: 1,2,3, ...}	
	SUBFRAME noon-midnight {noon-midnight: midi, minuit}	
	SUBFRAME minute {...}	

Fig. 4. Hierarchical case system with different levels. The symbol @ refers to lower-level frames, (from [46])

3.3.1 Rule-based Parsing Method.

In the rule-based understanding component, sentence parsing is done by first selecting the corresponding case frame with triggering reference words. Then the slots of the frame are instantiated using case markers which provide local syntactic constraints necessary to extract the meaning of the request. Higher level structures refer to lower level subframes. Sometimes the query may contain multiple concepts resulting in the generation of multiple semantic frames. The parser is recursively applied on the subframes of the case system until there are no suitable words left to fill in the slots. Once completed, the semantic frames represent the meaning of the input query.

Markers may have multiple functions in the same query: in the example *partir cet après midi avant dix-sept heures trente* (leaving this afternoon before five thirty pm), the word *heure* is used as a post-marker for the case **depart-hour** (*dix-sept*),

Q	*Je veux aller de Philadelphie à San Francisco avec escale à Dallas*
	(*I want to go from Philadelphia to San Francisco with a stop in Dallas*)
Q̂	*Je voudrais connaître les vols qui vont de Philadelphie à San Francisco avec une escale à Dallas*
	(*I would like to a s know the flights that go from Philadelphia to San Francisco that stop in Dallas*)

SF	\<flight\>
	from-city: philadelphie
	to-city: san-francisco
	stop-city: dallas
CS	SELECT airline_code, flight.flight_id, flight.departure_time,
	flight.arrival_time, stops, stop_airport
	FROM flight, flight_stop
	WHERE from-city=@from-city
	AND to-city=@to-city
	AND stop-city=@stop-city
R	*Voici les vols de Philadelphie à San Francisco faisant escale à Dallas*
	(*Flights from Philadelphia to San Francisco with a stop in Dallas*)

COMPANY	FLIGHT_NUM	DEPART	ARRIVE	STOP	STOP_CITY
DELTA	217/149	08h30	13h25	1	DALLAS/FORT-WORTH
AMERICAN	459	15h00	20h23	1	DALLAS/FORT-WORTH
DELTA	589/395	19h15	23h50	1	DALLAS/FORT-WORTH

Fig. 5. Example queries resulting in an identical semantic frame representation (F), SQL database request (DR) and formatted response (R) in L'ATIS, (from [43])

and also defined as a possible pre-marker for the case **depart-minute** (*trente*). The word *partir* (leaving) is used as a distant pre-marker of *dix-sept* (five pm). Therefore some words may have cumulative functions: adjacent or distant, pre- and/or post-marker(s).

The conceptual level is entirely task-dependent, whereas the basic levels may be task-independent to some extent: in the above example, they deal with lists of city names and times, which may be shared by different tasks.

Figure 5 illustrates the results obtained at different stages of the analysis: the case frame semantic representation, the SQL (System Query Language) command sequence and response generation. In the example, the reference word *aller* (to go) causes the parser to select the frame **flight**. The SQL command used for database access and information retrieval is built from the semantic frame using specific conversion rules. The retrieved information is then reformatted for presentation to the user along with an accompanying natural language response that can optionally be synthesized. The advantage of such a semantic grammar is to allow the user to formulate the same request differently and not to impose a strict syntax order.

The major work in designing the task model is defining the meaningful concepts along with the corresponding reference words and the cases, which requires the analysis of dialogue corpora recorded in realistic conditions [46].

• **Portability to different languages** In order to investigate language portability of the rule-based method for natural language understanding, the French prototype L'ATIS was ported to American English using the 10,718 answerable queries of the official ARPA ATIS corpus for iterative development and testing [43]. The porting consisted of translating the declarative language describing the rules for parsing and SQL command sequence generation. The understanding component was iteratively evaluated in order to monitor progress and to ensure the consistency of the changes.

The case grammar was found to be easily portable to a new language by translating the system of rules whilst considering some language specificities. L'ATIS had not been designed to cover the whole range of the American ATIS task, but rather to provide a simplified framework for a first French language prototype. Only five concepts with a set of 38 cases were identified. For the English version, the domain coverage was extended using the ARPA-ATIS data [48]. Supplementary concepts, such as **meal, ground-service, time-zone** were needed. For example in the subframe **flight-designation**, the French system only uses the cases **flight-number** and **company**. In the English version, **aircraft-type** and **capacity** are necessary. Table 1 shows the 13 semantic concepts with some example reference words that have been determined for the English version of L'ATIS.

Table 1. Concepts used in the American English version of L'ATIS, (from [43])

Semantic concept	Example reference words	Example query
abbreviation	*abbreviation*, explain	what does the **abbreviation** U S mean
quantity	capacity, **how many**	**how many** Delta flights leave from Washington
meal	eat, food, meal, **meals**	are **meals** available on Delta flight six eighty eight
grounds	transportation, **ground**	show me Boston **ground** transportation
airport	*airport*, airports	what is the name of the **airport** in Philadelphia
airfare	*fares*, cost, rate, ticket	I would like the **fares** for flight one oh two seven six nine
aircraft	description, kind, *type*	what **type** of airplane is an M eighty
flight-class	class, *classes*	how many different **classes** are there
restriction	restriction, **restrictions**	are there any **restrictions** on flight D L sixteen
airline	airline, **airlines**	what **airlines** go from Atlanta to Baltimore
city	city, **where**	**where** does flight one zero two six seven four stop
time-zone	*time zone*	what **time zone** is Denver in
flight	*flights*, operate, run	**flights** from Baltimore to Denver

Language specificities require case markers and reference words to be added, removed or their order to be changed in the grammar. For example, in French, *midi* (noon) and *minuit* (midnight) do not exactly behave like numbers, but unlike English can directly be followed by digits or numbers, such as in *demain midi trente* (tomorrow twelve thirty). Therefore they are only used in the French parser as a marker for the case **depart-minute** (*trente*). Another version of ATIS was developed for the Japanese language [49] and required pre-markers to be changed into post-markers.

• **Portability across application domains** Porting the natural language understanding component to the similar French MASK information retrieval task was also relatively straightforward. Compared to ATIS, most of the concepts and constraining values are also found in the train travel domain, albeit with slightly different significations: for example, the concept **type** corresponding to the aircraft type in ATIS corresponds to the type of train in MASK (TGV (Train à Grande Vitesse (High Speed Train)), TEE, etc.), whereas the concepts related to **arrival, departure times** and **fares** can be mapped directly. The MASK concepts determined by analysis of queries taken from a development corpora [50] are **train-time, fare, connection, type, book, service** and **reduction**.

3.3.2 Stochastic Parsing Method.

Establishing and maintaining a system of grammar rules for the semantic extraction is costly, as it requires human involvement. Alternatively a stochastic method can be used to automatically learn semantic concepts in function of their appearance from a large number of interactions. Statistical modelling techniques have been used with success for the acoustic and syntactic levels of speech recognition systems [51] [52], and more recently there has been interest in extending this approach to model the semantic content of the sentence. Stochastic methods for natural language understanding have been applied in the BBN-HUM [53], the AT&T CHRONUS [54] systems, by IBM [55] and LIMSI-CNRS [47] for the ARPA-ATIS task, and by Philips Research Laboratories [56] and LIMSI-CNRS [57] in the train travel application.

A stochastic method requires large amount of word sequences and corresponding semantic labels for training the parameter model. Figure 6 overviews the functional diagram of the stochastic parser developed at LIMSI-CNRS for the ATIS task [47]. The same set of semantic labels (concepts, markers and values) as in the rule-based case grammar is used. This allows for a direct comparison of both methods. The role of the query preprocessor is to reduce the model size. During training, the parameter estimator automatically establishes the model from the correspondences between the preprocessed word sequences and the corresponding semantic sequences. The semantic decoder, implemented as an ergodic Hidden Markov

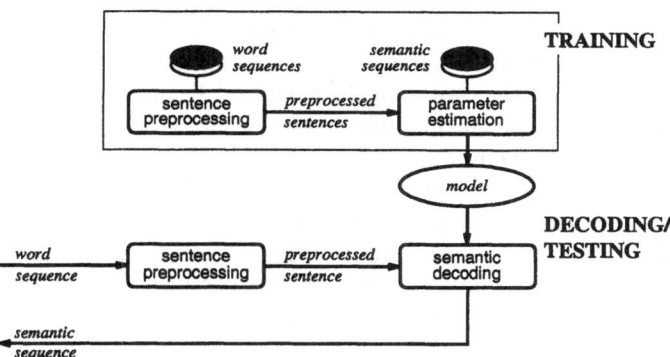

Fig. 6. Overview of the semantic analyser of a spoken language system using a stochastic method for natural language understanding, (from [57])

Model [51], outputs the most likely semantic sequence given the test input query. Using the token-value pairs, the template matcher directly converts the semantic sequence into a semantic frame for use by the database access and response generation components, which are shared by both methods.

• **Semantic data labelling** For ATIS, the semantic sequences were directly derived from the frame-based representations. However slight adaptations were necessary. In the rule-based method, the local syntactic constraints between the case markers and constraining cases were refined: case markers could be distinguished as pre- or post-markers. In the stochastic method, this notion of locality for case markers is implicitly contained in the semantic sequence. In the frame-based representation, the semantic labelling does not consider all the words of the query, but only those related to the concept and its cases. However, in order to estimate the model parameters, each word of the input query must have a corresponding semantic label. The additional semantic label **null** has been assigned to those words not used by the case frame analyser for the specific application. Therefore a semantic sequence consists of basic labels corresponding to the reference words concept, case values, case markers and irrelevant words (labelled as **null**). In the rule-based method, the markers and values are tightly coupled, and each marker corresponds to exactly one case value, e.g. the marker **from-city** to the value **from-city**, and the marker **from-airport** to the value **from-airport**. In the stochastic method, this relation is more relaxed. Therefore rather general markers can be defined: the marker **from** precedes either the value **from-city** or the value **from-airport**. This simplification helps to reduce the size of the parameter model.

For MASK, a straightforward method for developing a stochastic component was proposed by way of designing and annotating a corpus of semantic labels [57]. Since no previous knowledge other than the formalism was introduced, the semantic representation could be better adapted to the stochastic method. A common semi-automatic procedure was used to semantically label the queries. It reduces the effort substantially by integrating the stochastic component into the labelling process.

A portion of the MASK corpus containing 10,405 queries was divided into four subsets containing 488, 980, 2,937 and 6,000 queries. Using the notion of concepts, case markers and values, parses were manually determined for the first 488 sentences (*initialisation*). The model parameters were estimated on this initial subset. The *iterative* procedure consisted of using the model to label the next subset of data, to manually correct the labels and to estimate the new model parameters to be used in the next iteration. Typical errors at the initial stage of development arose from an increase in domain coverage when using new data. The labelled sets were merged and the model parameters recalculated for further query labelling. These steps were iterated until the complete training set was semantically labelled and corrected. Labelling became faster as more data were available for parameter estimation.

Stochastic methods require substantial amounts of data for the parameter estimation. Spoken language corpora are still limited in size, which is problematic in that events rarely observed in the training data are not adequately modelled. As a result, the parameter estimation may become unreliable. In addition to back-off techniques

[58], the data are preprocessed (categorisation and contextual information) in order to unify the input [57].

• **Model topology** Semantic decoding using a stochastic method consists of maximising the conditional probability $P(S|O)$ of some state (semantic label) sequence S given the observation (word) sequence O. Using Bayes' rule, this probability is reformulated as follows:

$$[S]_{\mathrm{opt}} = \underset{S}{\mathrm{argmax}}\{P(S)P(O|S)\}$$

Given the dimensionality of the sequence O, the estimation of the likelihood $P(O|S)$ is replaced by estimating the parameters of a Hidden Markov Model [51]. The parameters consist of bigram state transition probabilities $A = P\left(s_j(t)|s_i(t-1)\right)$ and of the observation symbol probability distribution $B = P\left(o_m(t)|s_j(t)\right)$ in state j.

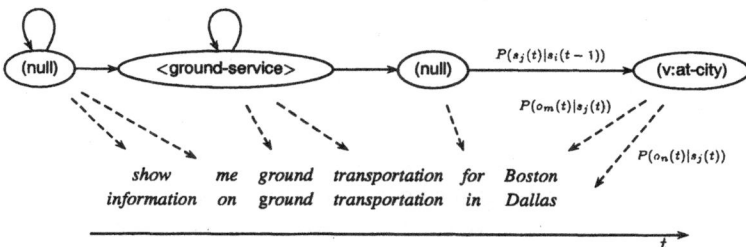

Fig. 7. Semantic decoding is progressing on a path through the Markov model and generating observation sequences, (from [37])

Figure 7 shows a particular path through the Markov model. The progression through the state sequence of semantic labels generates sequences of observations, each of which represents a word in the query. The temporal progression and sequence generation are guided by the state transition and observation probabilities, previously learned from a large number of correspondences between words and semantic labels in the training data. The example illustrates, that for a given progression through the model, several sequences of word hypotheses are generated, each of which yields different observation probabilities, e.g. $P(o_m = Boston)$ and $P(o_n = Dallas)$.

• **Training data characteristics** The stochastic models have been trained using 10,718 answerable queries of the ATIS and, as already specified, 10,405 queries of the MASK data along with their sequences of semantic labels. Even though the train travel application was represented in a more limited way than the air travel (74 semantic labels for MASK compared to 112 for ATIS), the main reason for the comparable lexicon sizes (1,449 versus 1,577) results from the higher inflectional characteristics of the French language as compared to English, providing a variety of words with identical root forms (especially for verbs). The MASK training corpus used for this study also contains more city names (86 compared to 46 in ATIS). For both applications, preprocessing was used to reduce the lexicon size to 164 lexical entries for MASK and to 293 for ATIS.

3.3.3 Comparative Evaluations.

Speech recognizers are typically evaluated in terms of word or sentence accuracy. The use of simple pattern matching for the semantic and dialogue evaluations is not so clearly appropriate, since natural language understanding may be assessed on different semantic levels. Applied on the completed semantic representation, a first evaluation stage gives an idea on system capability to ask the user for additional information and considers successive exchanges between the user and the system. Another stage consists of evaluating the system response, i.e. the information returned and the actions taken. MADCOW proposed a paradigm for spoken language systems response evaluation in the ATIS domain [42] [59].

Table 2. Semantic and response error rate (%) comparing stochastic and rule-based components for MASK and ATIS, (from [57])

	Semantic sequence (%)		Response (%)	
	Stochastic	Rule-Based	Stochastic	Rule-Based
MASK	7.2	13.8	8.3	9.4
ATIS	13.7	14.4	18.7	16.9

Performance evaluations have been carried out on the task model component using the stochastic and rule-based methods for MASK and ATIS (Table 2). For the MASK system, 15 travel scenarios containing 726 query transcripts were used as the test data. The ATIS evaluation was carried out on the 445 type A query (answerable without dialogue history) transcripts taken from the December 1994 Benchmark test data. The performance was assessed at the semantic sequence level, comparing the concepts and constraints with previously defined reference labels. The components were also evaluated on the accuracy of the system responses returned to the user, which are the retrieved database responses in ATIS and the natural response generated in MASK.

For MASK, the stochastic component obtains a 7.2% error rate compared to a 13.8% for the rule-based method. The difference in observed performance is much greater than for ATIS, as the independent system design enables a completely tailored semantic representation and does not limit the stochastic component by constraining it with the eventual shortcomings of the rule-based method. The stochastic implementation also profits from the mutual information between all the semantic labels. If an explicit marker is incorrectly decoded or does not exist, the surrounding labels yield the function of implicit markers. This makes a decoding of the associated constraints more robust. However the parser frequently fails to identify long distant marker-constraint relations.

For the rule-based MASK understanding component, 68% of the errors involve concept identification due to an incorrect triggering of reference words or due to multiple semantic concepts: the guidance through the declarative structure of the task model (Figure 4) normally limits the choice of the cases and consequently the risk of an incorrect case instantiation; however, should the identification of the concept fail, the underlying cases are in general incorrect. This is mainly due to the difficulty the rule-based decoding strategy has in coping with conflicting slots.

A priori we would expect the response evaluation to yield the highest performance, as even an incorrect semantic representation may yield a correct system reaction. However, this is only true for the rule-based implementation in MASK. The stochastic component was not integrated in an end-to-end spoken language system, where the semantic representation is also oriented so as to be able to respond appropriately to the user. The current decoder outputs the meaning of an isolated query regardless of the ongoing dialogue context. The response errors made by the stochastic component are partly due to ignoring dialogue specificities, which need to be addressed when the component is fully integrated in an end-to-end system. For ATIS, the performance loss of both implementations is attributed to the difficulty of matching the response generation to the reference answer strategy adopted by the ARPA community [59].

• **Discussion of the two methods** Figure 8 illustrates the main differences in developing rule-based and stochastic parsers. Given a pre-defined formalism such as the semantic case grammar, the linguistic expert needs to analyse and to annotate a large number of utterances for both methods (**level 1**). The data are explicitly labelled for the stochastic method in order to produce the training corpus. For the rule-based implementation the labelling usually goes along with developing a set of grammar rules (**level 2**). The data labelling gives an idea to the expert of what the semantic representation should look like. Therefore, rule-based systems are influenced twice by the human annotator, i.e. the data labelling and the elaboration of the system of rules. This second level represents a supplementary manual involvement and implies the risk of over-specialisation and error intrusion. In the stochastic method the labelled corpus is directly used to train the stochastic model, which in turn is used to directly decode the test query. The manual involvement is thus limited to the definition of semantic labels and to the data labelling itself which may be iterative. This

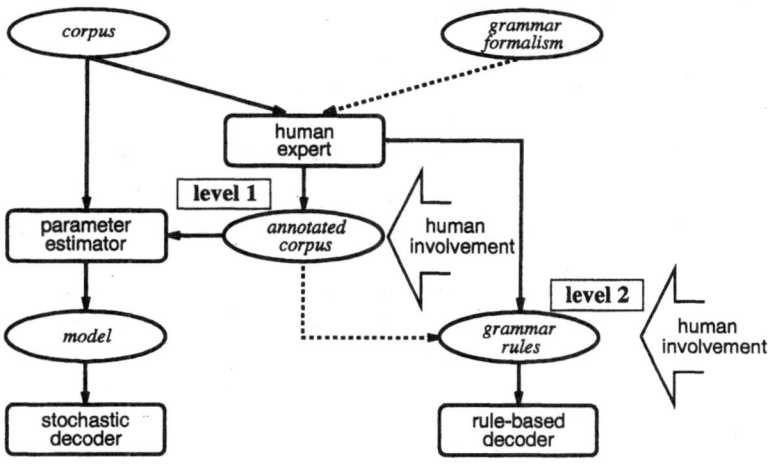

Fig. 8. Stochastic versus rule-based methods for natural language understanding

is much simpler than maintenance (and extension) of the grammar in a rule-based implementation. However, since the data are sparse, a major problem for developing a stochastic method is to find a good balance between the size of the parameter set and the performance of the semantic language model.

Parsing errors for both methods occur due to:

- the inappropriateness of the grammar formalism for a given application;
- the incorrect labelling of data, which may arise from a wrong judgement of the domain;
- the drawbacks of the decoding method, i.e. the way in which the grammar formalism is implemented;
- the different functionality of the semantic component in an end-to-end implementation compared to an out-of-system design.

Though limited, the first results obtained show that the stochastic outperforms the rule-based method, should the system be trained on a corpus using a correct set of labels. A global advantage of the stochastic method is that by learning from a large number of interactions, the system acquires an overall knowledge of the domain and of possible semantic structures, weighted by their occurrences. When developing a system of rules, the linguistic expert has not immediately this global picture of the domain. Misjudging the statistical significance of a request, i.e. its representativity, he risks to create complex, ambiguous and suboptimal structures and to make the component over-specialised and application-specific.

3.4 Task and Dialogue Model Integration

The rule-based task model only has been integrated in an end-to-end spoken language prototype [60], for both application domains (L'ATIS and MASK) using the French language. The semantic frame representation provided by the decoder is completed with a set of default and interpretative rules by the dialogue controller:

default rules assign values when necessary and when not specified by the user: e.g. if the departure day is not provided, the current day will be taken by default;

interpretative rules propagate values from one case to another: *this morning* provokes the instantiation of both **departure-day** and **departure-period** cases.

The semantic frame is also completed with the information gathered and rebuilt by the dialogue history from previous exchanges, which allows the system to process anaphora and ellipsis.

The dialogue model is structured into a hierarchy of subdialogues: each one has a particular label which depends on the subtask to be achieved, the state of the dialogue, or the metadialogue (dialogue about the dialogue). For both applications (L'ATIS and MASK), subdialogues were identified according to the abovementioned categories of dialogue acts:

informative: they concern the **task** and are application-dependent in so far as the information exchanged includes values directly related to the task objects characterising the application: e.g. *precision, explanation, contestation* or *discussion*.

communicative: they concern the **metadialogue** and are application-independent. They correspond to the parts that do not convey any specific information about the task objects, but that concern the way communication is handled through phatic messages: e.g. *hold-on, restart, repetition, confirmation, reformulation.*

interactive: they concern the **structure and organisation of the dialogue** and are more or less application-independent, since they mainly depend on social implication: e.g. *opening, closing formalities.*

Based on formal grammars and speech/dialogue act theories, the dialogue model is implemented as a set of rules [60]: the grammar non-terminal terms correspond to sub-dialogues, and the terminals correspond to dialogue acts. The model is represented by a recursive state transition network, in which exchanges between the user and the machine correspond to transitions in the network. One or several dialogue acts may correspond to one utterance.

The hierarchical structure of identified specific subdialogues enables the dialogue manager to control interaction more efficiently, and to vary response generation according to the state of the dialogue in the network, even if the user's request appears to be the same in its surface form: in case of repeated error, for instance, the system response may be formulated differently. The dialogue model enables mixed initiative: a non-complete frame causes the system to ask for precision, however the user may formulate a new request any time, corresponding to another semantic frame.

As a long-term perspective, such an approach based on act identification may provide utterance labels to automatically compute message occurrence probabilities, and allow for the design of a stochastic end-to-end system, including semantic and pragmatic levels.

4. Conclusion

The combination of several modalities requires modelling various types of knowledge sources. These knowledge sources enable interactive systems to interpret users' actions more accurately and to present information in the most appropriate way to the user. In a multimodal context, environment and system models play important roles. When speech is combined with other modalities, temporal aspects should be taken into account. Speech also implies complex cognitive processes which must be represented by specific strategies. The knowledge sources are application-dependent to some extent, especially for the task model. The results obtained with two different methods (rule-based and stochastic) used for representing the task model, show that the stochastic approach slightly outperforms the rule-based method. Furthermore, such a stochastic method should be more portable with respect to language and application, since the human involvement in developing the model is more limited. This may foster research on end-to-end systems entirely based on stochastic methods. Such an approach is already investigated for multimodality processing.

References

[1] Bernsen N.O., Dybkjaer L. and Dybkjaer H. A Dedicated Task-oriented Dialogue Theory in Support of Spoken Language Dialogue Systems Design. *Proceedings of the IC-SLP Conference*, Yokohama, Japan. 875-78. September 1994.

[2] Bellik Y., Ferrari S., Néel F., Teil D. Requirements for Multimodal Dialogue Including Vocal Interaction. *ESCA Workshop on Spoken Dialogue Systems. Theories and Applications*. Vigso, Denmark, May 30-June 2. pp.161-164. 1995,

[3] Bellik Y. La composante temporelle dans les interfaces multimodales. *Proceedings. Interfaces 97. 6th International Conference. Montpellier* May 28-30, 1997.

[4] Bellik Y. Media Integration in Multimodal Interfaces. *IEEE Workshop on Multimedia Signal Processing*, Princeton, New Jersey, 23-25 June 1997.

[5] Martin, J.-C.A Connectionist Model using Multiplexed Oscillations and Synchrony to Enable Dynamic Connections. *Proceedings of the Fourth International Conference on Artificial Neural Networks (ICANN'94)*, pp 755-758, vol. 1, 26-29 May, Sorrento, Italy. 1994.

[6] Martin, J.C. Towards "intelligent" cooperation between modalities. The example of a system enabling multimodal interaction with a map. IJCAI-97 workshop on Intelligent Multimodal Systems. August, 24th. Nagoya, Japan. 1997.

[7] Collet C., Finkel A. Gherbi R. CapRe: un système de capture du regard dans un contexte d'interaction homme-machine. *Proceedings Interfaces 97. 6th International Conference. Montpellier* May 28-30, 1997.

[8] Morel M-A. *Analyse linguistique d'un corpus de dialogues homme-machine* Tomes 1-2. Univ. Paris III Sorbonne Nouvelle. 1988.

[9] Néel, F. Etude lexicologique : vocabulaire propre à la tâche In: *Analyse linguistique d'un corpus de dialogues homme-machine* Tome 1. Univ. Paris III Sorbonne Nouvelle. 1988.

[10] Briffault X. et Denis M. Interactions pilote-copilote au cours de dialogues de navigation. *5èmes Journées Internationales Montpellier 96, L'Interface des Mondes Réels & Virtuels, Montpellier,* 21-24 mai 1996.

[11] Levinson S.E. and Shipley K.L. A Conversational-Mode Airline Information and Reservation System Using Speech Input and Output. *Bell Sys. Tech. Journ.*, Vol. 59, No. 1, pp. 119-137, 1980.

[12] Marque F., Bennacef S.K., Néel F., Trinh S. PAROLE : a Vocal Dialogue System for Air Traffic Control Training. *Joint ESCA-NATO/RSG.10 Tutorial and Research Workshop on Applications of Speech Technology*, Lautrach, 16-17 September 1993.

[13] Guyomard M. and Siroux J. Suggestive and Corrective Answers: A single Mechanism. In: *The Structure of Multimodal Dialogue*, M.M. Taylor, F. Néel, D.G. Bouwhuis (Eds). Elsevier Science Publ. pp. 361-374. 1989.

[14] Bellik Y., Ferrari S., Néel F., Teil D., Pierre E., Tachoires V. Interaction multimodale: concepts et architecture. *4èmes Journées Internationales sur l'Interface des Mondes Réels et Virtuels*. Montpellier, 26-30 juin 1995.

[15] Levinson S.C. *Pragmatics*. Cambridge University Press. 1983.

[16] Grice H.P. *Logic and Conversation*. In Cole & Morgan (eds) (1975: 41-58). (Part of Grice (1967)). 1975

[17] Grice H.P. *Further Notes on Logic and Conversation*. In Cole (1978: 113-28). (Part of Grice (1967)). 1978.

[18] Austin J.L., *How To Do Things With Words*. Oxford: Clarendon Press. 1962.

[19] Searle J.R., *Speech Acts, an essay in the Philosophy of Language*. Cambridge University Press. 1969.

[20] Bunt H.C. Information Dialogues as Communicative Actions in Relation to Partner Modelling and Information Processing. In: *The Structure of Multimodal Dialogue*, M.M. Taylor, F. Néel, D.G. Bouwhuis (Eds). Elsevier Science Publ. pp. 47-73. 1989.

[21] Taleb L. Communicational Deviation in Finalized Informative Dialogue Management. *Proceedings of the eleventh Twente Workshop on Language Technology (TWLT 11): Dialogue Management in Natural Language Systems*. LuperFoy S. Nijholt A. and Veldhuijzen van Zanten G. (eds). June 19-21 pp. 61-70 1996.

[22] Bilange E. An Approach to Oral Dialogue Modelling. *2nd Venaco Workshop. "The Structure of Multimodal Dialogue." ESCA ETRW*. Acquafredda di Maratea, Italy, September 16-20, 1991.

[23] Alexandersson J. Some Ideas for the Automatic Acquisition of Dialogue Structure. *Proceedings of the eleventh Twente Workshop on Language Technology (TWLT 11): Dialogue Management in Natural Language Systems*. LuperFoy S. Nijholt A. and Veldhuijzen van Zanten G. (eds). June 19-21 1996.

[24] Proctor C. and Young S. Dialogue Control in Conversational Speech Interfaces. In: *The Structure of Multimodal Dialogue*, M.M. Taylor, F. Néel, D.G. Bouwhuis (Eds). Elsevier Science Publ. pp. 385-398.1989.

[25] Sadek M.D., Bretier P., Cadoret V., Cozannet A., Dupont P., Ferrieux A., Panaget F. A Cooperative Spoken Dialogue System Based on a Rational Agent Model : A First Implementation on the AGS Application. ESCA Workshop on Spoken Dialogue Systems, Vigso, Denmark. pp. 145-148. May 1995,

[26] Maier E. Context Construction as Subtask of Dialogue Processing - the VERBMOBIL Case. pp. 113-122. Proceedings of the eleventh Twente Workshop on Language Technology (TWLT 11): Dialogue Management in Natural Language Systems. LuperFoy S. Nijholt A. and Veldhuijzen van Zanten G. (eds). June 19-21 1996.

[27] Briffault X. et Denis M., Analyse d'un Corpus de Dialogues de Navigation à Bord d'un Véhicule Automobile. *Notes et Documents LIMSI-CNRS* No. 95-28.1995

[28] Siroux J., Guyomard M., Jolly Y., Multon F., Remondeau C., Speech and tactile-based GEORAL system. *EUROSPEECH'95*, Madrid, pp. 1943-1946, September 1995.

[29] Montacié C., Projet AMIBE : Applications Multimodales pour Interfaces et bornes Evoluées; *GDR no. 39, rapport d'activité* 1994.

[30] Bellik Y. MEDITOR: a Multimodal Text Editor for Blind Users, *ACM UIST'96, Ninth Annual Symposium on User Interface Software*, Seattle, Washington, USA, November 6-8, 1996.

[31] Bourdot P., Krus M., Gherbi R. Gestion de périphériques non-standards pour des interfaces multimodales sous Unix / X11 : Application à un modeleur tridimensionnel, *4èmes Journées Internationales Montpellier 95 sur l'Interface des Mondes Réels et Virtuels*, juin 1995.

[32] Morel M-A. Computer Human-Communication. In: *The Structure of Multimodal Dialogue*, M.M. Taylor, F. Néel, D.G. Bouwhuis (Eds). Elsevier Science Publ., 1989. pp. 323-330.

[33] Beun R.J. Declarative Question Acts: Two Experiments on Identification. In: *The Structure of Multimodal Dialogue*, M.M. Taylor, F. Néel, D.G. Bouwhuis (Eds). Elsevier Science Publ., pp. 313-321. 1989.

[34] Beun R.J. The functions of repetitions in spoken information dialogues. *IPO Annual Progress Report, 20*, pp. 91-98. Eindhoven, The Netherlands: Institute for Perception Research.

[35] Fillmore Ch.J. The case for case, *Universals in Linguistic Theory* , Emmon Bach and Robert T. Harms, Holt, Rinehart and Winston Inc., pp. 1-90, 1968.

[36] Bruce B. Case Systems for Natural Language, *Artificial Intelligence,* Vol. 6, pp. 327-360, 1975.

[37] Minker W. Stochastic versus Rule-based Speech Understanding for Information Retrieval. *Speech Communication*, Vol 25 (4), October 1998, pp. 223-247.

[38] Hayes P., Hauptman A., Carbonnell J., and Tomita M. Parsing Spoken Language, a Semantic Caseframe Approach, *Proc. COLING*, 1986.

[39] Matrouf K., Néel F. Use of Upper Level Knowledge to improve Human-Machine Inter-action. *2nd Venaco Workshop. "The Structure of Multimodal Dialogue." ESCA ETRW.* Acquafredda di Maratea, Italy, September 16-20, 1991.

[40] Matrouf A.K., Gauvain J.-L., Néel F., Mariani J. An Oral Task-Oriented Dialogue for Air-Traffic Controller Training. *SPIE-IEEE.* Orlando, USA, April 1990.

[41] Price P. Evaluation of Spoken Language Systems: The ATIS Domain. *Proc. ARPA Human Language Technology*, June, 1990.

[42] Dahl D.A., Bates M., Brown M., Fisher W., Hunicke-Smith K., Pallett D., Pau C., Rud-nicky A., Shriberg E. Expanding the scope of the ATIS task: the ATIS-3 corpus. *Proc. ARPA Human Language Technology*, March 1994.

[43] Minker W. et Bennacef S.K Compréhension et Evaluation dans le Domaine ATIS. *Journées d' Etudes en Parole, JEP*, Juin, 1996.

[44] Bennacef S.K., Modélisation du dialogue oral Homme-Machine - Mise en oeuvre dans une application de demande d'informations. *PhD thesis, Université de Paris XI*, Orsay, 1995.

[45] Gauvain J.L., Bennacef S.K., Devillers L., Lamel L., and Rosset S. The Spoken Language Component of the MASK Kiosk. *Proc. Human Comfort Security Workshop.* 1995.

[46] Bennacef S.K., Bonneau-Maynard H., Gauvain J.L., Lamel L.F., and Minker W., A Spoken Language System For Information Retrieval. *Proc. ICSLP-94* , September 1994.

[47] Minker W., Bennacef S.K. and Gauvain J.L. A Stochastic Case Frame Approach for Natural Language Understanding. *Proc. ICSLP-96*, October, 1996.

[48] MADCOW Multi-Site Data Collection for a Spoken Language Corpus. *Proc. DARPA Speech and Natural Language Workshop*, February, 1992.

[49] Hayamizy S. Lively Communication with Spoken Dialogue Systems Utilizing Acoustic-Prosodic Information, *Internal Report, LIMSI-CNRS*, 1994.

[50] Lamel L.F., Rosset S., Bennacef S., Bonneau-Maynard H., Devillers L. and Gauvain J.L., Development of Spoken Language Corpora for Travel Information. *Proc. EUROSPEECH-95*, September 1995.

[51] Rabiner L.R. and Juang B.H. An introduction to Hidden Markov Models, *IEEE Transactions on Acoustics, Speech and Signal Processing*, 3(1):4-16, 1986.

[52] Jelinek F., Lafferty J., and Mercer R. Basic Methods of Probabilistic Context Free Grammars. *Speech Recognition and Understanding. Recent Advances*, Vol. 75, pp. 345-360, 1992.

[53] Schwartz R., Miller S., Stallard D. and Malkoul J. Language Understanding Using Hidden Understanding Models. *Proceedings of ICSLP*, pp. 997-1000, October 1996.

[54] Levin E. and Pieraccini R. Chronus - The Next Generation, *Proceedings ARPA Workshop on Human Language Technology*, January, 1995.

[55] Epstein M., Papineni K., Roukos S., Ward T. and Della Pietra S. Statistical Natural Language Understanding Using Hidden Clumpings. *Proceedings of ICASSP*, pp. 176-179, May 1996.

[56] Oerder M. and Aust H. A Realtime Prototype of an Automatic Inquiry System. *Proceedings of ICSLP*, pp. 703-706, 1994.

[57] Minker W. Stochastically-Based Natural Language Understanding Across Tasks and Languages. *Proceedings EUROSPEECH-93*, September 1997.

[58] Katz S.M. Estimation of Probabilities from Sparse Data for the Language Model Component of a Speech Recognizer. *IEEE Trans. on Acoustics, Speech and Signal Processing* , Vol. 35(3), pp. 400-401, 1987.

[59] Bates M., Boisen S., and Makhoul J. Developing an Evaluation Methodology for Spoken Language Systems. *Proc. DARPA Speech and Natural Language Workshop*, February 1992.

[60] Bennacef S.K., Néel F., Maynard H.B. An Oral Dialogue Model based on Speech Acts Categorisation. *ESCA Workshop on Spoken Dialogue Systems. Theories and Applications.* Vigso, Denmark, May 30-June 2, 1995, pp. 237-240.

Multimodal Interfaces for Multimedia Information Agents

Alex Waibel and Bernhard Suhm and Minh Tue Vo and Jie Yang

Interactive Systems Laboratories
Carnegie Mellon University 5000 Forbes Avenue, Pittsburgh PA 15217 (USA)
email: {waibel,suhm,tue,yang+}@cs.cmu.edu

Summary. When humans communicate they take advantage of a rich spectrum of cues. Some are verbal and acoustic. Some are non-verbal and non-acoustic. Signal processing technology has devoted much attention to the recognition of speech, as a single human communication signal. Most other complementary communication cues, however, remain unexplored and unused in human-computer interaction. In this paper we show that the addition of non-acoustic or non-verbal cues can significantly enhance robustness, flexibility, naturalness and performance of human-computer interaction. We demonstrate computer agents that use speech, gesture, handwriting, pointing, spelling jointly for more robust, natural and flexible human-computer interaction in the various tasks of an information worker: information creation, access, manipulation or dissemination.

1. Introduction

Human-computer interfaces today are limited and inflexible and do not take advantage of the many communication channels humans use to communicate verbal and non-verbal ways. Humans speak, point, gesture, write, fixate, use facial expressions, head-motion, eye-contact, etc. to express ideas, intentions and feelings. To build computer systems that operate more flexibly and robustly, and that are more intuitive to use by anyone, computer interfaces must be able to understand and process this multiplicity of human communication cues. Recently many research projects have been conducted to address this problem (e.g. [1, 2, 3, 4]). The need for more intuitive interfaces becomes all the more pressing as multimedia presentation, that is output, is becoming commonly available.

In this paper we describe information agents that use multimodal interfaces to access, manipulate, create and disseminate information. In particular, we seek effective methods by which human users can easily interact with a computer by speaking, pointing, drawing, spelling, etc. We show how cross-modal cues can be used effectively to recover from errors or miscommunications, since confusions in one modality usually are unambiguous in another. We describe how multiple modalities can cooperate and complement each other in various phases of the preparation of multimedia documents. Moreover, the added flexibility and robustness significantly increases reliability and naturalness, and with it the usability and acceptance of information systems. We present QuickDoc as a prototypical application that combines our multimodal subsystems in a simple, but powerful way.

2. Interpretation of Multimodal Input

2.1 Multimodal Components

Our labs have implemented recognizers and processing modules for various input modalities, most notably speech, pen input, and face tracking. These modality processors constitute the basic components of all our multimodal systems.

2.1.1 Speech. Our speech recognition subsystem is based on the recognition front-end of the JANUS speech translation system [5] which is capable of processing speaker-independent, spontaneous speech. The recognizer can be adapted to any task domain by retraining the language models and possibly tuning the acoustic models if the domain involves special vocabulary. We also have a high-performance, real-time continuous spelling recognizer for large lists of 100,000 or more names.

2.1.2 Gesture. Our approach to pen-based gesture recognition is to decompose pen strokes into sequences of basic shapes such as line, arc, arrow, circle, cross... [6]. The same gesture shape may mean different things depending on the surrounding context, hence each gesture component is augmented by gesture contexts indicating of spatial relationships between the gesture and nearby objects in the user interface.

2.1.3 Handwriting. Our MS-TDNN-based handwriting recognizer [7] is capable of processing writer-independent, continuous (cursive) handwriting at a recognition rate of 94% on a 20,000-word vocabulary. We employ simple heuristics to decide when to invoke handwriting recognition on pen input, e.g., when the gesture recognizer cannot identify the input strokes as basic shapes.

2.1.4 Face Tracking. We base our face-tracking subsystem on face color clustering and motion detection. The face tracker can control a pan-tilt-zoom camera to follow a freely moving person, producing a constant-sized image of the face area in real time. Another software layer built on top of this face tracker can identify the eyes and other face features, build a model of the face, and estimate the gaze direction to track the head pose [9].

2.2 Joint Interpretation

In order to make sense of input from all available sources, we need a multimodal interpreter capable of producing an interpretation of user intent (e.g., a command to execute in the application interface) from the output of the modality processors.

In our joint interpretation scheme, the user intent is represented by a frame consisting of slots specifying pieces of information such as the action to carry out or the parameters for that action. Recognition output from the modality processors are parsed into partially filled frames that are merged together to produce the combined interpretation as described in [6]. This technique leads to uniform handling of high-level information from all input sources, which is very important for modularity and extensibility. To add another input modality we need only provide a module to convert low-level recognizer output to a partially filled frame to be merged with others. In addition, context information can be retained across input events by merging with previous interpretation frames.

3. Multimodal Error Correction

Intensive research over recent years has boosted the performance of speech recognition technology significantly. Nevertheless it is widely believed that recognition performance will remain limited, at least for the foreseeable future. The potential for error with any speech interface will be balanced by the redundant and alternate ability to enter information or queries by spelled, handwritten or typed input. Rather than fighting a particular recognition error the user can therefore quickly circumvent the problem by choosing a different modality. Such alternates not only eliminate user frustration with potentially recurring errors, but also provide an avenue for background learning and adapting.

3.1 Multimodal Interactive Error Repair

Our approach is to have the user collaborate with the system recovering from interpretation errors [10]. First, errors have to be identified, either initiated by the system, for instance based on some confidence measure, or by the user. If a graphical user interface (GUI) is available, the user can simply highlight erroneous words (the reparandum) in the recognition hypothesis displayed. For error correction, the user provides additional input, choosing among different correction methods: repair by repeating the reparandum by respeaking, spelling out loud, or handwriting; repair by paraphrasing the reparandum; repair by pen gestures (e.g. to delete or insert), in addition to the standard repair by typing or selecting among N-best alternatives.

The rationale for this multimodal approach to error recovery is twofold. First, it exploits the fact that different input modalities are orthogonal: words which are confusable in one modality, can be disambiguated in a different modality (e.g. "road" and "rote" spoken versus spelled). Secondly, recent studies [11] show that switching modality after system misinterpretation alleviates user frustration.

3.2 Error Repair for Multimedia Information Agents

The design of speech user interfaces is constrained by the context of the application, and so is the design of error repair. The application context varies along several dimensions: available modalities (speech only versus GUI), what is repaired (isolated words, phrases, sentences), goal of repair is (get verbatim every word versus get the intended action), dialogue metaphor (command controls versus conversational). For multimedia information agents, we see the following two application contexts as particularly relevant:

- (Mixed initiative) Spoken dialogue, for example in user queries and command control of the application. In this context, a GUI may be available, repair will typically be performed at the level of phrases, the goal of repair is to initiate the intended action (semantic repair), and a conversational dialogue is desirable.
- Dictation, for example to entry in fields, or to dictate reports. In this context, a GUI is very appropriate, repair will occur from isolated word to sentence level, the goal of repair is to get each word correct (verbatim repair).

We have built prototypical speech user interfaces with error repair capabilities for dictation and form filling tasks. In this application context, requiring the user to identify errors by highlighting them is appropriate, since the user naturally focuses on the actual recognition hypothesis. For error correction, we offer to repeat the input (potentially switching to another modality), and pen gestures to indicate where to insert words, or which word to delete. In form filling tasks, knowledge of the current field can provide powerful constraints for recognition by restricting vocabulary and language model accordingly.

3.3 Evaluating Interactive Error Repair

Given a set of error recovery methods feasible with current technology in a given application context, a crucial issue is to predict which methods a user will prefer. Based on the assumption of a rational user [13], we use the time to complete some input, including the time needed to correct errors, as an objective and easily quantifiable measure for the effectiveness of different error recovery methods, and main predictor of user preference. By assuming stochastic independence of the various repair attempts, the expected accuracy after a certain number of repair attempts can be estimated as a geometric series. Given a high level of accuracy sufficient for the application (e.g. 99%), the number of repair attempts can be estimated, and thus the total time required to input including repair [12].

A pilot evaluation on a form filling task [12] suggests that given current technology, repair by spelling and handwriting can be very effective, and significantly better than standard choice from N-best alternatives. Using the context of a repair, e.g. in rescoring N-best lists obtained by decoding repair input, can substantially improve the accuracy of repair. Although results are still preliminary, they show that our multimodal approach to interactive error recovery is very promising.

4. Multimodal Information Agents

The objective for multimodal information agents is to quickly access, create, manipulate, and disseminate multimedia information. A multimodal agent can label multimedia information with appropriate classifications by voice or/and gesture annotations. All annotations are incorporated with original information. The agent will automatically attach hotlinks on the world-wide-web to broaden information sources. The agent then will generate an HTML format report. The agent can search all the information in the report including voice and gesture annotations and voice mails. The agent can repair errors and add more information. Finally, the agent can disseminate the report in electronic form or printed form.

4.1 Information Access

Multimodal information agents can speed up navigation of Information space and queries of a database. A city map, for example, can be queried by a combination of gesture, handwriting and speech. A query such as: "Show me all hospitals in this <circling-gesture> area," or "How far is from here to there <tracing-gesture>?" will access information that would otherwise have to e assembled and combined painstakingly in a series of steps. Multimodal agents will recognize gesture and speech concepts and map them onto a semantic representation. The semantic representation can be filled interchangeably by both gestures and verbal phrases using pattern recognition and chart parsing techniques. The parsing result is then sent to query the database. If interpretation errors occur, the user corrects them using multimodal correction techniques. The whole process is transparent to the user.

Searching multimedia data can be viewed as natural extension of multimodal database queries. For instance, voice annotations or video clips added during user interaction can be interpreted and indexed using multimodal interpretation components, and then accessed using database queries techniques.

4.2 Information Creation

A user often needs to add information to a database or modify existed information. Multimodal agents are also useful in such a process. For example, a doctor needs to mark his/her diagnoses on X-ray, CT, MR images and make comments on the case. He/she can use speech, gesture, and handwriting to do this. Similarly iconic gestures may be used on a touch sensitive display or writing pad, to quickly enter resources or objects in a database, such as a map, in non-verbal ways. Iconic gestures can be defined to represent objects that would be entered in the database. In the same way, handwriting symbols can be used to mark or tag objects on a screen. Multimodal input also need not always be interpreted, but can often be stored for later review. Thus a voice annotation, or a circling gesture can be simply stored to go along with an image for later review and retrieval. The efficiency of multimodal agents for information creation can illustrated by the following example. The scenario is an image analyst sitting in the front of a computer and classifying various targets. Images can be popped up quickly and in rapid succession and the analyst provides a quick classification. The image and its classification are entered into a database along with the analyst's spoken and gestured annotation. With the human in the loop to make decisions on intelligence relevance, but with the availability of a better (multimodal) interface, to do so quickly, the job of data creation and manipulation can be carried out with greater reliability and efficiency compared to a fully automated system.

4.3 Information Manipulation

Information will not only be requested to viewed, but will also need to be updated. A multimodal multimedia information system can offer new ways to maintain the underlying databases: the user can use speech, gesture, writing, spelling to update database information efficiently. manipulating visual representations on the display rather than having to learn and use a separate database manipulation system. Gestures play an important role in specifying object parameters (e.g. different iconic shapes to represent different object types) and spatial constraints (e.g., location and extent of objects), while speech is useful in specifying parameters not easily expressed visually. For example he/she might circle a bridge on a map and say "this bridge has been destroyed", or might cross out an object without words, etc.

In some situations, the answer for a query cannot be directly obtained from the database and the information from the database has to be future processed. For example, the query "Show me the nearest restaurant" requires not only database access but also other computations.

4.4 Information Dissemination

In addition to information access, creation and manipulation, the product has to be packaged and readied for later viewing. The world wide web can serve as a natural medium to disseminate multimedia information. The multimodal input signals are interpreted and compiled, and an HTML format report is automatically generated on the fly. It is then automatically available for viewing together with the original aligned speech waveform attached to it as hotlinks. Multimodal interaction can also be used to manipulate objects in a report, such as positioning objects on a page appropriately, or attaching appropriate hotlinks.

To quickly generate a multimedia report, speech dictation can be combined with point and click actions to arrange reports on the fly. Such a multimodal report generation would be faster than typing and can generate a hierarchy of facts and notions. The generation thus would consist of some dictation or typing, some multimodal object manipulation ("move this <point> here <point> and link it as an explanation for that <click>") and some handwriting or drawings that are to be attached. The result of these communicative acts will compile into HTML code, that is ready to be transferred. Time savings are possible in several ways: the report generation can be done more efficiently by speech, gesture, or handwriting, and by using references to already existing multimedia stickies, factoids, video footage or explanations, and the recipient of the report can probe and browse a hierarchical multimedia report in a non-linear fashion more efficiently, calling up aspects of the assessment as needed.

4.5 Controlling the Interface

Although the goal of a multimodal interface is to provide as transparent an access to functionality as possible, interaction to or control various aspects of the system can't be eliminated.

For example for navigation within visually presented image data, the user typically modifies the view by operations such as zooming and panning. The integration of speech and gesture increases flexibility by allowing different attributes of an operation to be specified in whatever modality is more appropriate for each attribute; for instance, the effect of a spoken "zoom in to this <circle> area" accompanied by a circle drawn around the desired area is much more difficult to achieve using a single input modality.

Additionally, multiple modalities can be used to enhance the reliability. For example, a gaze tracker can be used to detect the user's focus of attention. When the user is looking at a window on a screen, the window will be highlighted. No action is taken unless the user uses a voice command to confirm the selection. The voice commands could be "select this window" or "close window", etc. This can reduce unintended actions caused by the user randomly looking around.

5. The QuickDoc Application

QuickDoc is an example of an application that combines our multimodal subsystems in a simple but powerful way, letting the user perform a repetitive task with speed and convenience. The task is for a doctor to go through a series of images such as X-rays or computer-aided tomography scans, quickly identify an anomalous area, label the area with the name of a disease or condition, and attach relevant comments. The end product is an HTML report that summarizes the doctor's findings in a compact table listing the annotated images, the corresponding preliminary diagnoses, and automatically generated hotlinks to relevant sites based on the diagnoses.

The QuickDoc user labels each presented image by circling an area on the image (drawing directly on a touch-sensitive screen) and speaking a disease name as well as a percentage representing the confidence level of the diagnosis. The gesture recognizer identifies the circle and generates an area marker on the image, attaching other parts of the gesture as written annotations (a future version may run the handwriting recognizer to turn those into text). The user can issue a spoken command to initiate a voice annotation attachment; the recorded audio file is also run through the speech recognizer to produce an automatic transcription. When all the images have been processed, another spoken command causes the HTML report to be generated. Optionally the collected data can also be edited in a multimodal dictation and repair facility before being processed into HTML.

The generated report contains thumbnails of the annotated images accompanied by disease label, confidence level, location in the image (automatically extracted from the circling gesture), and voice annotation playback icon if applicable. The thumbnail is linked to a more detailed page containing the full-size image and the automatic voice transcription. The name of the disease is looked up in a database and turned into a hotlink to a page listing Web sites relevant to that disease. A keyword search mode allows to search spoken keywords in the text of the report and the automatic voice transcriptions; entries containing hits are then highlighted in the report.

QuickDoc thus embodies multimodal creation, manipulation, dissemination, and access of multimedia information in a simple yet surprisingly useful application.

6. Conclusions

In this paper we have described systems that combine speech, gesture, handwriting, pointing, spelling, and other communication modalities into interfaces that are robust, flexible, and intuitive to use. We described how multimodal error recovery can

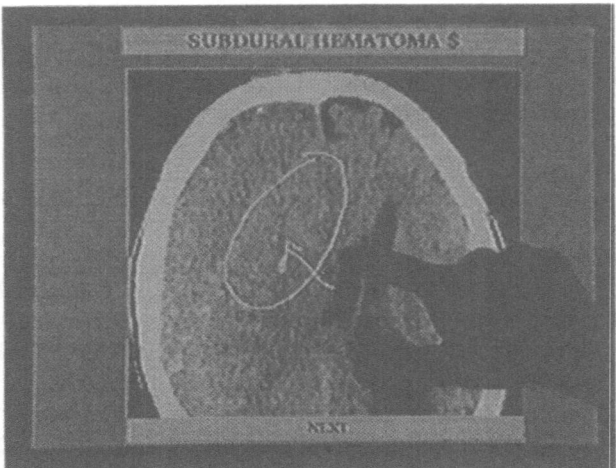

Fig. 1. Multimodal information creation in QuickDoc: labeling some area by a circle gesture and speech ("subdural hematoma")

ease the problem of unreliable automatic interpretation of communication modalities, in particular speech, thus removing a major obstacle in making multimodal interfaces truly usable. We demonstrated how these concepts can be combined for a more user friendly multimedia document production technology.

Acknowledgements. This research was sponsored by the DARPA under the Department of the Navy, Naval Research Office under grant number N00014-93-1-0806. Views and conclusions contained in this document are those of the authors and should not be interpreted as necessarily representing the official policies or endorsements, either expressed or implied, of the Navy or the U.S. Government.

References

[1] H. Ando, Y. Kitahara and N. Hataoka: "Evaluation of Multimodal Interface Using Spoken Language and Pointing Gesture on Interior Design System", Proc. ICSLP , 1994, Vol. 2, pp. 567-570.

[2] T. Nishimoto, N. Shida, T. Kobayashi, K. Shirai: "Multimodal Drawing Tool Using Speech, Mouse and Keyboard", Proc. ICSLP , 1994, Vol. 3, pp. 1287-1290.

[3] J. Wang: "Integration of Eye-gaze, Voice and Manual Response in Multimodal User Interface", Proc. ICSMC , 1995, Vol. 5, pp. 3938-42.

[4] A. Waibel, M.T. Vo, P. Duchnowski, and S. Manke: "Multimodal Interfaces," Artificial Intelligence Review , Special Volume on Integration of Natural Language and Vision . Processing , McKevitt, P. (Ed.), Vol. 10, Nos. 3-4, 1995.

[5] B. Suhm, P. Geutner, T. Kemp, A. Lavie, L.J. Mayfield, A. McNair, I. Rogina, T. Schultz, T. Sloboda, W. Ward, M. Woszczyna and A. Waibel: "JANUS: Towards Multilingual Spoken Language Translation," Proc. ARPA SLT Workshop 95 (Austin, Texas).

[6] M.T. Vo and C. Wood: "Building and application framework for speech and pen input integration in multimodal learning interfaces," Proc. ICASSP'96 (Atlanta, GA).

[7] S. Manke, M. Finke and A. Waibel: "The Use of Dynamic Writing Information in a Connectionist On-Line Cursive Handwriting Recognition System," Advances in Neural Information Processing Systems 6 , Morgan Kaufmann, 1994.

[8] J. Yang and A. Waibel: "A real-time face tracker," Proc. WACV'96 , pp. 142-147.

[9] R. Stiefelhagen, J. Yang and A. Waibel: "A Modelbased Gaze Tracking System," Proc. IEEE International Joint Symposia on Intelligence and Systems.- Image, Speech & Natural Language Systems , pp. 304-310, 1996.

[10] A.E. McNair and A. Waibel: "Improving Recognizer Acceptance through Robust, Natural Speech Repair", Proc. ICSLP , 1994, Vol. 3, pp. 1299-1302.

[11] S.L. Oviatt and R. VanGent: "Error Resolution during Multimodal Human-Computer Interaction", Proc. ICSLP, 1996.

[12] B. Suhm, B. Myers and A. Waibel: "Interactive Recovery from Speech Recognition Errors in Speech User Interfaces," Proceedings of the International Conference on Spoken Language Processing - ICSLP , 1996.

[13] A. Newell, S. Card and P. Moran: "The Psychology of Human Computer Interaction," Lawrence Earlbaum Associates, 1983.

Index

NATO ASI Series F

NATO ASI Series F